P9-AFZ-467

METHODS IN CELL BIOLOGY

VOLUME X

Contributors to This Volume

Lewis C. Altenburg
Paola Pierandrei Amaldi
James B. Boyd
Ph. Chevaillier
Bertil Daneholt
Elaine G. Diacumakos
Darrell Doyle
E. Anthony Evans
Joseph G. Gall
Michael J. Getz
Robert R. Klevecz
Bo Lambert
Judith Lengyel
Jürgen Maisenbacher
K. Marcinka
Y. Nozawa
Mary Lou Pardue
Sheldon Penman
M. Philippe
Hans Probst
Samuel Refetoff
Grady F. Saunders
Neal H. Scherberg
Edward L. Schneider
N. I. Shapiro
Robert H. Singer
Allan Spradling
Eric J. Stanbridge
John Tweto
José Uriel
N. B. Varshaver
David A. Wolff

Methods in Cell Biology

Edited by

DAVID M. PRESCOTT

DEPARTMENT OF MOLECULAR, CELLULAR AND
DEVELOPMENTAL BIOLOGY
UNIVERSITY OF COLORADO
BOULDER, COLORADO

VOLUME X

1975

ACADEMIC PRESS • New York San Francisco London
A Subsidiary of Harcourt Brace Jovanovich, Publishers

ACADEMIC PRESS, INC.
111 Fifth Avenue, New York, New York 10003

United Kingdom Edition published by
ACADEMIC PRESS, INC. (LONDON) LTD.
24/28 Oval Road, London NW1

LIBRARY OF CONGRESS CATALOG CARD NUMBER: 64-14220

ISBN 0-12-564110-9

PRINTED IN THE UNITED STATES OF AMERICA

CONTENTS

List of Contributors xi

Preface xiii

Contents of Previous Volumes xv

1. *Nucleic Acid Hybridization to the DNA of Cytological Preparations*
Mary Lou Pardue and Joseph G. Gall

Introduction 1
I. Procedure for *in Situ* Hybridization 2
II. Equipment and Reagents 8
III. Radioactive Nucleic Acids for Hybridization 10
IV. Conditions of Reaction 12
V. General Comments 14
References 16

2. *Microanalysis of RNA from Defined Cellular Components*
Bo Lambert and Bertil Daneholt

I. Introduction 17
II. Experimental Material 18
III. Labeling Conditions 20
IV. Preparation and Fixation 23
V. Microdissection 24
VI. Microextraction 27
VII. Microelectrophoresis in Gels 29
VIII. Microhybridization 34
IX. Combination of Micro- and Macrotechniques 44
X. Applications 45
References 46

3. *Staining of RNA after Polyacrylamide Gel Electrophoresis*
K. Marcinka

I. Introduction 49
II. Action of Fixatives and Stains on RNA 50

III. Procedure for RNA Staining 55
IV. Evaluation of the Stained RNA Pattern 62
V. Other Methods of RNA Detection 63
VI. Concluding Remarks 66
References 67

4. *Isolation and Fractionation of Mouse Liver Nuclei*
Ph. Chevaillier and M. Philippe

I. Fine Structure of Nuclei Isolated in Various Ionic Media 70
II. Fractionation of Mouse Liver Nuclei 77
III. Purification of Nuclear Membranes from Mouse Liver Nuclei 79
IV. Conclusions 82
References 83

5. *The Separation of Cells and Subcellular Particles by Colloidal Silica Density Gradient Centrifugation*
David A. Wolff

I. Introduction 85
II. Materials and Methods 87
III. Applications 93
IV. Conclusions 102
References 103

6. *Isolation of Subcellular Membrane Components from Tetrahymena*
Y. Nozawa

I. Introduction 105
II. Systematic Isolation Methods for Various Subcellular Membrane Components 108
III. Nonsystematic Isolation Methods for Various Subcellular Membrane Components 115
IV. Membrane Lipid Composition of Various Isolated Subcellular Components 130
References 132

7. *Isolation and Experimental Manipulation of Polytene Nuclei in Drosophila*
James B. Boyd

I. Introduction 135
II. Isolation of Nuclei and Nucleoli 136
III. Nuclear Purification in Sucrose Gradients 140

IV. Cytological Analysis of Isolated Nuclei 141
V. Manipulation and Incubation of Polytene Nuclei 141
VI. Related Studies 145
References 145

8. *Methods for Microsurgical Production of Mammalian Somatic Cell Hybrids and Their Analysis and Cloning*
Elaine G. Diacumakos

I. Introduction 147
II. Cell Culture 148
III. Cell Fusion 148
IV. Cell Isolation 151
V. Analyzing and Cloning Hybrid Cells 152
VI. Additional Comments 155
References 156

9. *Automated Cell Cycle Analysis*
Robert R. Klevecz

I. Introduction 157
II. Cell Cycle Analyzer 159
III. Optimizing Selection Conditions 164
IV. Calibrating the Cell Cycle 168
V. Limitations and Prospects 171
References 172

10. *Selection of Synchronous Cell Populations from Ehrlich Ascites Tumor Cells by Zonal Centrifugation*
Hans Probst and Jürgen Maisenbacher

I. Introduction 173
II. Materials 174
III. Zonal Centrifugation Run 177
IV. Results 179
V. Possible Applications 183
References 184

11. *Methods with Insect Cells in Suspension Culture I. Aedes albopictus*
Allan Spradling, Robert H. Singer, Judith Lengyel, and Sheldon Penman

I. Introduction 185
II. Adaptation of *A. albopictus* Cells to Desired Growth Conditions 186

III. Characteristics of the Cell Line 187
IV. Cell Fractionation and Other Procedures 189
V. Conclusion 193
References 194

12. *Methods with Insect Cells in Suspension Culture II. Drosophila melanogaster*
Judith Lengyel, Allan Spradling, and Sheldon Penman

I. Introduction 195
II. Adaptation of the Line to a New Medium and to Suspension Growth 197
III. Characteristics of the Cells 200
IV. Cell Fractionation 203
V. Summary and Conclusions 207
References 208

13. *Mutagenesis in Cultured Mammalian Cells*
N. I. Shapiro and N. B. Varshaver

I. Introduction 209
II. Conditions for Experiments on Mutagenesis 210
III. Spontaneous Mutagenesis 220
IV. Induced Mutagenesis 227
References 233

14. *Measurement of Protein Turnover in Animal Cells*
Darrell Doyle and John Tweto

I. Introduction 235
II. Methods for the Measurement of k_s 238
III. Measurement of k_d 243
IV. Conclusions 258
References 259

15. *Detection of Mycoplasma Contamination in Cultured Cells: Comparison of Biochemical, Morphological, and Microbiological Techniques*
Edward L. Schneider

I. Introduction 261
II. Microbiological Culture 262
III. Morphological Techniques 263
IV. Biochemical Techniques 265
V. Comparative Studies of Biochemical, Morphological, and

Microbiological Techniques for the Detection of Mycoplasma Contamination 268
VI. Sensitivities of Biochemical, Morphological, and Microbiological Techniques for Mycoplasma Detection 272
VII. Discussion 274
References 275

16. *A Simple Biochemical Technique for the Detection of Mycoplasma Contamination of Cultured Cells*
Edward L. Schneider and Eric J. Stanbridge

I. Introduction 278
II. Alterations in Exogenous Uridine and Uracil Incorporation into RNA in Mycoplasma-Infected Cells 278
III. Measurement of the Ratio of Uridine-^{3}H(Urd) to Uracil-^{3}H(U) Incorporated into Cellular RNA 279
IV. Evaluation of Results 282
V. Effect of Incubation Time, Cell Concentration, Antibiotics, and Serum Batch, Age, and Concentration on Urd/U Ratios 284
VI. Comparison of Parallel Single-Isotope and Double-Isotope Labeling for Urd/U Determinations 287
VII. Discussion 289
References 290

17. *Purity and Stability of Radiochemical Tracers in Autoradiography*
E. Anthony Evans

I. Introduction 291
II. Design of Tracer Experiments Using Autoradiography 293
III. Concepts of Purity 296
IV. Self-Decomposition of Radiochemicals—Observations and Control 308
V. Importance of the Specificity of Labeling 312
VI. Factors Associated with the Autoradiographic Techniques 320
VII. Concluding Remarks 321
References 321

18. *^{125}I in Molecular Hybridization Experiments*
Lewis C. Altenburg, Michael J. Getz, and Grady F. Saunders

I. Introduction 325
II. Iodination Methodology 327
III. Applications of ^{125}I-Labeled Nucleic Acids 335
IV. Discussion 341
References 342

19. *Radioiodine Labeling of Ribopolymers for Special Applications in Biology*

Neal H. Scherberg and Samuel Refetoff

I.	Introduction	343
II.	Special Considerations in Preliminary Preparation of the RNA	344
III.	Radioiodination of Polynucleotides	347
IV.	Applications of Iodine-Labeled RNA and Nucleotides	351
	References	358

20. *Autoradiographic Analysis of Tritium on Polyacrylamide Gel*

Paola Pierandrei Amaldi

I.	Introduction	361
II.	Procedure	362
	References	364

21. *A Radioautographic Method for Cell Affinity Labeling with Estrogens and Catecholamines*

José Uriel

I.	Introduction	365
II.	Procedure	366
III.	Results	369
IV.	Concluding Remarks	373
	References	374

Subject Index 375

Cumulative Subject Index Volumes I–X 381

LIST OF CONTRIBUTORS

Numbers in parentheses indicate the pages on which the authors' contributions begin.

LEWIS C. ALTENBURG, Medical Genetics Center, The University of Texas Graduate School of Biomedical Sciences, Health Science Center, Houston, Texas, and the Department of Biology, The University of Texas System Cancer Center, M. D. Anderson Hospital and Tumor Institute, Houston, Texas (325)

PAOLA PIERANDREI AMALDI, Laboratorio di Biologia Cellulare, Consiglio Nazionale delle Ricerche, Roma, Italy (361)

JAMES B. BOYD, Department of Genetics, University of California, Davis, California (135)

PH. CHEVAILLIER, Laboratoire de Biologie Cellulaire, Université Paris-Val de Marne, Créteil, France (69)

BERTIL DANEHOLT, Department of Histology, Karolinska Institutet, Stockholm, Sweden (17)

ELAINE G. DIACUMAKOS, Laboratory of Biochemical Genetics, The Rockefeller University, New York, New York (147)

DARRELL DOYLE, Department of Molecular Biology, Roswell Park Memorial Institute, Buffalo, New York (235)

E. ANTHONY EVANS, The Radiochemical Centre Limited, Amersham, Buckinghamshire, England (291)

JOSEPH G. GALL, Department of Biology, Yale University, New Haven, Connecticut (1)

MICHAEL J. GETZ, Department of Pathology, Mayo Clinic, Rochester, Minnesota (325)

ROBERT R. KLEVECZ, Department of Cell Biology, Division of Biology, City of Hope National Medical Center, Duarte, California (157)

BO LAMBERT, Department of Histology, Karolinska Institutet, Stockholm, Sweden (17)

JUDITH LENGYEL, Department of Biology, Massachusetts Institute of Technology, Cambridge, Massachusetts (185, 195)

JÜRGEN MAISENBACHER,[1] Physiologisch-chemisches Institut der Universität, Tübingen, West Germany (173)

K. MARCINKA, Institute of Virology, Slovak Academy of Sciences, Bratislava, Czechoslovakia (49)

Y. NOZAWA, Department of Biochemistry, Gifu University School of Medicine, Tsukāsamachi-40, Gifu, Japan (105)

MARY LOU PARDUE, Department of Biology, Massachusetts Institute of Technology, Cambridge, Massachusetts (1)

SHELDON PENMAN, Department of Biology, Massachusetts Institute of Technology, Cambridge, Massachusetts (185, 195)

M. PHILIPPE, Laboratoire de Biologie Cellulaire, Université Paris-Val de Marne, Créteil, France (69)

HANS PROBST, Physiologisch-chemisches Institut der Universität, Tübingen, West Germany (173)

SAMUEL REFETOFF, Department of Medicine, University of Chicago, Chicago, Illinois (343)

[1] *Present address:* E. Merck, Abteilung Klinische Forschung, Darmstadt, West Germany.

GRADY F. SAUNDERS, Department of Developmental Therapeutics, The University of Texas System Cancer Center, M. D. Anderson Hospital and Tumor Institute, Houston, Texas (325)

NEAL H. SCHERBERG, Department of Medicine, University of Chicago, Chicago, Illinois (343)

EDWARD L. SCHNEIDER, Gerontology Research Center, N.I.A., N.I.H., Baltimore, Maryland (261, 277)

N. I. SHAPIRO, Biological Department, Kurchatov Institute of Atomic Energy, Moscow, U.S.S.R. (209)

ROBERT H. SINGER,[2] Department of Biology, Massachusetts Institute of Technology, Cambridge, Massachusetts (185)

ALLAN SPRADLING, Department of Biology, Massachusetts Institute of Technology, Cambridge, Massachusetts (185, 195)

ERIC J. STANBRIDGE, Department of Medical Microbiology, Stanford University School of Medicine, Stanford, California (277)

JOHN TWETO, Department of Molecular Biology, Roswell Park Memorial Institute, Buffalo, New York (235)

JOSÉ URIEL, Institut de Recherches Scientifiques sur le Cancer, Villejuif, France (365)

N. B. VARSHAVER, Biological Department, Kurchatov Institute of Atomic Energy, Moscow, U.S.S.R. (209)

DAVID A. WOLFF, Department of Microbiology, The Ohio State University, Columbus, Ohio (85)

[2] *Present address:* Department of Anatomy, University of Massachusetts Medical School, Worcester, Massachusetts.

PREFACE

Volume X of *Methods in Cell Biology* continues the presentation of techniques and methods in cell research that have not been published or have been published in sources that are not readily available. Much of the information on experimental techniques in modern cell biology is scattered in a fragmentary fashion throughout the research literature. In addition, the general practice of condensing to the most abbreviated form materials and methods sections of journal articles has led to descriptions that are frequently inadequate guides to techniques. The aim of this volume is to bring together into one compilation complete and detailed treatment of a number of widely useful techniques which have not been published in full detail elsewhere in the literature.

In the absence of firsthand personal instruction, researchers are often reluctant to adopt new techniques. This hesitancy probably stems chiefly from the fact that descriptions in the literature do not contain sufficient detail concerning methodology; in addition, the information given may not be sufficient to estimate the difficulties or practicality of the technique or to judge whether the method can actually provide a suitable solution to the problem under consideration. The presentations in this volume are designed to overcome these drawbacks. They are comprehensive to the extent that they may serve not only as a practical introduction to experimental procedures but also to provide, to some extent, an evaluation of the limitations, potentialities, and current applications of the methods. Only those theoretical considerations needed for proper use of the method are included. Special emphasis has been placed on inclusion of much reference material in order to guide readers to early and current pertinent literature.

Finally, Volume X contains, in addition to the usual Subject Index for the volume, a Cumulative Index for Volumes I–X, which should facilitate the use of the series.

DAVID M. PRESCOTT

CONTENTS OF PREVIOUS VOLUMES

Volume I

1. Survey of Cytochemistry
   R. C. von Borstel

2. Methods of Culture for Plasmodial Myxomycetes
   John W. Daniel and Helen H. Baldwin

3. Mitotic Synchrony in the Plasmodia of *Physarum polycephalum* and Mitotic Synchronization by Coalescence of Microplasmodia
   Edmund Guttes and Sophie Guttes

4. Introduction of Synchronous Encystment (Differentiation) in *Acanthamoeba* sp.
   R. J. Neff, S. A. Ray, W. F. Benton, and M. Wilborn

5. Experimental Procedures and Cultural Methods for *Euplotes eurystomus* and *Amoeba proteus*
   D. M. Prescott and R. F. Carrier

6. Nuclear Transplantation in Ameba
   Lester Goldstein

7. Experimental Techniques with Ciliates
   Vance Tartar

8. Methods for Using *Tetrahymena* in Studies of the Normal Cell Cycle
   G. E. Stone and I. L. Cameron

9. Continuous Synchronous Cultures of Protozoa
   G. M. Padilla and T. W. James

10. Handling and Culturing of *Chlorella*
    Adolf Kuhl and Harald Lorenzen

11. Culturing and Experimental Manipulation of *Acetabularia*
    Konrad Keck

12. Handling of Root Tips
    Sheldon Wolff

13. Grasshopper Neuroblast Techniques
J. Gordon Carlson and Mary Esther Gaulden

14. Measurement of Material Uptake by Cells: Pinocytosis
Cicily Chapman-Andresen

15. Quantitative Autoradiography
Robert P. Perry

16. High-Resolution Autoradiography
Lucien G. Caro

17. Autoradiography with Liquid Emulsion
D. M. Prescott

18. Autoradiography of Water-Soluble Materials
O. L. Miller, Jr., G. E. Stone, and D. M. Prescott

19. Preparation of Mammalian Metaphase Chromosomes for Autoradiography
D. M. Prescott and M. A. Bender

20. Methods for Measuring the Length of the Mitotic Cycle and the Timing of DNA Synthesis for Mammalian Cells in Culture
Jesse E. Sisken

21. Micrurgy of Tissue Culture Cells
Lester Goldstein and Julie Micou Eastwood

22. Microextraction and Microelectrophoresis for Determination and Analysis of Nucleic Acids in Isolated Cellular Units
J.-E. Edström

Author Index—Subject Index

Volume II

1. Nuclear Transplantation in Amphibia
Thomas J. King

2. Techniques for the Study of Lampbrush Chromosomes
Joseph G. Gall

3. MICRURGY ON CELLS WITH POLYTENE CHROMOSOMES
H. Kroeger

4. A NOVEL METHOD FOR CUTTING GIANT CELLS TO STUDY VIRAL SYNTHESIS IN ANUCLEATE CYTOPLASM
Philip I. Marcus and Morton E. Freiman

5. A METHOD FOR THE ISOLATION OF MAMMALIAN METAPHASE CHROMOSOMES
Joseph J. Maio and Carl L. Schildkraut

6. ISOLATION OF SINGLE NUCLEI AND MASS PREPARATIONS OF NUCLEI FROM SEVERAL CELL TYPES
D. M. Prescott, M. V. N. Rao, D. P. Evenson, G. E. Stone, and J. D. Thrasher

7. EVALUATION OF TURGIDITY, PLASMOLYSIS, AND DEPLASMOLYSIS OF PLANT CELLS
E. J. Stadelmann

8. CULTURE MEDIA FOR *Euglena gracilis*
S. H. Hutner, A. C. Zahalsky, S. Aaronson, Herman Baker, and Oscar Frank

9. GENERAL AREA OF AUTORADIOGRAPHY AT THE ELECTRON MICROSCOPE LEVEL
Miriam M. Salpeter

10. HIGH RESOLUTION AUTORADIOGRAPHY
A. R. Stevens

11. METHODS FOR HANDLING SMALL NUMBERS OF CELLS FOR ELECTRON MICROSCOPY
Charles J. Flickinger

12. ANALYSIS OF RENEWING EPITHELIAL CELL POPULATIONS
J. D. Thrasher

13. PATTERNS OF CELL DIVISION: THE DEMONSTRATION OF DISCRETE CELL POPULATIONS
Seymour Gelfant

14. BIOCHEMICAL AND GENETIC METHODS IN THE STUDY OF CELLULAR SLIME MOLD DEVELOPMENT
Maurice Sussman

AUTHOR INDEX—SUBJECT INDEX

Volume III

1. Measurement of Cell Volumes by Electric Sensing Zone Instruments
R. J. Harvey

2. Synchronization Methods for Mammalian Cell Cultures
Elton Stubblefield

3. Experimental Techniques for Investigation of the Amphibian Lens Epithelium
Howard Rothstein

4. Cultivation of Tissues and Leukocytes from Amphibians
Takeshi Seto and Donald E. Rounds

5. Experimental Procedures for Measuring Cell Population Kinetic Parameters in Plant Root Meristems
Jack Van't Hof

6. Induction of Synchrony in *Chlamydomonas moewusii* as a Tool for the Study of Cell Division
Emil Bernstein

7. Staging of the Cell Cycle with Time-Lapse Photography
Jane L. Showacre

8. Method for Reversible Inhibition of Cell Division in *Tetrahymena pyriformis* Using Vinblastine Sulfate
Gordon E. Stone

9. Physiological Studies of Cells of Root Meristems
D. Davidson

10. Cell Cycle Analysis
D. S. Nachtwey and I. L. Cameron

11. A Method for the Study of Cell Proliferation and Renewal in the Tissues of Mammals
Ivan L. Cameron

12. Isolation and Fractionation of Metaphase Chromosomes
Norman P. Salzman and John Mendelsohn

13. Autoradiography with the Electron Microscope: Properties of Photographic Emulsions
D. F. Hülser and M. F. Rajewsky

14. Cytological and Cytochemical Methodology of Histones
James L. Pipkin, Jr.

15. Mitotic Cells as a Source of Synchronized Cultures
D. F. Petersen, E. C. Anderson, and R. A. Tobey

Author Index—Subject Index

Volume IV

1. Isolation of the Pachytene Stage Nuclei from the Syrian Hamster Testis
Tadashi Utakoji

2. Culture Methods for Anuran Cells
Jerome J. Freed and Liselotte Mezger-Freed

3. Axenic Culture of *Acetabularia* in a Synthetic Medium
David C. Shephard

4. Procedures for the Isolation of the Mitotic Apparatus from Cultured Mammalian Cells
Jesse E. Sisken

5. Preparation of Mitochondria from Protozoa and Algae
D. E. Buetow

6. Methods Used in the Axenic Cultivation of *Paramecium aurelia*
W. J. van Wagtendonk and A. T. Soldo

7. Physiological and Cytological Methods for *Schizosaccharomyces pombe*
J. M. Mitchison

Appendix (Chapter 7): Staining the *S. pombe* Nucleus
C. F. Robinow

8. Genetical Methods for *Schizosaccharomyces pombe*
U. Leupold

9. Micromanipulation of Ameba Nuclei
K. W. Jeon

10. Isolation of Nuclei and Nucleoli
Masami Muramatsu

11. The Efficiency of Tritium Counting with Seven Radioautographic Emulsions
Arie Ron and David M. Prescott

12. Methods in *Paramecium* Research
T. M. Sonneborn

13. Amebo-Flagellates as Research Partners: The Laboratory Biology of *Naegleria* and *Tetramitus*
Chandler Fulton

14. A Standardized Method of Peripheral Blood Culture for Cytogenetical Studies and Its Modification by Cold Temperature Treatment
Marsha Heuser and Lawrence Razavi

15. Culture of Meiotic Cells for Biochemical Studies
Herbert Stern and Yasuo Hotta

Author Index—Subject Index

Volume V

1. Procedures for Mammalian Chromosome Preparations
T. C. Hsu

2. Cloning of Mammalian Cells
Richard G. Ham

3. Cell Fusion and Its Application to Studies on the Regulation of the Cell Cycle
Potu N. Rao and Robert T. Johnson

4. Marsupial Cells *in Vivo* and *in Vitro*
Jack D. Thrasher

5. Nuclear Envelope Isolation
I. B. Zbarsky

6. Macro- and Micro-Oxygen Electrode Techniques for Cell Measurement
Milton A. Lessler

7. METHODS WITH *Tetrahymena*
L. P. Everhart, Jr.

8. COMPARISON OF A NEW METHOD WITH USUAL METHODS FOR PREPARING MONOLAYERS IN ELECTRON MICROSCOPY AUTORADIOGRAPHY
N. M. Maraldi, G. Biagini, P. Simoni, and R. Laschi

9. CONTINUOUS AUTOMATIC CULTIVATION OF HOMOCONTINUOUS AND SYNCHRONIZED MICROALGAE
Horst Senger, Jürgen Pfau, and Klaus Werthmüller

10. VITAL STAINING OF PLANT CELLS
Eduard J. Stadelmann and Helmut Kinzel

11. SYNCHRONY IN BLUE-GREEN ALGAE
Harald Lorenzen and G. S. Venkataraman

AUTHOR INDEX—SUBJECT INDEX

Volume VI

1. CULTIVATION OF CELLS IN PROTEIN- AND LIPID-FREE SYNTHETIC MEDIA
Hajim Katsuta and Toshiko Takaoka

2. PREPARATION OF SYNCHRONOUS CELL CULTURES FROM EARLY INTERPHASE CELLS OBTAINED BY SUCROSE GRADIENT CENTRIFUGATION
Richard Schindler and Jean Claude Schaer

3. PRODUCTION AND CHARACTERIZATION OF MAMMALIAN CELLS REVERSIBLY ARRESTED IN G_1 BY GROWTH IN ISOLEUCINE-DEFICIENT MEDIUM
Robert A. Tobey

4. A METHOD FOR MEASURING CELL CYCLE PHASES IN SUSPENSION CULTURES
P. Volpe and T. Eremenko

5. A REPLICA PLATING METHOD OF CULTURED MAMMALIAN CELLS
Fumio Suzuki and Masakatsu Horikawa

6. CELL CULTURE CONTAMINANTS
Peter P. Ludovici and Nelda B. Holmgren

7. ISOLATION OF MUTANTS OF CULTURED MAMMALIAN CELLS
Larry H. Thompson and Raymond M. Baker

8. Isolation of Metaphase Chromosomes, Mitotic Apparatus, and Nuclei
Wayne Wray

9. Isolation of Metaphase Chromosomes with High Molecular Weight DNA at pH 10.5
Wayne Wray

10. Basic Principles of a Method of Nucleoli Isolation
J. Zalta and J-P. Zalta

11. A Technique for Studying Chemotaxis of Leukocytes in Well-Defined Chemotactic Fields
Gary J. Grimes and Frank S. Barnes

12. New Staining Methods for Chromosomes
H. A. Lubs, W. H. McKenzie, S. R. Patil, and S. Merrick

Author Index—Subject Index

Volume VII

1. The Isolation of RNA from Mammalian Cells
George Brawerman

2. Preparation of RNA from Animal Cells
Masami Muramatsu

3. Detection and Utilization of Poly(A) Sequences in Messenger RNA
Joseph Kates

4. Recent Advances in the Preparation of Mammalian Ribosomes and Analysis of Their Protein Composition
Jolinda A. Traugh and Robert R. Traut

5. Methods for the Extraction and Purification of Deoxyribonucleic Acids from Eukaryote Cells
Elizabeth C. Travaglini

6. Electron Microscopic Visualization of DNA in Association with Cellular Components
Jack D. Griffith

7. AUTORADIOGRAPHY OF INDIVIDUAL DNA MOLECULES
D. M. Prescott and P. L. Kuempel

8. HELA CELL PLASMA MEMBRANES
Paul H. Atkinson

9. MASS ENUCLEATION OF CULTURED ANIMAL CELLS
D. M. Prescott and J. B. Kirkpatrick

10. THE PRODUCTION OF MASS POPULATIONS OF ANUCLEATE CYTOPLASMS
Woodring E. Wright

11. ANUCLEATE MAMMALIAN CELLS: APPLICATIONS IN CELL BIOLOGY AND VIROLOGY
George Poste

12. FUSION OF SOMATIC AND GAMETIC CELLS WITH LYSOLECITHIN
Hilary Koprowski and Carlo M. Croce

13. THE ISOLATION AND REPLICA PLATING OF CELL CLONES
Richard A. Goldsby and Nathan Mandell

14. SELECTION SYNCHRONIZATION BY VELOCITY SEDIMENTATION SEPARATION OF MOUSE FIBROBLAST CELLS GROWN IN SUSPENSION CULTURE
Sydney Shall

15. METHODS FOR MICROMANIPULATION OF HUMAN SOMATIC CELLS IN CULTURE
Elaine G. Diacumakos

16. TISSUE CULTURE OF AVIAN HEMATOPOIETIC CELLS
C. Moscovici and M. G. Moscovici

17. MEASUREMENT OF GROWTH AND RATES OF INCORPORATION OF RADIOACTIVE PRECURSORS INTO MACROMOLECULES OF CULTURED CELLS
L. P. Everhart, P. V. Hauschka, and D. M. Prescott

18. THE MEASUREMENT OF RADIOACTIVE PRECURSOR INCORPORATION INTO SMALL MONOLAYER CULTURES
C. R. Ball, H. W. van den Berg, and R. W. Poynter

19. ANALYSIS OF NUCLEOTIDE POOLS IN ANIMAL CELLS
Peter V. Hauschka

AUTHOR INDEX—SUBJECT INDEX

Volume VIII

1. Methods for Selecting and Studying Temperature-Sensitive Mutants of BHK-21 Cells
Claudio Basilico and Harriet K. Meiss

2. Induction and Isolation of Auxotrophic Mutants in Mammalian Cells
Fa-Ten Kao and Theodore T. Puck

3. Isolation of Temperature-Sensitive Mutants of Mammalian Cells
P. M. Naha

4. Preparation and Use of Replicate Mammalian Cell Cultures
William G. Taylor and Virginia J. Evans

5. Methods for Obtaining Revertants of Transformed Cells
A. Vogel and Robert Pollack

6. Monolayer Mass Culture on Disposable Plastic Spirals
N. G. Maroudas

7. A Method for Cloning Anchorage-Dependent Cells in Agarose
C. M. Schmitt and N. G. Maroudas

8. Repetitive Synchronization of Human Lymphoblast Cultures with Excess Thymidine
H. Ronald Zielke and John W. Littlefield

9. Uses of Enucleated Cells
Robert D. Goldman and Robert Pollack

10. Enucleation of Somatic Cells with Cytochalasin B
Carlo M. Croce, Natale Tomassini, and Hilary Koprowski

11. Isolation of Mammalian Heterochromatin and Euchromatin
Walid G. Yasmineh and Jorge J. Yunis

12. Measurements of Mammalian Cellular DNA and Its Localization in Chromosomes
L. L. Deaven and D. F. Petersen

13. Large-Scale Isolation of Nuclear Membranes from Bovine Liver
Ronald Berezney

14. A Simplified Method for the Detection of Mycoplasma
Elliot M. Levine

15. The Ultra-Low Temperature Autoradiography of Water and Its Solutes
Samuel B. Horowitz

16. Quantitative Light Microscopic Autoradiography
Hollis G. Boren, Edith C. Wright, and Curtis C. Harris

17. Ionic Coupling between Nonexcitable Cells in Culture
Dieter F. Hülser

18. Methods in the Cellular and Molecular Biology of Paramecium
Earl D. Hanson

19. Methods of Cell Transformation by Tumor Viruses
Thomas L. Benjamin

Author Index-Subject Index

Volume IX

1. Preparation of Large Quantities of Pure Bovine Lymphocytes and a Monolayer Technique for Lymphocyte Cultivation
J. Hinrich Peters

2. Methods to Culture Diploid Fibroblasts on a Large Scale
H. W. Rüdiger

3. Partition of Cells in Two-Polymer Aqueous Phases: A Method for Separating Cells and for Obtaining Information on Their Surface Properties
Harry Walter

4. Synchronization of Cell Division *in Vivo* through the Combined Use of Cytosine Arabinoside and Colcemid
Robert S. Verbin and Emmanuel Farber

5. The Accumulation and Selective Detachment of Mitotic Cells
Edwin V. Gaffney

6. Use of the Mitotic Selection Procedure for Cell Cycle Analysis: Emphasis on Radiation-Induced Mitotic Delay
D. P. Highfield and W. C. Dewey

7. Evaluation of S Phase Synchronization by Analysis of DNA Replication in 5-Bromodeoxyuridine
Raymond E. Meyn, Roger R. Hewitt, and Ronald M. Humphrey

8. Application of Precursors Adsorbed on Activated Charcoal for Labeling of Mammalian DNA *in Vivo*
George Russev and Roumen Tsanev

9. Growth of Functional Glial Cells in a Serumless Medium
Sam T. Donta

10. Miniature Tissue Culture Technique with a Modified (Bottomless) Plastic Microplate
Eliseo Manuel Hernández-Baumgarten

11. Agar Plate Culture and Lederberg-Style Replica Plating of Mammalian Cells
Toshio Kuroki

12. Methods and Applications of Flow Systems for Analysis and Sorting of Mammalian Cells
H. A. Crissman, P. F. Mullaney, and J. A. Steinkamp

13. Purification of Surface Membranes from Rat Brain Cells
Kari Hemminki

14. The Plasma Membrane of KB Cells; Isolation and Properties
F. C. Charalampous and N. K. Gonatas

15. The Isolation of Nuclei from *Paramecium aurelia*
Donald J. Cummings and Andrew Tait

16. Isolation of Micro- and Macronuclei of *Tetrahymena pyriformis*
Martin A. Gorovsky, Meng-Chao Yao, Josephine Bowen Keevert, and Gloria Lorick Pleger

17. Manipulations with *Tetrahymena pyriformis* on Solid Medium
Enore Gardonio, Michael Crerar, and Ronald E. Pearlman

18. The Isolation of Nuclei with Citric Acid and the Analysis of Proteins by Two-Dimensional Polyacrylamide Gel Electrophoresis
Charles W. Taylor, Lynn C. Yeoman, and Harris Busch

19. Isolation and Manipulation of Salivary Gland Nuclei and Chromosomes
M. Robert

Subject Index

METHODS IN CELL BIOLOGY

VOLUME X

Chapter 1

Nucleic Acid Hybridization to the DNA of Cytological Preparations[1]

MARY LOU PARDUE AND JOSEPH G. GALL

Department of Biology,
Massachusetts Institute of Technology,
Cambridge, Massachusetts
and Department of Biology,
Yale University, New Haven,
Connecticut

	Introduction	1
I.	Procedure for *in Situ* Hybridization	2
	A. Fixation and Squashing of Tissue	2
	B. Removal of Endogenous RNA	4
	C. Denaturation	5
	D. Hybrid Formation	5
	E. Removal of Nonspecifically Bound Nucleic Acid	6
	F. Autoradiography	7
	G. Staining Preparations after Autoradiography	8
II.	Equipment and Reagents	8
III.	Radioactive Nucleic Acids for Hybridization	10
IV.	Conditions of Reaction	12
V.	General Comments	14
	References	16

Introduction

Techniques for annealing two molecules of single-stranded nucleic acid and characterizing the resulting hybrid molecule make it possible to compare nucleic acid sequences even though the actual order of nucleotides

[1]Supported by research grants GB-35736 from the National Science Foundation and GM-12427 from the National Institute of General Medical Sciences.

in the sequences may not be known. When combined with cytological procedures that permit *in situ* detection of the cellular or chromosomal localization of the hybridized nucleic acid, these techniques can be used in several ways to study the organization and function of the genetic material in higher organisms. (1) *In situ* hybridization to condensed chromosomes can be used to map the sites of DNA sequences that have not been mapped by genetic breeding experiments. (2) Just as experiments with known nucleic acid sequences in the hybridizing solution can be used to identify chromosomal sites, hybridization to known chromosomal regions can be used in characterizing the nucleic acid sequences in the hybridizing solution. (3) Hybridization to cytological preparations gives information on the functional organization of particular sequences in the diffuse chromatin of interphase nuclei. This permits comparison of the organization in cells of different types. (4) Because cytological techniques permit analysis at the level of single cells, *in situ* hybridization can be used to obtain preliminary evidence on gene amplification or underreplication in single cell types in tissues composed of several cell types or cell stages. (5) The distribution of specific RNA species within a cell or in mixed cell populations can be studied by hybridization with DNA transcribed *in vitro* from that RNA by RNA-directed DNA polymerases.

Several years ago we published a general review of techniques for *in situ* hydridization (Gall and Pardue, 1971). Since that time we have continued to use and study *in situ* hybridization and have made several changes in the procedures described in our earlier article. Because there are a number of reviews on nucleic acid hybridization *in vitro* (Wetmur and Davidson, 1968; Walker, 1969; McCarthy and Church, 1970; Birnstiel *et al.*, 1972; Bishop, 1972; Britten *et al.*, 1974), and on techniques for *in situ* hybridization (Jacob *et al.*, 1971; Harrison *et al.*, 1973; Hennig, 1973; Jones, 1973; Wimber and Steffensen, 1973), this article describes only the procedures that have been used in our laboratories.

I. Procedure for *in Situ* Hybridization

A. Fixation and Squashing of Tissue

Many conventional cytological procedures can be used to make slides that give good results in *in situ* hybridization experiments. However, fixatives containing formaldehyde should not be used, since they apparently interfere with denaturation of the DNA. The best preparations are those that are well spread and very flat.

The following procedures are the ones we routinely use to make slides for *in situ* hybridization.

1. For Small Pieces of Tissue

Fresh tissue is teased into very small pieces not more than 5 mm in their longest dimension. It is then fixed for 5–10 minutes in freshly mixed ethanol–acetic acid (3:1).

Next a small drop of 45% acetic acid is placed on a siliconized 18-mm^2 cover slip. A small piece (less than 1 mm^3) is teased from the fixed tissue and transferred to this drop. Because addition of the fixative to 45% acetic acid causes violent mixing, the tissue is held for a few seconds in air to permit most of the fixative to evaporate before being placed in the acetic acid. However, the morphology may be ruined if the tissue becomes too dry in this step.

The tissue is thoroughly minced in the drop of acetic acid, and any large pieces that remain are removed. Then a "subbed" slide is carefully lowered onto the cover slip, and the cells are squashed. A simple way to squash the cells is to place the slide on filter paper with the cover slip side down and apply firm thumb pressure on the back of the slide over the cover slip. Some tough tissues will give a better preparation if they are allowed to soften on a 45°C warming plate for 3–5 minutes before being squashed. The flattest squashes tend to have the best preservation of morphology during the hybridization steps, so it is wise to use small amounts of tissue and to remove any bits of material that might interfere.

After squashing, the slides are placed on a flat surface of Dry Ice for a few minutes. When the preparation is completely frozen (5–10 minutes), the cover slip is flipped off with a razor blade and the slide is immediately placed in 95% ethanol. After the slide has been in ethanol for at least 5 minutes, it is air-dried and stored.

Slides can be stored dry for very long periods of time, although the level of hybridization tends to drop after several months. The level of hybridization drops much more rapidly if the slides are stored in ethanol.

2. For *Drosophila* Salivary Glands

The glands are dissected in insect Ringer's solution and transferred to a small drop of 45% acetic acid on a siliconized 18-mm^2 coverslip.

After the glands are allowed to fix briefly in the 45% acetic acid, a subbed slide is lowered onto the cover slip and the glands are squashed. There are many techniques for squashing salivary glands. The cover slip can be tapped with the point of a pencil, or with the eraser end, or it can be pressed with pointed forceps. The initial steps in squashing salivary glands are intended to spread out the chromosomes and should be checked in the phase micro-

When performing hybridization at high temperature, we ordinarily use an SSC or an SNB buffer at a concentration determined by the amount of sodium ion desired. For example, reactions can be done in 2× SNB at 65°C. Hybridizations can be done at lower temperatures by adding formamide to the reaction mixture. We frequently use 40% formamide (Matheson, Coleman, Bell; Norwood, Ohio) in 4 × SSC at 37°C.

Addition of a large excess of nonradioactive, noncompeting nucleic acid helps to eliminate nonspecific binding of radioactive nucleic acid to the slide.

The radioactive nucleic acid is placed over the preparation and covered with a cover slip. The slide is then placed in a moist chamber and incubated at the chosen temperature. Ordinarily we use 10–15 μl of solution under an 18-mm^2 cover slip. Much smaller volumes can be used with 9-mm^2 cover slips if the moist chamber is sealed. We routinely do 10- to 15-hour incubations at 65°C without noticeable loss of liquid from under the cover slip. Although it is possible to seal the cover slip over the liquid, it is simpler to seal the moist chambers. Such a procedure eliminates the possibility that the hybridization medium will extract some components from the sealing mixture.

Some brands of cover slips leach alkali into the incubation mix. It is useful to test a new brand of cover slip before using it. This can be done by doing a mock hybridization using the hybridization buffer plus phenol red under the cover slip. If this test shows that the cover slips are leaching alkali during the incubation, the cover slips can be boiled in 1 *N* HCl and rinsed well in pH 7 buffer before they are used.

E. Removal of Nonspecifically Bound Nucleic Acid

1. For RNA–DNA Hybrids

After the hybridization reaction the slides are removed from the oven. The cover slip and hybridization mix are immediately washed off by dipping the slide into a beaker of 2× SSC. Slides are then washed in 2× SSC at room temperature for 15 minutes before being incubated with pancreatic RNase (20 μg/ml in 2× SSC) for 1 hour at 37°C. Finally, the slides are rinsed twice for 10 minutes in 2× SSC and passed through 70 and 95% ethanol before being air-dried.

2. For DNA–DNA Hybrids

The cover slips are removed from the slides by dipping in 2× SSC at a temperature 5° less than the temperature used for the hybridization reaction. Nonspecifically bound DNA is then removed by three additional 10-minute washes in 2× SSC which is also 5° less than the hybridization temperature. Finally, the slides are washed in 70 and 95% ethanol and dried.

F. Autoradiography

The dried slides are coated with autoradiographic emulsion by standard procedures. We use Kodak NTB-2 liquid emulsion diluted 1:1 with distilled water. A thicker coating of emulsion on the slide would not increase the information gathered, since the beta particles emitted by ^3H are not energetic enough to reach the upper part of the film. However, a thicker film would increase the number of background grains produced extraneously.

Kodak NTB-2 is purchased in 4-oz bottles (112 ml) and is a solid at room temperature. It should be used only in absolute darkness or with a Kodak Wratten Series II or OA safelight. We heat each new bottle to 45°C for 30 minutes and then add it to 112 ml of warm, distilled water. The two solutions mix readily, and this step should be done carefully to avoid producing bubbles in the emulsion. Diluted emulsion is distributed into nylon scintillation vials in aliquots of about 10 ml. The vials are wrapped in aluminum foil and stored in a light-tight box in a refrigerator. Each vial contains enough emulsion for about 30 slides and is used only once. The emulsion will keep for many months. The caps used for the scintillation vials should not have cork inserts, since these can give off organic compounds which may produce background grains in the emulsion during storage.

If one suspects that the emulsion may have picked up some background grains before use, it is wise to dip a test slide, dry it for 20 minutes, and then develop and check it before dipping the experimental slides. A good emulsion has $0.1–0.2 \times 10^{-2}$ grains per square micrometer.

For use the emulsion is melted in a 45°C water bath for 10 minutes and then poured into a small plastic dipping chamber, taking care that no bubbles are formed. (A polyethylene two-place slide mailer available from most supply houses makes a good dipping chamber.)

The dipping chamber is held at 45°C. Slides are dipped into the chamber, withdrawn slowly, drained, and placed vertically in a rack. Slides should dry in the dark for 2 hours. A nonsparking fan can be used to reduce the drying time to 1 hour. Dried slides are stored in light-tight plastic boxes with a small container of silica gel included to maintain dryness. The boxes are sealed with electrician's tape and stored in lightproof containers in a refrigerator.

Autoradiographic exposures for *in situ* hybridization may range from a few hours to many days. It is wise to have several slides, so that test slides may be developed at intervals to determine the proper exposure time.

After the appropriate exposure time, slides are developed for $2\frac{1}{2}$ minutes in Kodak D-19, rinsed well in distilled water, and fixed for 2–5 minutes in

Kodak fixer. The slides are then rinsed through several changes of distilled water over a period of 30 minutes and stained before drying.

G. Staining Preparations after Autoradiography

There are several ways of staining cytological preparations through autoradiographic emulsions. We use Giemsa stain. A stock solution of Giemsa blood stain is diluted approximately 1:20 with 0.01 *M* phosphate buffer (pH 6.8) immediately before adding the slides. The staining time depends on the cytological preparation and on the batch of Giemsa stain. It is usually somewhere between 2 and 45 minutes. A metallic film forms rapidly over the surface of the Giemsa solution once it has been mixed with the buffer. Since this film will adhere to the slide if the slide is pulled out of the solution, the film is floated out of the staining dish with distilled water before removing the slides. The slides are rinsed with distilled water, air-dried, and covered with a cover-glass mounted in Permount. Slides may also be left uncovered and observed under immersion oil. Petroleum ether is used to remove oil from unmounted slides.

II. Equipment and Reagents

Subbed slides. Microscope slides are washed in either acid or detergent and thoroughly rinsed in distilled water. They are then immersed briefly in subbing solution and allowed to dry for several hours before being used.

Subbing solution. This is an aqueous solution of 0.1% gelatin and 0.01% chrome alum [chromium potassium sulfate, $CrK(SO_4)_2 \cdot 12H_2O$]. The gelatin is first dissolved in hot water, and the chrome alum is added after the solution cools.

Siliconized cover slips. Cover slips are immersed briefly in a 1% solution of Siliclad (Clay-Adams; Parsippany, N.J.), rinsed thoroughly in distilled water, and allowed to dry overnight at room temperature. They are stored dry. Before using, the cover slips can be rinsed with 95% ethanol and dried.

SSC. 0.15 *M* NaCl, 0.015 *M* sodium citrate; pH 7.0

SNB. 0.15 *M* NaCl, 10^{-3} *M* tris; pH 7.0.

Insect Ringer's solution. 7.5 gm NaCl, 0.35 gm KCl, and 0.21 gm $CaCl_2$ in 1 liter of water.

RNase. Pancreatic RNase (EC 2.7.7.16) is dissolved at 1 mg/ml in 0.02 *M* sodium acetate (pH 5). This solution is placed in a boiling water bath for for 5 minutes, cooled, and stored at $-20°C$.

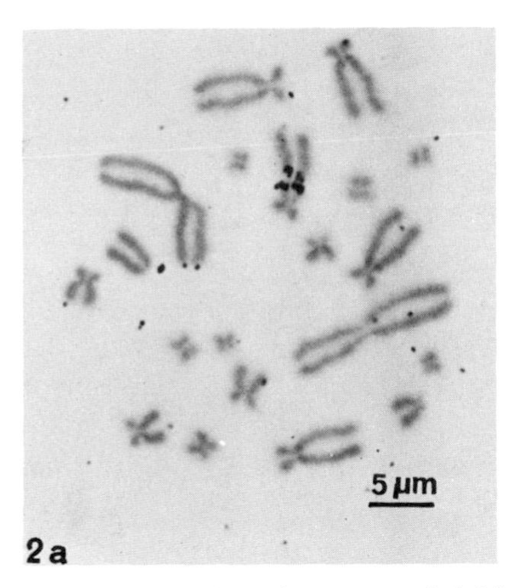

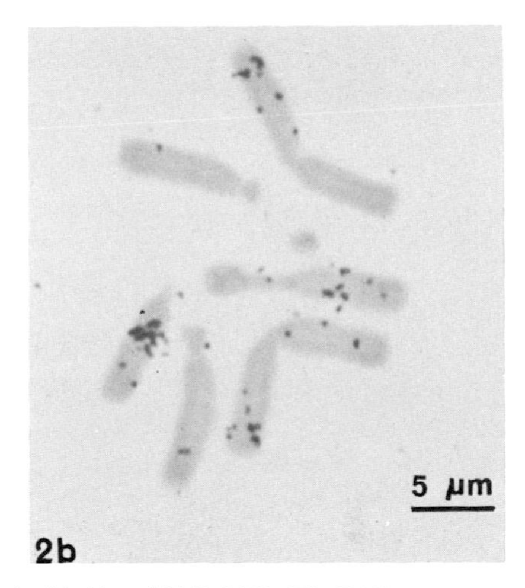

FIG. 2. Metaphase chromosomes hybridized with 18 and 28 S rRNA. The DNA sequences complementary to the rRNA are found at the secondary constrictions on the sex chromosomes in both karyotypes. The muntjac has an additional site of ribosomal sequences on the largest autosome. The specific activity of the rRNA-^{3}H was 5 $\times$ 10^6 dmp/μg. It was hybridized in 2$\times$ SNB at 65°C for 10 hours. (a) Bat, *Carollia perspicillata*. Male karyotype. rDNA only on X chromosome. Exposure 37 days. $\times$ 1445. (b) Indian mutjac, *Muntiacus muntjak*. Male karyotype. rDNA on X, Y, and one pair of autosomes. Exposure, 75 days. $\times$1870. (From T. C. Hsu, F. E. Arrighi, S. Spirito, and M. L. Pardue, unpublished.)

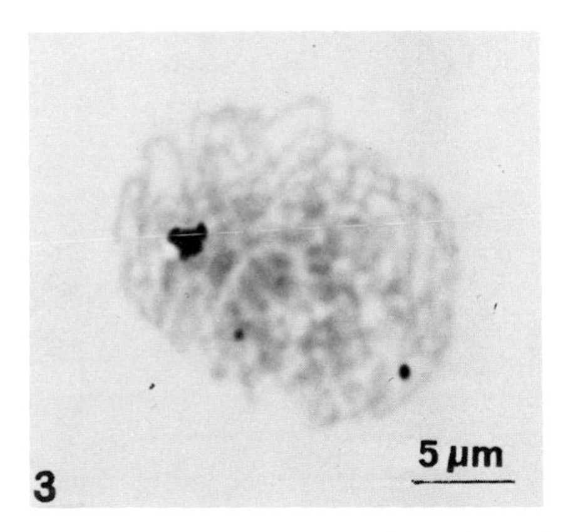

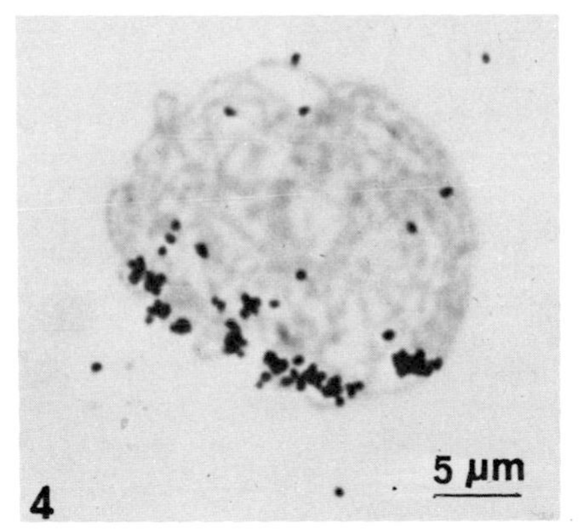

FIG. 3. Pachytene nucleus from the testis of the toad *X laevis* hybridized with cRNA-^{3}H transcribed *in vitro* from DNA coding for 18 and 28 S rRNA. In *X. laevis* the genes for 18 and 28 S rRNA are localized on one chromosome of the haploid complement, and thus the tightly paired homologous chromosomes of the pachytene nucleus show only one site of hybridization in this experiment. The cRNA-^{3}H had a calculated specific activity of 10^8 dpm/μg. Hybridization was carried out in 2$\times$ SSC for 10 hours at 65°C. Exposure, 41 days. $\times$2200.

FIG. 4. Pachytene nucleus from the testis of the toad *X. laevis* hybridized with cRNA-^{3}H transcribed *in vitro* from the DNA coding for 5 S rRNA. The ends of all of the chromosomes are oriented toward one side of the nucleus. The 5 S cRNA binds to DNA sequences located at the end of one arm of most, if not all, of the chromosomes. The cRNA-^{3}H had a specific activity of 4 $\times$ 10^7 dpm/μg. It was hybridized in 4$\times$ SNB for 8 hours at 65°C. Exposure, 26 days. $\times$1750. (From Pardue *et al.*, 1973.)

Moist chamber. (see Fig. 1). Moist chambers are made from 4 inch plastic petri plates. The bottom of the petri plate is lined with filter paper and contains 5–10 ml of the incubation buffer. The slide is supported above the liquid by two rubber grommets or a U-shaped glass rod. The solution in the bottom of the moist chamber must have the same salt concentration as the solution under the cover slip to prevent distillation and subsequent change in concentration. Glass petri plates do not make satisfactory moist chambers, because the moisture that condenses on the lid tends to coalesce into large drops which may fall onto the preparation. Moisture condensing on plastic lids remains in small droplets on the lid. Some types of plastic petri plates melt at temperatures used for hybridization reactions.

Giemsa stain. We use commercially prepared stock solutions of Giemsa blood stain.

III. Radioactive Nucleic Acids for Hybridization

The feasibility of detecting any particular DNA sequence by cytological hybridization depends on two factors, either of which may be the limiting one. The first factor is the sensitivity of detection. This depends on the amount of radioactivity in a given region and the efficiency of the autoradiographic procedure. The second factor is the rate of hybridization, which depends on the concentrations of the sequences being studied. Thus in choosing a source of radioactive nucleic acid for *in situ* hybridization experiments, one should consider both the specific radioactivity of the nucleic acid and the amounts of nucleic acid that can be obtained for use in the hybridization medium.

The minimum levels of radioactivity needed for a particular experiment are determined by the amount of nucleic acid bound at each chromosomal site. Repeated sequences clustered at a single chromosomal locus can be detected more easily than the same number of sequences scattered over sites on several chromosomes. In general, we find that 10^6 dpm/μg is the lower useful limit. There are several possible ways of obtaining nucleic acid with a specific activity greater than 10^6 dpm/μg.

1. *In Vivo* Labeled Nucleic Acid

Cultured cell lines grown in a medium containing tritiated nucleic acid precursors can yield both DNA and RNA with sufficient tritium label for use in *in situ* hybridization (Gall and Pardue, 1971). If culture and labeling conditions are optimized for the particular nucleic acid fraction desired, both stable and rapidly turning over fractions can be obtained with specific

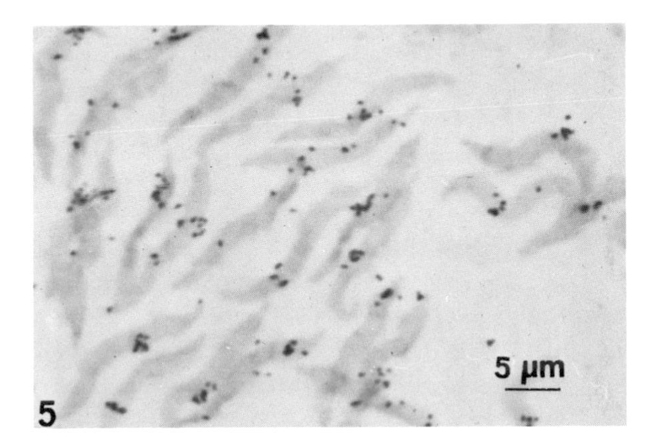

FIG. 5. Spermatozoa from *X. laevis* hybridized with cRNA-^{3}H transcribed *in vitro* from DNA coding for 18 and 28 S rRNA. The rRNA genes are apparently packed into the sperm head in a tight cluster, but the position of this cluster is not constant from sperm to sperm. The conditions for hybridization are those given in Fig. 3. Exposure, 31 days. $\times 1400$.

radioactivity of 10^6–10^7 dpm/μg. There are other biological systems that can be used to produce highly radioactive nucleic acids *in vivo*. For example, sea urchin embryos grown in sea water containing tritiated uridine during the early cleavage stages yield histone mRNA containing enough tritium label for *in situ* experiments (Pardue *et al.*, 1972).

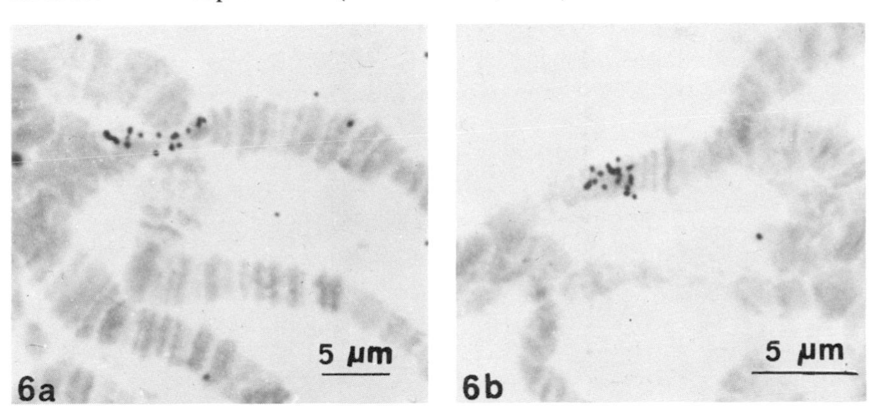

FIG. 6. Salivary gland chromosomes from *Drosophila melanogaster* hybridized *in situ* with 9 S mRNA-^{3}H from the sea urchin *Psammechinus milaris*. These stretched chromosomal preparations show hybridization of the putative histone mRNA fractions to bands 39D and 39E, as well as the intervening interband region. The 9-S mRNA-^{3}H was prepared from sea urchin blastulas grown in sea water containing uridine-^{3}H. The RNA was hybridized in $4\times$ SNB at 65°C for 10 hours. Exposure, 109 days. (a) $\times 1600$; (b) $\times 2600$. (From M. L. Pardue, L. H. Kedes, E. H. Weinberg, and M. L. Birnstiel, unpublished.)

2. *In Vitro* Transcription of Nucleic Acid

By far the most successful method for preparing extremely radioactive nucleic acid is by *in vitro* transcription. In cases in which a specific DNA fraction can be isolated, *Escherichia coli* RNA polymerase can be used *in vitro* to produce RNA copies of the DNA (Gall and Pardue, 1971). RNA fractions can be transcribed into DNA by RNA-directed DNA polymerases (Verma and Baltimore, 1974), or by DNA polymerase from *E. coli* (Loeb *et al.*, 1973). The product of an *in vitro* transcription should be characterized carefully, since any of the polymerases may transcribe certain nucleic acid sequences preferentially. When a mixture of nucleic acids is transcribed, the product may not be equally transcribed from all fractions of the template.

The specific radioactivity of nucleic acids obtained from *in vitro* transcription is limited only by the specific activity of the nucleotide precursors. The highest specific activity precursors currently available yield a product with greater than 10^8 dpm/μg.

3. Chemical Labeling of Preformed Nucleic Acids

In cases in which the desired nucleic acid cannot be labeled either *in vivo* or by *in vitro* transcription, radioactive atoms can be added chemically. Although there are several methods for chemical addition of ^{3}H to nucleic acids, none of the methods now in use yields the levels of radioactive labeling required for *in situ* hybridization. However, techniques have been developed for labeling nucleic acids with ^{125}I (Commerford, 1971; Prensky *et al.*, 1973). The specific radioactivity obtained with ^{125}I is about equal to that produced by *in vitro* transcription with ^{3}H precursors. Theoretically, ^{125}I-labeled nucleic acids of even higher specific activity can be produced, but the characteristics of hybridization may be affected by the increased iodination (Commerford, 1971).

The radiation emitted by ^{125}I is somewhat more energetic than that emitted by ^{3}H. For this reason ^{125}I is more efficient than ^{3}H for autoradiography, but gives less precise cytological localization.

IV. Conditions of Reaction

We have assumed that *in situ* hybridization shares some of the kinetic properties of the nitrocellulose filter hybridization technique. Our experimental evidence is consistent with this assumption, and we use filter

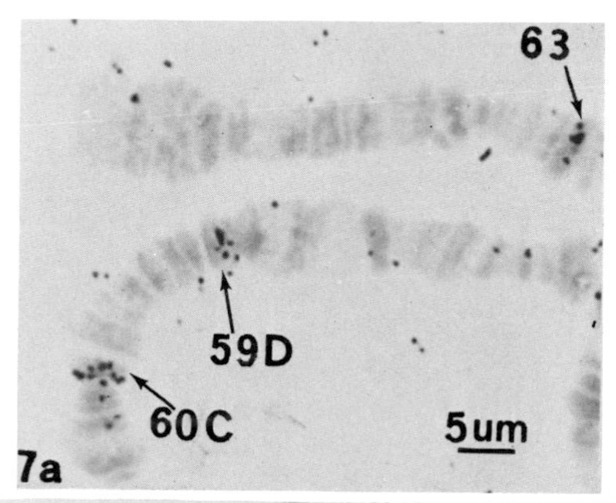

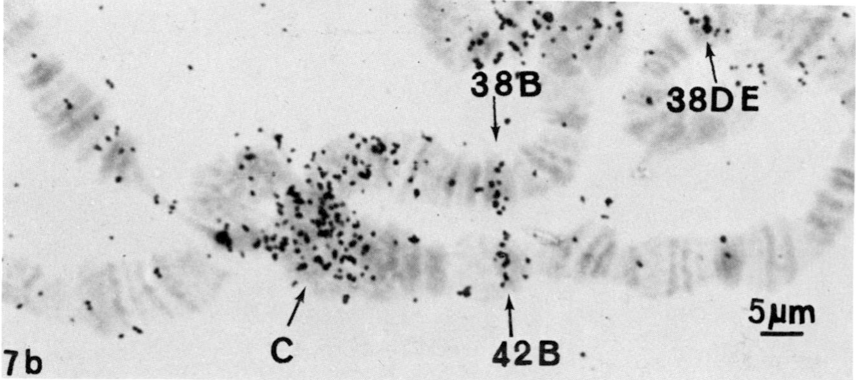

FIG. 7. Salivary gland chromosomes from *D. melanogaster* hybridized *in situ* with cytoplasmic RNA-^{3}H. The RNA was prepared from the cytoplasmic fraction of cultured *D. melanogaster* cells grown in a medium with uridine-^{3}H. These poly-A-containing sequences were isolated by oligo-dT chromatography and hybridized in $2\times$ SNB at 65°C for 10 hours. This RNA binds to a reproducible series of chromosomal bands and to the β-heterochromatin of the chromocenter (c). Exposure, 35 days. (a) $\times 1100$; (b) $\times 1400$. (From A. Spradling, S. Penman, and M. L. Pardue, unpublished.)

hybridization experiments when necessary to optimize reaction conditions for *in situ* hybridization.

In both *in situ* and filter hybridizations, one strand of the nucleic acid involved in the hybridization reaction is immobilized. We have not been able to detect either loss or renaturation of the DNA of the cytological preparation during the periods we have used for the annealing reaction, although there may be losses of DNA during the denaturation steps. *In situ* hybridization does differ from filter hybridization in having extremely high local concentrations of DNA sequences and in having some of the protein components of chromatin still associated with the DNA. We have not yet been able to assess the effect of either of these factors on the kinetics of hybridization *in situ*.

Precise quantitation of *in situ* hybridization experiments has proved difficult for several technical reasons. One source of variation is the degee of compaction of the DNA in different nuclei. Whether the compaction affects the amount of denaturation of the DNA, the access of the hybridizing nucleic acid to chromosomal sites, or the autoradiographic efficiency, it makes quantitative comparisons between different types of nuclei questionable in some cases. In many studies it is preferable to use information from *in situ* experiments to design filter hybridization experiments for precise quantitation.

Published *in situ* hybridization experiments have been calculated to have 5–15% efficiency of hybridization. Approximately the same efficiency is obtained whether the hybridization is to the diffuse amplified ribosomal genes in amphibian oocytes, to repeated sequences in heterochromatic regions of condensed chromosomes, or to repeated genes in euchromatic portions of chromosomes. The low efficiency of hybridization may be a reflection of the availability of the DNA in cytological preparations. However, it is also possible that none of the hybridizations were done with saturating concentrations of nucleic acid in solution.

It is unlikely that the concentrations and times now used for *in situ* hybridization reactions would permit the detection of unique sequences in the DNA of diploid cells of higher organisms. However, unique sequences can easily be studied in polytene chromosomes in which the close alignment of multiple chromatids provides high local concentrations of the sequences. Each metaphase chromosome in the toad *Xenopus laevis* has about 1000 copies of the sequences coding for 5 S rRNA. These can be detected by *in situ* hybridization using *in vivo* labeled 5 S RNA (Pardue *et al.*, 1973). The lateral repetitions of a unique gene in the polytene chromosomes of *Drosophila* salivary glands should be no more difficult to detect than the 5 S genes on a metaphase chromosome in *X. laevis*. A polytene band of 10^3 chromatids, each containing a single copy of a gene coding for a typical protein (a DNA sequence of 10^5–10^6 daltons, perhaps), should be a much better target for hybridization than the estimated 10^6–10^7 daltons of DNA detected when a *X. laevis* metaphase chromosome is hybridized with 5 S RNA.

V. General Comments

One of the first problems we encountered in our experiments on cytological hybridization was that of preserving nuclear morphology during the denaturation step. In our early experiments we dipped the slides into 0.5% agar just before the alkali denaturation. However, we have since found that, for preparations that are very flat and thoroughly dried to the slide, such protection is not necessary. Slides are dried several times during the hybridization procedure to ensure the adherence of the preparation. When we began our experiments we were concerned that basic proteins might either interfere with the hybridization reaction or cause nonspecific binding of the radioactive nucleic acid. However, a large fraction of the histones is extracted by the ethanol–acetic acid fixation and subsequent 45% acetic acid treatment (Dick and Johns, 1967), and this is sufficient to permit hybridization.

Much protein, of course, remains in a cytological preparation and is a possible source of nonspecific binding of radioactive nucleic acid. In our earlier article on the technical details of *in situ* hybridization (Gall and Pardue, 1971), we recommended treating preparations with 0.2 *N* HCl to further remove basic protein before hybridization. Since that time we have found that treating a preparation with 0.2 *N* HCl significantly reduces the amount of *in situ* hybridization that can occur. We have not determined whether the HCl treatment simply depurinates the DNA or whether it also disposes the DNA to loss during the subsequent steps of the hybridization procedure. We now omit the 0.2 *N* HCl treatment entirely. Instead we add an excess of nonradioactive, noncompeting nucleic acid to the hybridization mixture to reduce nonspecific binding to chromosomal proteins.

We originally avoided the use of RNase to remove endogenous RNA, because in filter hybridizations the enzyme binds to the filter and causes nonspecific attachment of radioactive RNA (Gillespie and Spiegelman, 1965). However, we found that treating *Xenopus* oocyte preparations with RNase before denaturation did not increase the background binding. Instead, it noticably improved the specific hybridization of ribosomal RNA, presumably by removing unlabeled ribosomal RNA from the tissue. We now routinely include the RNase step, although we have not investigated its effect on the hybridization of RNAs other than ribosomal RNA.

The genotypes of several organisms contain chromosomal pairs that are so similar that they can not be distinguished simply on the basis of size and centromere position. Recently, staining techniques have been developed that produce characteristic banding patterns on each pair of chromosomes in some organisms (reviewed in Hsu, 1973). We have found that it is possible to use at least one of these powerful techniques for chromosome identification, quinacrine banding, in conjunction with *in situ* hybridization (Evans *et al.*, 1974).

In our experiments human chromosomes were stained with quinacrine, and the fluorescence banding patterns of selected metaphase plates were photographed by the techniques of Evans *et al.* (1971). After the chromosomes had been photographed, the slides were washed for 5 or 6 hours with a gentle stream of running tap water to remove the quinacrine. The slides were dried and used for *in situ* hybridization experiments. By comparing the autoradiographs with the photographs of fluorescent-banded chromosomes, it was possible to identify unambiguously the chromosomes that had bound the radioactive RNA.

In choosing a staining technique for the identification of chromosomes before *in situ* hybridization, it is important to note that some chromosome stains can cause exposure of autoradiographic emulsions by chemical reactions. If such stains are used and not completely washed out, autoradiography may show chromosome regions that have a high affinity for the stain, as well as regions that bind the radioactive RNA.

References

Atherton, D., and Gall, J. G. (1972). *Drosophila Inform. Serv.* **49**, 131–133.

Birnstiel, M. L., Sells, B. H., and Purdom, I. F. (1972). *J. Mol. Biol.* **63**, 21–39.

Bishop, J. O. (1972). *Biochem. J.* **126**, 171–185.

Britten, R. J., Graham, D. E., and Neufeld, B. R. (1974). *In* "Nucleic Acids and Protein Synthesis" (L. Grossman and K. Moldave, eds.), Methods in Enzymology, Vol. 29, 363–418. Academic Press, New York.

Commerford, S. L. (1971). *Biochemistry* **10**, 1993–2000.

Dick, C., and Johns, E. W. (1967). *Biochem. J.* **105**, 46P.

Evans, H. J., Buckton, K. E., and Sumner, A. T. (1971). *Chromosoma* **35**, 310–325.

Evans, H. J., Buckland, R. A., and Pardue, M. L. (1974). *Chromosoma* **48**, 405–426.

Gall, J. G., and Pardue, M. L. (1971). *In* "Nucleic Acids Part D" (L. Grossman and K. Moldave, eds.), Methods in Enzymology, Vol. 21, 470–480. Academic Press, New York.

Gillespie, D., and Spiegelman, S. (1965). *J. Mol. Biol.* **12**, 829–842.

Harrison, P. R., Cronkie, D., Paul, J., and Jones, K. (1973). *FEBS* (*Fed. Eur. Biochem. Soc.*) *Lett.* **32**, 109–112.

Hennig, W. (1973). *Int. Rev. Cytol.* **36**, 1–44.

Hsu, T. C. (1973). *Annu. Rev. Genet.* **7**, 153–176.

Jacob, J., Todd, K., Birnstiel, M. L., and Bird, A. (1971). *Biochim. Biophys. Acta.* **228**, 761–766.

Jones, K. W. (1973). *New Tech. Biophys. Cell Biol.* **1**, 29–66.

Loeb, L., Tartof, K. D., and Travaglini, E. C. (1973). *Nature* (*London*), *New Biol.* **242**, 66–69.

McCarthy, B. J., and Church, R. B. (1970). *Annu. Rev. Biochem.* **39**, 131–150.

Pardue, M. L., Kedes, L. H., Weinberg, E. H., and Birnstiel, M. L. (1972). *J. Cell Biol.* **55**, 199a.

Pardue, M. L., Brown, D. D., and Birnstiel, M. L. (1973). *Chromosoma* **42**, 191–203.

Prensky, W., Steffensen, D. M., and Hughes, W. L. (1973). *Proc. Nat. Acad. Sci. U.S.* **70**, 1860–1864.

Verma, I. M., and Baltimore, D. (1974). *In* "Nucleic Acids and Protein Synthesis" (L. Grossman and K. Moldave, eds.), Methods in Enzymology, Vol. 29, 125–130. Academic Press, New York.

Walker, P. M. B. (1969). *Progr. Nucl. Acid Res. Mol. Biol.* **9**, 301–326.

Wetmur, J. G., and Davidson, N. (1968). *J. Mol. Biol.* **31**, 349–370.

Wimber, D. E., and Steffensen, D. M. (1973). *Annu. Rev. Genet.* **7**, 205–223.

Chapter 2

Microanalysis of RNA from Defined Cellular Components

BO LAMBERT AND BERTIL DANEHOLT

Department of Histology,
Karolinska Institutet,
Stockholm, Sweden

I. Introduction . . . 17
II. Experimental Material . . . 18
III. Labeling Conditions . . . 20
A. Incubation of Isolated Glands . . . 20
B. Bathing . . . 22
C. Injection . . . 22
IV. Preparation and Fixation . . . 23
V. Microdissection . . . 24
VI. Microextraction . . . 27
VII. Microelectrophoresis in Gels . . . 29
A. Preparation of Gels . . . 30
B. Application of RNA Sample . . . 31
C. Electrophoresis . . . 32
D. Recording . . . 33
VIII. Microhybridization . . . 34
A. Extraction and Specific Activity Determination of RNA . . . 36
B. Preparation and Immobilization of DNA on Microfilters . . . 38
C. Hybridization . . . 40
D. Sensitivity and Reliability of Microassays . . . 41
E. Measurements of Radioactivity . . . 43
IX. Combination of Micro- and Macrotechniques . . . 44
X. Applications . . . 45
References . . . 46

I. Introduction

In biochemical studies of RNA metabolism in eukaryotic cells, many investigators have found it necessary to adopt cell fractionation techniques.

With conventional procedures subcellular components can often be obtained reasonably clean and in large quantities, and can subsequently be characterized by standard biochemical methods. In many cases, however, this general approach is not feasible for various reasons, and other procedures have to be applied. For certain cell types microdissection can be the optimal fractionation technique. For example, it is possible to isolate defined giant chromosomes, or segments of such chromosomes, by microdissection but not by conventional cell fractionation methods. The small amounts of material obtained by microdissection can be further analyzed by biochemical microtechniques. Moreover, if RNA is labeled, macrotechniques can be applied as well.

Some of the microprocedures now available for RNA analysis of subcellular components in the salivary glands of *Chironomus tentans* are described in this article. Micromanipulation and tools used in such work (micromanipulators, dissection needles, micropipettes, etc.) have been treated extensively in earlier reviews (Edström, 1964; Diacumakos, 1973; Edström and Neuhoff, 1973) and are only briefly discussed. In our presentation of microtechniques we exemplify their application to *Chironomus* material, but it is evident that the techniques can be applied to other types of suitable biological material as well.

II. Experimental Material

The midge *C. tentans* can be preferably raised in the laboratory. A detailed description of suitable culture conditions have been given by Beermann (1952). Besides hibernated alder-tree leaves and cellulose tissue, as recommended by Beermann, nettle powder and sodium chloride (0.4 gm/liter) are also added.

The salivary glands of the larvae provide the most suitable cells for microwork. Each larva has a pair of salivary glands located in the second and third thoracic segments. The glands produce a secretion which forms a gelatinous, tubelike burrow for the larva during its life in water on the bottom of the rearing container. A single gland contains 30 to 40 cells (Fig. 1). These are unusually large, because the gland increases in size during larval development by growth of cells in the absence of cell division. Parallel to this process the chromatids replicate 12 to 13 times and remain side by side to form giant polytene chromosomes. Each cell has four, somatically paired chromosomes (Fig. 2); three of them (chromosomes I, II, and III) are about equally long (about 200 μm), while the fourth (chromosome

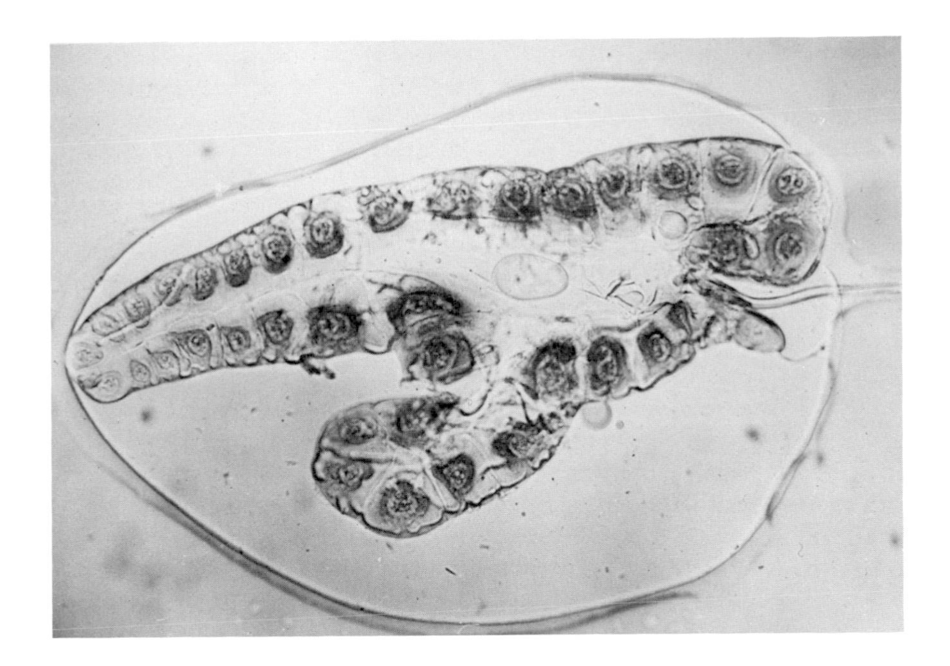

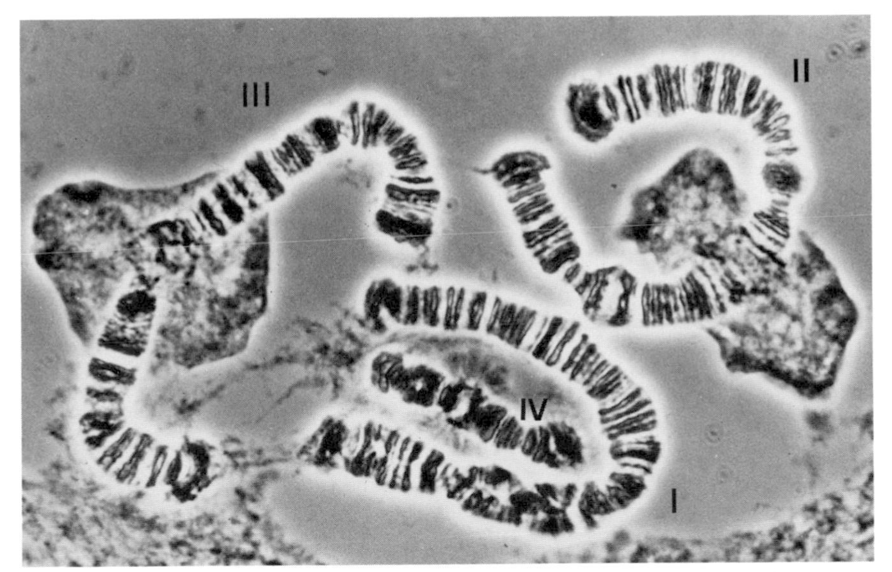

FIG. 1. Salivary gland of *C. tentans*. Magnification × 50.

FIG. 2. Polytene chromosomes (I to IV) from a salivary gland cell of *C. tentans*. Chromosomes II and III carry a nucleolus in the middle of the chromosome, while chromosome IV has a giant puff, Balbiani ring 2, located in about the center of the chromosome. Magnification × 400.

IV) is considerably shorter (about 80 μm). The chromosomes are banded transversally, and the distribution of the bands is constant and specific for each chromosome (Beermann, 1952). A band on a polytene chromosome consists of a large number of homologous chromomeres which are likely to be units of transcription and replication (see, e.g., Beermann, 1965).

Two of the chromosomes (II and III) have a prominent nucleolus. Furthermore, all four chromosomes have a certain number of so-called puffs, i.e., diffuse, swollen, RNA-containing bands. About 10% of all bands in a given cell are in this puffed condition. Puffs may vary in location in different tissues and during development. On the fourth chromosome there are three very large puffs, the so-called Balbiani rings (Balbiani rings 1, 2, and 3). These are likely to be responsible for the synthesis of mRNA for specific salivary polypeptides. The polytene chromosomes with their giant size, well-defined structure, and their puffs, especially the Balbiani rings, make them suitable material for studies on chromosomal structure and function (for reviews, see, e.g., Pelling, 1972; Daneholt 1974).

III. Labeling Conditions

RNA in the salivary glands of *C. tentans* can be efficiently labeled by the *incubation* of isolated glands in a medium containing radioactive RNA precursors (Edström and Daneholt, 1967). Alternatively, glands *in situ* can be exposed to labeled precursors. This can be achieved by addition of the precursors directly to the culture (*bathing*) (Daneholt and Hosick, 1973a), or by *injection* of the isotopes into the body cavity of the larvae (Ringborg *et al.*, 1970). These techniques have been modified in different ways since they were first published. At the present time they are performed as follows.

A. Incubation of Isolated Glands

Larvae (preferably 5–8 weeks old) are removed one by one from the culture and placed on a filter paper. The surplus of culture medium is adsorbed, and major contaminants from the culture (nettle powder, cellulose tissue, etc.) are removed from the larva with a forceps. Preferably, the larva is rinsed for a short time in a water solution and again placed on a filter paper. The larva is then transferred to a glass slide and cut with a scalpel through the first segment (prothorax). Usually the two glands, situated in the second and third segment, float out with the blood. If not, gentle pressure with a needle on the larval body close to the cut will extrude the glands into the pool of blood. The glands are immediately soaked with incubation medium (for composition see below) to prevent drying, and can then easily be lifted up on a steel or glass needle and transferred for incubation. The incubation medium is essentially of the composition described by Cannon (1964), but it has been modified to a certain extent. The inorganic salts and pH have been altered to correspond more closely to ionic composition, osmolarity, and pH of larval hemolymph in *Chironomus* (J.-E. Edström, unpublished). Methionine is added in a larger amount, according to Greenberg (1969). Besides penicillin, streptomycin is also included to prevent bacterial growth. Finally, the medium is supplied with glutamine immediately before use.

The composition of the incubation medium (in milligrams per 300 ml) is as follows.

Inorganic salts. $Na_2HPO_4 \cdot 2H_2O$, 356; $MgCl_2 \cdot 6H_2O$, 93.5; KCl, 82; NaCl, 420; $Na_2SO_4 \cdot 10H_2O$, 2000; $CaCl_2 \cdot 2H_2O$, 56.

Sugars. glucose, 140; fructose, 80; sucrose, 80, trehalose, 1000.

Organic acids. malic, 134; α-ketoglutaric, 74; succinic, 12; fumaric, 11.

Amino acids. L-arginine–HCl, 140; DL-lysine–HCl, 250; L-histidine, 500; L-aspartic acid, 70; L-asparagine, 70; L-glutamic acid, 120; L-glutamine, 120; glycine, 130; DL-serine, 220; L-alanine, 45; L-proline, 70; L-tyrosine, 10; DL-threonine, 70; L-methionine, 180; L-phenylalanine, 30; DL-valine; 40; DL-isoleucine, 20; DL-leucine, 30; L-tryptophan, 20; L-cystine, 5; cysteine–HCl; 16.

Vitamin B complex. thiamine hydrochloride, 0.004; riboflavin, 0.004; nicotinic acid, 0.004; pantothenic acid, 0.004; biotin, 0.004; folic acid, 0.004, inositol, 0.004; choline, 0.004.

Other compounds. cholesterol, 6; penicillin, 12; streptomycin sulfate, 30; phenol red, 20.

The solutes are dissolved in redistilled water. Inorganic salts, except $CaCl_2$, are dissolved in 60 ml; $CaCl_2$ in 15 ml; sugars and organic acids in 30 ml; amino acids (except L-glutamine) in 75 ml; cholesterol, penicillin, streptomycin, and phenol red in 5 ml; vitamins in 1000 ml (1 mg of each component), from which 4 ml is used. The solutions are mixed, $CaCl_2$ being added last. The pH is adjusted to 7.2 by addition of 1 *N* NaOH, and water is added to bring the total volume to 300 ml. The solution is filtered through a Millipore filter, divided into 2.5-ml portions, and stored frozen. When medium is needed, a single portion is rapidly thawed and L-glutamine is added (1 mg/2.5 ml). Medium left over is discarded. Fresh medium is made about every second month.

Prior to incubation solutions containing labeled RNA precursors [usually 100 μCi tritiated cytidine (>20.000 mCi/mmole; 1 mCi/ml) and 100 μCi tritiated uridine (40.000–60.000 mCi/mmole; 1 mCi/ml)] are evaporated on the bottom of an evaporating dish, the incubation medium is added (usually 25–50 μl), and the compounds are allowed to dissolve for 30 minutes at 18°C. The isolated glands are then added to the incubation medium (one to four glands per 25–50 μl). The dish is carefully sealed with a glass plate and silicone grease and kept at 18°C in a water bath (Heto, Birkeröd, Denmark).

B. Bathing

Larvae are maintained under standard cultivation conditions, and labeled RNA precursors are added directly to the culture medium. In most experiments 15 larvae are kept in 20 ml of culture medium supplemented with 400 μCi of cytidine and 400 μCi of uridine. However, if a very high incorporation is essential, three larvae can be bathed in only 1 ml of medium.

C. Injection

The labeled precursor solution (tritiated cytidine and/or uridine) is evaporated, and the compounds afterward dissolved in 0.67% NaCl to proper concentrations (usually 10–25 μCi/μl). A micropipette used for injection is loaded with 1 μl of this solution. [This is preferably done by placing the 1 μl as a hanging drop in an oil chamber as described below (Section V) and then sucking the solution into the micropipette with the aid of a micromanipulator. The pipette should contain no liquid paraffin.]

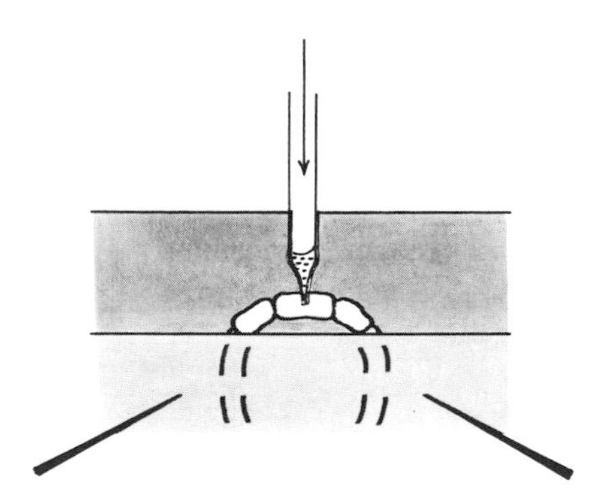

FIG. 3. Injection of labeled precursors into the larval body cavity. The larva is placed between two filter papers and kept in this position by two needles. Some of the intermediate segments are exposed, and the injection is made into one of them (usually the fourth or fifth).

The injection is preferably done under stereomicroscopic control. The larva is nailed down between two filter papers on a cork support, as shown in Fig. 3. A few of the intermediate segments are exposed, while the remainder of the larva is concealed between the two filter papers. To keep the larva wet, the filter papers are blotted with culture medium. The micropipette directed by the micromanipulator is then pushed through the skin in the middle of one of the exposed segments (usually segment 4 or 5), and the solution is expelled into the body cavity. If the larva is not fixed in a very tense state, there is as a rule no reflux of hemolymph when the pipette is withdrawn. The larvae are then kept under standard culture conditions for the incubation time required by the specific experiment.

IV. Preparation and Fixation

Salivary glands are isolated as described above (Section III,A), incubated if required, and then placed on a cover slip with a steel or glass needle in a flat, well out-stretched position. This can best be achieved if the glands are transferred into a droplet of a solution, e.g., incubation medium, deposited on the cover slip. The fiuid is then sucked away by a pipette, and the glands settle on the cover slip in an optimal way. This is important to facilitate later micromanipulation. Keeping the glands on a cover slip also simplifies their subsequent transfer through a series of solutions. In order to prevent evaporation, the cover slip with the glands is quickly placed in a fixation medium.

The glands are fixed, transferred through three rinsing media, and finally into a dissection medium (Edström and Daneholt, 1967). Media (about 15 ml) are kept cold (0°–4°C) in glass vials (25-ml Packard scintillation counting vials). In our routine procedure (Daneholt, 1972), fixation is carried out in a freshly prepared ethanol–acetic acid mixture (3:1) for 30 minutes. The rinsing step consists of three washes in 70% ethanol for 10 minutes each time. The glands are then finally placed in ethanol–glycerol (1:1), which is kept for about 60 minutes in the cold and then stored in a freezer (−20°C) until microdissection is carried out. If highly labeled giant RNA is studied, the sample should not be stored more than 24 hours prior to analysis because of the possibility of radioactive decomposition. This fixation procedure is essentially apolar and has proven very efficient in providing glands with optimal properties for subsequent dissection; it also keeps the RNA undegraded and prevents redistribution of RNA in the cell during the isolation. Buffers can be added to the media, but do not seem to improve the procedure.

The fixation medium can be varied to achieve particular properties of the glands and its components. Most important is the addition of formaldehyde to the fixation medium (final concentration 4%) (Edström and Daneholt, 1967). This treatment in some way stabilizes the chromosomes so that in a subsequent release of chromosomal RNA (Section VI) the giant chromosomes expand but still remain recognizable, while in the ordinary procedure they vanish. To keep the chromosomes visible and relatively compact is important, as the chromosomal material can then be separated mechanically from the extract.

V. Microdissection

After the glands have been fixed, different cellular components can be isolated by micromanipulation (Edström and Beermann, 1962; Edström and Daneholt, 1967). The cover slip with the two glands is removed from the ethanol–glycerol mixture, and the glands with some adhering medium are transferred to a clean cover slip with a steel needle. This cover slip is then inverted and placed on a thick glass plate with a transverse groove in such a way that it bridges the groove. Liquid paraffin is then added in the groove in between the glass plate and the cover slip; a so-called oil chamber is formed (Fig. 4). In this arrangement the glands are continuously bathed in a small remaining volume of ethanol–glycerol. Immersion of the glands and the dissection medium in liquid paraffin prevents a rapid change in the medium due to evaporation of ethanol.

The oil chamber is put on the stage of a Zeiss microscope (Standard GFL)

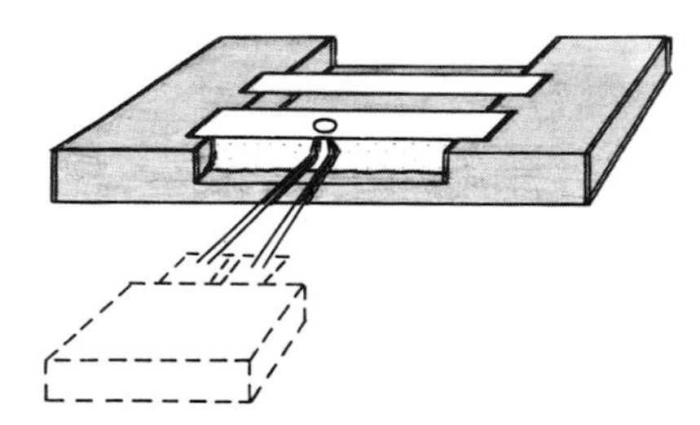

FIG. 4. Oil chamber. A cover glass (12 × 30 mm) is placed over a 25-mm-wide groove in a glass slide. Liquid paraffin fills the space under the cover glass. Glands for microdissection or incubation droplets for hybridization are kept on the lower surface of the cover glass. Glass needles for dissection or a micropipette for transfer of droplets are introduced in the groove, and their management is controlled by a micromanipulator. (Modified from Lambert *et al.*, 1973).

equipped with phase-contrast optics. The microdissection is performed on this stage at room temperature and can be controlled at all times through the microscope (for details, see Edström, 1964). Two glass needles connected to a de Fonbrune micromanipulator (Etablissements Beaudouin, Paris, France) are used to carry out the dissection and can easily be introduced into the oil chamber through one of the two open sides of the chamber (Fig. 4). During this process the gland sticks to the under surface of the cover slip and is constantly surrounded by liquid paraffin. One needle keeps the gland in position, while the other is used for the real dissection. Nuclear material can be isolated from a cell and a further subfractionation performed as demonstrated in Fig. 5. The individual chromosomes are freed from each other, and adhering nuclear sap is removed and collected. The nucleoli on chromosomes II and III are separated from the chromosome arms. Finally, Balbiani ring 2 is isolated from chromosome IV (Fig. 6). If cytoplasm is to be analyzed, it is feasible first to isolate whole cells. Pure cytoplasm can then be collected from the peripheral part of the cell, always some distance from the nuclear envelope.

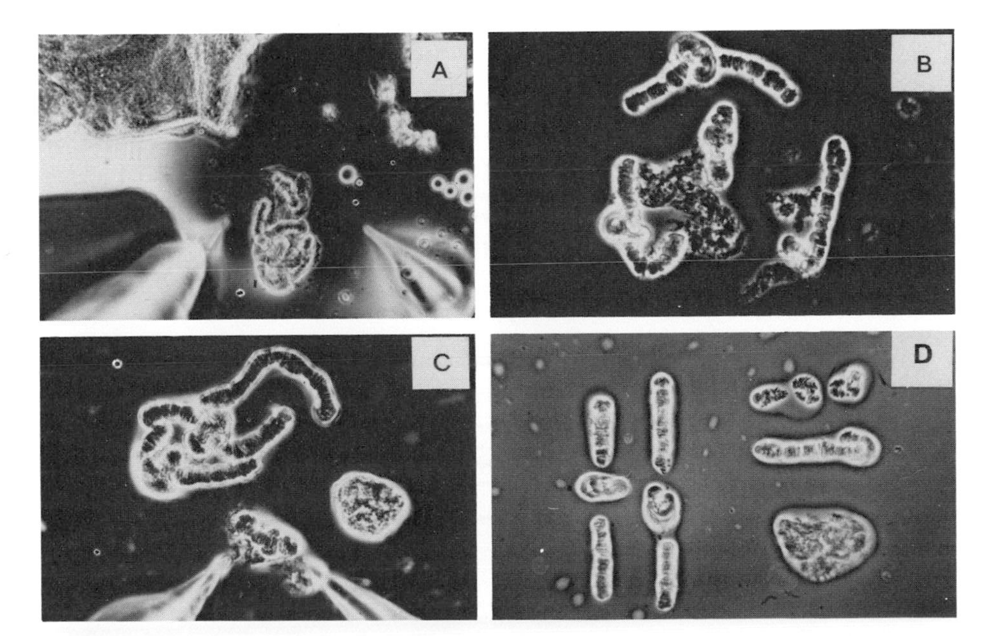

FIG. 5. Different steps in the microdissection procedure. The cell is opened up by two dissection needles, and the chromosome package pulled out from the nucleus (A). The four chromosomes are separated (B), and the nuclear sap is removed from the chromosomes (C). Finally, the nucleoli are isolated from the remainder of chromosomes II and III, and Balbiani ring 2 is isolated from chromosome IV. The nuclear sap is collected separately (D). Magnification: (A) $\times 80$; (B–D) $\times 125$.

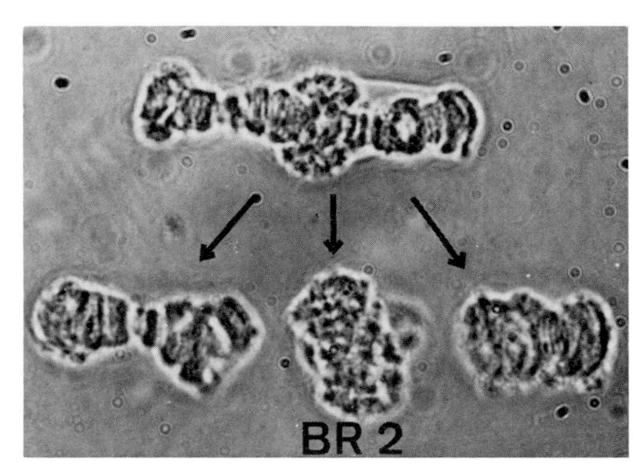

FIG. 6. Micromanipulatory isolation of Balbiani ring 2 from chromosome IV in *C. tentans*. The figure shows an intact chromosome IV (above) and a sectioned one (below) with Balbiani ring 2 in the middle segment. Magnification ×550.

If high-precision chromosome cuts are needed, two micromanipulators can be used simultaneously (Daneholt and Edström, 1969). Two dissection needles stretch the chromosome, and a third needle divides the chromosome between defined bands (Fig. 7).

There are certain aspects of micromanipulation that should be commented. First, it permits a series of well-defined components to be obtained

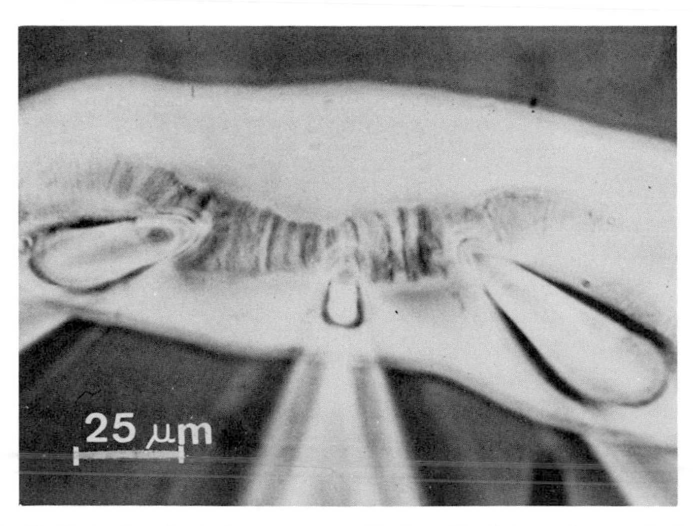

FIG. 7. Cutting of a giant chromosome with the aid of two micromanipulators. The chromosome is kept in a stretched condition with two dissection needles, while a third intermediate one is used for the cutting. (From Daneholt and Edström, 1969).

essentially free from contamination. This is true in particular for nucleoli, nuclear sap, and cytoplasm, whose purity can be directly checked in the microscope. The absence of nuclear contamination in the cytoplasmic sample can, for example, be easily avoided, as the intact nuclei can be continuously observed and only cytoplasm distant from the nuclear envelope collected. When chromosomes or parts of them are dissected, however, it is impossible to avoid minor amounts of nuclear sap material attached to the surface of the chromosomes, and above all between the individual chromatids. In most analyses, however, this contamination is negligible and, even in the case of Balbiani ring 2, an essentially clean product is obtained. Second, it is also important that redistribution of RNA species during isolation of material is minimized. Immediate fixation prior to dissection prevents redistribution to the extent that even the movements of low-molecular-weight RNA can be studied (Egyházi and Edström, 1972). Third, it should be pointed out that degradation due to nucleases is absent or very low. Isolation in apolar media, instead of the polar media conventionally used in biochemical studies, is likely to minimize the action of nucleases.

VI. Microextraction

RNA from fixed cellular components can preferably be extracted in a sodium dodecyl sulfate (SDS)–pronase solution (Pelling, 1970; Daneholt, 1972). The detergent SDS (Serva, Heidelberg) releases RNA from proteins, as has been known for a long time (Kay and Dounce, 1953; Kurland, 1960). The presence of pronase (free of nucleases; Calbiochem, Los Angeles, Calif.) has proven advantageous in minimizing degradation, particularly of the largest RNA molecules. Furthermore, if formaldehyde-containing fixatives are used, pronase is necessary in order to achieve a high yield of RNA.

Because of the small amounts of RNA available and the demand for relatively high RNA concentrations in a subsequent analysis of the sample by microtechniques, the RNA of the different components must be obtained in small volumes. This requirement can be fulfilled by carrying out the extraction in small droplets surrounded by liquid paraffin in an oil chamber (Egyházi *et al.*, 1968; Lambert *et al.*, 1973). The pooled components obtained by microdissection are placed as separate groups on the lower surface of a cover slip in an oil chamber. A droplet of the extraction medium, a buffered SDS–pronase solution (5 mg/ml SDS, 1 mg/ml pronase, pH 7.4) is applied to the sample with a micropipette so that the sample is extracted

A

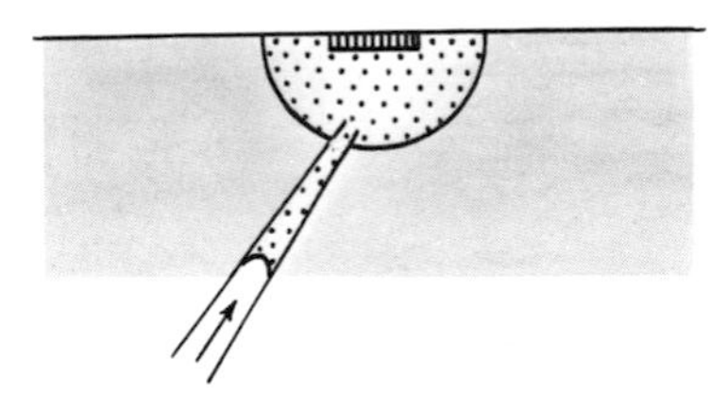

B

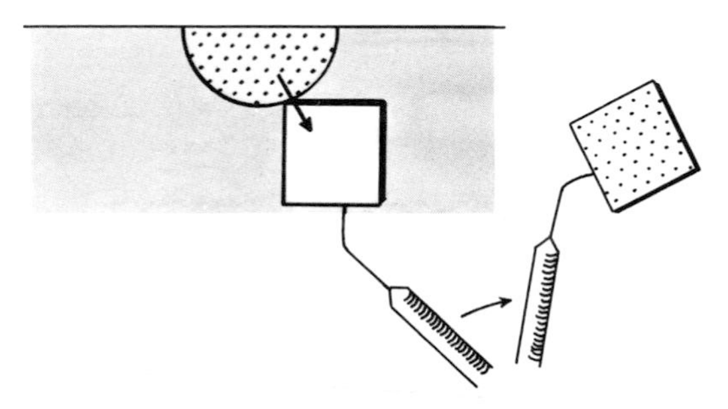

FIG. 8. Extraction of RNA. A sample is placed on the lower surface of a cover slip in an oil chamber arrangement. A drop of an SDS–pronase solution is added with a micropipette to extract the RNA (A). The extract is then absorbed on a filter paper and subsequently transferred on a glass needle (B).

in a hanging droplet in a liquid paraffin environment as depicted in Fig. 8A. It is important that the extraction solution is added in quantities giving rapid solubilization of the isolated components (volumes of the order of 20 to 50 times the volume of the component are necessary). (The extraction medium is preincubated at 37°C for at least 30 minutes prior to use.)

The digestion is carried out at room temperature or 37°C. The incubation time depends on the type of fixation used prior to the extraction. If formaldehyde is included in the fixative, the extraction is as a rule carried out for 3 hours at 37°C (Edström and Daneholt, 1967). In the microscope it can be observed that nucleoli and Balbiani rings disappear within 20 minutes, while the chromosomes expand only slowly and are still clearly visible after 180 minutes at 37°C. If formaldehyde is absent during fixation, extraction can be carried out at room temperature and a complete release of RNA is obtained after 3–5 minutes (Daneholt, 1972). The solubilization of nucleoli,

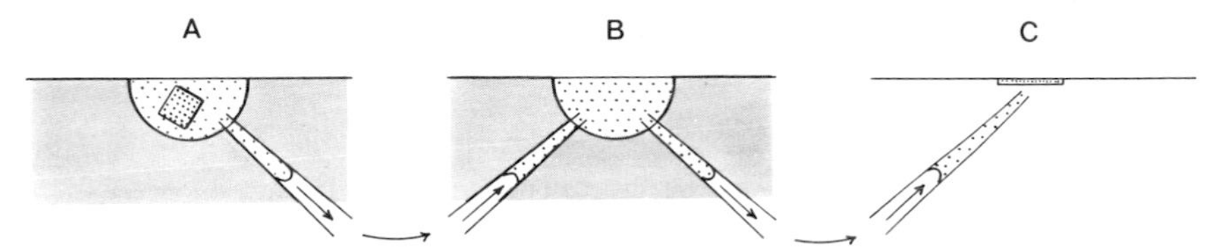

FIG. 9. Elution of RNA from filter paper. RNA is eluted from the filter paper in a hanging drop of distilled water surrounded by liquid paraffin (A). The eluate is transferred to another part of the chamber (B). The procedure is repeated five times. The total eluate is then brought to the under surface of a cover slip for evaporation (C).

Balbiani rings, and chromosomes is very rapid, and none of the nuclear components can be recognized after a few minutes.

After complete extraction a piece of filter paper (Munktell's Swedish filter paper 1F), is inserted with a steel needle through the oil into the chamber, and the digest is absorbed as shown in Fig. 8B. The size of the filter is 0.1–1 mm^2, depending on the volume of the droplet. The filter paper is removed from the oil chamber on a glass needle, rinsed for about 1 minute in chloroform to remove adhering liquid paraffin, dried in air, and transported to a great excess of 70% ethanol containing 0.2 *M* potassium acetate, where it is stored for a minimum of 30 minutes or, if convenient, overnight. This procedure removes the pronase–SDS, while the RNA remains precipitated on the filter. Hence the filter paper serves as an easily handled container. To elute RNA, the filter paper is removed from the potassium acetate–ethanol solution, rinsed for 30 minutes in 70% ethanol, and dried in air. It is then returned to the oil chamber, where repeated elutions are carried out in droplets of distilled water (Fig. 9). After five successive elutions less than 15% of the total radioactivity remains on the filter paper. The five droplets are pooled into one which after completed elution is transferred in small portions to a dry cover slip in the oil chamber and evaporated. This procedure serves to concentrate the RNA by evaporating it on a small spot. The RNA can then be dissolved again in a small volume of a suitable solution and further analyzed.

The extraction procedure as described above can be used in the range of 10^{-10}–10^{-6} gm of RNA.

VII. Microelectrophoresis in Gels

Electrophoresis can be scaled down over a wide range by adopting a system in which the overall dimensions are reduced, diffusion is restricted by greatly increased viscosity, and the migration rate kept sufficient by a very steep voltage gradient. In spite of such a high field strength, there is no problem of inadequate cooling because there is a favorable ratio of surface to volume. These principles were first worked out by Edström (1964), who applied them in the analysis of RNA base composition by microelectrophoresis of RNA hydrolysates on cellulose fibers. Gel electrophoresis has been similarly modified in a procedure described below, which permits fractionation of native RNA in amounts of 10^{-9}–10^{-10} gm (Daneholt *et al.*, 1968; Ringborg *et al.*, 1968). Different types of gels, such as agarose (Daneholt *et al.*, 1968), polyacrylamide (Egyházi *et al.*, 1968), and composite agarose–polyacrylamide (Ringborg *et al.*, 1968), can be used in gel microelectrophoresis. Agarose gel was chosen because of its firm but flexible consistency and its transparency to ultraviolet light. The fluid consistency of polyacrylamide gels of low percentage prevented their use on a microscale, but it was found that polyacrylamide–agarose composite gels could be handled on a small scale like agarose gels and polyacrylamide gels of high percentage. For analyses of high-molecular-weight RNA, we found that agarose (4%) and composite (2.75% polyacrylamide and 1% agarose) are most convenient, while high-percentage polyacrylamide gels (10%) are preferred in studies of low-molecular-weight RNA.

A. Preparation of Gels

Our standard gels are prepared as follows.

Agarose gel (4%). 1 gm of agarose is mixed with 25 ml of electrophoretic buffer (0.02 *M* NaCl, 0.002 *M* EDTA in 0.02 *M* tris–HCl, pH 8.0) and heated in a boiling water bath for 10 minutes. The solution is cooled slightly, poured into a Plexiglas mold (85 × 50 × 0.1 mm), and allowed to gel. From the resulting agarose block, thin strips (30 × 4 × 0.1 mm) are cut with a razor blade. These are immediately transferred to a glucose solution of high viscosity (4 gm liquid glucose per milliliter of buffer). The strips are equilibrated in this solution at room temperature for 6 hours or more, and then stored refrigerated for months if needed.

Polyacrylamide gel (10%). A stock solution of 15% recrystallized acrylamide and 1% recrystallized bisacrylamide is prepared in the electrophoretic buffer (see above). One milliliter of this stock solution is mixed with 400 μl of 1% bisacrylamide, 50 μl of 10% ammonium persulfate solution, and 50 μl of the electrophoretic buffer. Finally, 5 μl of *N*, *N*, *N′*, *N′*-tetramethylethylene-diamine is added and, after shaking, the mixture is poured into a Plexiglas mold as described above and left to polymerize for 2–3 minutes. Strips are cut, placed in a glucose-buffer solution, and stored as described for agarose gels.

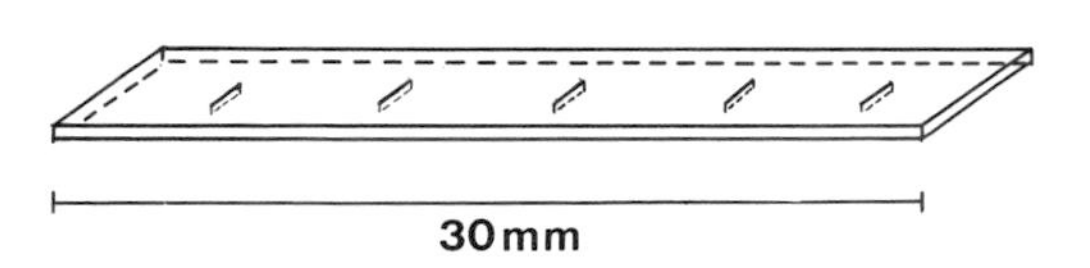

FIG. 10. A gel used for microelectrophoresis. The slits made for the applications are shown.

Composite gels (1% agarose–2.75% polyacrylamide). An agarose gel is prepared as described above by mixing 0.1 gm of agarose with 10 ml of electrophoretic buffer. The size of the Plexiglas mold is chosen to produce an agarose slice 0.3 mm thick (a 1% agarose slice, 0.1 mm thick, does not give acceptable stiffness). The slice is cut to 7.5 × 3.0 cm and transferred to a vial with 20 ml of electrophoretic buffer containing acrylamide (2.75%) and bisacrylamide (0.138%) monomers and equilibrated overnight at 4°C. The following day 10% ammonium persulfate (0.33 ml/gm acrylamide) and *N*,*N*,*N'*,*N'*-tetramethylethylenediamine (0.033 ml/gm acrylamide) are added, and the agarose slice is placed between two Plexiglas slides in the acrylamide solution. After 30–60 minutes the acrylamide polymerizes, and the composite gel is liberated and cut into strips (30 × 2 × 0.3 mm) and transferred to a solution of 4 gm of liquid glucose per milliliter of electrophoretic buffer. (If the gels adsorb considerably in ultraviolet, they can be washed for 24 hours in buffer at 4°C prior to immersion in glucose–buffer.) The procedure of Peacock and Dingman (1968) was also tried on a microscale, but gels of that type are difficult to handle because they are not firm enough to be manipulated.

Before the gel is used, excess moisture is removed from the surface by repeated transfers onto a clean glass surface; the gel is then placed on a quartz glass (38 × 25 × 0.5 mm; Zeiss, Oberkochen, West Germany) parallel to the long side. To permit application of RNA extracts, fine slits are cut in the gel with a piece of a razor blade (Fig. 10). The slit walls stick to each other so that the slit is well closed. The gel is immediately covered with liquid paraffin to avoid evaporation, as the gel cannot be exposed to air for more than 1 minute. The quartz glass with the gel slice is then inverted in an oil chamber and is ready for application of the sample.

B. Application of RNA Sample

RNA is extracted from the different cellular components as described above (Section VI). A micropipette is used to dissolve the sample (usually 500–1000 ng RNA) in a droplet of electrophoretic buffer (about 0.1 nl). [If only low-molecular-weight RNA is analyzed, these fractions are selectively dissolved from the total RNA extract with 1.5 *M* KCl in 0.02 *M* tris–HCl (pH 8.0), as described by Egyházi *et al.* (1968).] The RNA solution is sucked

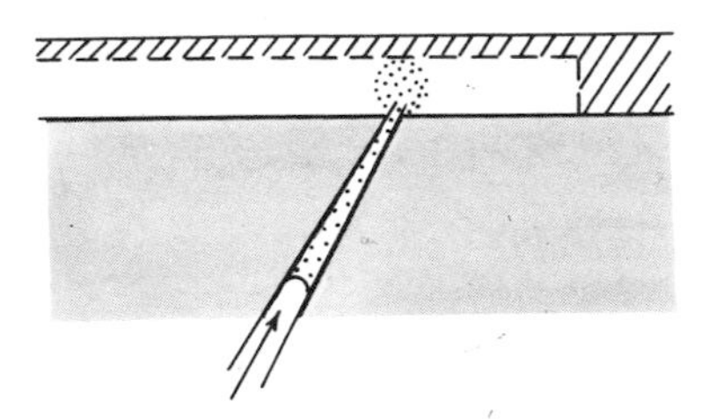

FIG. 11. Application of RNA sample for microelectrophoresis. The RNA sample is introduced into the slit with a micropipette. Several applications can be made in the same slit, which makes it possible to run different samples in parallel.

into the pipette and slowly introduced into a slit in the gel (Fig. 11). Usually the application is 100–200 μm long and not more than 5–10 μm thick. For effective applications we recommend a pipette with a pore diameter of 10–15 μm and an angle of 60°.

The slits are usually made considerably longer than is necessary for one RNA sample, so that several extracts can be applied in the same slit without mixing. This procedure is particularly useful when reference RNA is run parallel to the sample. Furthermore, as a rule slits are made at intervals along the gel; thus many separations can be run simultaneously in a single gel (Fig. 10).

C. Electrophoresis

The quartz glass with the gel strip is transferred to an electrophoretic chamber of the type described by Wieme (1965), but made smaller so that it fits the quartz glass used (Fig. 12). The gel strip forms a bridge between two blocks of 2% agarose in electrophoretic buffer. Each block is connected to

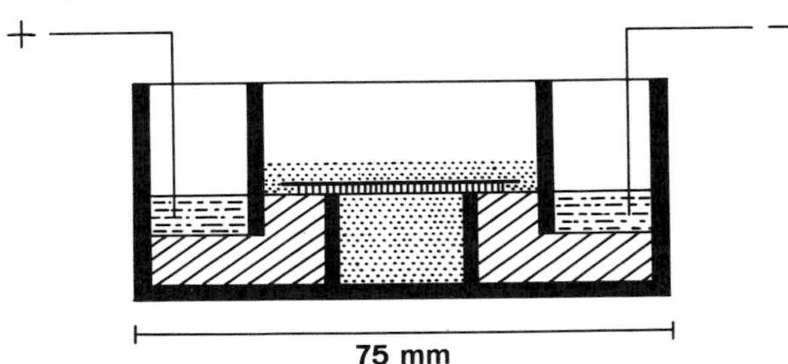

FIG. 12. Electrophoretic chamber for microelectrophoresis. A quartz glass with a gel strip (containing the applications of RNA samples) is inverted and placed in the electrophoretic chamber. The gel strip (vertical lines) bridges two gel blocks (hatched areas), each of which is connected to a buffer compartment (horizontal lines). The gel and the cover slip are immersed in liquid paraffin (dotted).

a buffer compartment containing a platinum electrode. The tension is obtained by a LKB power supply (Type 3290B, LKB-Produkter AB, Stockholm, Sweden). Because the demands on cooling are moderate, liquid paraffin instead of petroleum ether can be used to cover the gel during the run.

The separation is carried out in the cold with a constant field strength of about 150 V/cm and a constant current of about 0.1 mA. In standard agarose and composite gels, the run lasts for 1–2 hours, and in polyacrylamide gel (10%) for 2–3 hours. During this time 4 S RNA migrates 250–500 μm and 100–200 μm, respectively.

D. Recording

The separation of RNA species can be visualized and photographed during and after the run in an ultraviolet microscope at a wavelength of 265 nm with a total magnification of $\times 100$. The ultraviolet picture shows the position of the major bands. The photograph can be subjected to densitometric tracing with a Vitatron densitometer (Vitatron, Amsterdam, The Netherlands).

Improved contrast can be obtained by staining with acridine orange (C.I. 46005, S. No. 902; Merck, Darmstadt, West Germany) (Ringborg *et al.*, 1968). This procedure permits analysis of smaller amounts, which gives better resolution. The quartz glass with the gel is immersed in 10% trichloroacetic acid for 30 minutes at 4°C. The gel is then transferred to distilled water for 2 minutes, then to 10^{-4} *M* acridine orange in distilled water for 5 minutes, and finally to distilled water for 30 minutes. The gel is then examined in a Zeiss fluorescence microscope (with a mercury high-pressure lamp HBO 200 W, absorption filter Bg 12, transmission filter 53, and a total magnification of $\times 80$). Under these conditions RNA bands appear red against a dark background.

Alternatively, for inspection in a light microscope, staining can be done with methylene blue or toluidine blue. Before staining it is necessary to immerse the gel in 10% trichloroacetic acid for 30 minutes at 4°C to remove the glucose and precipitate the RNA in the gel. The gel is then dried on a glass slide, forming a 10 μm-thick film which can be stored.

Microelectrophoresis has a resolution not yet comparable to that of large-scale electrophoresis, but it is superior to that of sucrose density centrifugation. A separation of about 10^{-9} gm of RNA prepared from *C. tentans* is illustrated in Fig. 13, together with a densitometer curve of a photograph of a separation stained with acridine orange. The various bands have been identified in separate experiments using isolated ribosomal and 4 S RNA fractions obtained by sucrose density gradient centrifugation.

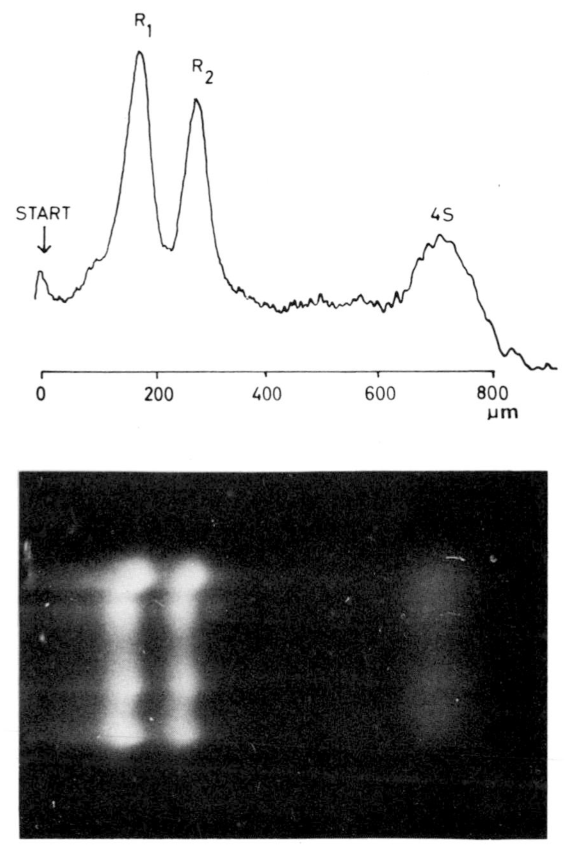

FIG. 13. Microelectrophoretogram of RNA from larval salivary glands of *C. tentans*. The separation was performed in a 4% agarose gel and stained with acridine orange and photographed in a fluorescent microscope. The densitometer scanning of the photograph is shown above the separation. (from Daneholt *et al.*, 1968).

VIII. Microhybridization

A quantitative microassay for RNA–DNA hybrids is useful in situations in which either DNA or RNA, or both reacting components, are available only in greatly limited amounts, as is often the case when microdissection is used for extracting RNA from defined chromosomal or cellular components. The microassay combines a scaled-down version of the membrane filter method for RNA–DNA hybridization (Gillespie and Spiegelman, 1965) with methods for the determination of microamounts of nucleic acids by ultraviolet microphotometry (Edström, 1964). The procedure, as described here, has been adopted for study of the salivary glands of *C. tentans* (Lambert, 1972; Lambert *et al.*, 1973), but should be applicable in any situation in which the amounts of RNA or DNA are limited, provided RNA of high specific activity can be obtained.

A diagram of the different steps of microhybridization is presented in Fig. 14. The preparation of labeled RNA has been described (Section III) and is only briefly commented on here. DNA is extracted by conventional biochemical methods, denatured, and immobilized on a cellulose nitrate filter. From this "parent" filter, small microfilters with a surface area of 0.2–0.4 mm^2 are cut. The DNA content per surface area is measured, and the microfilters used for hybridization. To determine the specific activity of the labeled RNA, a small sample of the original RNA extract is used. The RNA content of this sample is measured by ultraviolet microphotometry, and its total radioactivity is then measured in a scintillation counter.

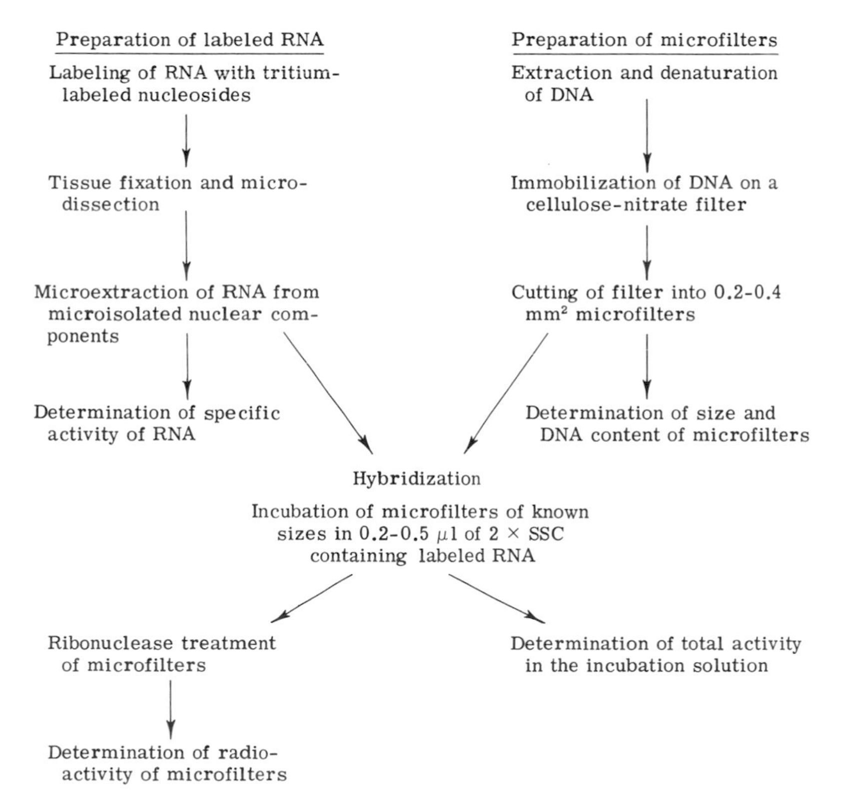

FIG. 14. Different steps in microhybridization. Explanations of the various procedures are given in the text. (from Lambert *et al.*, 1973).

therefore transferred to a dry cover slip, evaporated (Fig. 15F), and redissolved in a droplet of RNase (RAF, Worthington; Freehold, New Jersey) containing volatile buffer (100 μg RNase per milliliter of 0.2 *M* ammonium bicarbonate buffer, pH 7.6) (Fig. 15H). The digestion takes place in the oil chamber, which is incubated for 45 minutes at 37°C (Fig. 16A). The subsequent procedure essentially follows the technique developed by Edström (1964) for the determination of microamounts of RNA by ultraviolet microphotometry. The enzyme droplet is deposited on a quartz glass and evaporated (Fig. 16B), and the extract is redissolved in a droplet of a glycerol–phosphate buffer with a refractive index similar to that of the surrounding paraffin oil (Fig. 16C). The droplet is then photographed in the ultraviolet microscope at 265 nm and $\times$104 magnification, together with an optical density reference system (Fig. 16D). The photographic plate is used to prepare a photometer curve of the droplet and the reference system (Fig. 16E). The total absorption of the RNA in the droplet is determined from the photometer curve using an integrating device (Fig. 16F), and the total amount of RNA in the droplet is calculated, taking the optical density reference and the total magnification into account. After the RNA content has been determined, the droplet is absorbed onto a piece of filter paper, and the tritium activity of the RNA is determined by liquid scintillation counting (Section VIII,E).

B. Preparation and Immobilization of DNA on Microfilters

The method of choice for extracting DNA depends on the material used, as clearly stated in the explicit review on this subject by Travaglini (1973). DNA from frozen larvae of *Chironomus* is obtained by a detergent–phenol method (Lambert *et al.*, 1973). The DNA is further purified by centrifugation in CsCl, and dialyzed against 0.1 $\times$ SSC.

To prepare DNA-loaded microfilters, the DNA is first denatured at a concentration of 20 μg/ml in 0.1 $\times$ SSC by the addition of 0.1 *N* NaOH to a final pH of 12.5. After 10 minutes at room temperature, the DNA solution is neutralized by the addition of 0.1 *N* HCl. The absorption should be read before and after the denaturation to make sure that the hyperchromicity is about 30%. The denatured DNA solution is brought to 2 $\times$ SSC and filtered through a 30-mm cellulose nitrate filter (Sartorius MF 50, Sartorius Membranfilter, GmbH, West Germany). The filtering speed should not exceed 1 ml per minute, and the concentration of DNA in the solution should be calculated to yield a final amount of approximately 0.1 $\mu g/mm^2$ effective filtration area. The filter is washed by passing it through 50–100 ml 2 $\times$ SSC at moderate speed, and subsequently dried in air overnight, followed by "baking" in a vacuum for 2 hours at 80°C.

The border of the filter is removed, so as to leave only the effective filter area. This is then cut with a razor blade into small pieces differing in size between 0.1 and 0.3 mm^2. Each parent cellulose nitrate filter yields several hundred such microfilters which are then sorted into different size classes by measuring their surface areas in the microscope with the aid of an ocular micrometer.

Quantitative interpretation of the hybridization data requires exact knowledge of the amount of DNA on each microfilter. For practical reasons, however, it is desirable to avoid DNA determination for each microfilter, and more convenient to use the easily measured size of the microfilter as an indicator of its DNA content. For this purpose several microfilters of different sizes and originating from different parts of the parent filter are selected. Their surface areas are first measured as described above, and then they are washed in 70% ethanol to remove the salts, air-dried, and covered with a droplet of DNase solution [DPFF, Worthington, 100 μg/ml of 0.002 *M* ammonium bicarbonate buffer (pH 7.0) containing 0.003 *M* $MgCl_2$] in the oil chamber. Incubation then proceeds for 45 minutes at 37°C. This extraction is repeated twice, and additional extractions should be frequently performed to check that all DNA becomes released from the microfilter. The extracts are pooled on a quartz glass and evaporated, and their contents of DNA determined in the ultraviolet microscope as described for RNA Section VIII,A). The DNA content per unit filter area is determined from the average of 8 to 10 separate measurements. As shown in Table I, there is an insignificant variation in DNA content among groups of microfilters

TABLE I

DISTRIBUTION AND RETENTION OF DNA ON MICROFILTERS[a]

Sample number	Microfilter treatment	DNA content of microfilters[b]	
		Parent filter A	Parent filter B
1	Untreated	97.1 ± 9.4	68.5 ± 7.7
2	Untreated	98.8 ± 11.1	67.3 ± 11.1
3	Heated in 2 × SSC at 63°C for 12 hours	—	69.6 ± 9.8

[a] Two sets of four microfilters each (samples 1 and 2) were cut from different places on the parent cellulose nitrate filter, and the surface area of each microfilter measured in the microscope. The DNA content of each microfilter was determined as described in the text, and related to its surface area. A third set of microfilters (sample 3) was subjected to incubation under conditions similar to those of hybridization before the DNA contents of the microfilters were determined.

[b] Micrograms $\times 10^{-3}$ per square milliter of filter area; mean value ± S.D. ($n = 4$).

from different original positions on the parent filter. Included in Table I is the important control that the DNA is indeed retained on the microfilters under hybridization conditions, i.e., incubation for 12 hours at 63°C in 2 × SSC.

The amount of DNA per unit filter area is then used to estimate the content of DNA on microfilters used for hybridization, simply by measuring the size of the filter in the microscope. An even more rapid and probably more reliable method is to utilize different isotope labeling of RNA and DNA. The information required would then be only the specific activity of DNA, and the rather time-consuming microdetermination of DNA could be replaced by radioactivity measurement.

A control filter must be included in every hybridization reaction to serve as a check for unspecific reactions and background activity. Commerical bacterial DNA preparations may serve this purpose, and it is appropriate to use DNA of a guanine-plus-cytosine content similar to that being investigated. Control microfilters are prepared in exactly the same way as described above, but it is convenient to cut them in a different shape, to distinguish them easily from the experimental filters.

C. Hybridization

Incubation of microfilters with labeled RNA is carried out in the oil chamber. One square microfilter containing *C. tentans* DNA and one triangular microfilter containing bacterial DNA for control are placed close to each other on a cover slip, and a 0.1 to 0.5-μl droplet of 2 × SSC is applied on top of the microfilters with a Hamilton syringe (Fig. 17A). The cover slip is immediately transferred to an oil chamber. Part of the 2 × SSC solution is withdrawn into a micropipette and used to dissolve the evaporated RNA extract, which is contained in the same oil chamber but on a separate cover slip (Fig. 17B). The dissolved RNA is reinjected into the droplet, and the whole chamber is incubated at 63°C for 12 hours. After incubation the droplet is absorbed onto a 0.5 to 1-mm^2 piece of filter paper (Fig. 17C), and its total radioactivity is measured. The remaining activity on the microfilters after removal of the droplet is negligible compared to the activity in the droplet. Accordingly, the activity of the incubation droplet may serve as a measure of the total amount of RNA present during the reaction. The microfilters are removed from the cover slip on the tip of a glass needle (Fig. 17D), washed free from liquid paraffin with chloroform, dried, and rinsed in 2 × SSC. They are then treated with RNase (100 μg/ml of 2 × SSC preheated 5 minutes at 90°C) for 60 minutes at 37°C, followed by several washes in 2 × SSC. After drying in air the microfilters are analyzed for radioactivity (Section VIII,E).

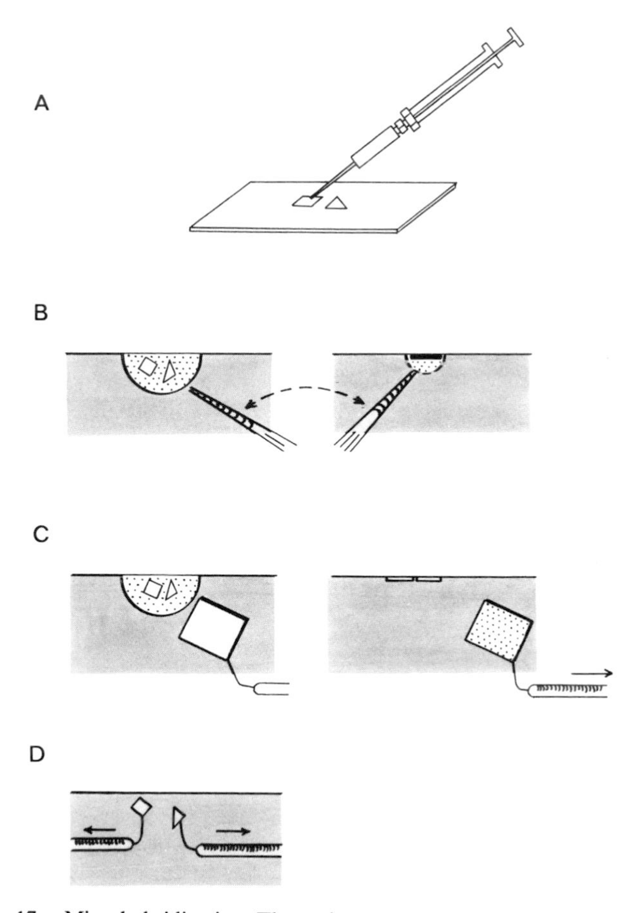

FIG. 17. Microhybridization. The various steps are presented in the text.

D. Sensitivity and Reliability of Microassays

The sensitivity of the hybridization, i.e., the minimum amount of hybridized RNA that can be detected, depends entirely on the specific activity of the RNA involved in the reaction. Microassays offers no special advantages over other hybridization techniques in this respect.

As shown in Table II, 25–30 pg of hybridized RNA can be detected at an input of 5–30 $\times$ 10^{-3} μg of RNA and DNA, provided the specific activity of the RNA is above 10^6 cpm. The sensitivity of the technique, as well as the minimum amounts of RNA and DNA required, can well be decreased below these values if the specific activity of the RNA is increased, and if the DNA is measured by isotope labeling instead of photometrically, as

TABLE II

MICROHYBRIDIZATION OF NUCLEOLAR RNA[a]

Experiment number	RNA specific activity (cpm/pg)	RNA input (cpm)	DNA input (μg $\times$ 10^{-3})	Input ratio RNA/DNA	Hybridized activity		
					Sample (cpm)	Control (cpm)	DNA in hybrid (%)
1	2.15	11,150	16.7	0.31	58	3	0.16
2	3.30	42,400	14.1	0.91	84	6	0.18
3	3.02	109,000	15.3	2.35	112	4	0.24

[a]Salivary glands of *C. tentans* were incubated for 90 minutes with ^{3}H-labeled nucleosides, fixed, and prepared for microdissection. The numbers of nucleoli isolated in experiments 1, 2, and 3 were 50, 150, and 300, respectively. The nucleolar RNA was liberated by pronase-SDS treatment, and its specific activity determined as described in the text. Incubation was in 0.3 μl droplets of 2 $\times$ SSC containing one microfilter loaded with *C. tentans* DNA and one microfilter (control) with *Micrococcus lysodeiktkius* DNA. The DNA contents of the microfilters were estimated from their surface areas, as described in the text. Hybridization was for 12 hours at 63°C. The microfilters were treated with RNase and washed in 2 $\times$ SSC before analysis of radioactivity.

suggested above. The size of the DNA-loaded microfilters has a lower limit of about 0.1 mm^2, at which size the cutting of the parent filters starts to become unreliable, as the thickness of the cellulose nitrate filter approaches the margin length of the microfilter. The size of the microfilters also limits the volume in which the reaction takes place to about 0.1 μl. This volume in turn limits the reaction, which is highly dependent on the concentration of RNA and its complementary binding sites in DNA. Precautions should be taken against overloading the cellulose nitrate filters with DNA by checking carefully for released DNA under incubation conditions (as shown in Table I). In our laboratory the upper limit of the binding capacity of the filters is in the range of 0.3 μg of DNA per square millimeter of effective filtration area.

The results of microassays have been found to agree with results obtained from comparative experiments with large-scale, standard filter hybridization procedures (consult Lambert *et al*., 1973, for details). As a further check on the reliability of the results, thermal dissociation studies should be carried out in each set of experiments. Analyses of thermal elution profiles are of considerable importance for the evaluation of results from hybridization studies, because partially mismatched hybrids are readily formed between sequences that contain complementary regions but which are not entirely identical. The midpoint of thermal elution (T_m) is reduced, and the elution profile becomes less steep if there is a pronounced decrease in stability due to mismatching within a set of hybrids (for discussion, see Kennel, 1971).

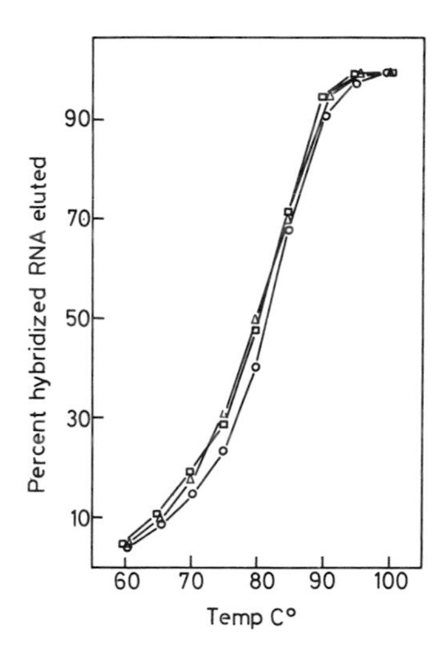

FIG. 18. Thermal stability of RNA/DNA hybrids. Nucleolar RNA was extracted from microisolated nucleoli of 90-minute labeled glands and hybridized by the microtechnique. Thermal elution was performed as described in the text. Ribosomal 18 and 28 S RNA were prepared from ribosomes of *in vivo* labeled larvae and hybridized by a conventional filter technique. Squares, nucleolar RNA, total eluted activity 81 cpm; circles, 28-S RNA, total eluted activity 230 cpm; triangles, 18-S RNA, total eluted activity 120 cpm. (from Lambert *et al.*, 1973).

The microfilters used in microassays are suitable for thermal elution analysis. After hybridization each microfilter is immersed in 25 μl of SSC in a test tube kept in a temperature-regulated water bath. After 5 minutes at each temperature interval of 5°C, the test tube is rapidly chilled in ice water and the SSC solution is absorbed onto a glass fiber filter (Whatman GF/B; Maidstone, England). The microfilter is washed with an additional 25 μl of SSC before continuing the elution at the next temperature increment. After the last elution at 98–100°C, the microfilter is checked for remaining radioactivity, which should not exceed 5% of the total eluted activity. The glass fiber filters, containing the combined eluates of each temperature step, are dried and analyzed for radioactivity. Figure 18 shows the outcome of an experiment in which a microfilter hybridized by microisolated nucleolar RNA was eluted, and the resulting elution profile is compared to that of 18S and 28S rRNA hybridized and eluted by a conventional large-scale procedure.

E. Measurements of Radioactivity

All measurements of radioactivity involve labeled RNA bound to or contained in different kinds of filters. The efficiency of the scintillation measurement is considerably increased if the radioactivity is released from the filters into the scintillation fluid. For this purpose the filters are placed in scintillation vials and treated with 0.5 ml of solubilizer (60 ml of Soluene, Packard, 40 ml of methoxyethanol, 8 ml of water) for 15 minutes at 60°C. The cellulose nitrate microfilters are completely dissolved by this treatment, but no quenching is observed because of the small filter areas. The 0.5- to 1-mm^2 pieces of filter paper (Munktell) used to determine the specific activity of RNA and the total radioactivity in the incubation droplet, as well as the glass fiber filters used in the thermal elution of hybrids, are not changed by the Soluene treatment, but their contents of radioactivity are completely released into the solubilizer. Immediately after Soluene treatment, 10 ml of scintillation fluid (5 gm of Permablend III, Packard, in 1000 ml of toluene) is added to each scintillation vial, and the activity measured in the liquid scintillation spectrometer at a background of 14 cpm and an efficiency of 33%.

IX. Combination of Micro- and Macrotechniques

Although this article is primarily devoted to microtechniques, it seems proper to point out briefly the advantages that can be derived from combining microtechniques with more conventional macrotechniques. Microdissection can undoubtedly provide well-defined and pure cellular components, but only in small physical amounts. If an accurate quantitation of RNA is required, microtechniques subsequently have to be applied. Very often, however, the quantitative aspect is not essential. As RNA can be efficiently labeled, it is often more convenient in such cases to combine microdissection with macrotechniques. The glands are labeled, the cellular components of interest are isolated by microdissection and dissolved in a SDS–pronase solution of proper volume together with carrier RNA, and the released labeled RNA is then further traced and analyzed by a convenient macrotechnique. Such an approach is usually preferred, as macrotechniques are as a rule simpler to carry out and give better and more reproducible results than the corresponding microtechniques (e.g., gel electrophoresis). Furthermore, only a limited number of microtechniques is available, and it is often quite a task to develop new ones. Microdissection has been successfully combined with several standard macrotechniques, such as sucrose gradient centrifugation (Edström and Daneholt, 1967; Daneholt, 1972) and gel electrophoresis of native RNA (Daneholt *et al.*, 1969; Egyházi *et al.*, 1969; Ringborg *et al.*, 1970; Daneholt, 1972), thin-layer chromatography of RNA nucleotides (Daneholt, 1970), biochemical DNA–RNA hybridization (Lambert *et al*., 1973), and cytological hybridization (Lambert *et al*., 1972); the reader is referred to these studies for further information on the combination of microdissection with macrotechniques.

X. Applications

The polytene chromosomes of *Chironomus* salivary gland cells present a distinct interphase morphology which offers a possibility for correlation of cytological features, such as bands, puffs, and nucleoli, with biochemical events. Recent work has mainly focused on the transcription process in defined chromosomal regions and the delivery of newly synthesized gene products through the nuclear sap into the cytoplasm. The microtechniques described in this article were developed for such studies so as to make possible the extraction of undegraded, high-molecular-weight RNA from various cellular components and its subsequent analysis by electrophoresis or hybridization.

Microelectrophoresis was used to demonstrate the main RNA species, high-molecular-weight RNA (Ringborg *et al*., 1968) and low-molecular-weight RNA (Egyházi *et al*., 1968), occurring on the chromosomes and in the nucleolus, nuclear sap, and cytoplasm. Microdissection in combination with analysis of labeled RNA by sucrose gradient centrifugation and large-scale electrophoresis has further clarified the synthesis and fate of preribosomal RNA in nucleoli, heterogeneous, high-molecular-weight RNA in chromosomes, and low-molecular-weight RNA. These results have been extensively treated in two recent reviews (Daneholt, 1974; Edström, 1974).

The microapproach has proven most useful in studies of the giant chromosome puff, Balbiani ring 2, which is likely to be the site of production of salivary polypeptide mRNA (for review, see, e.g., Daneholt, 1974). It has been demonstrated that only one main RNA product (75 S RNA) is synthesized in this defined chromosome region (Daneholt, 1972; Daneholt and Hosick, 1973b). In microhybridization experiments Balbiani ring 2 RNA was used to determine the amount of complementary DNA in the genome. Kinetic analyses of the hybridization reaction indicated that Balbiani ring 2 DNA contained repetitious nucleotide sequences (Lambert, 1972), which by cytological hybridization experiments were shown to be strictly limited within the puff (Lambert, 1973). The electrophoretic and hybridization results have led to the conclusion that Balbiani ring 2 RNA is transcribed as an internally repeated, giant RNA molecule (75 S RNA) and appears in the cytoplasm as a molecule of essentially the same size as in the nucleus (Daneholt and Hosick, 1973b; Lambert, 1973). These results may serve to illustrate the usefulness of microtechniques when applied to suitable material and, when possible, combined with the proper conventional biochemical techniques.

Acknowledgments

As is evident from the original publications, the microtechniques presented in this review have been worked out in close collaboration with Drs. J.-E. Edström, E. Egyházi, and U. Ringborg. Moreover, our co-workers have generously communicated to us their experiences with microprocedures and have provided us with additional unpublished information on the techniques. We also thank Miss Agneta Askendal for preparing the illustrations, and Miss Hannele Jansson for typing the manuscript. Our work was supported by the Swedish Cancer Society and Karolinska Institutet (Reservationsanslaget).

References

Beermann, W. (1952). *Chromosoma* **5**, 139.
Beermann, W. (1965). *Genet. Today, Proc. Int. Congr., 11th, 1963* Vol. 2, pp. 375–384.
Cannon, G. B. (1964). *Science* **146**, 1063.
Daneholt, B. (1970). *J. Mol. Biol.* **49**, 381.
Daneholt, B. (1972). *Nature (London), New Biol.* **240**, 229.
Daneholt, B. (1974). *Int. Rev. Cytol. Suppl.* **4**, 417.
Daneholt, B., and Edström, J.-E. (1969). *J. Cell Biol.* **41**, 620.
Daneholt, B., and Hosick, H. (1973a). *Proc. Nat. Acad. Sci. U.S.* **70**, 442.
Daneholt, B., and Hosick, H. (1973b). *Cold Spring Harbor Symp. Quant. Biol.* **38**, 629.
Daneholt, B., Ringborg, U., Egyházi, E., and Lambert, B. (1968). *Nature (London)* **218**, 292.
Daneholt, B., Edström, J.-E., Egyházi, E., Lambert, B., and Ringborg, U. (1969). *Chromosoma* **28**, 379.
Diacumakos, E. G. (1973). *In* "Methods in Cell Biology" (D. M. Prescott, ed.), Vol. 7, pp. 287–311. Academic Press, New York.
Edström, J.-E. (1964). *In* "Methods in Cell Physiology" (D. M. Prescott, ed.), Vol. 1, pp. 417–447. Academic Press, New York.
Edström, J.-E. (1974). *In* "The Cell Nucleus" (H. Busch, ed.), pp. 293–332. Academic Press, New York.
Edström, J.-E., and Beermann, W. (1962). *J. Cell Biol.* **14**, 371.
Edström, J.-E., and Daneholt, B. (1967). *J. Mol. Biol.* **28**, 331.
Edström, J.-E., and Neuhoff, V. (1973). *In* "Micromethods in Molecular Biology" (V. Neuhoff, ed.), pp. 215–256. Springer-Verlag, Berlin and New York.
Egyházi, E., and Edström, J.-E. (1972). *Biochem. Biophys. Res. Commun.* **46**, 1551.
Egyházi, E., Ringborg, U., Daneholt, B., and Lambert, B. (1968). *Nature (London)* **220**, 1036.
Egyházi, E., Daneholt, B., Edström, J.-E., Lambert, B., and Ringborg, U. (1969). *J. Mol. Biol.* **44**, 517.
Gillespie, D., and Spiegelman, S. (1965). *J. Mol. Biol.* **12**, 829.
Greenberg, J. R. (1969). *J. Mol. Biol.* **46**, 85.
Kay, E. R. M., and Dounce, A. L. (1953). *J. Amer. Chem. Soc.* **75**, 4041.
Kennel, D. E. (1971). *Progr. Nucl. Acid. Res. Mol. Biol.* **11**, 259.
Kurland, C. G. (1960). *J. Mol. Biol.* **2**, 83.

Lambert, B. (1972). *J. Mol. Biol.* **72**, 65.
Lambert, B. (1973). *Cold Spring Harbor Symp. Quant. Biol.* **38**, 637.
Lambert, B., Wieslander, L., Daneholt, B., Egyházi, E., and Ringborg, U. (1972). *J. Cell Biol.* **53**, 407.
Lambert, B., Egyházi, E., Daneholt, B., and Ringborg, U. (1973). *Exp. Cell Res.* **76**, 369.
Peacock, A. G., and Dingman, C. W. (1968). *Biochemistry* **7**, 668.
Pelling, C. (1970). *Cold Spring Harbor Symp. Quant. Biol.* **35**, 521.
Pelling, C. (1972). *In* "Results and Problems in Cell Differentiation" (W. Beermann, J. Reinert, and H. Ursprung, eds.), Vol. 4, pp. 87–99. Springer-Verlag, Berlin and New York.
Ringborg, U., Egyházi, E., Daneholt, B., and Lambert, B. (1968). *Nature* (*London*) **220**, 1037.
Ringborg, U., Daneholt, B., Edström, J.-E., Egyházi, E., and Lambert, B. (1970). *J. Mol. Biol.* **51**, 327.
Travaglini, E. C. (1973). *In* "Methods in Cell Biology" (D. M. Prescott, ed.), Vol. 7, pp. 105–127. Academic Press, New York.
Wieme, R. J. (1965). "Agar Gel Electrophoresis." Elsevier, Amsterdam.

Chapter 3

Staining of RNA after Polyacrylamide Gel Electrophoresis

K. MARCINKA

Institute of Virology,
Slovak Academy of Sciences,
Bratislava, Czechoslovakia

I. Introduction . . . 49
II. Action of Fixatives and Stains on RNA . . . 50
A. Binding of Fixatives . . . 52
B. Binding of Dyes . . . 54
III. Procedure for RNA Staining . . . 55
A. Preparation of Gel and Electrophoresis . . . 55
B. Staining of RNA in Gels with Pyronine Y . . . 58
C. Destaining of Gel Background after RNA Staining with Pyronine Y . . . 59
D. Staining with Other Dyes . . . 60
E. Staining of RNA in Indirect Detection of RNases . . . 62
IV. Evaluation of Stained RNA Pattern . . . 62
V. Other Methods of RNA Detection . . . 63
A. Direct Photometric Scanning . . . 63
B. Fluorescence Detection . . . 64
C. Elution from Sliced Gels . . . 66
D. Use of Radioisotopes . . . 66
VI. Concluding Remarks . . . 66
References . . . 67

I. Introduction

The early rapid expansion of polyacrylamide gel electrophoresis concerned primarily proteins, inasmuch as this method (Davis, 1962) was originally developed for analysis of this material. It should be stressed that the rapid acceptance and expansion of this technique and the continual interest in it can be ascribed to this procedure being rapid, simple, sensitive, and reproducible. Furthermore, because an appropriate sieve density can be selected depending on the concentration of the monomer, polyacrylamide gel electrophoresis can be employed for the analysis of macromolecules of a broad range of molecular weights. The small samples required for analysis can be further minimized by the use of ultramicromethods (e.g., Pun and Lombrozo, 1964; Neuhoff, 1968). It is not surprising therefore that, since the first application of this method for nucleic acids (Richards *et al.*, 1965), its expansion in this direction has also been rapid.

A prerequisite for full utilization of any technique is the gaining of expertise in the various steps involved. One of the critical steps in polyacrylamide gel electrophoresis is detection of the substance analyzed—in our case electrophoresed RNA. Staining appears to be the simplest and a very sensitive method of RNA detection.

The requirements for RNA staining techniques are: (1) ability to make possible the visualization of very small amounts of RNA (0.01 μg and less), and (2) minimal binding of the stain to the gel (i.e., minimal background). Further requirements, although not of such importance, are (3) proportionality of staining depending on the amount of RNA present in an equal volume unit, especially if semiquantitative or quantitative evaluation is desired, and (4) persistence of binding of the dye to the nucleic acid, especially when analyzed samples must be stored for prolonged periods; this persistence should manifest itself under normal laboratory conditions, i.e., at room temperature and under normal illumination.

We are concerned mainly with staining techniques for RNA; other detection methods are mentioned only briefly. Fluorescence detection, developed in our laboratory and not yet described in the literature, is discussed in some detail. Mention of it in this chapter is useful, because this method is very rapid and permits RNA detection within 5 minutes after electrophoresis.

II. Action of Fixatives and Stains on RNA

Generally, to explain the mode of action of fixatives and stains on RNA, the principles of binding of a substance possessing a small molecule (this is the case with all fixatives and dyes used thus far) to polymers can be applied. Under suitable conditions complexes are formed. Evidently, various kinds of bonds are involved; their strength may vary, depending primarily on the properties of the substances between which the binding occurs (structure of RNA, character of the binding of the fixative or the dye), but also on the conditions under which the binding occurs and is maintained (pH, temperature, ionic strength, and ionic composition of the solution, presence of other active groups, etc.). These factors affect the value of total free energy of reactive groups involved in the formation of the complex.

The nature of the binding of either a fixative or a dye to RNA has been widely examined (Lerman, 1961; Lamm *et al.*, 1965; Santilli, 1966; Blake and Peacocke, 1968; Popa and Repanovici, 1969; Löber, 1969; and others). Because of chemical differences between RNA and fixatives or dyes, various steric configurations of the complexes can be expected. The mechanism by which binding occurs between the two partners is explained by the generally accepted theory of Lerman (1961). According to this theory, two types of complexes are gradually formed. Complex I arises by intercalation of one molecule of fixative or dye between a pair of nucleic acid bases; such binding is rather strong. Furthermore, it apparently makes the bond between the bases even stronger. Complex I is formed to a certain degree, in principle until all pairs of bases with disposable free energy are occupied by the binding sites of the substance. In the presence of excess molecules of the substance, complex II is formed. In this complex, binding of the substance to the surface of the RNA molecule is involved. Naturally, this binding is more labile and so far has been studied only a limited extent.

A brief attempt at a physicochemical interpretation of the complex formation theory follows. If a fixative or dye with single type of site is bound to RNA, the equivalence between free substance and RNA-bound substance follows the Guldberg-Waage law which can be expressed by Scatchard's (1944) equation written in the form

$$\frac{r}{c} = \frac{n}{k} - \frac{r}{k}$$

where r is the number of molecules of substance bound per nucleotide, c is the molar concentration of unbound substance, k is the dissociation constant of the complex, and n is the number of binding sites per nucleotide. For the type of interaction RNA–substance, the plot of r/c versus r is characteristic. If expressed graphically and r/c versus r is linear, interactions of a single type of site are involved. But if the binding process involves more than a single type of site, or binding at one site affects the interaction at neighboring sites, the relationship of r/c versus r is not linear.

For the relationship between the binding of a fixative and/or the binding of a dye to the nucleic acid, with respect to subsequent detection, it is required that (1) the binding forces of the fixative confer on the RNA higher free energy at the free reactive sites, thus allowing stronger binding of the dye. Otherwise, fixation would be meaningless. Another important condition is that (2) after fixation the dye disposes of a sufficient number of reactive groups. For example, RNA fixation by formalin is being abandoned because it blocks so many reactive sites that it strongly reduces RNA stainability. This principle applies both in histology and in the use of formalin in the fixative stain solution for RNA in polyacrylamide gel electrophoresis in which, in a 0.2% concentration, formalin strongly inhibits the binding of Pyronine Y to RNA (K. Marcinka, unpublished).

A. Binding of Fixatives

Fixatives of any type induce changes in the physical state of the RNA molecule. This applies to both fixatives in which the metal component is linked to the RNA molecule (e.g., mercury, lanthanum, chromium salts) and which form stronger bonds, the changes induced thus being more persistent, and fixatives that are active without being bound to RNA (methanol, acetic acid, etc.) in which the changes induced persist only as long as the RNA remains in such fixatives. As mentioned above, the purpose of all changes is to induce stronger binding of the dye to the RNA. For RNA staining therefore almost exclusively so-called fixative stain solutions are used; these contain, in addition to the dye, one or a combination of several fixatives. In the methylene blue staining method (Peacock and Dingman, 1967), in which a solution of the dye in acetate buffer is used, the dye is bound to the RNA weakly. Consequently, this results in complete destaining in a short time (Marcinka, 1972; see also Section III,D).

The mechanisms of the action of the fixative stain solution have not yet been fully elucidated. Investigations of these actions are complicated by the simultaneous presence of fixative and dye. The reaction proceeds at room temperature for 8–16 hours (Marcinka, 1972). Apparently, the fixative which is bound by relatively strong binding forces occupies some of the reaction sites on the RNA molecule to the disadvantage of the dye. It was found (K. Marcinka, unpublished) that all types of RNA isolated from plant tissue after fractionation in 4% polyacrylamide gel became more intensively stained with Pyronine Y (for the method see Section III,B) in a fixative stain solution of acetic acid–methanol–water (1:1:8) without lanthanum acetate (Fig. 1A) than in the presence of 1 or 3% of this fixative (Fig. 1B and C). When parallel gels were left in the same mixture without lanthanum acetate for 24 hours before staining, massive diffusion from the gels of 4 and 5 S RNA occurred (Fig. 1D). In parallel samples containing lanthanum acetate, no difference was observed in those stained in the same mixtures immediately (cf. Figs. 1B and C and 1E and F). The presence of lanthanum acetate in the fixation mixture prevented diffusion of RNA from the gel even after 1 week (Fig. 1H and J), while in its absence diffusion of all RNA

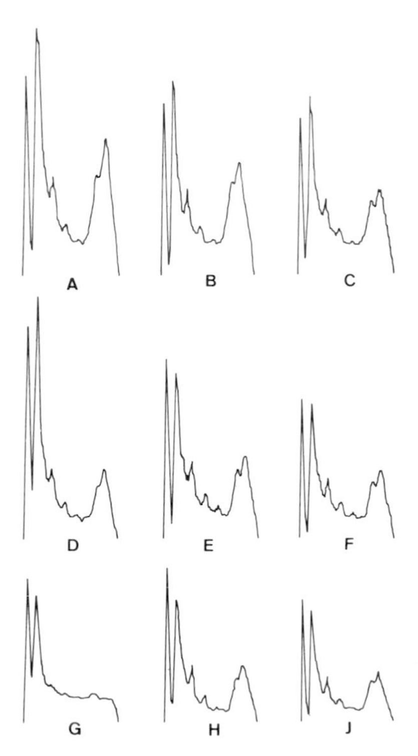

FIG. 1. Densitometric evaluation of RNA pattern. RNA was isolated from pea (*Pisum sativum* L. cv. Raman) plants by phenol extraction. Electrophoresis was performed in 4% polyacrylamide gel for 65 minutes at 90 V. Parallel samples were stained in fixing stain solution according to the procedure given in Section III. 1. Immediately after electrophoresis: (A) without lanthanum acetate (LaAc); (B) with 1% LaAc; (C) with 3% LaAc. 2. After 24-hour storage in the fixing solution: (D) without LaAc; (E) with 1% LaAc; (F) with 3% LaAc; staining was the same as in (A)–(C). 3. After 1-week storage in the fixing solution: (G) without LaAc; (H) with 1% LaAc; (J) with 3% LaAc. Staining was the same as in (A)–(C). Migration from left to right. A densitometer Joyce-Loebl Chromoscan was used.

species occurred (Fig. 1G). These results indicate that lanthanum acetate should be used if gels are left unstained for some time after the completion of electrophoresis.

It is interesting that, of the fixatives having in their molecule a metal component, lanthanum acetate has been used exclusively for RNA in polyacrylamide gels. Possibly, this is due to the fact that lanthanum acetate was used by the first investigators who reported RNA fractionation by polyacrylamide gel electrophoresis (Richards *et al.*, 1965). Their argument for the use of lanthanum acetate was that it provides specificity for RNA staining with acridine orange. In addition, a further reason for using lanthanum acetate is that it blocks RNase activity.

The binding of fixatives to RNA has been studied mainly in mercury(II)–RNA. Such studies have been based on changes in ultraviolet absorption spectra of RNA at different molar ratios (r): mercury (II)/nucleotide or mercury(II)/P-RNA. The results of experiments of this type made with tobacco mosaic virus (TMV) RNA (Santilli, 1966), in accordance with the complex formation theory (see above), indicated a two-phase course of the reaction of mercury(II) with RNA. The concentration of mercury(II), up to one-half the molar ratio with respect to nucleotides ($0 < r \leqq 0.5$), results in the formation of the bond base mercury(II)–base (first complex). For TMV RNA, the formation of this complex can be interpreted as being a result of the known high degree of hydrogen bonding. By further addition of mercury(II), the cross-linked structure is broken, with a simultaneous formation of the second base–mercury(II) complex until $r = 1.0$, when the saturation of each RNA base by one mercury(II) is reached.

B. Binding of Dyes

As in the binding of the metal ion of a fixative, the complex formation theory also applies to the binding of dyes to nucleic acids. Under appropriate conditions these processes can be studied in the visible region of the spectrum—in the absorption region of the stain. Marked changes occur in the absorption spectrum of the dye during its binding to the nucleic acid: (1) a shift in the absorption maximum (the so-called metachromic effect), and (2) a decrease in absorbancy. Changes in absorption maxima, under appropriate concentration conditions, allow quantitative estimates in the region of complex formation with RNA (and DNA). For example, with toluidine blue it was found that complex I is formed at r in the range $0 < r \leqq 0.15$ to 0.20, i.e., two to three base pairs are saturated by one molecule of the stain. Complex II is formed until $r = 1$ is reached, i.e., one molecule of toluidine blue is bound to one nucleotide. Apparently, this basic dye is fixed to the negatively charged phosphate group of the nucleotides, but these linkages are weaker. It is assumed that, e.g., in DNA, such surface-bound molecules of the dye are released on heating, before the melting temperature is attained.

It has been known for a long time that certain soluble dyes form aggregates in solution. The absorption maximum of such aggregates, with respect to the monomer, is shifted to a shorter wavelength. Based on this phenomenon it was found that, formation of complex II, in addition to nucleic acid–dye binding, interactions also occur between neighboring molecules of the dye bound to nucleic acids in the formation of dimers or even larger aggregates. Concentrations of dyes used for the staining of RNA in polyacrylamide gels are so high that such aggregates undoubtedly are formed, although methanol, which prevents aggregation, is frequently used in the fixative stain solution.

Most of the published reports suggest that the free energy of dye–nucleic acid binding differs with respect to different bases. It seems that dyes show a greater affinity for adenylic acid or for the base pairs AU and AT. But this cannot be accepted as a generalized postulate, since extreme variability exists among the types of nucleic acids and the milieus in which the binding takes place.

III. Procedure for RNA Staining

This section is concerned mainly with the staining of RNA and the destaining of the gel background (Section III,B and C). To achieve good reproducibility of staining, is important to bring out the method of gel preparation, the staining procedure employed, and the conditions for electrophoresis.

A. Preparation of Gel and Electrophoresis

The staining procedure described in Section III,B can be used with any type of polyacrylamide gel, including the polyacrylamide–agarose gels reported in the literature. Only a brief description of Loening's (1967) method is given. According to our experience, this method is the most suitable for fractionation of all RNA species, as concerns both the purity of the gel and the possibility of densitometric evaluation of RNA fractions either directly or after staining.

For gel polymerization we used glass tubes 7 cm high with an inner diameter of 0.5 cm. The height of the gel columns was mostly 5.0 cm. Depending on the molecular weight of the RNA to be separated, we used gels in the concentration range 2.5–7.5%, calculated on acrylamide monomer only. Concentrations lower than 2.5% were unnecessary, even for the separation of viral nucleic acids with a molecular weight of about 2 million daltons. Acrylamide and *N*,*N'*-methylene bisacrylamide were used after recrystallization from benzene and from water, respectively (Loening, 1967). The gels and reservoir buffer contained tris–sodium acetate–EDTA buffer in the same concentration (a continuous system). Contaminating substances, particularly ammonium persulfate, were removed from gels by preelectrophoresis at 5 mA per tube before loading of the samples. The time period necessary for quantitative removal of ammonium persulfate from the gels

TABLE I
DYES FOR STAINING OF RNA IN POLYACRYLAMIDE GEL[a]

Color Index number[b]	Name of dye and synonyms	Chemical name and empirical formula	Molecular weight	References[c]
46005	Acridine orange (acridine orange 2G, rhodulinorange N)	3,6-Bis(dimethylamino)acridine hydrochloride (double salt with zinc chloride), $C_{17}H_{19}N_3 + HCl + ZnCl_2$	438.11	Richards *et al.* (1965)
52005	Azure A (MacNeal)	7-(Dimethylamino)-3-imino-3*H*-phenothiazine hydrochloride, $C_{14}H_{14}ClN_3S$	291.80	
51030	Gallocyanine (alizarine navy blue AT; brilliant chrome blue P; ultrabrilliant blue P)[d]	7-(Dimethylamino)-3*H*-4-hydroxy-3-oxophenoxazine-1-carboxylic acid, $C_{15}H_{12}N_2O_5$	300.27	Grossbach and Weinstein (1968)
52015	Methylene blue (Swiss blue; basic blue 9; solvent blue 8)	3,7-Bis(dimethylamino)phenazathionium chloride, $C_{16}H_{18}ClN_3S$	319.86	Peacock and Dingman (1967)

42585	Methyl green (Paris green; double green SF; basic blue 20)	Heptamethyl-*p*-rosaniline chloride (double salt with zinc chloride), $C_{26}H_{33}Cl_2N_3$ + $ZnCl_2$	594.78	
45005	Pyronine Y (Pyronine G)	3,6-Bis(dimethylamino)xanthylium chloride, $C_{17}H_{19}ClN_2O$	302.81	Marcinka (1972)
—	Stains-all	4,5,4′,5′-Dibenzo-3,3′-diethyl-9-methylthia-carbocyanine bromide, $C_{30}H_{27}BrN_2S_2$	559.60	Dahlberg *et al.* (1969); Marcinka (1972)
52000	Thionine (Lauth's violet)	3,7-Diaminophenothiazonium Chloride, $C_{12}H_{10}ClN_3S$	263.75	
52040	Toluidine blue O (tolonium chloride, blutene chloride, tolazul)	3-Amino-7-dimethylamino-2-methylphenazathionium chloride, $C_{15}H_{16}ClN_3S$	305.85	Konings and Bloemendal (1969). Marcinka (1972)

[a]This list is not complete. Dyes tested for RNA staining in polyacrylamide gel have been included.
[b]The numbers refer to those given in the second edition (1956).
[c]Only basic references, containing methods or results, are listed.
[d]Grossbach and Weinstein (1968) used gallocyanine in a mixture with chrome alum (chromic potassium sulfate, $CrK(SO_4)_2 \cdot 12H_2O$) prepared according to de Boer and Sarnaker (1956).

was determined according to Bennick (1969). Then the gels were loaded with RNA samples (60–120 μg per tube), in amounts depending on how many fractions were expected. The RNA sample was mixed with reservoir buffer (diluted 2- to 4-fold) containing 5–10% sucrose. Electrophoretic separation of the sample lasted for usually 65 minutes at 90 V (5 mA per tube). For further details of the procedure, see Loening (1967). The book by Maurer (1971) is recommended for a detailed study of polyacrylamide gel electrophoresis.

B. Staining of RNA in Gels with Pyronine Y

Gels removed from the tubes were transferred to the fixative stain solution. The best results were obtained when staining was done in a solution of Pyronine Y (for chemical characteristics see Table I) in a 1:1:8 acetic acid–methanol–water mixture. Acetic acid and methanol were tested either separately or in combination, in concentrations of from 0 to 50%. The mixture mentioned proved to be the most efficient. Unlike in the original

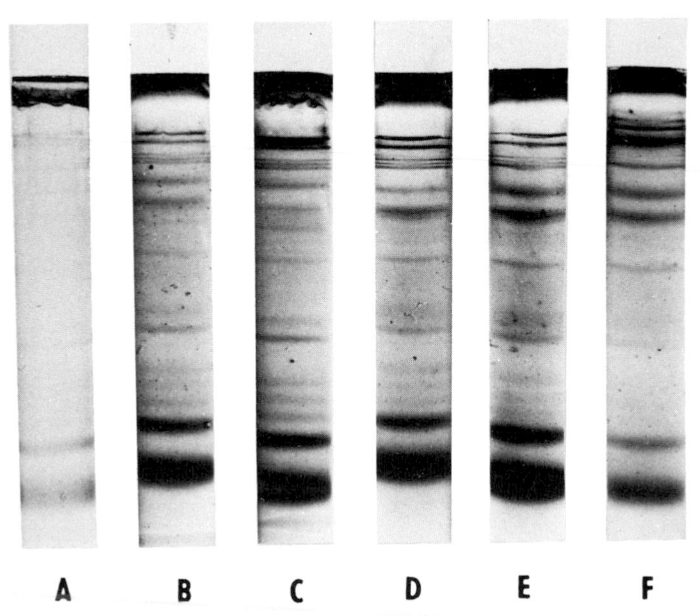

FIG. 2. Patterns of RNA from ascites cells after separation by polyacrylamide gel electrophoresis. Stained as described in Section III,B with Pyronine Y for 1 hour (A), 8 hours (B), and 16 hours (C). Parallel samples were stained with methylene blue (D) and Stains-all (E), as described in Section III,D. Staining with acridine orange (F) was the same as with Pyronine (16 hours).

method, lanthanum acetate was omitted as the fixative in the fixative stain solution. According to recent results (see Section II,A), the presence of lanthanum acetate decreases the intensity of RNA staining by blocking some of the binding sites on the RNA molecule to which the dye can be bound. Therefore addition of lanthanum acetate is not recommended, provided that staining is carried out immediately after completion of electrophoresis (Fig. 1). However, if staining takes place after a longer period of time, a fixative solution with 0.7% lanthanum acetate to prevent diffusion of RNA from the gel is recommended. Pyronine Y concentrations lower than 0.5% in the mixtures mentioned gave less intensive staining of the RNA bands, and concentrations higher than 0.5% had no effect on the quality of staining. The time period for staining was also examined. Gels A, B, and C in Fig. 2 illustrate gels after parallel electrophoresis of the same RNA sample, stained for different lengths of time. Staining for 1 hour (Fig. 2A) was quite insufficient, but even after 8 hours (Fig. 2B) not all bands that became clearly evident after 16 hours of staining (Fig. 2C) were detectable. Staining for longer periods did not result either in an increased intensity of staining or in the detection of additional RNA bands.

For staining it is sufficient to use the stain solution in a volume approximately 2.5-fold that of the gel. Occasional stirring of the fixative stain solution during the course of staining is recommended. The solution can be used repeatedly without impairing the quality of RNA staining.

C. Destaining of Gel Background after RNA Staining with Pyronine Y

After the gels are placed in the staining solution, penetration of the stain into the entire gel occurs, including those parts that are devoid of RNA. By destaining we mean removal of stain not bound to the RNA. Many destaining mixtures have been proposed, including destaining with running tap water. We tested many destaining solutions. A 0.5:1:8.5 acetic acid–methanol–water mixture proved to be the most satisfactory.

Destaining can be accomplished by electrophoresis or by simple leaching.

1. Electrophoresis

As the term implies, this destaining technique is similar to that used for fractionation, the difference being that destaining solutions are employed instead of electrophoresis buffers. At the beginning destaining was done in the same apparatus as fractionation, in tubes variously modified. One of the most frequently used modifications consisted of placing the stained gel in a larger tube with a short gel plug at the bottom to prevent the gel rod from slipping. In this case the destaining direction was the same as in the case of electrophoretic separation. This method is still being used. Soon, however, special apparatuses were developed. In these the gel rods are oriented perpendicularly to the separation direction. We found the latter technique of destaining very efficient. A simple and suitable instrument was described by Maurer (1966). The destaining solution is placed in the apparatus so that it completely covers the gels. With 5-mm gels and a current density of 20 mA per gel, the gels can be destained within 40–60 minutes. The main drawback of electrophoretic destaining is that some of the dye bound to the RNA is freed, resulting in less intensively stained bands. This may result, especially for minor RNA fractions, in undetectability. However, this method is of great utility, because results are available within a short time. A compromise solution seems to be the best: to start with electrophoretic destaining at lower current densities and then proceed to complete destaining of the background by simple leaching.

2. Simple Leaching

Destaining by simple leaching or by diffusion is preferred when leakage of dye from weaker binding sites on the RNA in the gel should be avoided to visualize even very small amounts of RNA if such amounts are expected. A disadvantage of this method is that it is rather time-consuming. However, it is still used very often because of its simplicity and lack of deleterious effects.

The procedure is as follows. The gel is removed from the staining solution, rinsed with water, and placed in a 200 to 300-ml Erlenmeyer flask containing about 100 ml of the destaining mixture. Diffusion of excess stain can be accelerated by continuous stirring of the destaining solution, e.g., on a rotatory shaker. Destaining is stopped after an equilibrium between the concentrations of stain in the gel and in the destaining mixture is reached, which takes about 8–10 hours. The destaining mixture should then be replaced by a fresh one. Destaining is usually completed after three to five changes, as indicated by the destaining mixture remaining colorless.

The destaining mixture can be used repeatedly after removal of the stain by absorption, e.g., by passage through a column of granulated activated charcoal. The whole process of destaining can be mechanized by continuous circulation of the destaining mixture from the gel through a column of activated charcoal and back to the gel.

D. Staining with Other Dyes

In addition to Pyronine Y, several other dyes, summarized in Table I, can be used for staining RNA in polyacrylamide gels. Our experiences with some of them are described below.

1. Toluidine Blue O

All criteria stated above for Pyronine Y, including those for destaining, are also valid for staining RNA with toluidine blue. The only difference concerns the concentration of dye in the fixative stain solution, the optimum concentration being 0.7%. As with Pyronine, lanthanum acetate should be omitted from the fixative stain solution if staining is carried out immediately after electrophoresis (this is in contrast to the statement in a previous article—Marcinka, 1972). The quality of toluidine blue used is of great importance (for details on this point, see Section VI).

2. Methylene Blue

The method proposed by Peacock and Dingman (1967) is suitable. When carrying out electrophoresis according to Loening (1967), staining time was 16 hours; destaining proceeded in running tap water for 20–24 hours. In this way all RNA bands, detectable by Pyronine according to the method described above, were demonstrated, but they were more diffuse (Fig. 2D). A great drawback of methylene blue staining was the weak binding of the stain to RNA. Washing for 48 hours resulted in total destaining of both the background and RNA.

The method described above for Pyronine staining proved unsuitable for methylene blue.

3. Stains-all

This compound is a photosensitive dye. We tested the conditions suitable for the staining of RNA. The staining solution was prepared according to Dahlberg *et al.* (1969) from a stock stain solution (0.1% dye in 100% formamide) by mixing 10 ml of stock stain with 90 ml of 100% formamide and 100 ml of water, in that order. We found that with gels of 5-mm diameter it was necessary to use a volume of staining solution at least seven times that of the gel and to stain the gels for 10–15 hours. This staining solution can be used only once. Destaining by running tap water was not very efficient, the background remaining heavily stained. Only destaining for 16 hours with 7% acetic acid proved effective. As compared with Pyronine staining, all RNA bands were present (Fig. 2E). The fact that the entire procedure must be conducted in the dark remains a distinct disadvantage of this method.

Gels kept in 7% acetic acid in the dark remained practically unchanged for 9 months (the longest period tested).

4. Acridine Orange

The results obtained by both the original method (Richards *et al.*, 1965) and our Pyronine staining proved unsatisfactory, because it was impossible to reduce the heavy staining of the background. This apparently was also the reason why some bands of RNA could not be detected as compared with the Pyronine staining technique (Fig. 2F).

E. Staining of RNA in Indirect Detection of RNases

RNases fractionated electrophoretically in polyacrylamide gels can be detected based on their enzymatic activity. This method is mentioned here because it is based on RNA staining and the procedure of staining differs from that described for detection of RNA patterns. The detection of RNases proceeds as follows. The gel, after fractionation of RNases, is placed for a short period of time in a solution of low-molecular-weight RNA. Under suitable conditions splitting of RNA occurs at those places in the gel where RNases are present. Thereafter the gel is stained for 1 minute in a 0.2% solution of toluidine blue in 1% acetic acid. Subsequent destaining of the gel in 1% acetic acid results after 1–2 hours in destaining of those parts of the gel in which RNA was enzymatically split, i.e., where RNase was present. A measure of the amount of RNase and its activity is the degree of destaining of the gel in the RNase band. It must be stressed, however, that the selection of suitable conditions for the manifestation of RNase activity (preincubation in buffer, incubation with RNA, postincubation in buffer, selection of suitable buffers, and their concentrations, pH, incubation time and temperature, quality of RNA, etc.) is of great importance for successful detection of RNases.

For details concerning methods for detecting RNases, see Wilson (1971).

IV. Evaluation of the Stained RNA Pattern

It is often sufficient to evaluate the stained patterns of electrophoresed RNA visually. This subjective evaluation needs no commentary.

For an objective evaluation densitometers are employed. Among the commercially available equipment are filter photometers (e.g., Chromoscan, Joyce Loebl, Gateshead, U.K.) and spectrophotometers with special equipment for gel scanning (e.g., spectrophotometer Model 240, Gilford Instruments, Oberlin, Ohio, with Model 2410 linear transport equipment). Adesnik (1971) has compared the two types of instrumenns in detail in regard to sensitivity of detection.

In principle, there are two methods of densitometric evaluation of gels with stained RNA patterns: (1) direct scanning of the gel, and (2) scanning of a photographic record (less frequently used at present).

Of the various problems encountered in densitometric evaluation, only those concerning quantitative evaluation are briefly dealt with here. Except for pointing out possible errors, no attempt is made to reduce the value of the quantitative densitometric evaluation itself. By contrast, we consider densitometry to be very useful and frequently to offer a more correct view of the results. At present it plays a substantial part in the evaluation of numerous results. However, great caution should be taken in quantitative evaluation, especially when the results from different laboratories are compared.

The sources of differences in densitometric records are as follows.

1. *Instrument*. Type of densitometer; size of slit; location of ray bundle across the gel; wavelength of radiation used for scanning; sensitivity of the detector; proportionality of the record or of values of the integrating counter with respect to concentration; speed of sample transport.

2. *Gel*. Purity of chemicals (differences may occur even between two purification procedures in one laboratory); concentration of monomers and other ingredients; thickness of the gel; height of the gel; damaged gel; air bubbles; mechanic impurities.

3. *Conditions of electrophoresis*. Source; voltage used; time of preelectrophoresis; time of electrophoretic separation; composition of reservoir buffers; temperature.

4. *RNA, fixative, and stain*. Type of RNA; space distribution of RNA in gel (bandwidth); use of fixatives and their chemical purity; stain used; chemical purity of dye; proportionality of RNA staining with respect to the amount of RNA in the gel.

5. *Conditions of destaining*. Mode and time of destaining; composition of destaining solution.

6. *Photographic record*. Quality of both negative and positive material (e.g., different sensitivity); method of processing (e.g., exposure time, development, quality of the developer, etc.).

In each case it is necessary to take into consideration whether efforts toward quantitative evaluation of the phenomena are supported by the data, and what degree of importance to attach to the conclusions thus obtained.

V. Other Methods of RNA Detection

A. Direct Photometric Scanning

Polyacrylamide gel can be scanned directly, without previous staining. After electrophoresis a bundle of ultraviolet rays in the region of RNA absorption (at about 260 nm) is allowed to pass through the gel. Radiation is absorbed in those regions of the gel in which RNA is present. This detection method has expanded rapidly since its first introduction by Loening (1967). A sufficiently low gel background in the given spectral region is a prerequisite. According to Loening (1967), minimal background can be achieved by purification of the monomers of recrystallization. Recrystallization of acrylamide from chloroform results in better transparency of gels than recrystallization from benzene. But at wavelengths shorter than 260 nm, the absorbancy of the gel (i.e., background) increases rapidly, because acrylamide absorbs the radiation in this region. Therefore densitometry of RNA patterns at wavelengths shorter than 260 nm is impossible.

Equipment necessary for this method of detection is similar to that described in Section IV. However, it must be adapted to radiation and detection in the ultraviolet region; filter photometers must be equipped with appropriate spectral filters.

B. Fluorescence Detection

Very rapid detection of RNAs after electrophoresis in polyacrylamide gel is sometimes essential, especially before subjecting fractions (from parallel samples) to further investigations (e.g., Lane and Kaesberg, 1971; Hull, 1972). As a result of such problems arising in our laboratory, the fluorescence detection method for RNA was developed (Marcinka, 1974). Its principle is based on the fact that purine and pyrimidine derivatives, hence nucleic acids, intensively absorb ultraviolet radiation, especially at 260 nm. This property, in relation to fluorescence quenching, was used by Wieland and Bauer (1951) for the detection of purines in paper chromatography.

The procedures is as follows. After completion of electrophoresis, carried out according to Loening (1967), the gel is removed from the tube and immersed in a 0.02% fluorescein solution for 3 minutes. Then the gel is rinsed with tap water and placed on a black, frosted-glass support. Excess fluid around the gel is sucked off with a Pasteur pipette. The gel is then examined in a darkroom under a source of ultraviolet radiation at about 260 nm (e.g., Mineralight, UltraViolet Products, Inc., San Gabriel, Calif., at a distance of 5–20 cm from the gel). The presence of RNA is revealed by a dark band showing little or no fluorescence (depending on the amount of RNA present); the rest of the gel exhibits yellow-green fluorescence (Fig. 3A). All RNA species can be detected by this method. The position of the RNA is marked appropriately. The fluorescein solution can be used repeatedly.

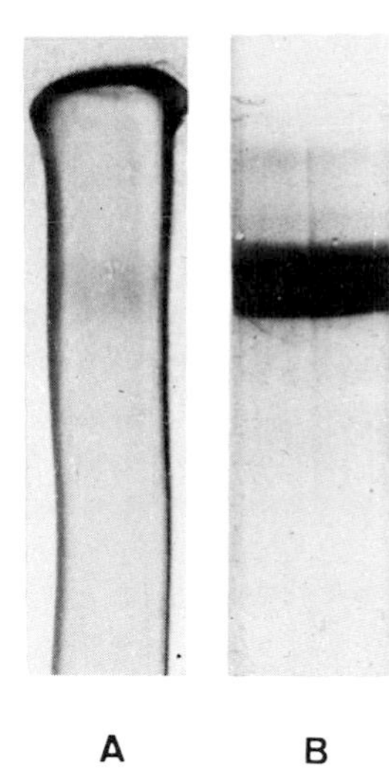

FIG. 3. Detection of RNA in polyacrylamide gel. (A) By fluorescence in ultraviolet radiation according to the method described in Section V,B and (B) by staining with Pyronine Y. In both cases 10 μg of TMV RNA was placed on the gels.

The 0.02% aqueous solution of fluorescein is obtained by 10-fold dilution of the stock solution with distilled water. The stock solution is prepared by adding to 5 ml of distilled water first 20 mg of fluorescein and then 0.1 *N* NaOH dropwise until complete solution of the dye occurs; then distilled water is added to make 10 ml. At lower fluorescein concentrations the intensity of fluorescence of the gels was reduced, and the areas of fluorescence quenching became less distinct. Higher fluorescein concentrations did not improve RNA detection.

Fluorescence detection of RNA was compared with its detection by staining with Pyronine Y. In parallel samples the position of RNA was identical (see Fig. 3).

By employing this method it was possible to detect RNA fractionated on gels containing 2.5–10% acrylamide monomer (the range tested).

The sensitivity limit of fluorescence detection when determined with TMV RNA on a gel prepared from 2.5% acrylamide monomer, was about one dex[1] lower than with Pyronine staining.

The fluorescence detection method is simple and does not require elaborate equipment. After it has become a routine procedure, detection may be completed within 5 minutes after the completion of electrophoresis.

So far, no objective method for fluorescence detection of RNA has been proposed.

[1]One order of magnitude. See Haldane (1960). *Nature* (*London*) **187**.

C. Elution from Sliced Gels

In this detection method a discontinuous record of RNA patterns is obtained. After electrophoresis the gel is sliced with appropriate equipment into slices usually 1–2 mm thick. Each slice is then separately eluted into a suitable buffer with continuous stirring. With high-molecular-weight RNA, the elution should be longer, especially if the RNAs are present in gels with a higher monomer concentration. After elution the absorbancy at 260 nm is measured.

This method is used in preparative separation of RNA in isoelectric focusing and, in particular, in separating labeled RNA.

D. Use of Radioisotopes

Many biological experiments can hardly be conducted without using labeled compounds. It is therefore quite natural that analysis of labeled biological material by electrophoresis in polyacrylamide gel has been employed since its very inception. This system of detection of radioactive substances, including labeled nucleic acids, has been elaborated widely and deserves discussion in a separate chapter. For details on the detection of labeled RNA in polyacrylamide gels, see, e.g., Adesnik (1971).

VI. Concluding Remarks

It seems unnecessary to note that, as in the preparation of RNA, in all electrophoresis steps up to the removal of gels from the tubes, care should be taken to avoid splitting the RNA strand, in particular by the action of ubiquitous RNases. It is necessary therefore that all equipment be free of RNases and that contamination by this enzyme be prevented throughout the experiment. Boiling of the equipment in distilled water containing 0.1% diethylpyrocarbonate (e.g., Baycovin, Bayer AG, Leverkusen, West Germany) for 20 minutes is recommended. Only after electrophoresis, if no further analysis of RNA is planned, are these precautions not so critical. The gels can be removed from the tubes using tap water and manipulated with unprotected hands. Similarly, stain solutions and the destaining mixture need not be sterilized, especially if the fixative stain solution contains lanthanum acetate which protects RNA from the action of RNases.

For the analysis of RNA with a large molecule (e.g., ribosomal or viral RNAs), usually large-pore gels, i.e., gels containing about 2.5% acrylamide monomer, are employed. With Loening's (1967) method, i.e., polyacrylamide without agarose, the gels have to be handled with considerable care.

Last but not least, it must be stressed that all chemicals, in particular dyes, must conform to rigid standards. Not all brands are satisfactory, and pretesting is recommended. The quality of a dye may differ from one batch to another, even from the same manufacturer. This was confirmed in our experiments. We used Pyronine and toluidine blue samples that, without previous purification, stained RNA in gels to less than half the intensity of that of other products. Bands containing small amounts of RNA were thus undetectable. In consequence, when carrying out experiments requiring extreme accuracy in the dosage of pure substance, commercial dyes must be purified. Semmel and Huppert (1964) purified acridine orange, alcian blue, toluidine blue, and Pyronine by shaking out aqueous solutions of the dyes with chloroform until the latter was colorless. Lamm *et al.* (1965) purified toluidine blue dissolved in ethanol by passing it through an acid-washed alumina (Merck) column. The fraction of pure dye, after precipitation with diethyl ether, was dried at 100° C.

In staining RNA in our laboratory, Pyronine produced by E. Merck AG, Darmstadt, West Germany (supplied as Pyronine G), proved satisfactory.

ACKNOWLEDGMENTS

All experiments mentioned above were carried out with the excellent technical assistance of Mrs. M. Gregorová. I am indebted to Dr. O. P. Sehgal, Department of Plant Pathology, University of Missouri, Columbia, Mo., for reading and correcting the English text. I thank Mr. P. Kvíčala for the photographs. Thanks are also due to SERVA Feinbiochemica GmbH, Heidelberg, West Germany, for the gift of a sample of Stains-all.

REFERENCES

Adesnik, M. (1971). *In* "Methods in Virology" (K. Maramorosch and H. Koprowski, eds.), Vol. 5, pp. 125–177. Academic Press, New York.

Bennick, A. (1969). *Anal. Biochem.* **26**, 453.

Blake, A., and Peacocke, A. R. (1968). *Diopolymers* **6**, 1225.

Dahlberg, A. E., Dingman, C. W., and Peacock, A. C. (1969). *J. Mol. Biol.* **41**, 139.

Davis, B. J. (1962). "Disc Electrophoresis." Distillation Prod. Div., Eastman Kodak Co., Rochester, New York (reprint).

de Boer, J., and Sarnaker, R. (1956). *Med. Proc.* (*S.Afr.*) **2**, 218.

Grossbach, U., and Weinstein, I. B. (1968). *Anal. Biochem.* **22**, 311.

Hull, J. (1972). *J. Gen. Virol.* **17**, 111.

Konings, R. N. H., and Bloemendal, H. (1969). *Eur. J. Biochem.* **7**, 165.

Lamm, M. E., Childers, L., and Wolf, M. K. (1965). *J. Cell Biol.* **27**, 313.

Lane, L. C., and Kaesberg, P. (1971). *Nature* (*London*), New Biol. **232**, 40.

Lerman, L. S. (1961). *J. Mol. Biol.* **3**, 18.

Löber, G. (1969). *Z. Chem.* **9**, 252.

Loening, U. E. (1967). *Biochem. J.* **102**, 251.

Marcinka, K. (1972). *Anal. Biochem.* **50**, 304.

Marcinka, K. (1974). *Fed. Eur. Biochem. Soc. Symp. 9th* (abstr.), p. 440.

Maurer, H. R. (1966). *Z. Klin. Chem.* **4**, 85.
Maurer, H. R. (1971). "Disc Electrophoresis and Related Techniques of Polyacrylamide Gel Electrophoresis." de Gruyter, Berlin.
Neuhoff, V. (1968). *Arzneim-Forsch.* **18**, 35.
Peacock, A. C., and Dingman, C. W. (1967). *Biochemistry* **6**, 1818.
Popa, L. M., and Repanovici, R. (1969). *Biochim. Biophys. Acta* **182**, 158.
Pun, J. Y., and Lombrozo, I. (1964). *Anal. Biochem.* **9**, 9.
Richards, E. G., Coll, J. A., and Gratzer, W. B. (1965). *Anal. Biochem.* **12**, 452.
Santilli, V. (1966). *Biochim. Biophys. Acta* **120**, 239.
Scatchard, G. (1944). *Ann. N.Y. Acad. Sci.* **51**, 660.
Semmel, M., and Huppert, J. (1964). *Arch. Biochem. Biophys.* **108**, 158.
Wieland, T., and Bauer, L. (1951). *Angew. Chem.* **63**, 511.
Wilson, C. M. (1971). *Plant Physiol.* **48**, 64.

Chapter 4

Isolation and Fractionation of Mouse Liver Nuclei[1]

Ph. CHEVAILLIER AND M. PHILIPPE

Laboratoire de Biologie Cellulaire,
Université Paris-Val de Marne,
Créteil, France

I. Fine Structure of Nuclei Isolated in Various Ionic Media 70
A. General Methods 70
B. Preliminary Observations 70
C. Action of Various Cations on the Fine Structure of Isolated Nuclei . . 73
D. Action of Nuclease and Protease Inhibitors on Nuclear Fine Structure. . 74
II. Fractionation of Mouse Liver Nuclei 77
III. Purification of Nuclear Membranes from Mouse Liver Nuclei 79
IV. Conclusions 82
References 83

The chromatin of eukaryotes assumes two different configurations in interphase nuclei, a compact form or heterochromatin, and a diffuse form or euchromatin. Comparative studies of the molecular and functional properties of the two chromatin fractions necessitate isolation and fractionation methods that preserve, as much as possible, the structural characteristics the nuclear constituents possess *in situ.*

Numerous investigators tried to demonstrate differences in the chemical composition of the DNA and associated proteins of the two chromatin fractions. Their results were often contradictory, probably because methods of nuclear fractionation were unsuitable for the nuclei studied.

Our work was designed to obtain a medium for the isolation and fractionation of mouse liver nuclei that would satisfactorily preserve the intranuclear structures.

[1]This investigation was supported by grants from the C.N.R.S. (ERA 400; RCP 283).

I. Fine Structure of Nuclei Isolated in Various Ionic Media

A. General Methods

Nuclei were isolated from the livers of normal adult mice according to the method of Chauveau *et al.* after homogenization of the livers in 19 vol of a 2.2 *M* sucrose medium of known ionic strength. The homogenate was filtered through several thicknesses of gauze and then centrifuged at 40,000 *g* for 1 hour.

The various media generally used for the preparation of nuclei and eventually for their fractionation are compared in Table I. In our experiments different ionic media were used in addition to the medium of Chauveau *et al.* (1957) and that of Frenster *et al.* (1963) (Chevaillier and Philippe, 1973). These media contained in varying combinations $MgCl_2$, $CaCl_2$, KCl, a polyamine (spermidine or spermine), tris buffer, and sometimes β-mercapto-ethanol.

Isolated nuclei were studied with an electron microscope to test their purity and to determine their morphological components. These observations were made after a double classic fixation with 2.5% glutaraldehyde for 15 minutes and 2% osmium tetroxide for 1 hour.

The respective phospholipid, RNA, and DNA contents of each nuclear fraction were determined after their extraction according to a variation (Moulé, 1953) of the method of Schmidt and Tannhauser (1945). Phosphorus was determined according to the method of Delsal and Manhouri (1958). RNA was measured separately by a variation of the Mejbaum technique (Moulé, 1953). Proteins were determined according to Lowry *et al.* (1951).

B. Preliminary Observations

In media containing small quantities of bivalent cations (medium of Chauveau et al., 5×10^{-5} *M*; medium of Frenster et al., 5×10^{-3} *M*), nuclei were evenly dispersed, and their structure was relatively homogeneous as compared to intact nuclei *in situ*. Euchromatin and heterochromatin could no longer be distinguished. After fractionation in the medium of Chauveau *et al.*, only the nucleoli were identifiable, but their structure appeared less compact than that of intact nucleoli.

To preserve the morphological constituents of the nuclei correctly, the isolation and fractionation medium must therefore have an ionic composition that is both sufficient and balanced (Chevaillier, 1971).

TABLE I

SALINE MEDIA USED FOR ISOLATION OR FRACTIONATION OF NUCLEI BY DIFFERENT INVESTIGATORS

Method	Buffer	$CaCl_2$	$MgCl_2$	KCl	NaCl	Amines	Other salts
Chauveau *et al*. (1957)	—	0.05 m*M*	—	—	—	—	—
Anderson and Norris (1960)	—	—	—	—	—	1 mM–0.5 *M*	—
Frenster *et al*. (1963)	—	3.3 m*M*	—	—	—	—	—
Blobel and Potter (1966)	50 m*M* tris (pH 7.5)	—	5 m*M*	25 m*M*	—	—	—
Sadowski and Steiner (1968)	—	—	5 m*M*	—	—	—	—
Schildkraut and Maio (1968)	10 m*M* tris (pH 7.6)	3 m*M*	—	—	—	—	—
Utakoji *et al.* (1968)	—	1 m*M*	1 m*M*	—	—	—	1 m*M* $ZnCl_2$
Haussler *et al*. (1969)	50 m*M* tris (pH 7.5)	—	5 m*M*	25 m*M*	—	—	—
Dolbeare and Koenig (1970)	—	1–3 m*M*	1–3 m*M*	0–50 m*M*	—	—	—
Dounce and Ickowicz (1970)	—	—	—	—	—	—	0.3 m*M* $InCl_3$ 0.2 m*M* $Pb(CH_3COO)_2$ 0.2 m*M* $CdSO_4$
Yasmineh and Yunis (1970)	—	1.5 m*M*	—	—	—	—	—
Brasch *et al*. (1971)	—	—	—	—	15 m*M*	—	—
Yasmineh and Yunis (1971)	—	0.5 mM	3 m*M*	—	—	—	—
Chevaillier and Philippe (1973)	1 m*M* tris (pH 7.5)	0.9 m*M*	0.9 m*M*	25 m*M*	—	0.14 m*M* Spermidine	—
Laval and Bouteille (1973)	1 m*M* phosphate (pH 6.8)	—	5 m*M*	—	—	—	—

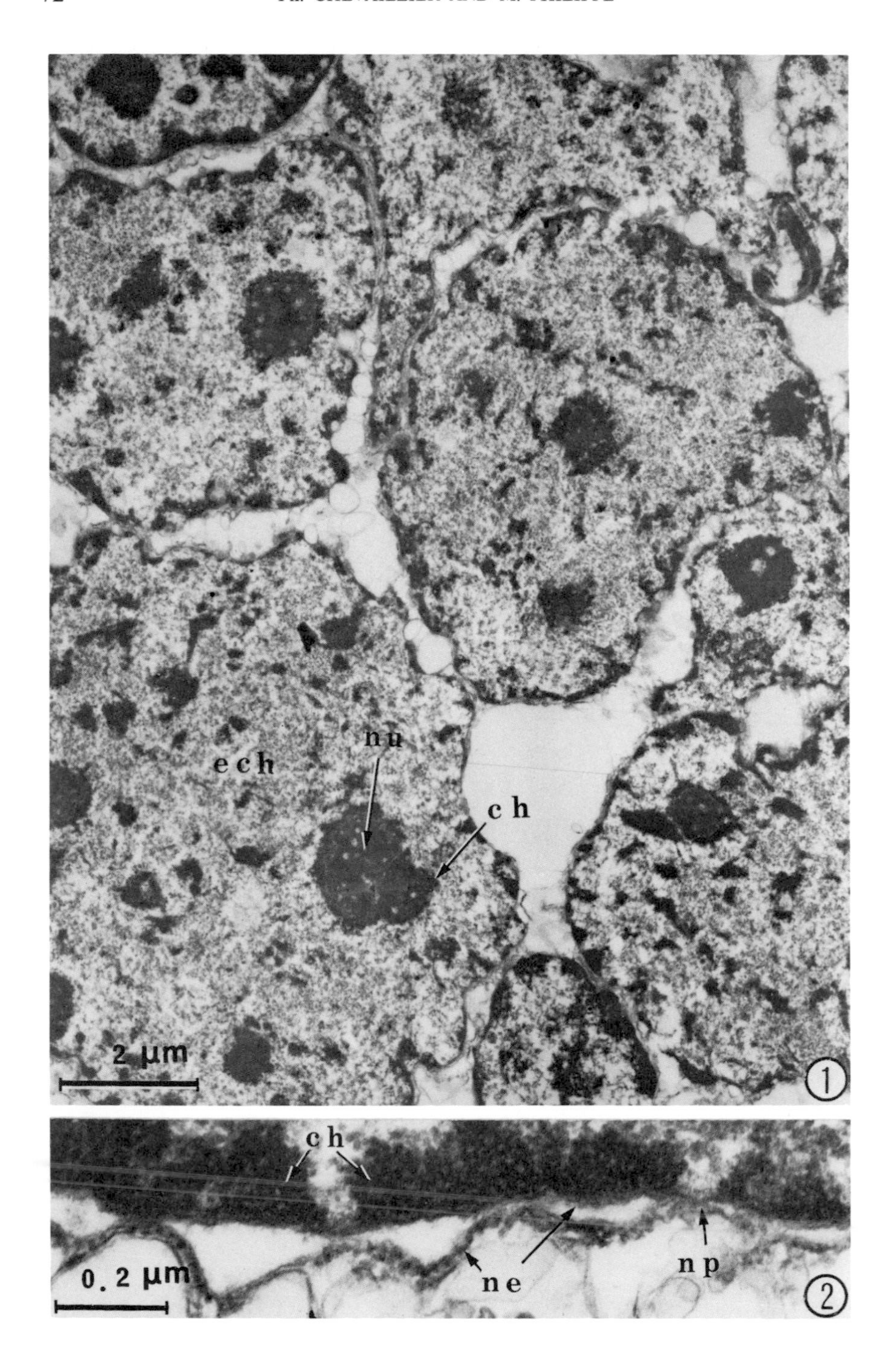
ech
nu
ch
2 μm
1
ch
ne
np
0.2 μm
2

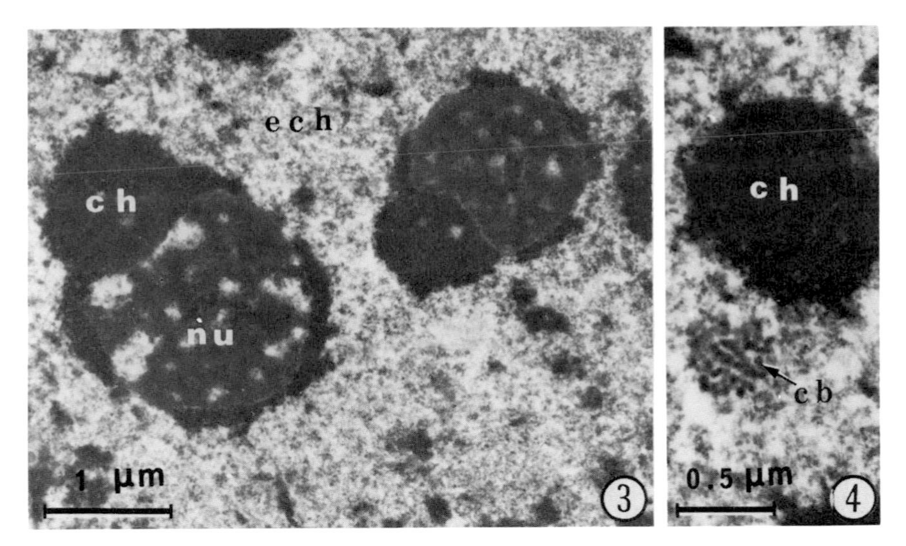

FIGS. 1–8. Nuclei isolated (Figs. 1–4) and subfractionated (Figs. 5–8) in a medium containing 2.2 *M* sucrose, 1 m*M* tris (pH 7.5), 25 m*M* KCl, 0.9 m*M* $CaCl_2$, 0.9 m*M* $MgCl_2$, and 0.14 m*M* spermidine.

FIGS. 1–4. Fine structure of isolated nuclei.

FIG. 1. General view. ×9400.

FIG. 2. Detail of the structure of the nuclear envelope (ne) and of peripheral heterochromatin (ch) lying under the inner nuclear membrane. Note the persistence of nuclear pores (np). ×78,000.

FIGS. 3 and 4. Detail of nucleolar fine structure (nu) and of the dense perinucleolar chromatin (ch); note the presence of a coiled body in Fig. 4. Note also the well-preserved diffuse chromatin (ech) in Fig. 3. Fig. 3: ×16,800. Fig. 4: ×25,500.

C. Action of Various Cations on the Fine Structure of Isolated Nuclei

Three different mineral cations (K^+, Ca^{2+}, and Mg^{2+}) and two polyamines (spermidine and spermine) were tried at different concentrations, most often in a 1 m*M* tris buffer (pH 7.5). This concentration was ultimately retained to prevent extraction of soluble nuclear constituents in the 10 m*M* buffer. However, absence of the buffer did not greatly modify nuclear structure, even though euchromatin appeared less homogeneous. K^+ ions were necessary for the proper preservation of chromatin, particularly euchromatin, but they alone did not suffice. When the medium contained only KCl in tris buffer, the chromatin and the nucleoli were totally dispersed and the nuclei appeared homogeneous. Moreover, the nuclear envelope was very fragile, and numerous nuclei were broken during the initial homogenization.

Mineral cations were found to be indispensable for stabilization of the chromatin and the nuclear envelope. The following ions were therefore proposed: Ca^{2+}, Mg^{2+}, Pb^{2+}, Cd^{2+}, Zn^{2+}, or In^{3+}. Among the bivalent mineral cations, we chose Ca^{2+} and Mg^{2+}, which were less likely to inhibit the nuclear enzymatic activities most often studied. The concentrations utilized varied between 0.5 and 2.5 m*M*. At the most elevated concentrations, chromatin and nuclear structures were more compact, and euchromatin was barely visible. With concentrations of 1.25 m*M* for Ca^{2+} and Mg^{2+}, we observed a marked dispersal of heterochromatin; the nucleoli were, however, well preserved and, with the method of Frenster *et al.*, they could be isolated after sonication of the nuclei.

Utilization of only Mg^{2+} and Ca^{2+} cations at different concentrations did not permit us to preserve satisfactorily the intranuclear structures. We therefore added a polyamine, either spermine or, in most cases, spermidine, to the medium for the isolation and fractionation of the nuclei. By trying different concentration combinations of this polyamine and Ca^{2+} or Mg^{2+} in the presence of K^+ ions, we were able to prepare a medium that permitted good preservation of the intranuclear structures and, in particular, of the different chromatin fractions. For the liver of normal adult mice, the medium used (medium O) contained: 1 m*M* tris (pH 7.5); 25 m*M* KCl; 0.9 m*M* $MgCl_2$; 0.9 m*M* $CaCl_2$; 0.14 m*M* spermidine.

This medium, which was used for the isolation of nuclei and the separation of intranuclear constituents, maintained the heterochromatin and the nucleoli in a condensed form (Fig. 1). Euchromatin remained in its diffuse state in the nucleoplasm, even though the diameter of its fibers was a little larger than that of those in the nuclei *in situ*. Nucleolus-associated chromatin was clearly visible (Fig. 3); the heterochromatin located under the inner nuclear membrane remained associated with this membrane (Fig. 2). A coiled body whose structure was well preserved could also be observed in some isolated nuclei (Fig. 4). The quantity of DNA recovered in the nuclear fraction corresponded to about 60% of the total liver DNA.

Among the isolation media used, those containing β-mercaptoethanol yielded very small amounts of DNA. We verified that these small values of DNA were not due to interference of this compound with the colorimetric determination of DNA, and consequently abandoned the media containing β-mercaptoethanol.

D. Action of Nuclease and Protease Inhibitors on Nuclear Fine Structure

Even though the chromatin and nucleoli were well preserved by the isolation medium described above, perichromatin granules were, in con-

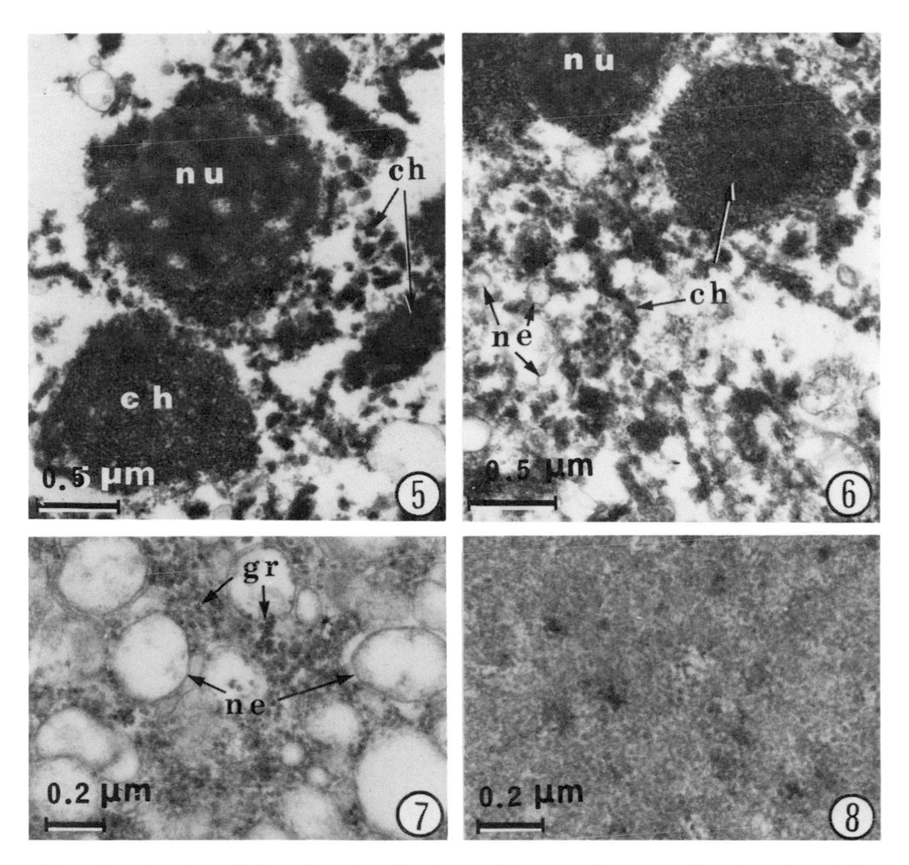

FIG. 5. Fraction 1000 *g* (fraction I) obtained by centrifugation of sonicated nuclei at 1000 *g* for 10 minutes. ×22,000.

FIG. 6. Nuclear fraction obtained by centrifugation at 3500 *g* for 20 minutes (fraction II). × 22,000.

FIG. 7. Nuclear fraction obtained by centrifugation at 78,000 *g* for 1 hour (fraction III). ×40,000.

FIG. 8. Light nuclear fraction obtained by centrifugation at 105,000 *g* for 16 hours (fraction IV). ×47,000.

trast, never identified with certainty in the isolated nuclei. However, interchromatin granules were quite visible.

Since the intranuclear structures are probably ribonucleoprotein in nature (Monneron and Bernhard, 1969), we tried several different isolations of the nuclei in the above ionic medium to which was added an inhibitor of RNases or proteases. The isolations were therefore made either in the presence of potassium polyvinyl sulfate (PVS), or phenylmethylsulfonyl fluoride

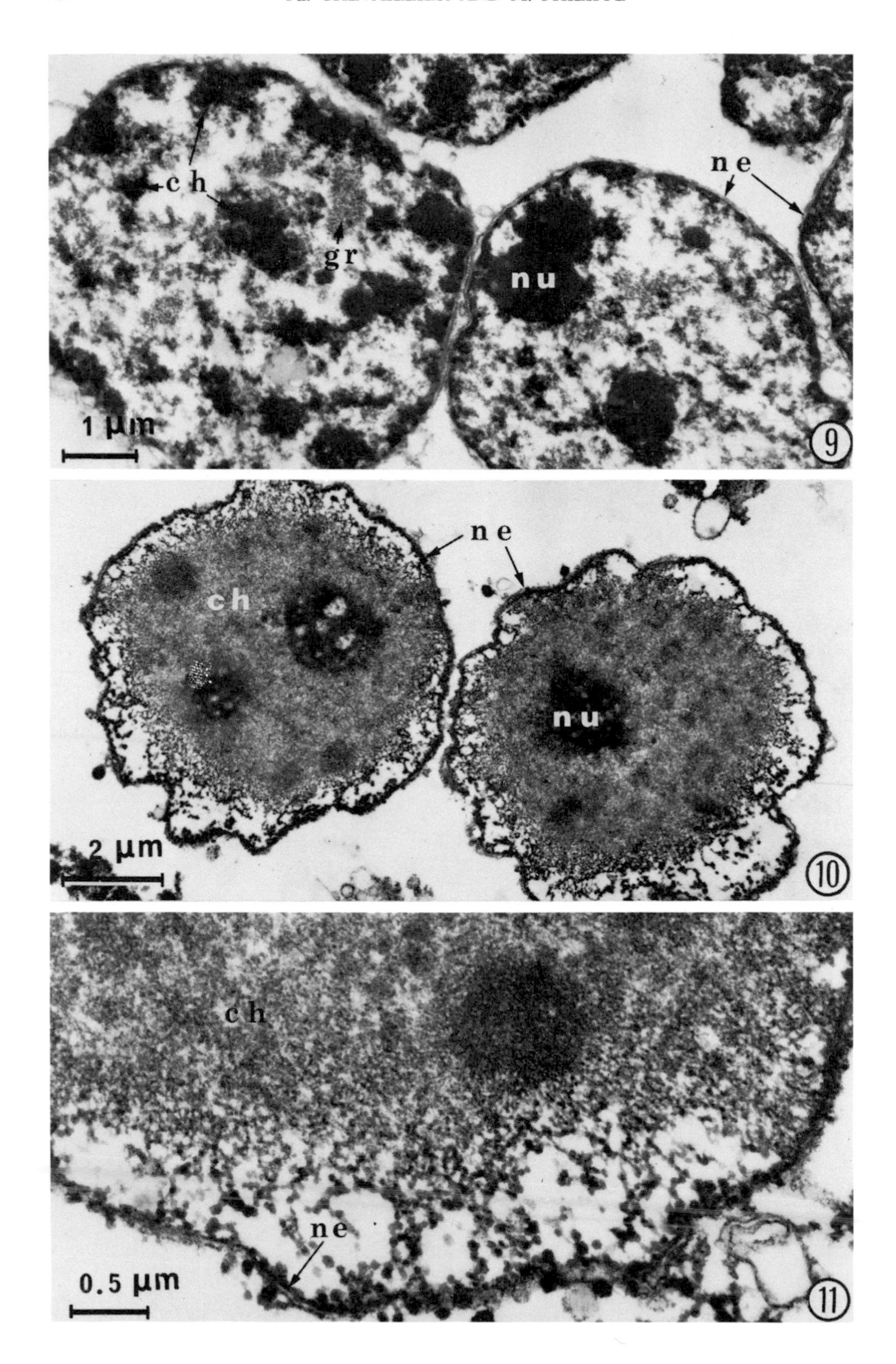
ch
gr
ne
nu
1 µm
9
ne
ch
nu
2 µm
10
ch
ne
0.5 µm
11

(PMSF), or in the presence of both inhibitors simultaneously. PMSF produced only slight modifications of intranuclear structures (Fig. 9); by contrast, PVS alone (Fig. 10) or associated with PMSF (Fig. 11) profoundly altered the nuclei. The nucleoli were partially diffuse. Chromatin was completely dispersed, except at the nuclear periphery where it appeared aggregated in reticular clumps very similar to the clumps of bacterial DNA formed after precipitation by fixatives. These observations were similar to those made by investigators treating isolated nuclei with other polyanions (Arnold *et al.*, 1972). In no case did we preserve perichromatin granules with the concentrations utilized. These organelles are therefore structures particularly sensitive either to the action of nuclear enzymes or to the ionic conditions of nuclear isolation methods; their disappearance was noted in all the media used during this work.

II. Fractionation of Mouse Liver Nuclei

Isolated nuclei were resuspended in 10 vol of a 0.25 *M* sucrose solution in the same ionic medium used for isolation of the nuclei. They were sonicated for 30 seconds to 4 minutes, depending on the medium utilized. Disintegration of the nuclei was controlled with a phase-contrast microscope.

The disintegrated nuclei were first centrifuged for 10 minutes at 1000 *g*; the pellet obtained was designated fraction I. The supernatant was centrifuged again for 20 minutes at 3500 *g*, providing the pellet that was fraction II. Centrifugation of this supernatant for 1 hour at 78,000 *g* produced the pellet designated fraction III. The last supernatant was again centrifuged for 18 hours at 105,000 *g*, producing the pellet that was fraction IV.

The nuclear fractions prepared in this manner were studied biochemically and structurally by utilizing the same methods as for whole nuclei.

After isolation of the nuclei in the medium of Chauveau *et al.*, we obtained morphologically similar fractions containing chromatin fibers and membrane vesicles derived from the nuclear envelope. We did not observe the nucleoli that were dispersed during sonication. The DNA was divided almost equally into heavy fractions (I and II) and light fractions (III and IV).

FIGS. 9–11. Mouse liver nuclei isolated in the same medium as in Fig. 1, but in the presence of PMSF (Fig. 9), PVS (Fig. 10), or both (Fig. 11). Note the alteration of nuclear fine structure in the presence of PVS. Fig. 9: ×10,600. Fig. 10: ×7300. Fig. 11: ×22,000.

Using the medium of Frenster *et al.*, we obtained in fractions I and II the nucleoli and their associated chromatin, as well as the nuclear membranes. Fraction III contained chromatin fibers with diameters from 150 to 200 Å. The presence of calcium in quantities greater than in the medium of Chauveau *et al.* stabilized the nucleoli and some of the heterochromatin.

With the more complex media containing Ca^{2+} and Mg^{2+} ions and spermidine, fractionation of nuclei in the same medium allowed us to prepare two heavy fractions containing nucleoli, nucleoli-associated chromatin, extranucleolar heterochromatin, and several membranes (fractions 1000 *g* and 3500 *g*). Two other nuclear fractions (fractions III and IV), called light fractions, were also obtained. Fraction III contained most of the nuclear membranes in the form of smooth or granular vesicles. Fraction IV was essentially composed of fine euchromatin fibrils.

The quantities of DNA recovered in each of the nuclear subfractions varied according to the medium used. We observed that the amount of DNA recovered in the heavy fractions varied from 45% to more than 95%; the more diffuse the nuclear structure, the greater the quantity of DNA found in the light chromatin fractions. In the medium chosen for isolation and fractionation of mouse liver nuclei, we recovered an average of 90% of the nuclear DNA in fraction I, 5% in Fraction II, 1% in fraction III, and 0.25% in fraction IV. A small fraction of total nuclear DNA remained in the supernatant of the last centrifugation; this quantity varied between 1 and 10% according to the fractionation medium utilized.

The compositions of the different nuclear fractions isolated in medium O are indicated in Table II.

TABLE II

CHEMICAL COMPOSITION OF THE NUCLEI AND ISOLATED NUCLEAR SUBFRACTIONS[a]

		Nuclear fraction			
Component	Nuclei	I	II	III	IV
DNA	60.4	91.3	6.5	1.1	0.2
RNA	3.8	65	7.8	12.7	3.3
Proteins	3.1	44	13.5	0.8	17
Phospholipids	2.1	23.8	19	33.3	4.8
RNA/DNA	0.21	0.14	0.25	2.47	3.21
DNA/proteins	0.20	0.40	0.09	0.25	0.002

[a] The quantities of DNA, RNA, proteins, and phospholipids in the nuclei are expressed in percent of the corresponding quantities in the total homogenate. The quantities present in each nuclear fraction are expressed in percent of the corresponding amounts in isolated nuclei.

Most of the nuclear DNA was contained in the large clumps of extra-nucleolar heterochromatin and heterochromatin associated with the nucleoli. Fraction IV, consisting almost exclusively of fine chromatin fibrils, contained a small percentage of nuclear DNA; it contained, by contrast, appreciable quantities of RNA and especially of proteins. Although free of nuclear membranes, phospholipids accounted for 5% of its composition. Fractions I and II together contained more phospholipids than fraction III, even though the nuclear membranes were greatly reduced in the heavy fractions. The majority of nuclear RNA was found in fraction I which contained most of the nucleoli and was the largest in size. However, fractions III and IV contained large quantities of RNA as compared with their content of DNA.

III. Purification of Nuclear Membranes from Mouse Liver Nuclei

This purification was made essentially on fraction III, which contained the largest portion of nuclear membranes. Fraction III was resuspended in an ionic medium identical to that used for the isolation and fractionation of nuclei. It was placed at the top of a discontinuous gradient composed of four layers of sucrose of the following densities: 1.18, 1.24, 1.26, and 1.28. The gradients were centrifuged for 40 hours at 130,000 *g*. After centrifugation each gradient was fractionated with an ISCO gradient fractionator. Different subfractions corresponding to the same peak were pooled, centrifuged at 130,000 *g* for 2 hours, and used for structural or biochemical studies.

Five subfractions, one of which was a pellet, were therefore obtained with this method (Fig. 12). Electron microscope analysis of the different sub-

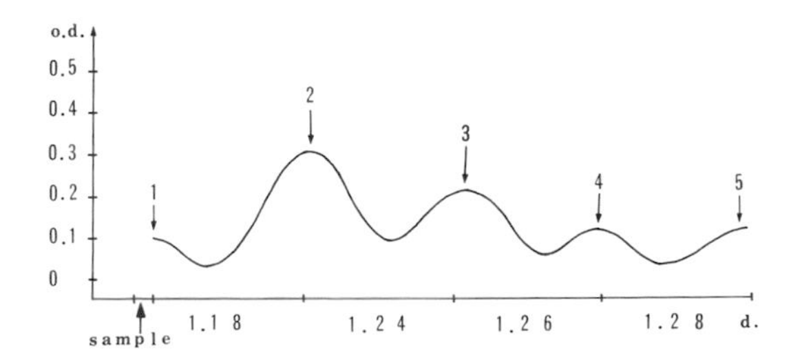

FIG. 12. Gradient profile obtained by centrifugation of nuclear fraction III for 40 hours at 135,000 *g* on a discontinuous gradient of four layers of sucrose (density from 1.18 to 1.28). Five subfractions are separated.

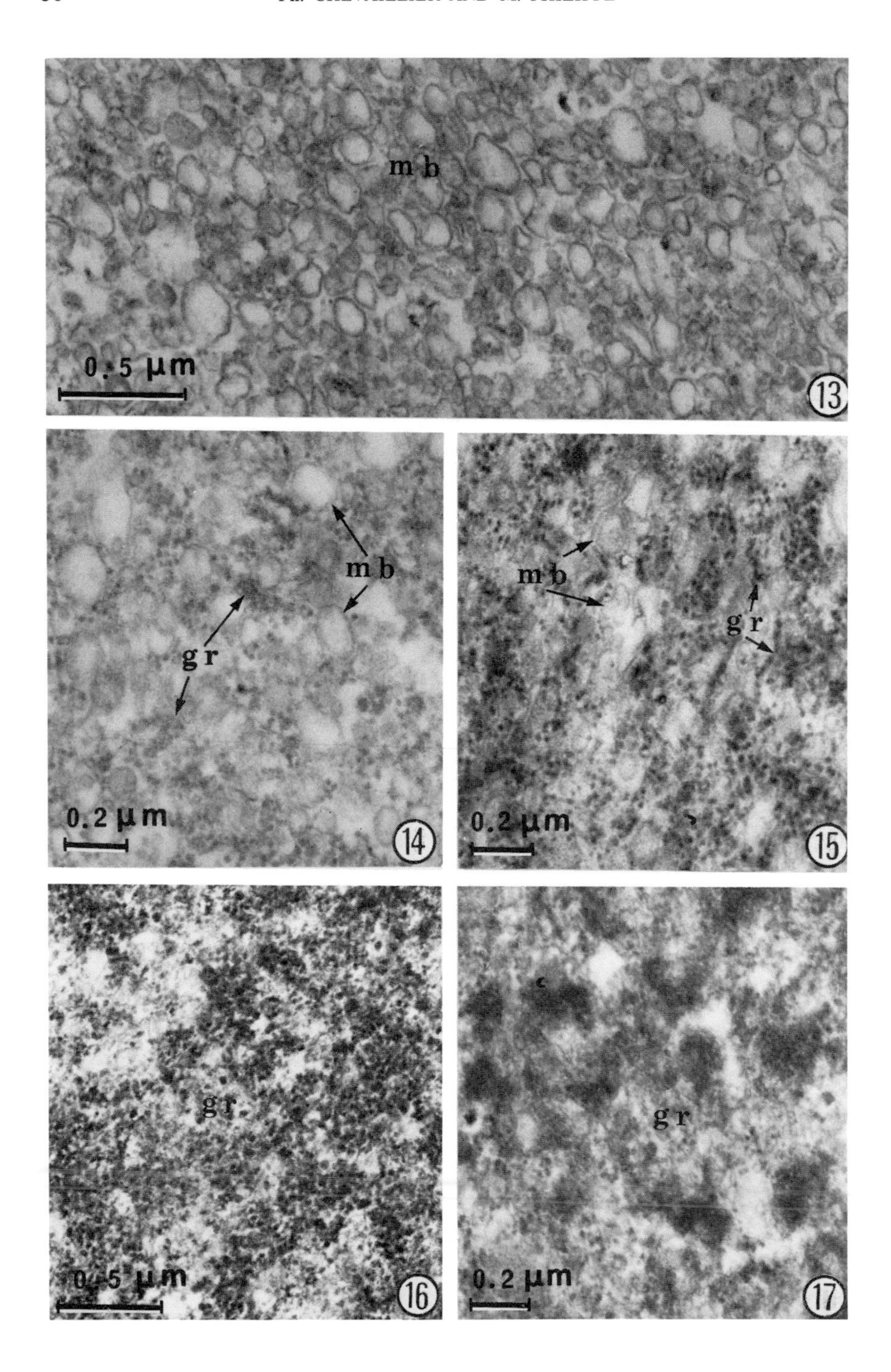
m b
0.5 µm
13
m b
g r
0.2 µm
14
m b
g r
0.2 µm
15
g r
0.5 µm
16
g r
0.2 µm
17

fractions shows that the first peak contained almost exclusively inner nuclear membranes; only a few membranes with attached ribosomes and free particles were found (Fig. 13). The second peak was composed mostly of granular membranes; a few particles not linked to the membranes were noted (Fig. 14). Subfraction 3 contained essentially free particles, as well as granular membranes (Fig. 15). The last two subfractions contained essentially particles dispersed or grouped in small clumps (Figs. 16 and 17). The distribution of DNA, RNA, proteins and phospholipids among the different subfractions is given in Table III.

The phospholipids of fraction III were distributed essentially in the first three peaks. These results agree with electron microscope observations which show the presence of membranes in the three subfractions; the first peak contained inner nuclear membranes. In the two following peaks, outer nuclear membranes with attached ribosomes were found. In contrast, the RNA of the 78,000 *g* fraction was found especially in subfractions 2 to 5, and principally in peaks 2 and 3; peak 1 contained only a very small quantity

TABLE III

RELATIVE COMPOSITION OF THE SUBFRACTIONS OBTAINED AFTER CENTRIFUGATION OF FRACTION III (78,000 *g*) FOR 40 HOURS AT 27,000 rpm ON A DISCONTINUOUS GRADIENT COMPOSED OF FOUR LAYERS OF SUCROSE (DENSITIES: 1,28, 1.25, 1.24, 1.18)

	Subfraction[a]				
Component	1	2	3	4	5
Phospholipids	22	38	24.5	10.5	5
RNA	5	42	23.5	11	18.5
DNA	4	19.5	45	24.5	7
Proteins	3.5	40.5	25	22.5	8.5

[a]The amount of each component recovered in each subfraction is expressed in percent as compared to those found in fraction III (78,000 *g*).

FIGS. 13–17. Fine structure of nuclear subfractions obtained from the gradient of Fig. 12.

FIG. 13. Subfraction 1, consisting of smooth vesicles derived from the inner nuclear membrane, ×36,000.

FIGS. 14 and 15. Subfractions 2 and 3, containing rare smooth vesicles (mb), ribosomelike particles (gr), and granular vesicles derived from the outer nuclear membrane, Fig. 14: ×45,000. Fig. 15: ×46,000.

FIGS. 16 and 17. Subfractions 4 and 5 correspond to ribosomelike granules existing either in a dispersed or an aggregated state. Fig. 16: ×36,000. Fig. 17: ×44,000.

of RNA. Peaks 2 to 4 contained most of the DNA of fraction III, especially peak 3, which contained almost half of the total DNA of that fraction. The significance of the presence of DNA in subfraction 3 is still under study to determine the relation of this DNA to the membrane systems also present in this fraction.

The densities of outer and inner nuclear membranes prepared by this method are greater than those obtained by other investigators. Zbarsky (1972), by gradient centrifugation of osmotically broken nuclei, isolated two membrane fractions: one with a density of 1.16 composed of outer nuclear membranes, and the other with a density of 1.18 composed of inner nuclear membranes.

In addition, Kashnig and Kasper (1969) and Mizuno *et al.* (1971) isolated two membrane fractions from sonicated nuclei by sucrose gradient centrifugation in the presence of sodium citrate. The first fraction had a density between 1.16 and 1.18 and contained more membranes with attached ribosomes; the second had a density between 1.18 and 1.20. The differences in density observed, especially for outer nuclear membranes, were probably caused by the loss of material provoked by the osmotic shock or by the potassium citrate. Pure nuclear membranes prepared according to Monneron *et al.* (1973), by treating nuclei with high ionic concentrations, most often lack ribosomes and have densities varying between 1.17 and 1.18, depending on the molarity of the extraction solutions. The advantage of the method described here as compared to other methods of preparation and purification of nuclear membranes is that it permits separation of inner nuclear membranes from outer nuclear membranes and nuclear particles. It avoids the use of the very acid media of certain techniques (Bornens, 1968; Kashnig and Kasper, 1969) capable of denaturing membrane proteins. This method may limit the extraction of membrane constituents to a greater extent than when use is made of solutions of low (Franke, 1966; Mizuno *et al.*, 1971; Zbarsky, 1972) or high ionic strength (Monneron *et al.*, 1973), or enzymes capable of degrading components bound to nuclear membranes (Berezney *et al.*, 1972; Kay *et al.*, 1972).

IV. Conclusions

A medium for the isolation and fractionation of mouse liver nuclei that preserves the fine structure of nuclear constituents has been proposed. This medium contains, in addition to 2.2 *M* sucrose, 1 m*M* tris (pH 7.5), 25 m*M* KCl, 0.9 m*M* $MgCl_2$, 0.9 m*M* $CaCl_2$, and 0.14 m*M* spermidine. After sonication and the rupture of the nuclei in this medium, two heavy fractions (1000 *g* and 3500 *g*), an intermediate fraction (78,000 *g*) and a light fraction (105,000 *g*) can be isolated by differential centrifugation. The two heavy fractions contain essentially nucleoli, nucleoli-associated chromatin, and extranucleolar heterochromatin. The fraction sedimenting at 78,000 *g* contains dense particles and nuclear membranes. The light fraction is composed of euchromatin fibrils. The heterogeneous 78,000-*g* fraction can be subfractionated on a discontinuous sucrose gradient with densities varying from 1.18 to 1.28. A light subfraction composed of inner nuclear membranes is thus obtained, along with two subfractions of intermediate density containing outer nuclear membranes and ribosomelike particles, and two heavy subfractions composed of dense, dispersed or associated particles which are probably ribosomes detached from the outer membranes during sonication.

The object of these studies was to develop a technique that preserves the fine structure of the different intranuclear organelles in isolated nuclei or in nuclear subfractions. We found that the presence of different cations in specified quantities and proportions is necessary during the fractionation of nuclei. Solutions of weak or high ionic strength or too acid pH conditions must be avoided.

REFERENCES

Anderson, N. G., and Norris, C. B. (1960). *Exp. Cell Res.* **19**, 605–618.

Arnold, E. A., Yawn, D. H., Brown, D. G., Wyllie, R. C., and Coffey, D. S. (1972). *J. Cell Biol.* **53**, 737–757.

Berezney, R., Macaulay, L. K., and Crane, F. (1972). *J. Biol. Chem.* **17**, 5549–5561.

Blobel, G., and Potter, V. R. (1966). *Science* **154**, 1662–1665.

Bornens, M. (1968). *C. R. Acad. Sci.* **266**, 596–599.

Brasch, K., Seligy, V. L., and Setterfield, G. (1971). *Exp. Cell Res.* **65**, 61–72.

Chauveau, J., Moulé, Y., and Rouiller, C. (1957). *Bull. Soc. Chim. Biol.* **34**, 1521–1533.

Chevaillier, P. (1971). *Exp. Cell Res.* **67**, 466–471.

Chevaillier, P., and Phillippe, M. (1973). *Exp. Cell Res.* **82**, 1–14.

Delsal, J. L., and Manhouri, H. (1958). *Bull. Soc. Chim. Biol.* **40**, 1623–1636.

Dolbaere, F., and Koenig, H. (1970). *Proc. Soc. Exp. Biol. Med.* **135**, 636–641.

Dounce, A. L., and Ickowicz, R. (1970). *Arch. Biochem. Biophys.* **137**, 143–155.

Franke, W. W. (1966). *J. Cell Biol.* **33**, 619–623.

Frenster, J. H., Allfrey, V. G., and Mirsky, A. (1963). *Proc. Nat. Acad. Sci. U.S.* **50**, 1026–1032.

Haussler, M. R., Thomson, W. W., and Norman, A. D. (1969). *Exp. Cell Res.* **58**, 234–242.

Kashnig. D. M., and Kasper, C. B. (1969). *J. Biol. Chem.* **244**, 3786–3792.

Kay, R. R., Fraser, D., and Johnston, I. R. (1972). *Eur. J. Biochem.* **30**, 145–154.

Laval, M., and Bouteille, M. (1973). *Exp. Cell Res.* **76**, 337–348.

Lowry, O. H., Rosebrough, N. J., Farr, A. L., and Randall, A. J. (1951). *J. Biol. Chem.* **193**, 265–275.

Mizuno, N. S., Stoops, C. E., and Sinha, A. A. (1971). *Nature (London), New Biol.* **229**, 22–24.

Monneron, A., and Bernhard, W. (1969). *J. Ultrastruct. Res.* **27**, 266–288.
Monneron, A., Blobel, G., and Palade, G. E. (1973). *J. Cell Biol.* **55**, 104–125.
Moulé, Y. (1953). *Arch. Sci. Physiol.* **7**, 161–187.
Sadowski, P. D., and Steiner, J. W. (1968). *J. Cell Biol.* **37**, 147–162.
Schildkraut, C. L., and Maio, J. J. (1968). *Biochim. Biophys. Acta* **161**, 76–93.
Schmidt, G., and Thannhauser, S. J. (1945). *J. Biol. Chem.* **161**, 83–89.
Utakoji, T., Muramatsu, M., and Sugano, H. (1968). *Exp. Cell Res.* **53**, 447–458.
Yasmineh, W. G., and Yunis, J. J. (1970). *Exp. Cell Res.* **59**, 69–75.
Yasmineh, W. G., and Yunis, J. J. (1971). *Exp. Cell Res.* **64**, 41–48.
Zbarsky, I. B. (1972). *In* "Methods in Cell Physiology" (D. M. Prescott, ed.), Vol. 5, pp. 167–198. Academic Press, New York.

Chapter 5

The Separation of Cells and Subcellular Particles by Colloidal Silica Density Gradient Centrifugation

DAVID A. WOLFF

Department of Microbiology,
The Ohio State University,
Columbus, Ohio

I. Introduction . . . 85
II. Materials and Methods . . . 87
A. Properties of Colloidal Silica . . . 87
B. Preparing the Gradient Medium . . . 88
C. Gradient Formation . . . 91
III. Applications . . . 93
A. Cultured Cell Separation . . . 93
B. Separation of Cells from Tissues . . . 97
C. Separation of Blood Cells . . . 98
D. Separation of Marine Organisms . . . 98
E. Separation of Subcellular Particles . . . 99
F. Purification of Viruses . . . 101
IV. Conclusions . . . 102
References . . . 103

I. Introduction

The use of density gradient centrifugation has played a major role in the recent advances made in cellular and molecular biology. Various gradient media such as sucrose, CsCl, and Ficoll have been developed for specific purposes to take advantage of certain properties such as proper density range, ease of preparation, and low cost. Gradient media are usually chosen that do not alter the particles to be separated, and most investigators select familiar and inexpensive media, such as sucrose, that provide a useful density range for separation. The most common problems encountered with gradient media are osmotic pressure changes and chemical toxicity. Low-molecular-weight materials such as sucrose, CsCl, potassium tartrate, and NaBr have been useful, but create high osmotic pressure, making them unsuitable for the separation of osmotically fragile particles. The use of large-molecular-weight substances such as Ficoll (MW 400,000) (polymer of sucrose) and dextran T40(MW 40,000) enables the investigator to adjust the osmotic pressure to suit his needs, but these materials are quite viscous.

This chapter does not presume to review all density gradient procedures, nor to provide a thorough critique of the various gradient media available. The purpose here is to describe the practical use of colloidal silica as a gradient medium, explaining procedures for its use and the many advantages of this system.

The usefulness of colloidal silica as a gradient medium has come largely from the work of Håkan Pertoft and his collaborators at the Institute of Medical Chemistry in Uppsala, Sweden. Pertoft and his coworkers applied this gradient medium for purifying or separating viruses, cells, and subcellular particles, and determined a variety of medium components necessary for application to these specific purposes. The medium has been used for rate zonal and isopycnic or equilibrium separation in tubes and zonal systems.

The use of colloidal silica was first reported in 1959 by Mateyko and Kopac who compared colloidal silica with various polysaccharides, methyl celluloses, gum arabic, proteins, and other polymers. In their 1959 paper, and more extensively in 1963, they report osmotic pressure effects, ability to separate cells, and cell toxicity for the above-mentioned compounds, as well as the criteria desired of an ideal density gradient medium. These criteria were similar to those presented by deDuve, Berthet, and Beaufay (1959), and include isoosmolarity, production of a useful density range for separation, and absence of toxicity. In addition, they list stability at a wide variety of temperatures and pH values, lack of permeation into the particles, and solubility or dispersibility in aqueous solutions. According to Mateyko and Kopac (1959, 1963), no substance comes closer to providing all these desired characteristics than colloidal silica. In 1966, Juhos described the use of colloidal silica to separate *Escherichia coli* from a contaminating bacteriophage. The bacteria retained viability, and the phages were sedimented and trapped in the silica pellet, providing a convenient way to maintain separation after centrifugation.

Pertoft (1966) improved colloidal silica gradient media by using polysaccharides to stabilize the gradients. He found the polymer dextran to be useful by acting on the colloidal silica as a molecular sieve, enabling him to produce nearly linear self-forming gradients during relatively short runs of 30–60 minutes at 35,000 rpm (Beckman rotor 40.2). The rapid, self-forming features made colloidal silica very convenient and, with the addition of dextran, made it easier to obtain a linear gradient. Other modifications are presented in a review by Pertoft and Laurent (1969).

II. Materials and Methods

A. Properties of Colloidal Silica

The colloidal silica most commonly used in forming density gradients is a product of Du Pont (Wilmington, Delaware) called *Ludox*, although other brands are available such as Nyacol (Nyanza, Inc., Ashland, Mass.). The material is produced and used commercially in large batches and is available in tank cars, tank trucks, and 55-gallon drums, quantities larger than those needed in the average laboratory. Smaller quantities may be obtained by writing directly to Du Pont.

Colloidal silica is an aqueous suspension of colloidal particles formed by polymerization of monosilicic acid from SiO_2 dissolved in water (Iler, 1955; Nash *et al.*, 1966). The colloidal suspension is most stable for storage at pH 8–10, and the colloidal particles have a net negative charge. Ludox HS is more completely described, since it has been most widely used for density gradient formation. It is a sodium-stabilized silica sol; the particles have diameters between 8 and 25 nm, with an average of 15 nm (Pertoft, 1969). The viscosity of a 40% solution (SiO_2 in water) is only 27 cps, and Ludox HS 40 has a specific gravity of 1.295 gm/ml (note that separated particles have much different apparent specific gravities (density) when centrifuged to equilibrium in colloidal silica than in sucrose or CsCl). Other colloidal silica preparations have been used, such as Ludox AM which is aluminum- and sodium-stabilized and is the most stable of the Ludox preparations (Du Pont, 1967). Ludox SM is chemically similar to Ludox HS but has particle diameters of 7–8 nm and a specific gravity of 1.093 gm/ml. Ludox LS is also chemically similar to HS and SM but has particle diameters of 15–16 nm and a specific gravity of 1.209 gm/ml. Colloidal silica solutions are irreversibly precipitated on freezing and form gels under certain ionic conditions such as above 0.1 *M* NaCl at pH 5–7 (see Section II,B). Certain alcohols, ketones, ethers, and other organic solvents also cause gel formation, and are therefore to be used with care if needed.

Commercial colloidal silica preparations such as Ludox are used in floor coatings, foam rubber preparation, and paper coatings, among other things. As such, it is considered "technical" grade and must often be dialyzed, sterilized, or prepared in some manner to be compatible with a biological or chemical system. Silica is toxic, causing silicosis on inhalation of the dust. Colloidal silica was also shown to be toxic on parenteral administration (Gye and Purdy, 1922) and to hemolyze red blood cells (Harley and Margolis, 1961). The use of colloidal silica as a laboratory reagent, however, presents no problem, as external contact has not been shown to be hazardous (Du Pont); it is not irritating to the skin and is easily removed by washing.

B. Preparing the Gradient Medium

The gradient may be formed by conventional techniques such as the layering of various concentrations, or by using a gradient-forming machine. An attractive feature of colloidal silica as a gradient medium, however, is that it is possible to self-form a gradient in a short time by high-speed centrifugation. Pertoft (1966) demonstrated the formation of gradients varying in density from 1.0 to 1.25 gm/ml in 30 minutes, using 120,000 *g*, the density and shape being determined by the colloidal silica content, time of centrifugation, and polymers added. The gradient is presumably formed by differential sedimentation of the polydisperse, heterogeneous, colloidal silica particles. Larger, heavier particles sediment faster, creating greater density.

In use the gradient mixture must be assembled to meet specific particle separation requirements. For whole-cell separation it is often important to maintain viability and to provide isoosmotic conditions and a pH value near neutrality, in addition to providing some of the nutritional requirements such as supplied in Eagle's minimal essential growth medium (MEM) (Wolff and Pertoft, 1972b). Other particles do not need all these components, but the separation of living cells with a minimum of alteration represents one of the more difficult tasks of a gradient medium.

1. Colloidal Silica Preparation

For cell separation dialyze the colloidal silica against 20 vol of distilled water with five changes during a 6-hour period in conventional dialysis tubing. This dialysis results in dilution of the colloidal silica of from 52% (w/v) to about a 42% (w/v) concentration.

2. Polymer Preparation

The polymers used include dextran, polyethylene glycol (PEG), and polyvinyl pyrrolidone (PVP), the last-mentioned being the least toxic for use with cultured HeLa and L cells and necessary to protect the cells from the toxic action of the colloidal silica. PVP (A. H. Thomas Co., Philadelphia, Pa.; MW 40,000) was prepared as a 33% (w/v) solution in distilled water and dialyzed as the colloidal silica (Section II,B,1), resulting in approximately a 20% (w/v) final concentration.

3. Adjusting the Osmotic Pressure and pH

Osmotic pressure and pH can be adjusted by modifying the buffer system used. An osmometer is useful to measure the osmotic pressure, and models using freezing-point depression (such as that from Advanced Instruments, Newton Highlands, Mass.) can be used with colloidal silica. Osmotic pressures of 290–330 mosmoles are isoosmotic for L, HeLa, and HEp-2 cells. For more details on osmotic pressure, see Weymouth (1970). Other methods, such as microscope observation of red blood cells or the cells to be separated suspended in the medium, can give one a crude idea of the osmotic pressure. When using a mixture of salts and buffer in the gradient medium, such as Eagle's MEM, it is convenient to prepare a solution more concentrated than that used for cell cultures and, by adding various amounts of a diluent such as water, one can easily adjust the osmotic pressure. To adjust the pH, 1 *N* HCl is added with constant mixing, to prevent gel formation. Another method of pH adjustment is by the addition of 1 *N* HCl and Dowex AG50 W or Amberlite IR-120, strong cationic sulfonic acid resins. Back-titration with NaOH or other basic solutions is not successful without production of a gel.

4. Density Measurements

Pertoft (1966) modified the Miller and Gasek (1960) procedure of producing a density column of organic solvents. After standardization drops of the aqueous samples are allowed to drift to their own density level, and then from a standard curve one can read the density directly. This is a very convenient and accurate method for measuring density to a difference of 0.001 gm/ml ($\pm$0.001), and is especially useful for measuring a large number of samples in a short time.

To form an organic solvent density column, a 500-ml graduated cylinder is filled by layering four different mixtures of kerosene (density 0.770 gm/ml) and carbon tetrachloride (density 1.595 gm/ml). The lightest mixture is added first through a funnel and tube leading to the bottom of the cylinder, followed by increasingly more dense mixtures. When the column is full, one or two gentle turns of the glass filling tube followed by gentle removal suffice to initiate diffusion of the four layers so that 24 hours after standing a linear gradient is produced.

Stock standard density solutions are prepared from sucrose in water by weighing. The gradient and standards are equilibrated to room temperature, and a standard curve is prepared as follows. Use a Pasteur pipette and allow a drop of the standard sucrose solution to fall into the center of the column from about 3–5 mm above the liquid surface. Two to three drops of each standard solution are used and allowed to stop drifting, usually within 10–15 seconds after being applied to the column. The graduation marks on the cylinder are convenient to use, and a standard curve is constructed to display density against the level in the column. At least five standard sucrose solutions are used to extend over the expected ranges of densities to be measured. The columns are quite stable, being useful for weeks, but must be standardized shortly before each use because of some evaporation, mixing, and accumulation of drops of samples in the column. Samples are added dropwise as were the standards, and the density read from the standard curve. Dual columns are used to ensure accuracy, and it is important to have all solutions at room temperature or a predetermined constant temperature before standardization and density measurement.

For use with cultured cells, which have a density range of 1.03–1.07 gm/ml, the following four mixtures of kerosene and carbon tetrachloride, respectively, are used: 75–25, 70–30, 65–35, and 60–40. We have successfully used nonreagent grade kerosene and carbon tetrachloride, although odorless kerosene is available from Fisher Chemical Co., Pittsburgh, Pa. Bromobenzene (density 1.513 gm/ml) may be substituted for carbon tetrachloride.

5. Mixing the Gradient Medium

It has been found by experience that colloidal silica has a tendency to gel irreversibly, especially between pH 5 and 7 and at salt concentrations above 0.1 *N* (Du Pont, 1967). The gelling tendency also increases at lower temperatures and higher concentrations of colloidal silica, and is the problem most often met by users of colloidal silica. It is important to mix ingredients in the order suggested and to prepare the mixture just before use, since the conditions that promote gel formation tend to reduce the gelling time. We have used gradient media prepared the previous day, and have used colloidal silica for centrifugation at 4°C without gellation, but the investigator should be aware of the conditions that promote gel formation.

For our use in cultured-cell separation, one gradient mixture described here has been useful, however, several variations are discussed in subsequent sections on specific procedures. Contamination of cells has not been encountered in using materials without sterilization, since the bacteria have densities different from those of the cells after separation. However, if desired, each of the ingredients can be presterilized and mixed under sterile conditions. Colloidal silica, water, and PVP may be steam-sterilized by the usual method (121°C, 15 minutes). Although approximately 10–15% of the water is lost in the process, so readjustment of concentrations or calculations will be necessary. The Eagle's MEM should be filter-sterilized (0.22-μm

TABLE I

INGREDIENTS OF A COLLOIDAL SILICA SELF-FORMING DENSITY GRADIENT OF VARIOUS DENSITIES (PEL GRADIENT)

Ingredients[a]	Parts required to obtain a density of		
	1.055 gm/ml	1.058 gm/ml	1.060 gm/ml
Colloidal silica (Ludox HS-40, dialyzed 42% w/v)	18	21	24
Eagle's MEM (2× concentrated)	42	42	42
PVP (dialyzed, 20% w/v in water)	25	25	25
1 *N* HCl (add to obtain pH 7.6)	2.5	2.5	2.5
Double-distilled water	12.5	9.5	6.5

[a]These are added in order with hand mixing between each addition.

pore diameter). Colloidal silica and PVP solutions may also be boiled separately for 10 minutes to inactivate vegetative bacterial cells (not spores), most viruses, and mycoplasma.

To construct a PEL gradient medium useful for cell separation, the ingredients listed in Table I should be added in order. Each ingredient should be thoroughly mixed before the next is added, and all additions made with agitation by active hand swirling. Magnetic stirrers have not been as satisfactory in our use as hand mixing.

After mixing the density is determined on the organic solvent density gradient column, and the osmotic pressure recorded.

C. Gradient Formation

1. DISCONTINUOUS GRADIENT

A step or discontinuous gradient may easily be formed by choosing concentrations of colloidal silica that produce a gradient both lighter and heavier than the material to be separated. The least dense medium is added first through a tube or large needle leading to the bottom of the centrifuge tube. Subsequently, heavier layers are added through the same tube by using a stopcock or pinch clamp to stop the flow between concentrations, otherwise air bubbles enter the tube and mix the gradient. After layering, the sample can be added to the top.

This type of gradient often gives good separation of particles, but is not as sensitive or selective as a continuous gradient. Erroneous results can be obtained as a result of clumps of particles forming at density interfaces which form streams, disrupting the gradient. Discontinuous gradients are especially useful when separating only a few different particle types of known density.

2. Continuous Gradients

These are either pumped or mixed with gradient-forming machines or are self-formed by high-speed centrifugation. Only a relatively few materials lend themselves to the latter method, colloidal silica being one.

Employing a gradient-mixing device (Britten and Roberts, 1960), one uses somewhat more than the minimum and maximum density media needed for separation. These are mixed and fed into the centrifuge tube. (ISCO, Lincoln, Nebr., and Beckman Instruments, Fullerton, Calif., produce various devices for making these gradients.) The formation of gradients by these methods is quite reproducible, and one can produce linear, convex, concave, or S-shaped gradients, depending on the need and availability of equipment. Linear gradients are usually most useful for separation mixture particles having many different densities. S-Shaped gradients are useful for producing steep gradients at either end of the tube, thus eliminating from the mixture particles that may not be desirable to harvest, and providing a shallow or gradual density change through the center of the gradient, resulting in the greatest sensitivity for separation of particles having nearly the same density.

The so-called self-forming gradient media are few, and include CsCl and colloidal silica. Under high forces of gravity the molecules or colloidal particles are rearranged in solution so a density gradient is formed. CsCl gradients are usually formed in 24 hours, while colloidal silica gradients can be formed in 30 minutes at equivalent or lower forces of gravity. In forming a gradient with colloidal silica, one chooses a density at the center of the tube that is the average density of the particles to be separated, and places the gradient mixture under centrifugal force. The length of the tube, desired

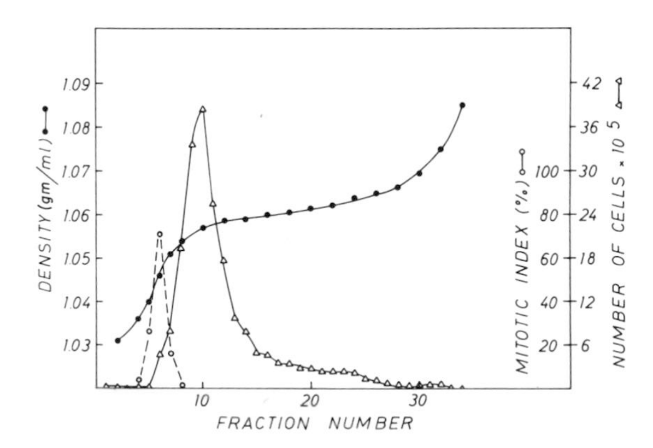

FIG. 1. The distribution and mitotic index of HeLa cells after a 20-minute centrifugation at 800 g on a preformed PEL gradient in a 95-ml test tube (taken with permission from Wolff and Pertoft, 1972b).

shape of the gradient, volume of the tube, type of rotor, and density of the gradient determine the time and speed to be used for forming the gradient.

As an example, for a gradient of 95 ml having an average density of 1.055 gm/ml, the following can be used: 3.8 × 10.1 cm tube, using a Beckman 42 or 21 angle-head rotor with a gradient medium having a density of 1.055 gm/ml, centrifuged at 20,000 rpm for 20 minutes (53,664 *g*). This forms an S-shaped gradient similar to that in Fig. 1.

It has been shown by Flamm *et al.* (1966) that better resolution of density gradients is obtained when an angle-head rotor is used rather than a swinging-bucket type. We find that, with a self-formed colloidal silica gradient, the central linear portion of the gradient is nearly flat when formed in a swinging-bucket rotor and not as satisfactory as that formed in an angle-head rotor (see Fig. 1).

The gradients formed by centrifugation are quite stable and have been found to remain stable overnight, although it is preferable to prepare the gradient just before use.

III. Applications

A. Cultured Cell Separation

1. Cell Viability

The maintenance of viability throughout the procedure is a prerequisite for successful cell separation. This is probably the best indication that the cells have not been altered and so is a good marker even if living cells are not needed for a particular application.

Several techniques exist for the evaluation of viability. One of the easiest methods is to form nearly confluent monolayers of the cells to be studied and then add the gradient mixture for the period of time the cells would be in the gradient for separation and harvesting. Following this, the gradient medium can be washed off and the cell culture incubated in its regular growth medium. If the cells are unaltered by the medium, they will look normal, remain attached to the plastic or glass, and continue to form confluent monolayers. In our studies of various ingredients for gradient media (Wolff and Pertoft, 1972b), we found that colloidal silica without a polymer, colloidal silica with PEG, and colloidal silica with Methocel were able to separate cells but were also toxic to HeLa and L cells. The polymer PVP was protective and caused no cell detachment or granulation, and the cells grew like the untreated controls after the gradient medium was removed.

Other investigators (Harley and Margolis, 1961; Allison *et al.*, 1966) had shown colloidal silica to be toxic to cells such as red blood cells and macrophages, and Nash *et al.* (1966) showed protection by PVP and PVP *N*-oxide. Apparently, the polymer actually coats the colloidal silica particles and neutralizes their toxic action.

Another good method for measuring toxic activity is to follow the growth pattern of cultured cells after exposure to the gradient medium. This is especially useful when the cells to be studied cannot grow attached to a substrate such as glass or plastic but can grow or be maintained in suspension. Such cells are exposed to the gradient mixture, like monolayer cells, and then returned to the culture and counted periodically. Our results using PEL media on HeLa cells show little difference in growth rate from interacted controls (Fig. 2).

Cell cloning is also a well-established procedure for evaluating toxicity, and the details of this procedure are well described elsewhere (Puck *et al.*, 1956; Cooper, 1973). Dye exclusion tests have been useful for estimating cell viability (Hanks and Wallace, 1949; Phillips, 1973), but their use can lead to misleading results, either the false security of apparent nontoxicity or an erroneous indication that all cells have been killed. We have used Erythrocin B stain with some success in evaluating the percent of cells damaged by poliovirus infection and suggest the use of this procedure only as a preliminary indication of toxic level in studying gradient media. A 0.05% solution of Erythrocin B is dissolved in phosphate-buffered saline

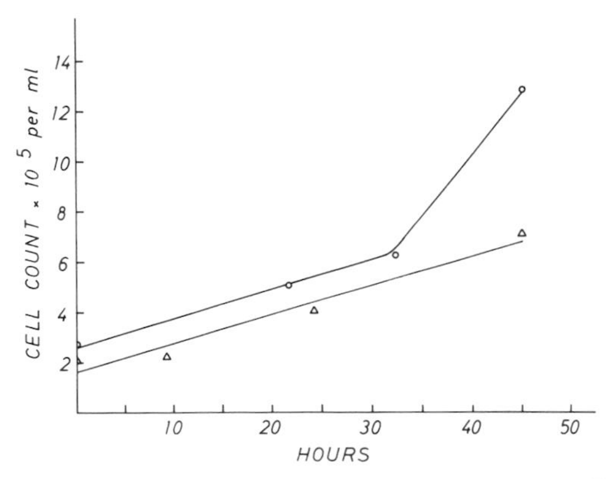

FIG. 2. HeLa cell growth rate after exposure to PEL medium. Cells were placed on a preformed PEL gradient and centrifuged for 20 minutes at 800 *g*. The major cell band was harvested, diluted with an equal volume of 2× Eagle's MEM and pelleted at 100 *g* for 8 minutes. The cells were washed once in Eagle's MEM spinner medium and a spinner culture established at a cell density of 3.3×10^5 cells per milliliter. Time of exposure was approximately 1 hour. Growth rates of control, untreated cells (triangles) and PEL-treated cells (circles) are shown (taken with permission from Wolff and Pertoft, 1972b).

(8.0 gm NaCl, 3.0 gm KCl, 0.073 gm Na_2HPO_4, 0.02 gm KH_2PO_4, and 2.0 gm glucose to 1 liter with water), and a drop added to one drop of cells to be studied on a microscope slide; then cover glass is added. The cells are observed microscopically for the number taking up the red stain (these are nonviable), and the percent that stain within a defined period of time is determined.

2. Cell Dispersion

In order to separate cells they must not be allowed to aggregate or clump together. Some gradient media enhance the clumping of cells. Various cell types respond in different ways so various methods, from adjustment of ionic strength to enzymatic treatment, have been used. It is useful to test the gradient medium with cells to be used by mixing them with the medium in a test tube, allowing them to interact for approximately 30 minutes, and observing them under the microscope for formation of cell clumps. Several gradient media may be tested simultaneously, and the best chosen without the need for use of larger numbers of cells.

In our experience HeLa and L cells remain dispersed in the gradient medium described in Table I, which includes Eagle's MEM as a buffered salt solution. Using HEp-2 cells, K. and D. Bienz, Basel University (unpublished personal communication), obtained good cell separation and maintained viability by substituting tris buffer for double-strength Eagle's MEM. They used double-strength tris prepared by combining 8.0 gm of NaCl, 0.38 gm of KCl, 0.1 gm of Na_2HPO_4, 3.0 gm of tris, and distilled water to 500 ml. In formulating the gradient medium, 39 parts of the double-strength tris buffer is used and, after adjusting the pH to 7.6, 5 parts of calf serum added.

In the separation of chick embryo fibroblast cells, we used the enzyme collagenase (Worthington Biochemical Co., Freehold, N.J.), prepared to 0.25% in single-strength tris buffer and added to 2× Eagle's MEM (1 part to 10 parts 2× MEM). This resulted in better dispersion for these fibroblasts than Eagle's MEM or tris alone. We found that trypsin, often used to separate cells from tissue in the preparation of cell cultures, promotes cell aggregation under conditions needed for gradient separation.

3. Centrifugation Separation Procedure

Once a suitable gradient has been formed, the cells to be separated are resuspended in a small volume of fluid taken from the top of the gradient, dispersed with two or three strokes of a Dounce homogenizer (type A or loose pestle) (Blaessig Glass Co., Rochester, N.Y., or Kontes Glass Co., Vineland, N.J.), and then gently layered on the gradient with a pipette. The number of cells that can be applied depends on the volume of the gradient and the properties of the cells. We found that by using a PEL gradient of 95 ml in a 3.8 × 10.1 cm tube we obtained good separation with up to 5 × 10^8 HeLa cells; we have also used the smaller (1.6 × 7.6 cm, 13.5 ml) tube in the Beckman 40 rotor with 1 × 10^8 cells.

The separation after gradient formation and layering is best carried out in a swinging-bucket rotor and can be done at relatively low speeds. We obtained satisfactory separations at 800 *g*, 2000 rpm, for 20–30 minutes in a Wifung Model X-3 (Stockholm, Sweden) or in a Model CS International centrifuge (Needham Heights, Mass.). This centrifuge force for this time period brought the cells into a near-equilibrium condition. This was established by centrifuging the cells for various periods of time and measuring cell densities; HeLa cells reached their own density level in 20 minutes and did not change with increased time.

Harvesting after centrifugation can be carried out by any of the standard techniques. One we find simple and effective for the collection of separated cells is shown in Fig. 3. By introducing undiluted colloidal silica under the gradient, the contents are forced up and out of the collection tube. The use of a peristaltic pump is convenient, however, good control is also obtained by flow of the displacement fluid from a flask elevated at the proper level above the gradient to provide the desired flow rate.

Figure 1 shows the results of a separation of HeLa cells on a 95-ml gradient with the collection of 3-ml fractions. The majority of the HeLa cells have densities of 1.054–1.058 gm/ml. When the mitotic index of these fractions was determined, a sharp peak at 1.046 gm/ml was observed, thus showing the ability of the gradient to separate cells of slightly different densities. In our laboratory this has now become a standard means of collecting mitotic phase cells for direct use or for synchronized growth of HeLa cells (Fig. 4).

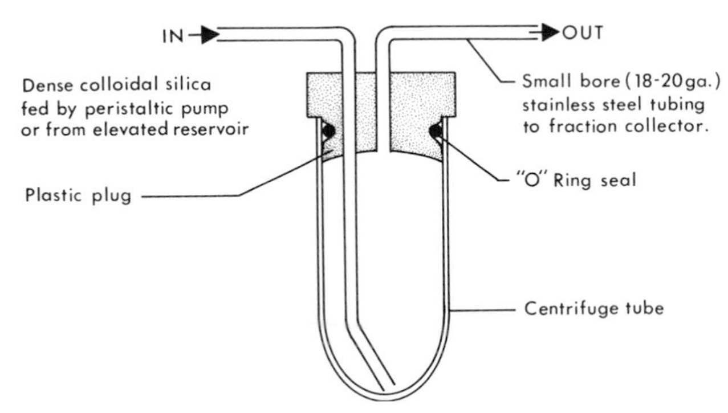

FIG. 3. A simple, inexpensive system for the controlled collection of fractions from a density gradient tube.

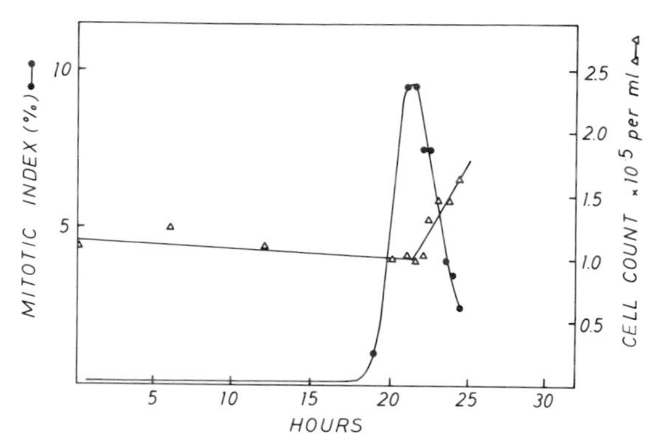

FIG. 4. A synchronous culture of HeLa cells. The culture was established by separating mitotic phase cells on a PEL density gradient (see Fig. 1) and growth in a spinner culture (taken with permission from Wolff and Pertoft, 1972b).

By using the same technique, we separated normal chick embryo fibroblast cells (density 1.0394 gm/ml) from vaccinia virus–infected cells (density 1.041 gm/ml), as well as cultured LM cells from chick embryo fibroblasts.

On removing the fractionated cells from the gradient medium by washing in growth medium (100 *g*, 10-minute pelleting and resuspending in growth medium), we found that the selected mitotic cells grow in synchrony (Wolff and Pertoft, 1972b) and the remainder grow and divide with continued culturing.

B. Separation of Cell from Tissues

Pertoft used a colloidal silica gradient for the separation of rat liver cells (1969; Peterson *et al.*, 1973) and cells from a mast cell tumor (Pertoft, 1970c).

In the separation of rat liver cells, a mixture of colloidal silica and PEG (MW 4000) was used to separate cells having an average density of 1.08–1.10 gm/ml from heavier cells with greater phagocytic activity. The cells were separated from tissues by mincing and treatment with a modified Potter-Elvehjem homogenizer with a rubber pestle in a solution containing 10–16% PEG. The cells were then added to colloidal silica gradient medium containers and layered on either continuous or discontinuous preformed gradients. The gradient medium contained 40 ml of 46.8% (w/v) Ludox HS colloidal silica after neutralization to pH 7.0 with 1 *N* HCl, 10 ml of a solution containing 60% (w/v) PEG (MW 4000) and 5.4% PEG (MW 300), and 38 ml of the above mixed with 57 ml of the cell suspension [cells in 10% (w/v) PEG 4000].

For the separation of mast cells, the tumor was taken from the musculature of a mouse and homogenized like the liver (above). After centrifugation of the cells on gradients prepared as for liver cell separation, Pertoft separated four different cell populations having a density range of 1.06–1.15 gm/ml. He found most of the tumor-inducing cells to have a density of 1.08 gm/ml.

Both cell separations showed the efficacy of the colloidal silica density gradient system. The separation of cells from tumors that can induce other tumors is an exciting finding, since the separation of malignant from non-malignant cells in a tumor has been a problem thus far.

C. Separation of Blood Cells

Thrombocytes, monocytes, polymorphonuclear cells, and red blood cells were successfully separated by the use of colloidal silica gradients (Evrin and Pertoft, 1973).

The gradient medium was prepared as two solutions, and the gradient formed by a Britten and Roberts (1960) gradient-mixing device. Solution A was prepared by mixing 166 ml of 7.5% PVP in 0.83% NaCl with 69 ml of 52.4% (w/v) Ludox HS and adjusting the pH to 7.5 with 9.2 ml of 1 *N* HCl. Solution B contained 30 ml of solution A plus 24 ml of 0.946% NaCl. The gradient was formed in the mixing chamber by adding equal volumes of solutions A and B. Whole blood was diluted 2:1 with a mixture of 51.0 m*M* NaCl, 4.7 m*M* KCl, 2.5 m*M* $CaCl_2$, 1.2 m*M* KH_2PH_4, 1.2 *m*M $MgSO_4$, 24.4 m*M* Na acetate, 25.0 $NaHCO_3$, 16.7 m*M* glucose, 42 m*M* G[1] acid and 1.0% gelatin; approximately 25 ml of this mixture can be layered on the top of a 95-ml preformed gradient. The gradient is centrifuged at 800 *g* for 50 minutes, and on fractionation the thrombocytes are found at the top of the tube with mononuclear cells at density 1.073 gm/ml, polymorphonuclear leukocytes at 1.060–1.073 gm/ml, and red blood cells at greater than 1.090 gm/ml. The separated cells may be washed with a solution of 0.25 *M* sucrose adjusted to pH 7.5 and, if aggregation is a problem, 0.1% Na EDTA may be added.

D. Separation of Marine Organisms

Bowen *et al.* (1972) reported the success of colloidal silica gradient in separating zooplankton, fish eggs, and fish larvae. They compared the use of sucrose, dextran, and NaBr gradient media with colloidal silica (Ludox AM) and found no effect on the zooplankton and good separation of various classes within this group from fish eggs and larvae.

[1] 2-napthol, 6, 8, disulphonic acid (dipotassium salt), Eastman Organic Chemicals.

The gradient medium used by Bowen *et al.* (1972) consisted of Ludox AM mixed with tap water. They formed 0–10% (w/v) gradients by the use of a gradient pump in 85-ml tubes and obtained separation after layering the cells on the gradient by centrifugation for 60 minutes at 280 *g*.

E. Separation of Subcellular Particles

1. Lysosomes

Since lysosomes are osmotically fragile subcellular particles, colloidal silica density gradients are well suited to their separation and purification. We developed a procedure for the purification of lysosomes isolated from HeLa cells, which resulted in separation of the lysosomes from mitochondria and peroxisomes, both having density and size similar to that of lysosomes (Wolff and Pertoft, 1972a).

To prepare purified lysosomes, 8–16 × 10^8 HeLa cells are pelleted at 140 *g* for 10 minutes and then resuspended in cold (4°C), 0.125 *M* sucrose to create a slightly hypoosmotic condition and repelleted. The cells are homogenized in 5 ml of 0.125 *M* sucrose using a 15-ml-size Dounce homogenizer with a tight or type-B pestle (Blaessig Glass Co., Rochester, N. Y., or Kontes Glass Co., Vineland, N. J.). Homogenization is carried out until 95% of the nuclei are released from the cells as observed by phase-contrast microscopy. The homogenate is centrifuged at 1085 *g* for 10 minutes at 4°C to pellet nuclei and unbroken cells. The supernatant is harvested and centrifuged at 20,200 *g* for 20 minutes at 4°C to pellet the lysosomes. The pellet contains lysosomes, mitochondria, peroxisomes, and cell debris of similar size. The separation of these takes place on one of two colloidal silica density gradients. One, designated PEL (density 1.055 gm/ml, is described in Table I; the other, designated PSL, uses sucrose to provide the proper osmotic pressure (290–310 m smoles) in place of Eagle's MEM. The PSL gradient medium of density 1.057 gm/ml is prepared by adding 6 parts of 52.4% (w/v) HS-40 Ludox colloidal silica (undialyzed) to 44 parts of distilled water with mixing, and then adding 50 parts of 0.48 *M* sucrose in 10% (w/v) PVP (A. H. Thomas Co., Philadelphia, Pa.) and adjusting the pH to 7.5 with 1 *N* HCl. The above are added in the order listed, with constant hand mixing to prevent gel formation.

The lysosome preparation from the 20,200 *g* pellet can be resuspended in the gradient medium with a few strokes of a Dounce homogenizer pestle, and the lysosome–gradient medium mixture placed in the centrifuge tubes or the zonal rotor. Our best separations were obtained with an angle-head rotor on a zonal centrifuge. When a 95-ml tube (3.8 × 10.1 cm) is used for a Beckman 21 or 42 rotor, the tube is filled with the lysosome–gradient mixture and centrifuged at 40,000 *g* for 20 minutes at speed. The lysosome suspension can be layered on the gradient medium, but this is not necessary since the same results are obtained by mixing. On harvesting, as in a system shown in Fig. 3, fractions can be assayed for marker enzymes for lysosomes, such as β-glucuronidase, aryl sulfatase, or β-glucosaminidase. Results

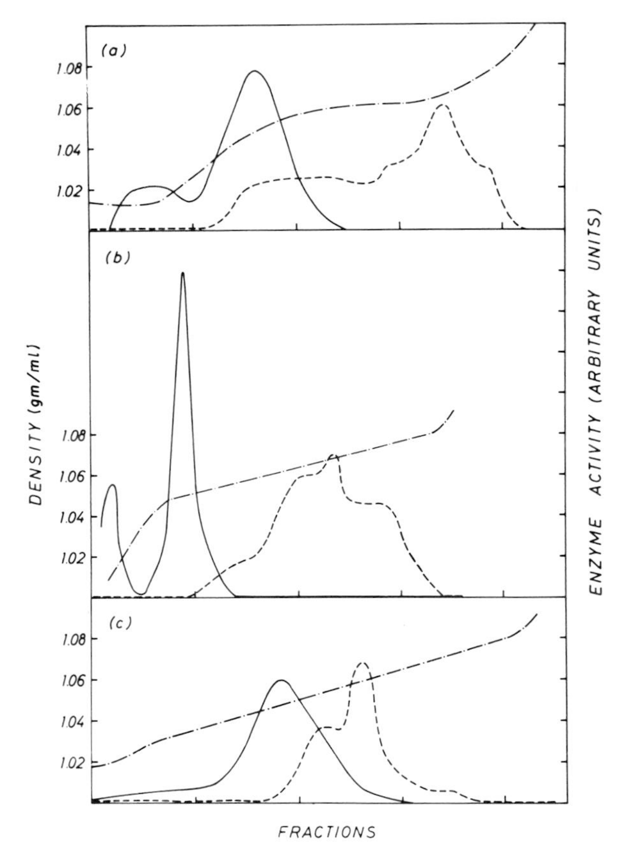

FIG. 5. Separation of lysosomes and mitochondria from HeLa cells by centrifugation on a PEL gradient. Separation was performed in three separate rotors: (a) an angle-head Beckman rotor 21 at 40,000 *g* for 90 minutes, (b) a zonal Beckman rotor B-XIV at 66,000 *g* for 20 minutes, and (c) a swing-out Beckman SW26 rotor at 97,000 for 20 minutes, all operated at 4°C. Density (·—·—), β-glucosaminidase as a lysosome marker (——), and cytochrome C oxidase as a mitochondrial marker (– – –) were measured in each fraction. The gradients were produced as follows. (a) A 5-ml sample was layered onto 85 ml of PEL gradient medium, density 1.055 gm/ml, and the gradient formed during centrifugation; (b) the zonal rotor was loaded by using a gradient-forming machine with a linear PEL gradient, density 1.035–1.095 gm/ml, the cell homogenate (25 ml) was layered on the gradient; (c) 2-ml sample was layered on a PEL linear preformed gradient, density 1.035–1.095 gm/ml (taken with permission from Wolff and Pertoft, 1972a).

comparing angle-head, zonal, and swinging-bucket centrifugation are shown in Fig. 5 and demonstrate that the best resolution was obtained in the angle-head and zonal rotors.

2. Nuclei, Catacholamine Granules, Synaptosomes, and Chloroplasts

To indicate the range of application of a colloidal silica density gradient, we have included references to several diverse separation procedures.

Both plant and animal nuclei have been purified by the use of a colloidal silica gradient. Hendriks (1972) improved on existing procedures of separating tobacco plant cell nuclei by using colloidal silica (Ludox TM) and obtained a very clean separation from cell debris without shrinkage as when sucrose was used. Loir and Wyrobek (1972) used colloidal silica in density gradient separation of spermatid nuclei from mice. They obtained a good separation of the nuclei which were in different phases of spermatogenesis, making a harvest of homogenous preparations possible.

In the separation of particles from nerve tissue, Lagercrantz *et al.* (1970) used colloidal silica to separate catecholamine granules, and Lagercrantz and Pertoft (1972) used this method in the purification of synaptosomes.

Colloidal silica density gradients have also been used in the separation of chloroplasts from spinach leaves by tube-type centrifugation (Lyttleton, 1970) and by zonal centrifugation (Price, 1973).

F. Purification of Viruses

While the methodology is not included in this chapter, a brief review of the application of colloidal silica gradients to virus purification is presented. The first report on the use of colloidal silica to purify viruses was published by Pertoft *et al.* (1967), in which they described the purification of tobacco mosaic virus, poliovirus type 1, and adenovirus type 2. This procedure was rapid and gave good separation, but the viruses remained bound to the silica particles after centrifugation. This problem was overcome by precipitation of the colloidal silica using polyamines, after harvesting the gradient fraction containing purified virus. A 5% solution of colloidal silica can be completely precipitated by 3.9×10^{-3} *M* spermine or 4.8×10^{-3} *M* spermidine, however, 10^{-3} *M* EDTA plus 10^{-2} *M* Cleland's reagent was also required to prevent inactivation of the polio- and adenoviruses.

The herpesvirus group has been difficult to purify by the use of CsCl or sucrose density gradients, presumably because the envelope is modified by these compounds to cause virus inactivation. Within this virus group, infectious bovine rhinotracheitis virus (IBRV), equine abortion virus (EAV), and herpes simplex virus (HSV) have been successfully purified.

IBRV was purified by tube-type density gradients of colloidal silica–PEG (Pertoft, 1970a) and zonal centrifugation (Pertoft, 1970b), offering rapid and low-cost separation of the viruses from cell debris. Using a colloidal silica–PEG density gradient, Klingeborn and Pertoft (1972) purified EAV and found that, by placing concentrated virus bands on an 80% (w/v) cushion of sucrose, the colloidal silica could be separated from the EAV at 250,000 *g* for 3 hours. The virus remained on top, while the silica penetrated the sucrose cushion. A more efficient procedure for separating viruses from gradient media is to chromatograph on a short Sepharose 2B column. The intact viruses are eluted with the void volume, while the colloidal silica is retained on the column (B. Klingeborn and H. Pertoft, personal communication).

HSV was purified by Vahlne and Blomberg (1974), using a gradient of colloidal silica followed by a T70 dextran gradient, and entire virions were separated from nucleocapsid (deenveloped) material. The virus was purified on a linear colloidal silica isopycnic gradient, and the dextran T70 gradient was used under rate zonal conditions to sediment the silica at 41,000 *g* for 2 hours, leaving the virus banded and purified 1250 to 2000 times relative to protein concentration and infectivity.

Virus concentration and purification is facilitated by the use of colloidal silica density gradients in many circumstances, improving on retention of infectivity of herpesvirus and providing a low-cost means of processing large amounts of viruses in a zonal ultracentrifuge.

IV. Conclusions

The use of colloidal silica in the formation of a density gradient offers several advantages to the biologist. The ability to control the osmotic pressure to match the osmolarity of the particles to be separated seems to be one of the major advantages. This allows harvesting of cells and particles under isoosmotic conditions. With the use of PVP, cell viability can be maintained and mitotic cells separated and grown in synchronous culture. Similarly, the preparation of purified subcellular particles and viruses is facilitated by the maintenance of isoosmotic and nontoxic conditions.

Other advantages of this gradient medium include low cost and rapid gradient formation during centrifugation.

If the precautions included in this chapter are observed, gel formation can be avoided and colloidal silica becomes the medium of choice for many separation procedures.

ACKNOWLEDGMENTS

I express my deep appreciation to Dr. Håkan Pertoft and Professor Torvard Laurent of Uppsala University for their willingness to instruct and collaborate. Part of the work described here was performed at the Department of Medical Chemistry, Uppsala University, and supported by the Swedish Medical Research Council and the Swedish Cancer Society. I also thank Dr. Kurt Bienz, Denise Egger, and Professor Hans Loeffler of Basel University for their advice and hospitality during the preparation of this chapter. Some of the unpublished work was carried out by Ms. Beth Wichman at Ohio State University, and I thank her for allowing its inclusion.

REFERENCES

Allison, A. C., Harington, J. S., and Birbeck, M. (1966). *J. Exp. Med.* **124**, 141.
Bowen, R. A., St. Onge, St. J. M., Colton, J. B., Jr., and Price, C. A. (1972). *Mar. Biol.* **14**, 242.
Britten, R. J., and Roberts, R. B. (1960). *Science* **131**, 32.
Cooper, J. E. K. (1973). *In* "Tissue Culture: Methods and Applications" (P. Kruse, Jr. and M. K. Patterson, Jr., eds.), pp 266–269. Academic Press, New York.
deDuve, C., Berthet, J., and Beaufay, H. (1959). *Progr. Biophys. Biophys. Chem.* **9**, 326.
Du Pont (1967). "Product Information Bulletin on Ludox Colloidal Silica." Wilmington, Delaware.
Evrin, P. E., and Pertoft, H. (1973). *J. Immunol.* **111**, 1147.
Flamm, W. G., Bond, H. E., and Burr, H. E. (1966). *Biochim. Biophys. Acta* **129**, 310.
Gye, W. E., and Purdy, W. J. (1922). *Brit. J. Exp. Pathol.* **3**, 75.
Hanks, J. H., and Wallace, R. E. (1949). *Proc. Soc. Exp. Biol. Med.* **71**, 196.
Harley, J. D., and Margolis, J. (1961). *Nature (London)* **189**, 1010.
Hendriks, A. W. (1972). *FEBS (Fed. Eur. Biochem. Soc.) Lett.* **24**, 101.
Iler, R. K. (1955). "The Colloidal Chemistry of Silica and Silicates." Cornell Univ. Press, Ithaca, New York.
Juhos, E. (1966). *J. Bacteriol.* **91**, 1376.
Klingeborn, B., and Pertoft, H. (1972). *Virology* **48**, 618.
Lagercrantz, H., and Pertoft, H. (1972). *J. Neurochem.* **19**, 811.
Lagercrantz, H., Pertoft, H., and Stjärne, L. (1970). *Acta Physiol. Scand.* **78**, 561.
Loir, M., and Wyrobek, A. (1972). *Exp. Cell. Res.* **75**, 261.
Lyttleton, J. W. (1970). *Anal. Biochem.* **38**, 277.
Mateyko, G. M., and Kopac, M. J. (1959). *Exp. Cell Res.* **17**, 524.
Mateyko, G. M., and Kopac, M. J. (1963). *Ann. N.Y. Acad. Sci.* **105**, 185.
Miller, G. L., and Gasek, J. M. (1960). *Anal. Biochem.* **1**, 78.
Nash, T., Allison, A. C., and Harington, J. S. (1966). *Nature (London)* **210**, 259.
Pertoft, H. (1966). *Biochim. Biophys. Acta* **126**, 594.
Pertoft, H. (1969). *Exp. Cell Res.* **57**, 338.
Pertoft, H. (1970a). *Virology* **41**, 368.
Pertoft, H. (1970*b*). *Anal. Biochem.* **38**, 506.
Pertoft, H. (1970c). *J. Nat. Cancer Inst.* **44**, 1251.
Pertoft, H., and Laurent, T. (1969). *Progr. Separ Purif.* **2**, 71–90.
Pertoft, H., Philipson, L., Oxelfelt, P., and Höglund, S. (1967). *Virology* **33**, 185.
Peterson, P., Rask, Östberg, L., Anderson, L., Kamwendo, F., and Pertoft, H. (1973). *J. Biol. Chem.* **248**, 4009.
Phillips, H. J. (1973). *In* "Tissue Culture Methods and Applications" (P. Kruse, Jr. and M. K. Patterson, Jr., eds.), pp. 406–408. Academic Press, New York.
Price, C. A. (1973). *Eur. Symp. Zonal Centrifugation Density Gradient*, 4, p. 162.

Puck, T. T., Marcus, P. I., and Cieciura, S. J. (1956). *J. Exp. Med.* **103**; 273.
Vahlne, A. G., and Blomberg, J. (1974). *J. Gen. Virol.* **22**, 297.
Weymouth, C. (1970). *In Vitro* **6**, 109.
Wolff, D. A., and Pertoft, H. (1972a). *Biochim. Biophys. Acta* **286**, 197.
Wolff, D. A., and Pertoft, H. (1972b). *J. Cell Biol.* **55**, 579.

Chapter 6

Isolation of Subcellular Membrane Components from Tetrahymena

Y. NOZAWA

Department of Biochemistry,
Gifu University School of Medicine,
Tsukasamachi-40, Gifu, Japan

I. Introduction 105
II. Systematic Isolation Methods for Various Subcellular Membrane Components 108
A. Isolation of Mitochondria and Microsomes 108
B. Isolation of Mitochondria, Peroxisomes, and Lysosomes 109
C. Isolation of Cilia, Pellicles, Mitochondria, and Microsomes 111
III. Nonsystematic Isolation Methods for Various Subcellular Membrane Components 115
A. Isolation of Cilia and Their Subfractions 115
B. Isolation of Oral (Buccal) Apparatus 118
C. Isolation of Pellicles 119
D. Isolation of Kinetosomes (Basal Bodies) 120
E. Isolation of Mitochondria 121
F. Isolation of Nuclei and Nuclear Membranes 124
G. Isolation of Ribosomes 127
H. Isolation of Peroxisomes 130
I. Other Subcellular Components 130
IV. Membrane Lipid Composition of Various Isolated Subcellular Components . 130
References 132

I. Introduction

A variety of methods has served for isolating subcellular organelles from tissues of higher animals and much information about mophological and biochemical properties has been obtained. But in some cases difficulties are encountered in interpreting data, since many factors are involved in metabolic interactions among individual cells in a tissue. Efforts have been made to use cell cultures, *i.e.*, HeLa, fibroblast, and L cells, instead of the complicated intact mammalian system, but there is still difficulty in growing and collecting cells on a large scale. In contrast to the cells listed above, the free-living, unicellular, ciliated protozoan *Tetrahymena* has been widely used for diverse biochemical investigations because of ease in growing cells in large quantities in media that are axenic or even completely defined. Especially for membrane studies, *Tetrahymena* is a potentially convenient model system, since in addition to ease in cell growth it has several other advantages: well-defined development of subcellular organelles, strikingly different distribution of specific lipids among the various functionally distinct membranes within the cell, and sensitive adjustment to environmental factors such as starvation, temperature, and chemicals.

Tetrahymena pyriformis cells are most commonly pear-shaped, 30 × 50 μm, and have many cilia on the cell surface, as conveniently seen in a scanning electron micrograph (Fig. 1). The subcellular membrane components

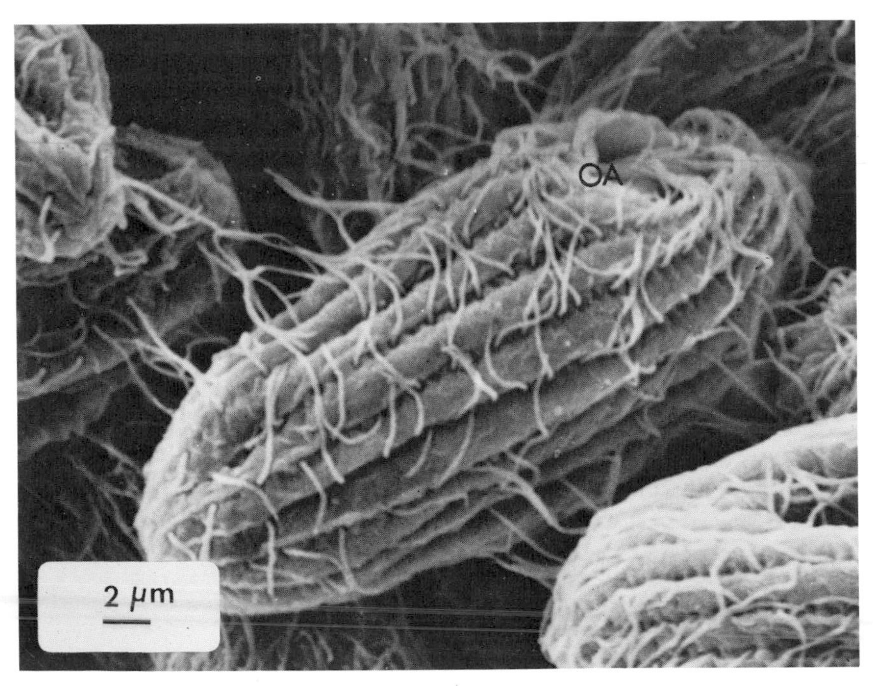

FIG. 1. Scanning electron micrograph of *T. pyriformis* E. Fixed with 3% glutaraldehyde in 50 m*M* sodium phosphate buffer. OA, Oral apparatus.

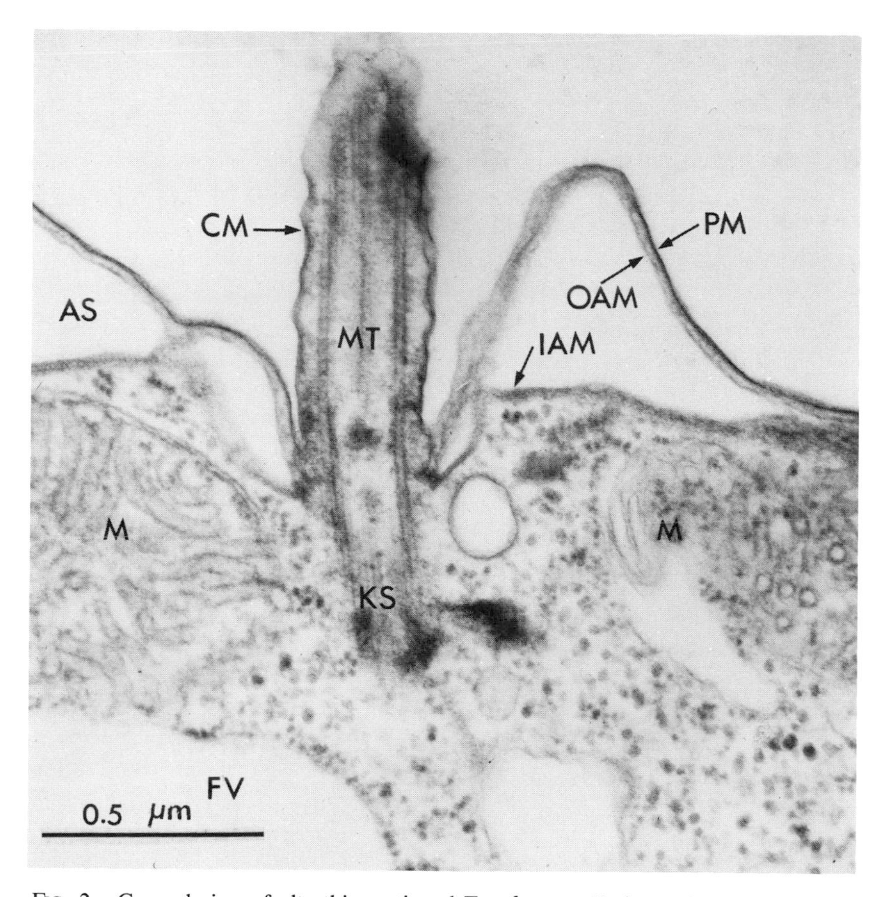

FIG. 2. General view of ultrathin-sectioned *Tetrahymena*. Various subcellular membrane components are seen. PM, Plasma membrane; OAM, outer alveolar membrane; IAM, inner alveolar membrane; CM, ciliary membrane; M, mitochondria; KS, kinetosome; MT, microtubule; AS, alveolar space; FV, food vacuole.

include nuclei, mitochondria, lysosomes, endoplasmic reticulum, cilia, pellicles, oral apparatus, and food vacuoles, some of which are shown in Fig. 2.

This article aims to detail and review procedures for isolation of various membrane components from *T. pyriformis*. For convenience, the isolation methods are divided into two large groups: (1) systematic isolation methods by which several different membrane components can be isolated from the same batch of culture, and (2) nonsystematic isolation by which individual organelles are separately isolated.

II. Systematic Isolation Methods for Various Subcellular Membrane Components

For studies requiring various membrane components from the cells at the same metabolic level, the systematic isolation method must be employed.

A. Isolation of Mitochondria and Microsomes

Smith and Law (1970) isolated mitochondria and microsomes from *T. pyriformis* WH-4 by the procedure shown in Fig. 3. The washed cells suspended in 0.25 *M* sucrose–0.1 *M* tris–HCl (pH 7.4) are mechanically disrupted and centrifuged at 500 *q* for 10 minutes to remove the unbroken cells.

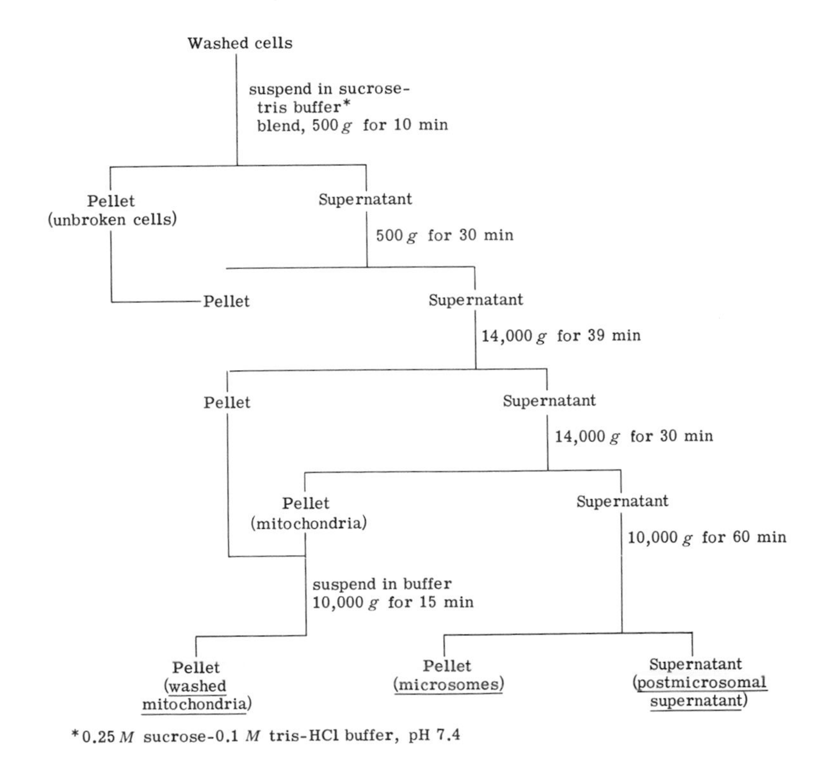

FIG. 3. Scheme for the systematic isolation of mitochondria and microsomes. (Smith and Law, 1970).

B. Isolation of Mitochondria, Peroxisomes, and Lysosomes

The first attempt at subcellular fractionation of *Tetrahymena* by use of zonal centrifugation was made by Müller *et al.* (1968). Cell disruption is achieved by passage of a chilled cell suspension in 0.25 *M* sucrose through a fritted-glass filter (pore size 10–15 μm) under light suction. The homogenate is fractionated by zonal differential sedimentation through discontinuous sucrose gradients in a B-XIV rotor (Anderson *et al.*, 1967). The rotor is

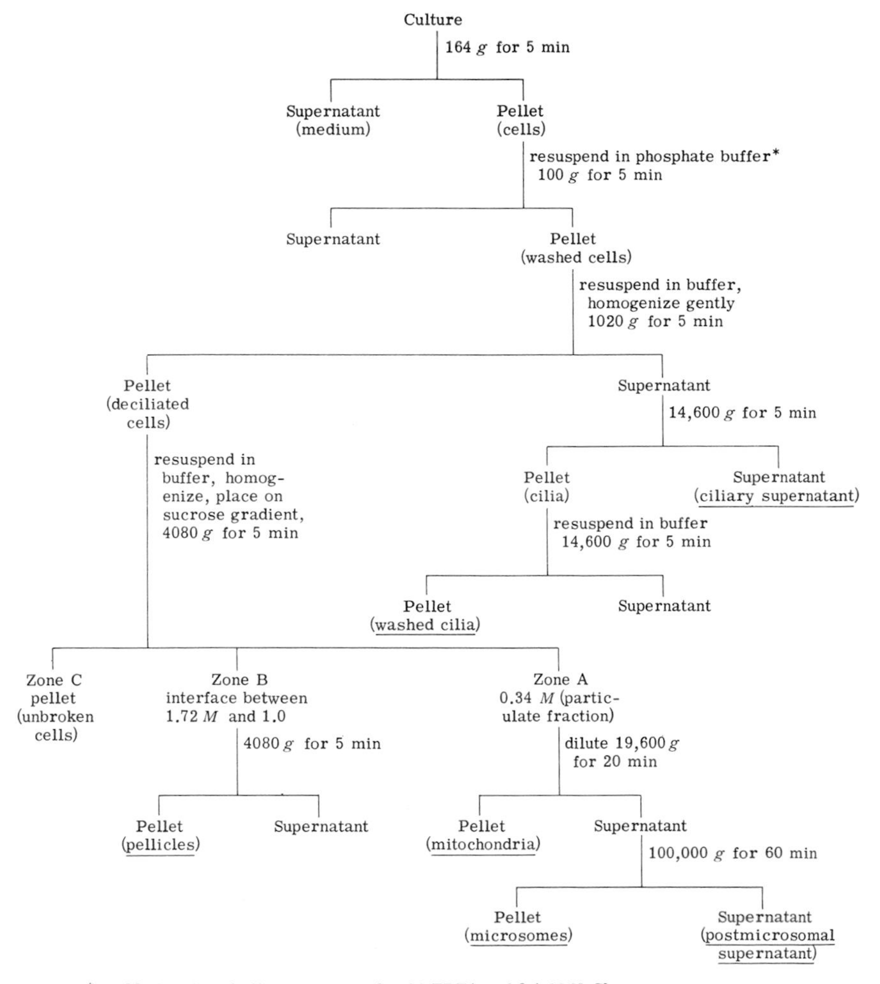

FIG. 4. Scheme for the systematic isolation of various organelles from *Tetrahymena*. (Nozawa and Thompson, 1971a).

loaded at 2800 rpm, establishing the following initial conditions: 90 ml of 0.2 *M* sucrose overlay; 40 ml of 0.25 *M* sucrose homogenate; 50 ml each of 0.30, 0.35, 0.40, 0.45, 0.50, and 0.55 *M* sucrose; and 220 ml of 60% (w/w) sucrose cushion. The rotor is accelerated to 5000 rpm for 11 minutes, following which the fractions are collected at 2800 rpm through the flow cell. Thus mitochondria, peroxisomes, and lysosomes are separated.

A similar method has been employed by Lloyd *et al.* (1971) and Poole *et al.* (1971), except that the preparation medium is 0.32 *M* sucrose containing 24 m*M* tris–HCL and 10 m*M* EDTA, and cell disruption is carried out in a glass homogenizer with 10 to 30 gentle strokes. All investigators have reported that the marker enzymes are reasonably distributed among these isolated membrane fractions.

Such fractionation by zonal centrifugation has an advantage of large-scale preparation of membrane components, but further improvement in purity will be required.

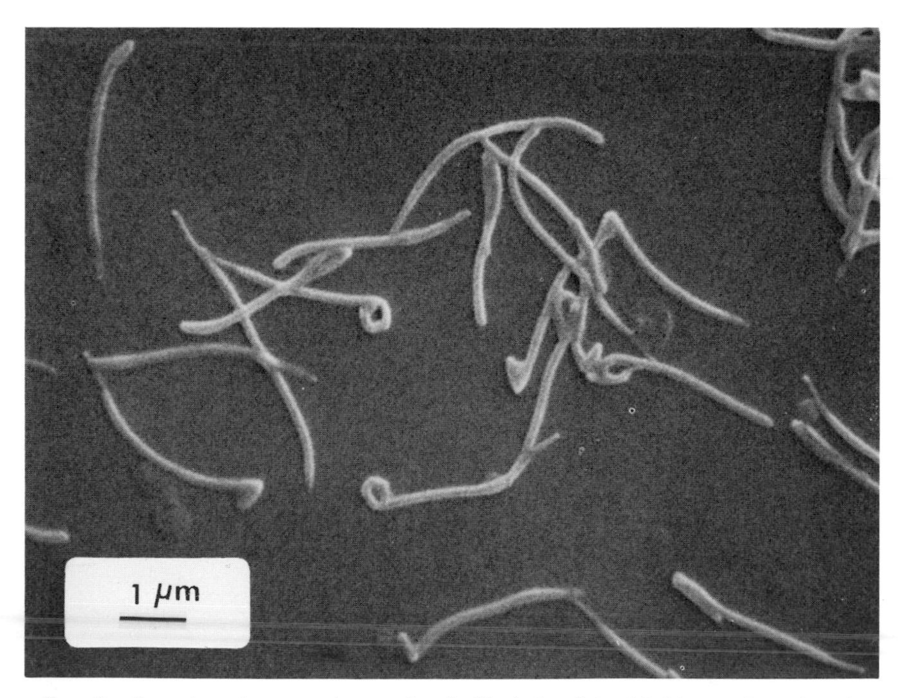

FIG. 5. Scanning electron micrograph of cilia isolated in 0.2 *M* potassium phosphate buffer (pH 7.2) containing 3 m*M* EDTA–0.1 *M* NaCl.

C. Isolation of Cilia, Pellicles, Mitochondria, and Microsomes

Nozawa and Thompson (1971a) have developed a procedure for isolating several membrane components including surface membrane fractions such as cilia and pellicles. The overall procedure is presented in Fig. 4. The harvested cells of *T. pyriformis* E are resuspended in cold 0.2 *M* phosphate buffer (pH 7.2) containing 0.1 *M* NaCl and 3 m*M* 2Na EDTA and then centrifuged at 100 *g* for 5 minutes. This cooling step is important for rendering the cells well shrunken. Unless good shrinkage is obtained, cells are easily broken, and isolation of pure membrane components is no longer possible. The shrunken cells are resuspended in 12–15 ml of the cold phosphate buffer to make a suspension of approximately 6–8 $\times$ 10^6 cells per milliliter and gently homogenized by hand (four to six strokes) in a loose-fitting glass

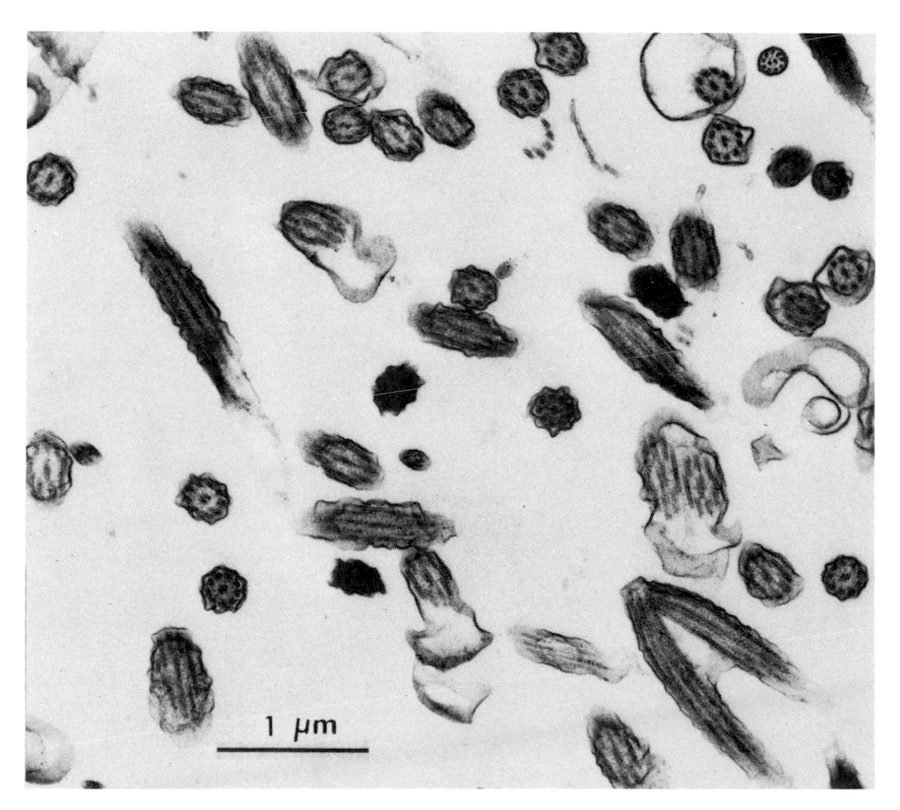

FIG. 6. Ultrathin section of isolated cilia.

homogenizer (Arthur H. Thomas Co., Philadelphia, Pa.) until most of the cilia have been detached from the cells as determined by phase-contrast microscopy. The homogenate is then centrifuged at 1020 *g* for 5 minutes, and the resulting supernatant is recentrifuged at 14,600 *g* for 5 minutes. The pellet of cilia thus obtained is washed once with the buffer (Figs. 5 and 6).

The deciliated cells are resuspended in 4–5 ml of buffer and homogenized vigorously by hand in a tight-fitting glass homogenizer (Arthur H. Thomas Co., Philadelphia, Pa.). The homogenate is layered on a discontinuous buffered sucrose gradient (0.34 *M*, 10 ml; 1.0 *M*, 15 ml; 1.72 *M*, 15 ml) and centrifuged at 4080 *g* for 5 minutes. Three major zones are separated: zone A, a top band down through the 0.34 *M* layer; zone B, a discrete band of pellicles at the interface between 1.0 *M* and 1.72 *M*; zone C, a small pellet of unbroken cells at the bottom.

The particulate (zone A) and pellicle (zone B) fractions are carefully removed with a syringe. The pellicle fraction is centrifuged at 4080 *g* for 5 minutes to form a pellet. The plasma membrane and outer and inner

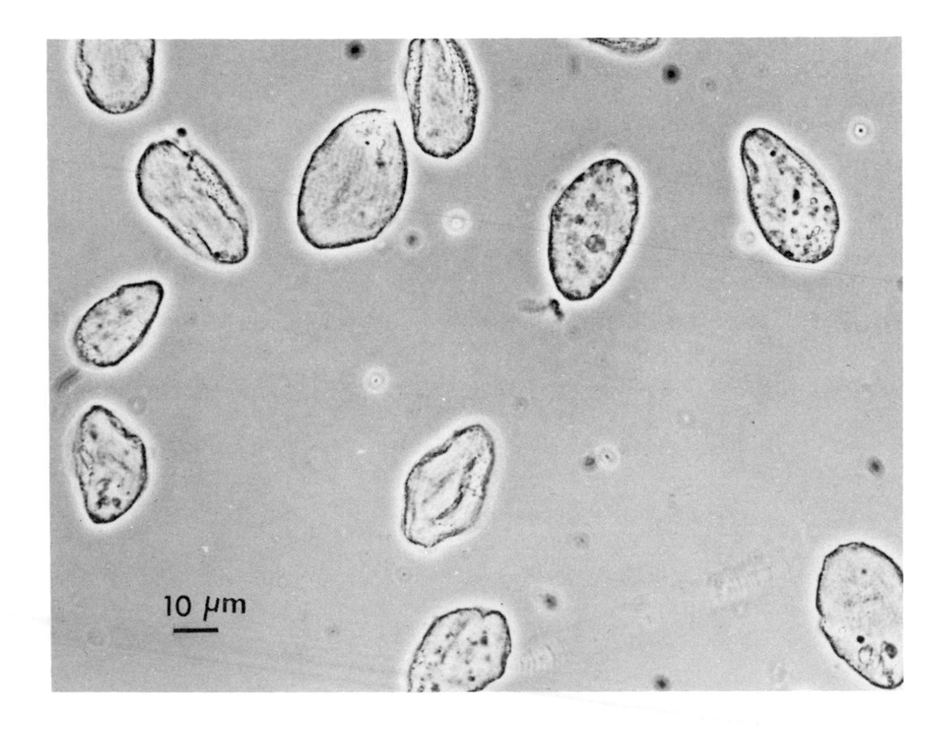

FIG. 7. Phase-contrast micrograph of isolated pellicle ghosts.

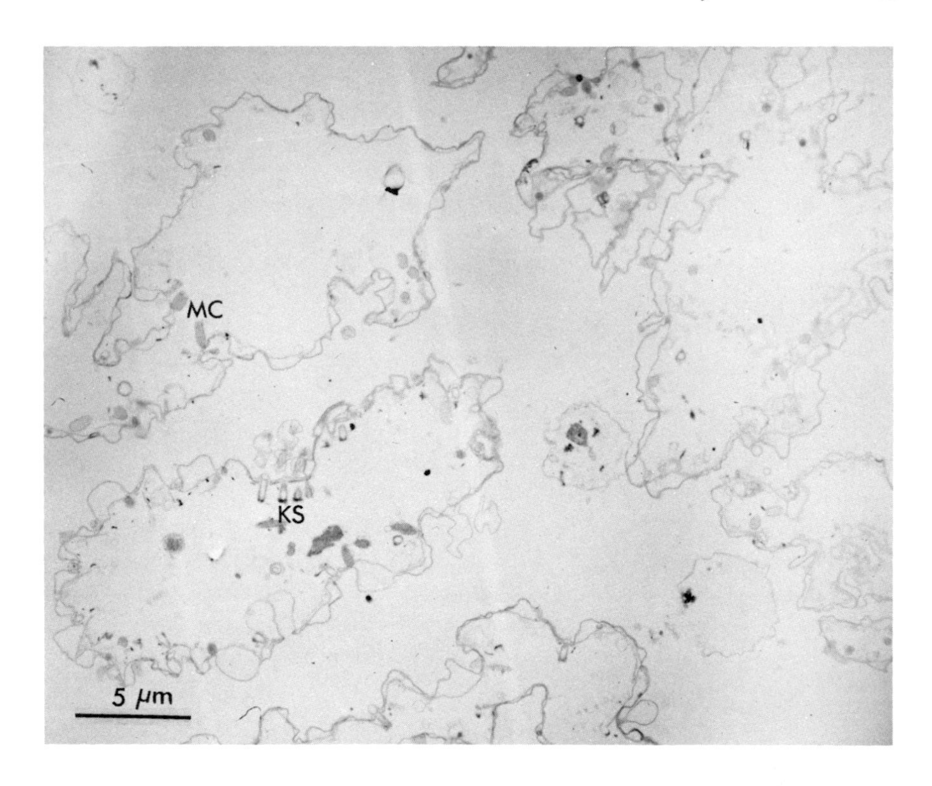

FIG. 8. Ultrathin section of isolated pellicles. Plasma membrane, and outer and inner alveolar membranes, are well preserved. A few kinetosomes (KS) and mucocysts (MC) are also seen.

alveolar membranes are well preserved (Figs. 7 and 8). For further purification gentle homogenization and washing are repeated. The particulate fraction (zone A) composed of mitochondria and microsomes is diluted with buffer and spun at 19,600 *q* for 20 minutes to sediment mitochondria (Fig. 9). Some of them are irregular in shape and size. This might be due to exposure to high-phosphate buffer. The resulting supernatant is further centrifuged at 100,000 *g* for 60 minutes, yielding a pellet comprised of microsomes (Fig. 10). Thus four different membrane components (cilia, pellicles, mitochondria, and microsomes) are isolated in a highly pure state. In fact, with this procedure we have isolated various membrane components and performed radioisotope labeling experiments to follow membrane biosynthesis (Nozawa and Thompson, 1971b, 1972), as well as other studies (Thompson *et al.*, 1971, 1972; Nozawa *et al.*, 1974).

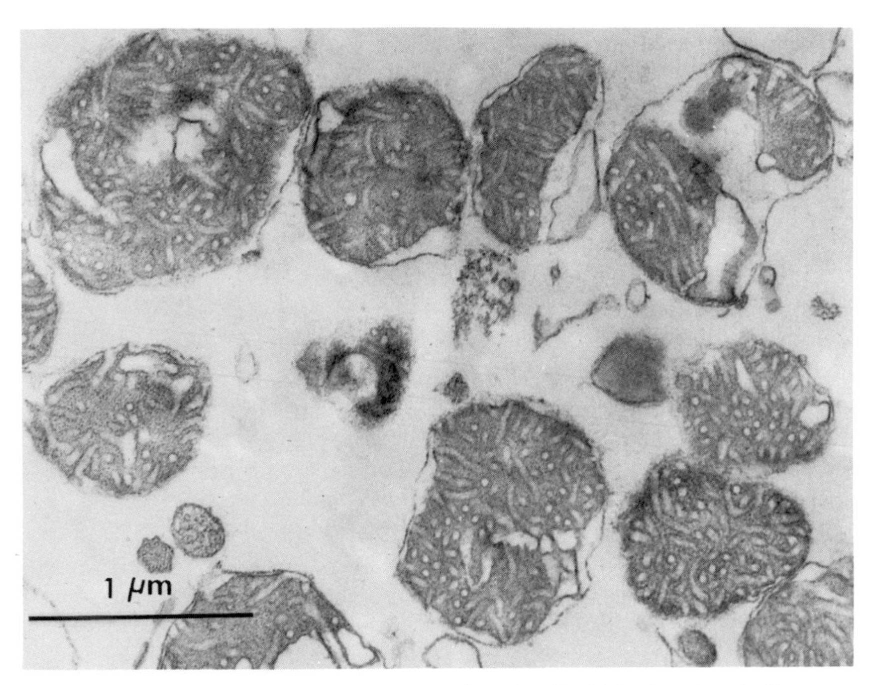

FIG. 9. Ultrathin section of mitochondria isolated in high-phosphate buffer.

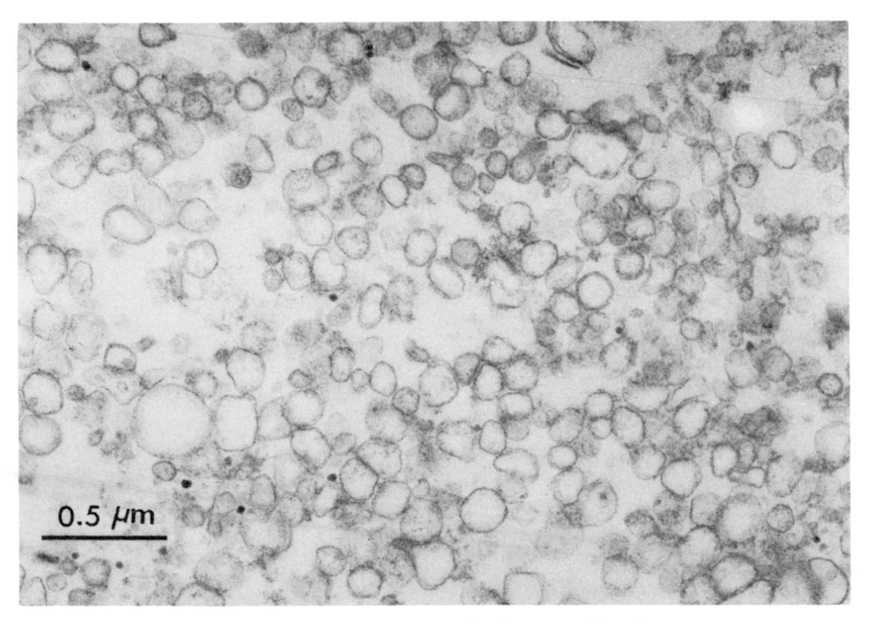

FIG. 10. Ultrathin section of microsomal fraction.

III. Nonsystematic Isolation Methods for Various Subcellular Membrane Components

A. Isolation of Cilia and Their Subfractions

1. WHOLE CILIA

The cilia of *Tetrahymena* can be detached from the cells by mild shearing forces after several different chemical treatments. Blum (1971) specified the transitional region between the kinetosome and the ciliary shaft as a breaking point of a cilium.

a. Use of Ethanol. Cilia were first isolated from *Tetrahymena* cells by Child (1959), using an ethanol–KCl solution. Cells are suspended in 40% ethanol prechilled to −10°C while stirring vigorously, and kept at −10°C for 12–24 hours. After centrifugation at 3000 *g* for a few minutes, the supernatant ethanol is decanted and the packed cells suspended in 0.1 *M* KCl at pH 7.0 and stirred vigorously for 10 minutes in an ice bath. The suspension is centrifuged at 4000 *g* for 10 minutes, and the supernatant fluid containing the cilia is carefully decanted and recentrifuged at 12,000 *g* for 15 minutes to sediment the cilia.

Watson and Hopkins (1962) used a similar procedure for isolating cilia. The concentrated cells are suspended in 150 ml of cold 25 m*M* sodium acetate and diluted with 750 ml of cold 12% ethanol in 25 m*M* sodium acetate containing 0.1% EDTA, followed by immediate addition of 25 ml of cold 1.0 *M* $CaCl_2$ solution. The suspension is allowed to stand in the ice bath for 10 minutes with occasional stirring, and the cell bodies are removed by centrifugation at 1000 *g* for 10 minutes. The resulting supernatant contains only cilia.

b. Use of Glycerol. Gibbons (1965) has described an alternative procedure using glycerol instead of ethanol. To 10 ml of a concentrated cell suspension is added 100 ml of 70% glycerol solution containing 50 m*M* KCl, 2.5 m*M* $MgSO_4$, and 20 m*M* tris–thioglycolate buffer (pH 8.3). As soon as it is throughly mixed, the suspension of cells in glycerol is cooled to −20°C and maintained at this temperature. Vigorous agitation of the suspension of a vortex mixer for 1 minute causes the majority of the cilia to become detached. The supernatant obtained after centrifugation at 12,000 *g* for 10 minutes contains pure cilia. He showed that cilia isolated by this method can be reactivated by ATP, unlike the cilia isolated by the ethanol–calcium method of Watson and Hopkins (1962) described above. Raff and Blum (1966, 1969) isolated cilia by slight modifications of Gibbons' glycerol method for the study of reactivation of cilia by ATP.

c. Use of EDTA–$CaCl_2$. Rosenbaum and Carlson (1969) have described a procedure for the amputation of cilia, in which the cells remain viable and regenerate cilia. At zero time 2.5 ml of concentrated cells is added to 5.0 ml of medium A [10m*M* 2Na EDTA, 50 m*M* sodium acetate (pH 6.0)]. At 30 seconds 2.5 ml of cold distilled water is added, followed by the addition at 90 seconds of 2.5 ml of 0.2 *M* $CaCl_2$. At 3.5 minutes the suspension of cells is subjected to two to four shearings with a glass syringe fitted with an 18-gauge needle, the cilia becoming detached from the cell bodies.

d. Use of Dibucaine. For regeneration experiments with deciliated cells, an alternative procedure for cilia amputation has been described by Thompson *et al.* (1974). *Tetrahymena pyriformis* E cells are suspended in 10 ml of the fresh medium, and 0.55 ml 2.5 m*M* dibucaine–HCl is added with mixing. Cilia are almost completely detached within 3–5 minutes, but these investigators have suggested that use of dibucaine for various strains requires its optimum concentration.

e. Use of High Phosphate. To obtain a pure preparation of isolated cilia, it is essential that there is no lysis of cells during the deciliation procedure. Any lysed cells give rise to serious contamination of the cilia with cytoplasmic particles. Therefore we used 0.2 *M* phosphate buffer containing 0.1 *M* NaCl and 3 m*M* 2 Na EDTA to cause the cells to shrink and to prevent cell lysis. After the cells are well shrunken three to four gentle strokes by hand in a loose-fitting glass homogenizer or several shearings with a glass syringe give efficient deciliation on a large scale.

2. Subfractions of Celia

Gibbons (1963) has described a procedure for the isolation of microtubules and central fibers from isolated cilia. This method involves selective solubilization of the ciliary membrane with a 0.5% solution of digitonin containing 2.5 m*M* tris–HCl (pH 8.3) for microtubule isolation, and the subsequent dialysis of digitonin-extracted cilia against tris–EDTA solution [0.1 m*M* EDTA, 1 m*M* tris–Chl (pH 8.3)] for separating the outer fibers.

Renaud *et al.* (1968) analyzed in great detail the protein forming the outer fibers of *Tetrahymena* cilia by using disc electrophoresis and analytical ultracentrifugation, and demonstrated a close similarity between outer fiber protein and actin. The isolation of the outer fibers must be done as follows. Cilia isolated by an ethanol–calcium procedure are resuspended in tris–EDTA solution as mentioned above, dialyzed against this same solution, and then centrifuged. Although the supernatant contains the matrix, the central fibers, and most of the dynein, the resulting pellet consists largely of ciliary membranes and outer fibers. For removal of membranes the mixture of ciliary membranes and outer fibers is centrifuged, and the pellet is resuspended twice in 0.5% digitonin and 1 m*M* tris–HCl (pH 8.3) at 0°C. The digitonin solubilizes the membranes, and a pure outer fiber fraction is obtained.

Rubin and Cunningham (1973) isolated the axonemal microtubule fraction by the method of Stephens (1970), originally designed for sea urchin flagella, in which the ciliary membranes are removed by solubilization in 1% Triton X-100 detergent in 30 m*M* tris–HCl, 3 m*M* $MgCl_2$, and 0.1 m*M* dithiothreitol. A preparation of microtubules is shown in Fig. 11. For the isolation of ciliary membranes, the suspension containing ciliary membranes and outer fibers, which is prepared by the method of Renaud *et al.* (1988), is resuspended in tris–EDTA–0.6 *M* KCl solution, dialyzed against this solution overnight, and centrifuged at 19,600 *g* for 20 minutes to pellet the ciliary membranes.

Recently, Subbaiah and Thompson (1974) isolated the ciliary membrane fraction with the method using Triton X-100 solution developed by Gibbons and Gibbons (1972) for use on sea urchin flagella. This preparation medium consists of 0.05% (w/w) Triton X-100, 0.15 *M* KCl, 2 m*M* $MgSO_4$, 0.5 m*M* 2Na EDTA, 0.5 m*M* mercaptoethanol, and 2 m*M* tris–HCl buffer (pH 8.0).

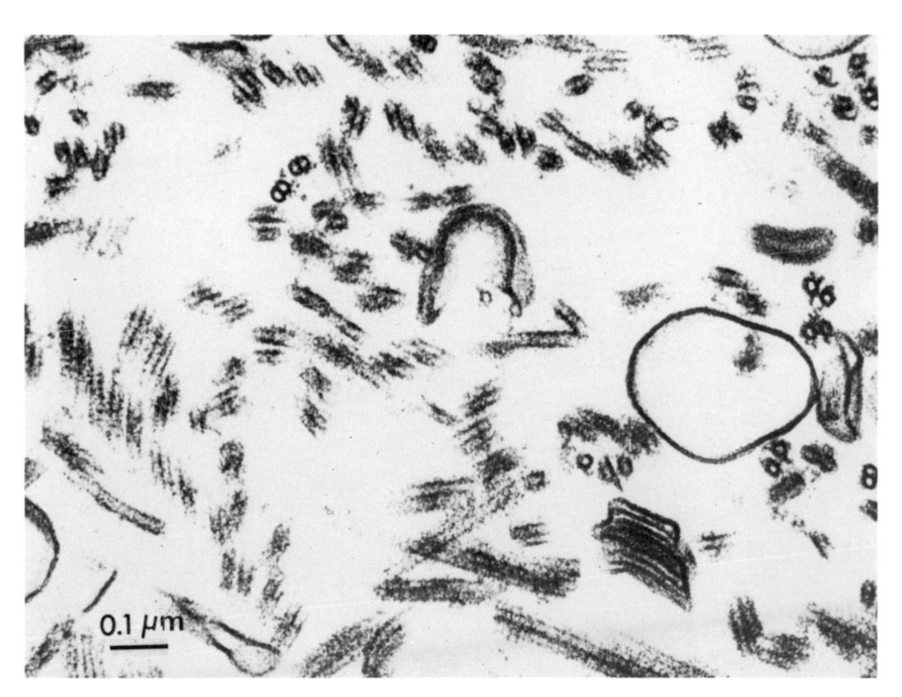

FIG. 11. Ultrathin section of a demembranated axonemal microtubule fraction. (Kindly provided Dr. William P. Cunningham.)

B. Isolation of Oral (Buccal) Apparatus

The oral apparatus is composed of kinetosomes, microtubules, and microfilaments. Therefore preparations of this apparatus are a good source of kinetosomes (see Section III,D). For the isolation of pure oral apparatus, efficient cell lysis is a prerequisite. Basically, there are three methods.

1. Method with Butanol

Williams and Zeuthen (1966) have described a procedure for isolating oral apparatuses. The cells are lysed in 1.5 *M* tertiary butanol by stirring on a vortex mixer, and centrifuged at 2000 *g* for 5 minutes to obtain a pellet of oral apparatuses.

2. Method with Indole

Whitson *et al.* (1966a) performed a large-scale isolation of oral apparatuses by zonal centrifugation from cells lysed by the addition of indole made up in 10 m*M* tris buffer (pH 7.5).

3. Method with Triton X

Wolfe (1970) used a Triton X solution to disrupt cells. To 200 ml of resuspended cells in 0.12 *M* sucrose is added 1 liter of a solution consisting of 1.0 *M* sucrose, 1 m*M* EDTA, 0.1% 2-mercaptoethanol, and 10 m*M* tris (pH 9). Then, before swelling occurs, 100 ml of 10% Triton X-100 is added in order to lyse the cells. The lysate is spun at 12,000 *g* for 30 minutes to obtain a loosely packed pellet. The pellet is resuspended in the isolation medium without detergent and homogenized in a glass homogenizer. The homogenate is recentrifuged at 12,000 *g* for 15 minutes to remove small particles from the oral apparatus fraction.

Rannestad and Williams (1971) have described an alternative isolation procedure for the oral apparatus for protein analysis by disc electrophoresis. To the pelleted cells 20 ml of 0.1% Triton X-100 solution is added, and the mixture is swirled rapidly for 2 minutes with a spatula to lyse the cells. The lysate is mixed with 180–190 ml of cold distilled water and homogenized in a Logeman homogenizer (Scientific Products, Evanston, Ill.). The homogenate is centrifuged at 2000 *g* for 10–15 minutes in a centrifuge tube containing a 1 m*M* sucrose cushion made with 0.01% Triton. The pelleted oral apparatus is resuspended in 0.01% Triton X and recentrifuged. The oral apparatus fraction thus isolated is highly pure, and the three membranelles and the undulating membrane are well preserved (Fig. 12).

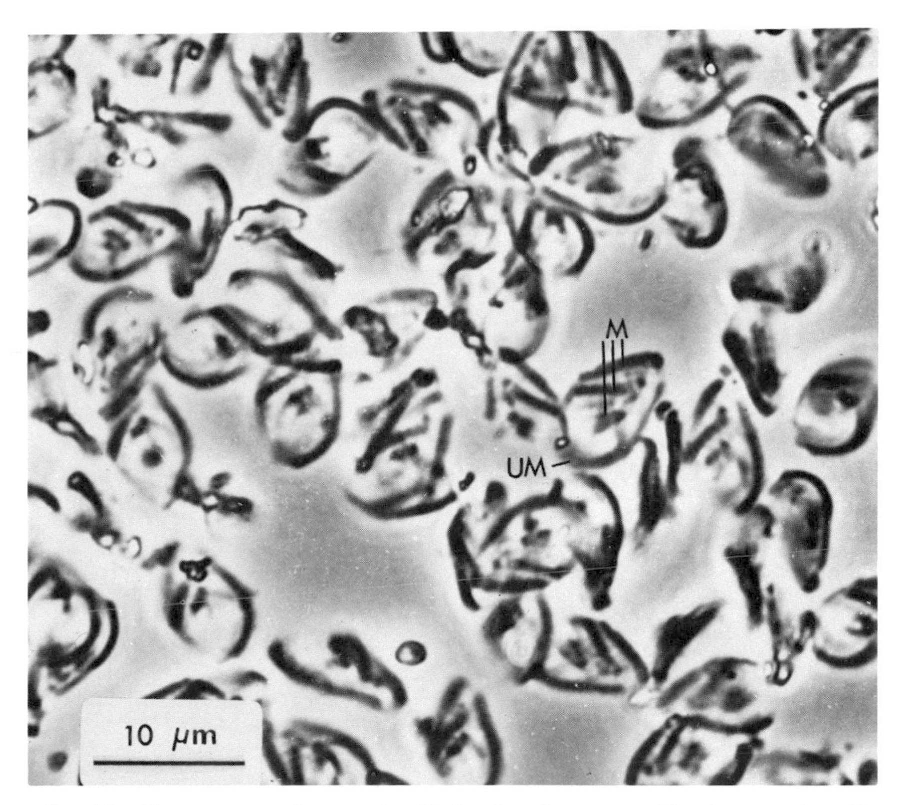

FIG. 12. Phase-contrast photograph of isolated oral apparatus. The three membranelles (M) and the undulating membrane (UM) of an oral apparatus are indicated. (Kindly provided by Dr. Norman E. Williams.)

C. Isolation of Pellicles

The term *pellicle* has been used to describe the surface structures of ciliates, and includes cilia, surface membranes, kinetosomes, and tubular components (Everhart, 1972). However, in this chapter we use the term pellicle to designate the surface membranes, which include the plasma membrane and the outer and inner alveolar membranes. Until recently, all methods have been developed to isolate pellicle fragments as the source material of kinetosomes (see Section III,D).

1. USE OF DIGITONIN

Seaman (1960) first isolated pellicle fragments from *Tetrahymena*. The cell suspension cooled at $-15°$ C in 40% ethanol is stripped of cilia by centrifugation at 250 *g* for 10 minutes and resuspended in 1% digitonin solution made up in 0.4 *M* KCl. The resulting pellicle shells are pelleted by centrifugation at 1000 *g* for 10 minutes.

Other investigators (Argetsinger, 1965; Hoffman, 1965; Satir and Rosenbaum, 1965) isolated the fragments of pellicles, prior to separation of kinetosomes, by modifications of Seaman's method.

2. Use of High Phosphate

An isolation procedure for the intact pellicle membrane complex that preserves the tripartite membranes well was first developed by Nozawa and Thompson (1971a), as described previously (Section I,C). For nonsystematic isolation of the pellicles, cells suspended in 0.2 *M* phosphate buffer containing 0.1 *M* NaCl and 3 m*M* 2Na EDTA (pH 7.2) are vigorously homogenized by hand in a tight-fitting glass homogenizer until almost all the cells become pellicle "ghosts." The homogenate is then centrifuged at 1465 *g* for 5 minutes to remove cilia and small cytoplasmic particles. The resultant pellet, consisting mostly of pellicle ghosts, is resuspended in the buffer, loaded on a discontinuous sucrose density gradient (1.0 *M*, 10 ml; 1.72 *M*, 15 ml), and centrifuged at 4080 *g* for 5 minutes. The highly purified pellicle membranes are removed by a syringe from the interface between the two sucrose phases. Attempts to separate the plasma membrane from the alveolar membranes have met with little success.

3. Use of Tween 80

Hartman *et al.* (1972) have developed an alternative procedure in which *Tetrahymena* (GL or W) are suspended in 1% Tween 80–0.1 *M* EDTA (pH 9) and homogenized with a Virtis 45 homogenizer at 6000 rpm for 20 minutes. They reported that in comparison with our pellicle preparations their pellicles lost some membrane components. We have analyzed proteins from both pellicle preparations by disk gel electrophoresis and found that some membrane protein bands are lacking in the Tween–EDTA-extracted pellicles.

D. Isolation of Kinetosomes (Basal Bodies)

1. From Pellicles

As described above (Section III,C,1), pellicle fragments have been isolated as a source of kinetosomes. In fact, Argetsinger (1965), Hoffman (1965), and Satir and Rosenbaum (1965) attempted to isolate kinetosomes from pellicle fragments using modifications of Seaman's procedure. All these methods involve homogenization of pellicles leaving kinetosomes intact, but each uses a different approach. For example, Hoffman (1965) used three different methods: a quartz method, a fixation-quartz method, and an ether method. Whitson *et al.* (1966a) used zonal centrifugation for isolating kinetosomes from digitonin-extracted pellicles.

Recently, Rubin and Cunningham (1973) reported that none of these procedures produces basal bodies of sufficient purity for chemical analysis in their laboratory, and therefore developed a better technique for kinetosome isolation. The pellet obtained after treatments with ethanol, digitonin, and KCl as described earlier (Section III,A,2) is resuspended in 0.2 *M* sucrose, homogenized with a Polytron 10-ST (Brirkman Instruments, Inc., Westbury, N.Y.), and centrifuged at 7000 *g* for 8 minutes. The supernatant is spun at 50,000 *g* for 10 minutes. The resulting pellet is resuspended in 0.2 *M* sucrose, placed over sucrose gradients (1.4, 1.7, 1.8 *M*), and centrifuged at 100,000 *g* for 90 minutes. The top zone of the 1.8 *M* interface is the basal body–rich fraction.

2. From Oral Apparatuses

Wolfe (1970) isolated oral apparatuses by using 10% Triton X-100 as a supplier of kinetosomes (Section III,B,3). The oral apparatus fraction prepared as described above is a purely fibrous structure composed of microtubules and microfilaments. Membranes have been stripped off, while the triplet structure of the basal bodies remains intact.

E. Isolation of Mitochondria

Although *Tetrahymena* has been one of the most widely used materials for biochemical investigation of isolated mitochondria, there are some difficulties in isolating pure mitochondria because of their irregular shape and fragility.

Byfield *et al.* (1962) first attempted to isolate mitochondria from *Tetrahymena* cells. The homogenate suspended in 0.25 *M* sucrose after passing through a French press is centrifuged at 5000 *g*, and then the resulting supernatant is spun at 100,000 *g*. The pellet is rich in active mitochondria.

Kobayashi (1965) and Suyama and Preer (1965) independently developed procedures for the isolation of mitochondria. By the procedure of Kobayashi, the cells are suspended in a solution of 0.35 *M* mannitol, 0.05% bovine serum albumin (BSA), 0.1 m*M* EDTA, and 1 m*M* tris–HCl (pH 7.0–7.2), gently homogenized with a Teflon probe in a tight-fitting glass homogenizer, and immediately centrifuged at 100,000 *g* for 3 minutes. The pellet consists of two layers: an upper one, which is white-grey, jellylike, loosely packed, and highly viscous, and a lower one, which is yellow-brown and tightly packed. The upper layer is carefully removed. The bottom layer is centrifuged at

200 *g* for 2 minutes to remove unbroken cells. The supernatant is then centrifuged at 9000 *g* for 5 minutes to form a mitochondrial pellet, which has a high respiratory control ratio.

Conklin and Chou (1972) used a modification of Kobayshi's method in which the preparation medium is made up with 5 m*M* tris–HCl (pH 7.4), 1 m*M* phosphate, 0.25% BSA, 4 m*M* $MgCl_2$, and 0.25 *M* sucrose.

Suyama and Preer (1965) have described a procedure for mitochondrial isolation for characterization of DNA in *Tetrahymena* ST mitochondria. Cells suspended in a medium of 0.2 *M* raffinose, 0.25% BSA, and 1 m*M* potassium phosphate buffer (pH 6.2) are disrupted in a milk homogenizer. The homogenate is centrifuged at 5000 *g* for 6 minutes, and the resulting

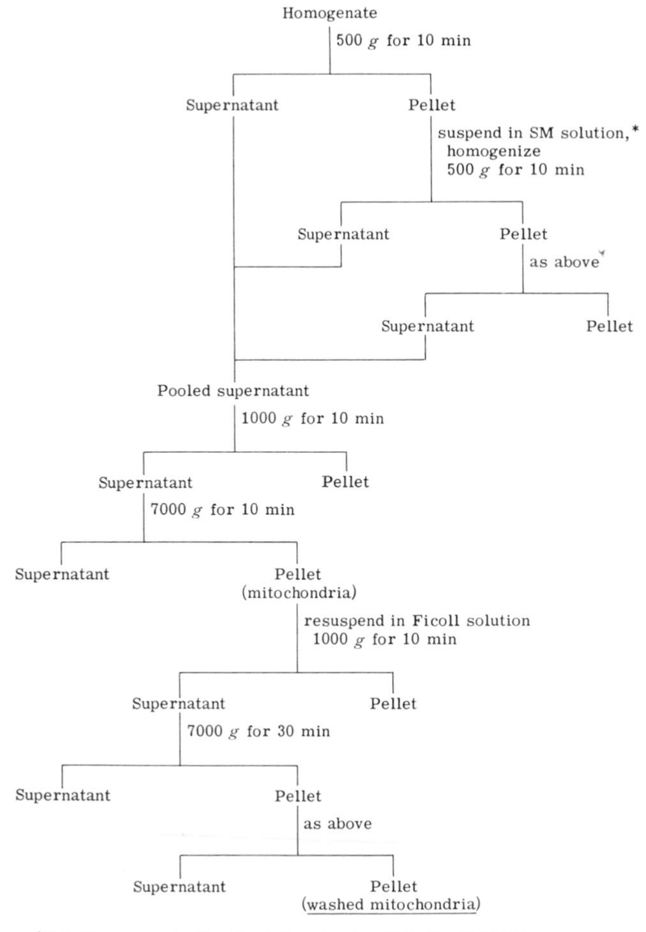

FIG. 13. Scheme for the isolation of mitochondria (Schwab-Stey *et al.*, 1971).

pellet is resuspended in the same medium. The suspension is centrifuged again under the same conditions, and the final pellet contains mitochondria with some contaminating cilia. However, since it was found that this method could not be applied to all other strains of *Tetrahymena*, Suyama (1966) attempted to improve his technique by the use of various media: (1) 0.5 *M* sucrose, 10 m*M* tris–HCl (pH 7.4), 1 m*M* EDTA: (2) 30 m*M* mannitol, 50 m*M* tris–HCl (pH 7.4) 1 m*M* EDTA, 0.05% cysteine, 0.1% BSA; (3A) 0.2 *M* raffinose, 1 m*M* potassium phosphate buffer (pH 6.2), 0.25% BSA; (3B) 0.2 *M* raffinose, 1 m*M* potassium phosphate buffer (pH 7.0), 0.25% BSA. He reported that, except for (3A), these media gave little success.

Later, other workers (Westergaard *et al.*, 1970; Flavell and Jones, 1970) followed essentially the method of Suyama and Preer (1965) for mitochondrial DNA studies.

Schwab-Stey *et al.* (1971) isolated mitochondria by modification of the method of Lloyd *et al.* (1968), which is presented in Fig. 13. Cells are suspended in a medium of 0.3 *M* sucrose in 10 m*M* triethanolamine–HCl and 2 m*M* EDTA, and homogenized in a Potter-Elvehjem homogenizer. The isolated mitochondria are highly pure, as shown in Fig. 14.

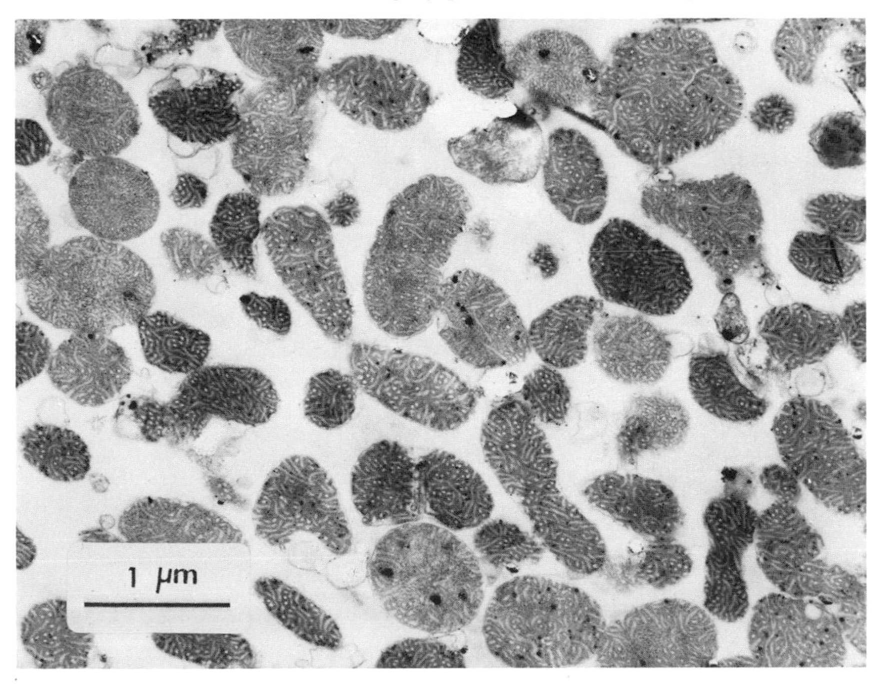

FIG. 14. Ultrathin section of mitochondria isolated in 0.3 *M* sucrose solution containing 10 m*M* triethanolamine–HCl and 2 m*M* EDTA. (Kindly provided by Dr. Hiltrud Schwab-Stey.)

Several other methods have been described previously by Smith and Law (1970), Müller *et al.* (1968), Lloyd *et al.* (1971), and Nozawa and Thompson (1971a) (see Section II,A).

F. Isolation of Nuclei and Nuclear Membranes

1. Isolation of Macronuclei

There have been many methods for the isolation of macronuclei from *Tetrahymena*, but one must chose a suitable procedure depending on what is required for biochemical analysis.

a. Methods with Detergents. Lee and Scherbaum (1966) described a procedure using the nonionic detergent Triton X-100 for isolation of macronuclei. This method has been modified by Lee and Byfield (1970). Cells are washed with buffer [10 m*M* tris–HCl (pH 7.4) containing 2 m*M* $CaCl_2$ and 1.5 *M* $MgCl_2$] and suspended in the same buffer. To the suspension is added Triton X-100 to a final concentration of 0.1%. After complete cell lysis within 5 minutes, polyvinylpyrrolidone (PVP) is added to a final concentration of 2%, and the mixture is filtered through a cotton filter. The filtrate is then layered over 0.5 *M* sucrose in buffer containing 2% PVP, and spun down by success stepwise 5-minute accelerations to 70, 250, and 800 *g*. The macronuclei are packed softly at the bottom of the tube. Engberg and Pearlman (1972) have isolated macronuclei by the use of a similar method and determined the amount of nuclear rRNA genes.

Mita *et al.* (1966) used another nonionic detergent, Nonidet P-40, for lysis of cells. Cells are washed twice with 0.25 *M* sucrose containing 10 m*M* tris–HCl (pH 7.5), 0.1 m*M* $MgCl_2$, and 3 m*M* $CaCl_2$. To 100 ml of the cell suspension in the same buffer is added 30 ml of 1% (v/v) Nonidet P-40 made up with buffer. The suspension is shaken by hand for several seconds, placed on 0.3 *M* sucrose, and centrifuged at 1200 *g* for 5 minutes. The macronuclei are collected as a pellet and washed by centrifuging at 400 *g* with 0.25 *M* sucrose buffer.

Prescott *et al.* (1966) have described a similar method using Triton X-100 and spermidine. The cells are washed in 0.15 *M* KCl and suspended in a medium of 0.1% Triton X-100, 0.001% spermidine, and 0.25 *M* sucrose. After lysis of cells by expulsion through a Pasteur pipette, the lysate is centrifuged at 700 *g* for 30 minutes to pellet macronuclei.

b. Methods without Detergents. Gorovsky (1970) has described a method that uses *n*-octanol and gum arabic. Nozawa *et al.* (1973) isolated macronuclei by a modification of Gorovsky's method for the separation of macronuclear membranes. The cells are suspended in a solution of 0.1 *M* sucrose, 1.5 m*M* $MgCl_2$, and 4% gum arabic (pH 6.7), and washed by centrifugation.

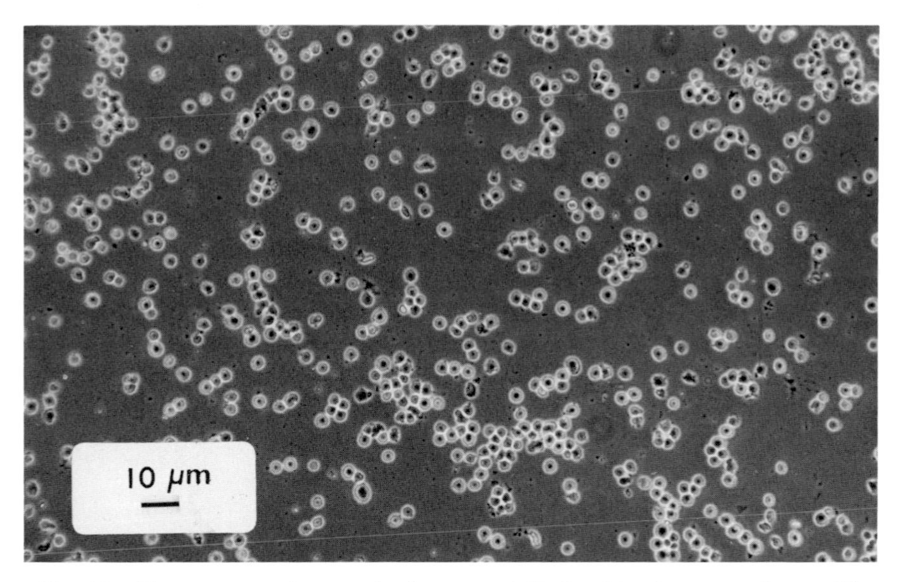

FIG. 15. Phase-contrast photograph of macronuclei isolated by the *n*-octanol–spermidine method.

The washed cells are then resuspended in the same medium, containing 24 m*M* *n*-octanol–0.01% spermidine, and homogenized gently with several stroked by hand in a loose-fitting Teflon probe homogenizer. To the cell lysate is immediately added 3 vol of the $MgCl_2$–sucrose (MS) solution. Great care is taken not to expose the isolated macronuclei too long to octanol. The diluted suspension is centrifuged at 365 *g* for 5 minutes to form a pellet of macronuclei. The pellet is resuspended in the MS solution, layered over a 1.0 *M*, 1.5 *M* discontinuous sucrose gradient, and centrifuged at 10,400 *g* for 5 minutes. The highly purified macronuclei are sedimented at the bottom (Fig. 15).

2. ISOLATION OF MICRONUCLEI

Muramatsu (1970) has described a procedure for isolating pure micronuclei, which is based on the lysis of cells with Nonidet P-40 and differential centrifugation to separate micro- and macronuclei, followed by elimination of contaminating macronuclei by sonic oscillation.

Gorovsky (1970) used the octanol procedure described above, and macro- and micronuclei were isolated together. Such a macronuclear fraction containing micronuclei is resuspended in *n*-octanol solution made up of 0.1 *M* sucrose, 4% gum arabic, and 1.5 m*M* $MgCl_2$ (pH 6.75), and is homogenized in a Waring Blendor to destroy the macronuclei. The homogenate is then centrifuged at 25 *g* for 10 minutes, and the resulting supernatant is spun again at 250 *g* for 10 minutes. The pellet is suspended in the *n*-octanol solution and centrifuged at 1000 *g* for 30 minutes to pellet the micronuclei.

3. ISOLATION OF NUCLEAR MEMBRANES

Purity of the initial isolated nuclei is a prime prerequisite for the isolation of clean nuclear envelopes.

France (1967) first worked out a method to isolate nuclear envelope fragments from *Tetrahymena*. To the isolated nuclei, which are suspended in a few drops of a medium [4% gum arabic, 0.1 *M* sucrose, 4 m*M* *n*-octanol, 20 m*M* tris–HCl (pH 7.5)], is added distilled water or 20 m*M* of sucrose solution. The nuclear envelope ruptures at one or several sites, leaving ghosts. The suspension of nuclear ghosts is gently sonicated, layered over 62% sucrose, and centrifuged at 3000 *g* for 30 minutes to sediment the nuclear membranes. Wunderlich (1969) isolated the macronuclear envelope from cells in different physiological states by a modification of Franke's method.

Nozawa *et al.* (1973) described the following procedure for isolating macronuclear membranes, which preserves full integrity. To the pellet of macronuclei isolated as mentioned earlier (Section III,F,1,b) is added a small amount of 0.2 *M* phosphate buffer in 0.25 *M* sucrose containing 1 *M* NaCl (pH 7.2). By such hypertonic shock almost all the naked nuclei are

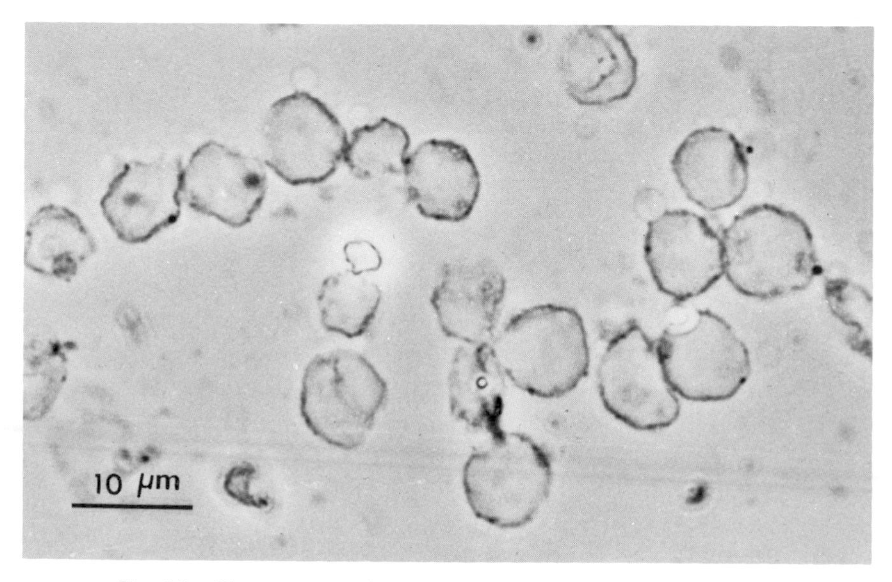

FIG. 16. Phase-contrast photograph of isolated macronuclear envelopes.

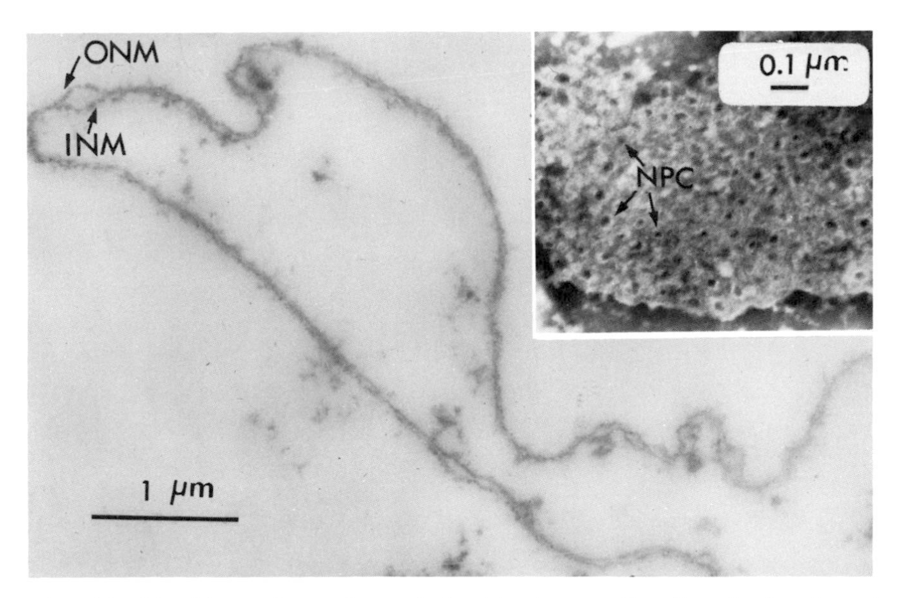

FIG. 17. Ultrathin section and negatively stained preparation (insert) of a piece of isolated macronuclear envelope. Both outer (ONM) and inner (INM) nuclear membranes are well preserved. Nuclear pore complexes (NPC) are also observed (insert).

immediately ruptured, leaving the intact nuclear ghosts. Then the suspension is layered on 2.0 *M* sucrose in phosphate buffer and centrifuged at 3000 *g* for 30 minutes. The top layer, containing the nuclear membranes, is carefully removed, subsequently overlayered on 1.6 *M* sucrose in buffer, and spun at 3000 *g* for 20 minutes. The membrane fraction in the top layer is diluted with 5 vol if 0.2 *M* phosphate buffer–1 *M* NaCl and centrifuged at 10,000 *g* for 10 minutes. Thus the purified nuclear membranes are pelleted at the bottom of the tube (Figs. 16 and 17). Both outer and inner nuclear membranes are observed. The nuclear pore complexes are also preserved as shown by the negatively stained sample (Fig. 17, insert).

G. Isolation of Ribosomes

Leick and Plesner (1968) described a procedure for the isolation of ribosomes. Cells are washed in 0.01 *M* tris–HCl containing 0.15 *M* sucrose, 0.1 m*M* magnesium acetate, and 10 m*M* KCl. The washed cells are subjected to freezing and thawing, and homogenized in a Potter-Elvehjem-type homogenizer. The homogenate is centrifuged at 15,000 *g* for 15 minutes. The resulting supernatant is spun at 100,000 *g* for 60 minutes to pellet the ribosomes.

Kumar (1969, 1970) isolated ribosomes for RNA extraction by the following method. The cell pellet is suspended in TKM buffer [15 m*M* tris–HCl (pH 7.5), 1.5 *M* $MgCl_2$, and 5 m*M* KCl] supplemented with 0.01% spermidine–HCl and 0.25 *M* sucrose, and is homogenized by passing through a Logeman hand mill. The homogenate is centrifuged at 16,000 *g* for 10 minutes. Ribosomes are pelleted from the supernatant at 105,000 *g* for 90 or 120 minutes. Byfield *et al.* (1969) used a similar method with some modifications, in which the isolation medium is 10 m*M* tris–HCl buffer (pH 7.3) containing 10 m*M* $MgCl_2$.

Polyribosomes have been prepared by Whitson *et al.* (1966b) and by Hartman and Dowben (1970). The procedure of the latter involves cell disruption by nitrogen cavitation and linear 15–30% (w/w) sucrose gradient centrifugation.

Chi and Suyama (1970) attempted to isolated *Tetrahymena* mitochondrial and cytoplasmic ribosomes for comparative studies of their physicochemical properties. The lysis of isolated mitochondria and cytoplasmic extracts is

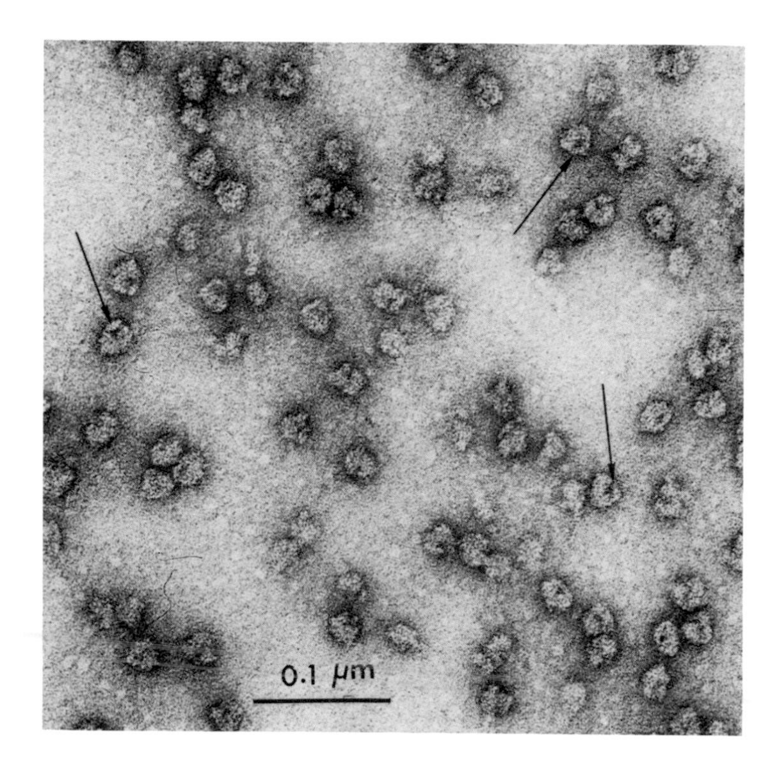

FIG. 18. Negatively stained preparation of isolated cytoribosomes. The majority of profiles show an electron-opaque spot (arrows). (Kindly provided by Dr. Jean-Jacques Curgy.)

carried out by addition of 1% Triton X-100 or 0.5% sodium deoxycholate.

Quite recently, Curgy *et al.* (1974) compared mito- and cytoribosomes by gel electrophoresis and electron microscopy. To prepare cytoribosomes, cells are broken by an emulsion homogenizer (Arthur H. Thomas Co., Philadelphia, Pa.) and centrifuged at 500–800 *g* for 6 minutes. The supernatant is then centrifuged at 5000 *g* for 6 minutes, and the resulting supernatant is spun at 56,000 *g* for 20–30 minutes. Triton X-100 is added to the supernatant (0.02 ml of 1.4% Triton X-100 per milliliter of supernatant). The mixture is layered on 1.5 *M* sucrose in TMK buffer containing 10 m*M* tris–HCl (pH 7.4), 10 m*M* $MgCl_2$, 100 m*M* KCl, and 0.25 *M* sucrose (Chi and Suyama, 1970), and centrifuged at 140,000 *g* for 150 minutes to pellet cytoribosomes (Fig. 18). For mitoribosome isolation, the mitochondrial pellet is resuspended in TMK buffer and lysed with Triton X-100 and sodium deoxycholate at final concentrations of 2.5 and 0.4%, respectively, and maintained for 30 minutes. The lysate is layered on a linear sucrose gradient (0.3–1.4 *M*)

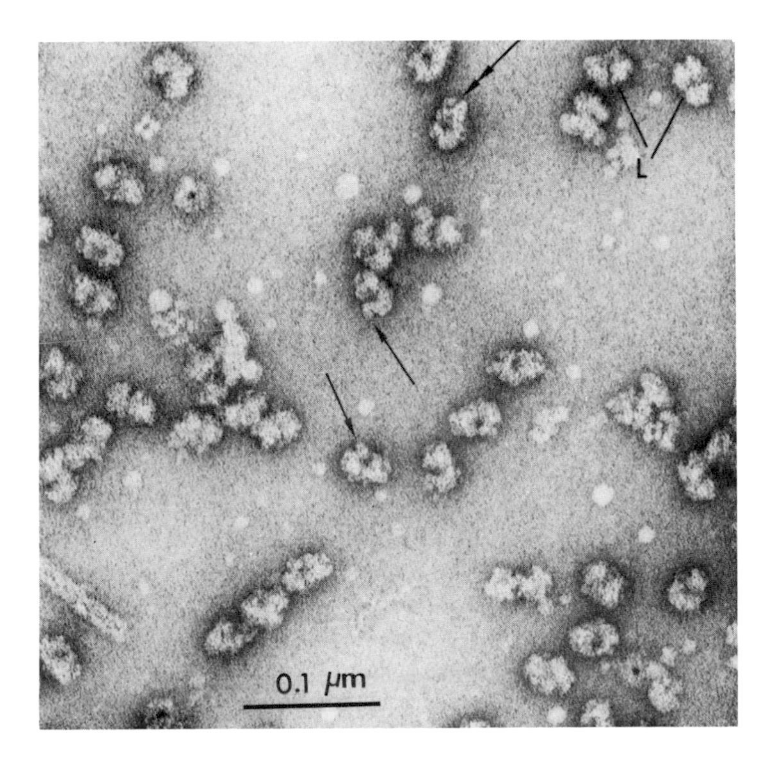

FIG. 19. Negatively stained preparation of isolated mitoribosomes. A dense line appears to divide some profiles into equal-size subunits (L). (Kindly provided by Dr. Jean-Jacques Curgy.) One end shows a definite lobe (arrows).

in TMK buffer. Centrifugation at 35,000 rpm for 180–210 minutes is carried out to obtain mitoribosomes (Fig. 19).

H. Isolation of Peroxisomes

Levy (1970) has described a preliminary procedure for isolating peroxisomes. Cells are disrupted with a Teflon pestle homogenizer in 0.25 *M* sucrose and centrifuged at 8000 *g* for 5 minutes. The thightly packed layer is layered over a gradient consisting of 3.0 *M*, 2.0 *M*, and 0.9 *M* sucrose, and centrifuged at 6000 *g* for 10–15 minutes. Three bands are obtained, and the highest activities of lactate oxidase and isocitrate lyase are localized in the top layer where peroxisomes are concentrated.

Hokama *et al.* (1971) improved Levy's procedure, using the same sucrose gradient system. Since peroxisomes localize between the 0.9 and 2.0 *M* sucrose layers, or between the 2.0 and 3.0 *M* layers, these fractions are collected and diluted with 0.25 *M* sucrose, and then centrifuged at 25,000 *g* for 5 minutes. For further purification several centrifugations are repeated. However, the final preparation consists mostly of peroxisomes with few or no mitochondria.

The procedures of Müller *et al.* (1968) and Poole *et al.* (1971) have been described previously (Section II,B).

I. Other Subcellular Components

Lysosomes have been prepared by zonal centrifugation (Poole *et al.*, 1971), but better purity will be required for detailed biochemical analysis.

Weidenbach (1973) attempted to establish a procedure for the isolation of food vacuoles by using ferric oxide particles.

IV. Membrane Lipid Composition of Various Isolated Subcellular Components

It is well known that functionally distinct membranes within a cell have markedly different lipid compositions. The localization of certain lipids in particular membranes is especially striking in *T. pyriformis*, as presented in Table I. It is of great interest to note that the phospho lipids containing direct a carbon–phosphorus bond and the triterpene alcohol tetrahymanol are localized in surface membranes such as those of cilia and pellicles.

TABLE I

PHOSPHOLIPID COMPOSITION OF VARIOUS SUBCELLULAR COMPONENTS FROM *T. pyriformis* CELL[a]

Cell fraction	C—P bond (% of lipid phosphorus)	Total phospholipids (mole %)[b]						Tetrahymanol (moles/mole lipid phosphorus)	Glyceryl ethers (moles/100 moles of lipid phosphorus)
		LysoPC	PC	LysoAEPL and LysoPE	PE	AEPL	Cl		
Whole cells	29	2	33	0	37	23	5	0.057	29.7
		17.6[c]	21.2[c]		19.v[c]	31.7[c]	8.7[c]		
Cilia	67	1	28	9	11	47	1	0.30	52.6
		20.2[c]	8.6[c]		14.7[c]	56.2[c]	0[c]		
Ciliary supernatant	44	8	19	13	16	35	1	0.16	23.1
Pellicles	42	5	25	3	34	30	2	0.084	32.8
Mitochondria	26	2	35	0	35	18	10	0.048	24.7
		17.1[c]	30.2[c]		19.4[c]	16.1[c]	16.2[c]		
Nuclear membranes[d]	—	6	31	6	26	23	3	0.036	—
Microsomes	33	1	35	3	34	23	1	0.041	18.3
Postmicrosomal supernatant	26	5	34	4	30	22	2	0.016	27.4

[a]All data from Nozawa and Thompson (1971a) except as otherwise noted.
[b]PC, Phosphatidylcholine; PE, phosphatidylethanolamine; AEPL, 2-aminoethylphosphonolipids; Cl, cardiolipin.
[c]Data from Jonah and Erwin (1971).
[d]Data from Nozawa *et al.* (1973).

REFERENCES

Anderson, N. G., Waters, D. A., Fisher, W. D., Cline, G. E., Nunley, C. E., Flord, L. H., and Rankin, C. T. (1967). *Anal. Biochem.* **21**, 235–252.

Argetsinger, J. (1965). *J. Cell. Biol.* **24**, 154–157.

Blum, J. J. (1971). *J. Theory. Biol.* **33**, 257–263.

Byfield, J. E. Chou, S. C., and Scherbaum, O. H. (1962). *Biochem. Biophys. Res. Commun.* **9**, 226–230.

Byfield, J. E., Lee, Y. C., and Bennett, L. R. (1969). *Biochem. Biophys. Res. Commun.* **37**, 806–812.

Chi, J. C. H., and Suyama, Y. (1970). *J. Mol. Biol.* **53**, 531–556.

Child, F. M. (1959). *Exp. Cell Res.* **18**, 258–267.

Conklin, K. A., and Chou, S. C. (1972). *Comp. Biochem. Physiol*, **41**, 45–54.

Curgy, J. J., Ledoigt, G., Stevens, B. J., and Audié, J. (1974). *J. Cell Biol.* **60**, 628–640.

Engberg, J., and Pearlman, R. E. (1972). *Eur. J. Biochem.* **26**, 393–400.

Everhart, L. P. (1972). *In* "Methods in Cell Physiology" (D. M. Prescott, ed.), Vol. 5, pp. 219–288. Academic Press, New York.

Flavell, R. A., and Jones, I. G. (1970). *Biochem. J.* **116**, 811–817.

Franke, W. W. (1967), *Z. Zellforsch. Mikrosk. Anat.* **80**, 585–593.

Gibbons, B. M., and Gibbons, I. R. (1972). *J. Cell Biol.* **54**, 75–97.

Gibbons, I. R. (1963). *Proc. Nat. Acad. Sci. U.S.* **50**, 1002–1010.

Gibbons, I. R. (1965). *J. Cell Biol.* **25**, 400–402.

Gorovsky, M. A. (1970). *J. Cell Biol.* **47**, 619–630.

Hartman, H., and Dowben, R. M. (1970). *Biochem. Biophys. Res. Commun.* **40**, 964–967.

Hartman, H., Moss, P., and Gurney, T. (1972). *J. Cell. Biol.* **55**, 107a.

Hoffman, E. J. (1965). *J. Cell Biol.* **25**, 217–228.

Hokoma, Y., Nishimura, E. T., and Chou, S. C. (1971). *Res. Commun. Chem. Pathol. Pharmacol.* **2**, 899–918.

Jonah, M., and Erwin, J. A. (1971). *Biochim. Biophys. Acta* **231**, 80–92.

Kobayashi, S. (1965). *J. Biochem. (Tokyo)* **58**, 444–457.

Kumar, A. (1969). *Biochim. Biophys. Acta* **186**, 326–331.

Kumar, A (1970). *J. Cell. Biol.* **45**, 623–634.

Lee, Y. C., and Byfield, J. E. (1970). *Biochemistry* **9**, 3947–3959.

Lee, Y. C., and Scherbaum, O. H. (1966). *Biochemistry* 5, 2067–2075.

Leick, U., and Plessner, P. (1968). *Biochim. Biophys. Acta* **169**, 398–408.

Levy, M. R. (1970). *Biochem. Biophys. Res. Commun.* **39**, 1–6.

Lloyd, D., Evans, D. A., and Venables, E. (1968). *Biochem. J.* **109**, 897–907.

Lloyd, D., Brightwell, R., Venables, S. E. Roach, G. I., and Turner, G. (1971). *J. Gen. Microbiol.* **65**, 209–223.

Mita, T., Shiomi, H., and Iwai, K. (1966). *Exp. Cell Res.* **43**, 696–699.

Müller, M., Hogg, J. F., and de Duve, C. (1968). *J. Biol. Chem.* **243**, 5385–5395.

Muramatsu, M. (1970). *In* "Methods in Cell Physiology" (D. M. Prescott, ed.), Vol. 4, pp. 195–230. Academic Press, New York.

Nozawa, Y., and Thompson, G. A. (1971a). *J. Cell Biol.* **49**, 712–721.

Nozawa, Y., and Thompson, G. A. (1971b). *J. Cell Biol.* **49**, 722–730.

Nozawa, Y., and Thompson, G. A. (1972). *Biochim. Biophys. Acta* **282**, 93–104.

Nozawa, Y., Fukushima, H., and Iida, H. (1973). *Biochim. Biophys. Acta* **318**, 325–344.

Nozawa, Y., Iida, H., Fukushima, H., Ohki, K., and Ohnishi, S. (1974). *Biochim. Biophys. Acta* **367**, 134–147.

Poole, R. K., Nicholl, W. G., Turner, G., Roach, G. I., and Lloyd, D. (1971). *J. Gen. Microbiol.* **67**, 161–173.

Prescott, D. M., Rao, M. V. N., Evenson, D. P., Stone, G. E., and Thrasher, J. D. (1966). *In* "Methods in Cell Physiology" (D. M. Prescott, ed.), Vol. 2, pp. 131–142. Academic Press, New York.
Raff, F. C., and Blum, J. J. (1966). *J. Cell Biol.* **31**, 445–453.
Raff, E. C., and Blum, J. J. (1969). *J. Biol. Chem.* **244**, 366–376.
Rannestad, J., and Williams, N. E. (1971). *J. Cell Biol.* **50**, 709–720.
Renaud, F. L., Rowe, A. J., and Gibbons, I. R. (1968). *J. Cell Biol.* **36**, 79–90.
Rosenbaum, J. L., and Carlson, K. (1969). *J. Cell Biol.* **40**, 415–425.
Rubin, R. W., and Cunningham, W. P. (1973). *J. Cell Biol.* **57**, 601–612.
Satir, B., and Rosenbaum, J. L. (1965). *J. Protozool.* **12**, 397–405.
Schwab-Stey, H., Schwab, D., and Krebs, W. (1971). *J. Ultrastruct. Res.* **37**, 82–93.
Seaman, G. R. (1960). *Exp. Cell Res.* **21**, 292–302.
Smith, J. D., and Law, J. H. (1970). *Biochim. Biophys. Acta* **202**, 141–152.
Stephens, R. E. (1970). *J. Mol. Biol.* **47**, 353–363.
Subbaiah, P. V., and Thompson, G. A. (1974). *J. Biol. Chem.* **249**, 1302–1310.
Suyama, Y. (1966). *Biochemistry* **5**, 2214–2221.
Suyama, Y., and Preer, J. R. (1965). *Genetics* **52**, 1051–1058.
Thompson, G. A., Bambery, R. J., and Nozawa, Y. (1971). *Biochemistry* **10**, 4441–4447.
Thompson, G. A., Bambery, R. J., and Nozawa, Y. (1972). *Biochim. Biophys. Acta* **260**, 630–638.
Thompson, G. A., Baugh, L. C., and Walker, L. F. (1974). *J. Cell Biol.* **61**, 253–257.
Watson, M. R., and Hopkins, J. M. (1962). *Exp. Cell Res.* **28**, 280–295.
Weidenbach, A. (1973). Ph.D. Dissertation, University of Texas, Austin.
Westergaard, O., Marcker, K. S., and Keiding, J. (1970). *Nature (London)* **227**, 708–710.
Whitson, G. L., Padilla, G. M., Canning, R. E., Cameron, I. L., Anderson, N. G., and Elrod, L. H. (1966a). *Nat. Cancer Inst., Monogr.* **21**, 317–321.
Whitson, G. L., Padilla, G. M., and Fisher, W. D. (1966b). *Exp. Cell Res.* **42**, 438–446.
Williams, N. E., and Zeuthen, E. (1966). *C. R. Trav. Lab. Carlsberg* **35**, 101–118.
Wolfe, J. (1970). *J. Cell. Sci.* **6**, 67–700.
Wunderlich, F. (1969). *Exp. Cell Res.* **56**, 369–374.

Chapter 7

Isolation and Experimental Manipulation of Polytene Nuclei in Drosophila

JAMES B. BOYD

Department of Genetics,
University of California,
Davis, California

I. Introduction . . . 135
II. Isolation of Nuclei and Nucleoli . . . 136
A. Small-Scale Nuclear Isolation . . . 137
B. Large-Scale Nuclear Isolation . . . 137
C. Nonaqueous Preparation of Polytene Nuclei . . . 138
D. Isolation of Nucleoli . . . 140
III. Nuclear Purification in Sucrose Gradients . . . 140
IV. Cytological Analysis of Isolated Nuclei . . . 141
V. Manipulation and Incubation of Polytene Nuclei . . . 141
A. Manipulation . . . 141
B. RNA Synthesis in Isolated Polytene Nuclei . . . 142
C. DNA Synthesis in Isolated Polytene Nuclei . . . 143
VI. Related Studies . . . 145
References . . . 145

I. Introduction

Isolated nuclei offer unique advantages for the investigation of chromosome metabolism. Many nuclear functions appear to be sensitive to mechanical disruption and are either distorted or absent in cellular homogenates (Goulian, 1971). In isolated nuclei, however, the integrity of chromosome metabolism can be maintained under the proper incubation conditions. Purified nuclei therefore permit the investigation of normal nuclear metabolism in the absence of complicating cytoplasmic influences. The use of polytene nuclei in such studies further increases the potential of this approach. Since the heterochromatic portion of the diploid genome is grossly underrepresented in polytene nuclei (Rudkin, 1972), this material is particularly suited for studying the structure and function of euchromatin. In addition, polytene chromosomes represent an extensive magnification of the euchromatic portion of the genome. As such, they permit the detailed localization of sites of transcription, or puffs, which usually arise from single chromosomal bands (Beermann, 1967). Since a single band houses one essential genetic function (Beermann, 1972), the puffing phenomenon makes it possible to monitor the activity of individual functional units of the genome. Furthermore, the effect of altering the activity of specific genetic functions can be studied routinely in this material, because the puffing pattern itself is readily subjected to experimental manipulation (Berendes, 1972). The extensive size of polytene chromosomes has also permitted the identification and investigation of multiple sites of DNA synthesis (Rudkin, 1972).

These unique cytological properties of polytene chromosomes have stimulated efforts to procure the quantities required for biochemical investigations. The possibility of pursuing large-scale studies with polytene chromosomes was realized with the development of methods for the mass isolation of salivary glands from dipteran larvae (Ristow and Arends, 1968; Boyd *et al.*, 1968; Zweidler and Cohen, 1971). Although a salivary gland from *Drosophila hydei*, for example, weighs only about 0.1 mg, it is now feasible to obtain gram quantities of this tissue on a routine basis (Boyd, 1975).

This chapter documents a few relatively simple procedures for the isolation and incubation of polytene nuclei from salivary glands of *D. hydei*. These methods were originally developed from a procedure described by Prescott *et al.* (1966). Our methods also rely heavily on procedures developed by Ristow and Arends (1968) for the isolation of polytene nuclei from salivary glands of *Chironomus*. Procedures for the mass isolation of polytene nuclei from salivary glands of *Drosophila melanogaster* have been described by Zweidler and Cohen (1971) and Cohen and Gotchel (1971).

II. Isolation of Nuclei and Nucleoli

All nuclear isolation procedures are carried out at 0–4°C with either siliconized glass or plastic equipment. The glassware is treated with 1 *N* NaOH for 30 minutes prior to treatment with Siliclad (Clay Adams, Parsippany, N. J.).

A. Small-Scale Nuclear Isolation

Small amounts of nuclei can be prepared quickly by repeated passage of 100 to 1000 salivary glands through the tip of a pipette. The product of this procedure consists of about 95% polytene nuclei.

1. Suspend isolated glands in 1–5 ml of nuclear isolation buffer [0.11 *M* NaCl, 0.002 *M* KCl, 0.01 *M* tris(hydroxymethyl)aminomethane, 0.0025 *M* $MgCl_2$ at pH 7.2 (20°C)] containing 0.001% spermidine.
2. Draw the suspension in and out of a Pasteur pipette until each gland has been reduced to small fragments.
3. Replace the pipette with one whose tip has been slightly constricted in a flame and bring the solution to 0.2% Triton X-100. Continue pipetting until about 95% of the nuclei are free in solution. Steady, even pressure with the pipette over a period of about 5 minutes is necessary to achieve optimum results. Care must be taken to avoid placing the tip of the pipette directly on the bottom of the tube. If pipetting is too vigorous, the free nuclei will rupture and the chromatin will immediately be drawn into fibers which appear as long, black threads when observed with low-power phase-contrast optics. As soon as fibers begin to appear, pipetting is terminated, even though some gland fragments remain intact.
4. Pour the nuclear suspension through a nylon screen (Nitex, 53-μm opening, Tet/Kressilk, Monterey Park, Calif.) previously saturated with isolation buffer. The screen is held taut in a round two-piece glass funnel (diameter 2.4 cm) which provides a reservoir above the screen and a funnel beneath.
5. Rinse the tube and pipette with nine additional volumes of isolation buffer. Pour this gently over the screen in order to free nuclei trapped among the gland ducts and membranes.
6. Spin the combined filtrate and rinse in chilled adapters of a tabletop centrifuge. A force of 40 *g* for 5 minutes pellets the dense polytene nuclei but leaves most of the cellular debris and any small nuclei in suspension.

B. Large-Scale Nuclear Isolation

The above procedure becomes tedious and inefficient for processing more than 0.2 gm of glands. To obtain nuclei from gram quantities of glands, we employ a procedure similar to that of Helmsing and Van Eupen (1973).

1. Allow isolated glands to settle through insect Ringer's (Hennig, 1972) to the bottom of a loose (A clearance) Dounce homogenizer (20-ml stem

capacity, Kontes Glass Co. Vineland, N.J.). Replace the Ringer's with isolation buffer containing 0.001% spermidine and 0.2% Triton X-100.

2. Homogenize the glands with slow, complete strokes in 5 ml of the isolation buffer per gram of glands. Allow the gland fragments to settle; transfer the supernatant, which contains about one-half of the nuclei, to a separate tube.

3. Repeat the entire operation with the remaining tissue fragments. Combine the nuclear suspensions and bring them to 0.5% Triton X-100.

4. Establish the nuclear concentration with a hemacytometer after shaking the solution vigorously for 90 seconds.

5. Filter the nuclear suspension through a 53-μm screen supported by a two-piece glass funnel (5.5-cm diameter) into 50-ml conical centrifuge tubes.

6. Wash the screen gently to a final volume of 100 ml per gram of glands with isolation buffer containing 0.5% detergent.

The product obtained by centrifugation of the nuclei at 40 *g* for 5 minutes contains 80–90% polytene nuclei (Fig. 1a). Short sections of the chitinous gland duct and the tough basement membrane of the gland make up a major portion of the contamination. The nuclear yield, based on a value of 131 nuclei per gland (Berendes, 1965), is 80–90%. Because the salivary glands are fragile and their nuclei so large, it is not necessary to start the preparation with absolutely pure gland samples. Other tissues are not as easily disrupted by these procedures, nor do their nuclei sediment as rapidly as large salivary gland nuclei.

C. Nonaqueous Preparation of Polytene Nuclei

Although this procedure has not been developed as extensively as the aqueous methods, it is outlined here because of the potential importance of nuclei prepared in this way (Busch, 1967). The product of this procedure is enriched in nuclei that exhibit normal polytene chromosome morphology following hydration in isolation buffer. The retention of morphologically normal chromosomes in lyophilized glands makes long-term storage and shipment of this material feasible.

1. Pellet the salivary glands in Ringer's and completely remove the supernatant.

2. Freeze the tissue rapidly by pouring liquid nitrogen directly into the tube, while placing the tube itself in a Dry Ice–acetone bath.

3. Lyophilize the tissue from a frozen state, suspend it in dried petroleum ether (boiling point 30°–60°C), and homogenize it in a loose glass-Teflon homogenizer.

4. Suck off the petroleum ether after centrifugation for 10 minutes at 500 *g* and allow the remaining solvent to evaporate.

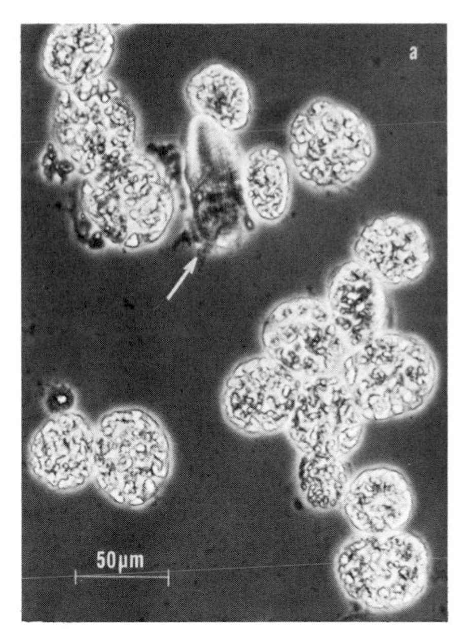

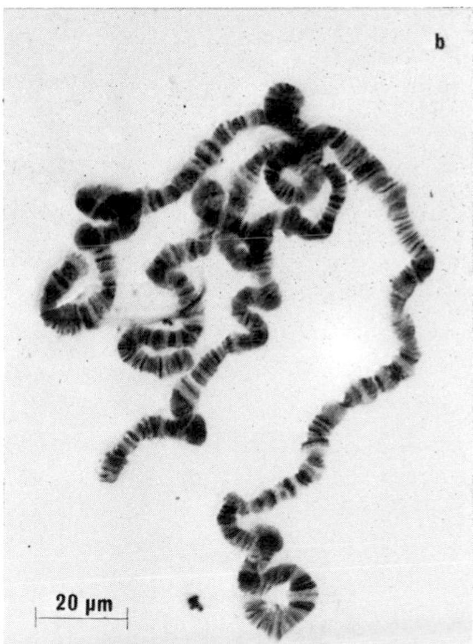

FIG. 1. Isolated polytene nuclei. (a) Phase-contrast micrograph of nuclei prepared according to Section II, B of the text. A section of contaminating salivary gland duct is indicated by arrow. (b) Chromosome morphology following *in vitro* incubation of isolated polytene nuclei. Nuclei were isolated according to Section II, B, incubated according to Section V, B, and prepared for bright-field observation as described in Section V.

5. Suspend the tissue in roughly equal volumes of dried, cold benzene and carbon tetrachloride and centrifuge at 12,000 *g* for 15 minutes. The nuclear pellet can be dried and hydrated if desired.

D. Isolation of Nucleoli

The details of this procedure were provided by H. D. Berendes. A similar protocol has been employed successfully in the author's laboratory. Salivary glands are homogenized in 5 ml of buffer-A (0.08 *M* NaCl; 0.005 *M* KCl; 0.01 *M* $MgCl_2$; 0.005 *M* PIPES and 1% NPT, pH 6.8–7.2), in an all-glass homogenizer. The homogenate is filtered through nylon gauze with a pore diameter of 80 μm. The gauze is rinsed with 5 ml of buffer A without NPT. The filtrate is centrifuged for 5 minutes (750 *g*) in a polyallomer tube. The pellet is suspended in 2 ml of buffer A without NPT and sonicated 3 × 1 second (Branson Sonifier, step 2). The sonicate is filtered through gauze with a pore diameter of 31 μm. The gauze is rinsed with 8 ml of buffer A without NPT and the filtrate centrifuged for 10 minutes at 900 *g*. The pellet is resuspended in 1 ml of buffer A without NPT and filtered again through gauze with a pore diameter of 31 μm and centrifuged. The pellet consists of clean nucleoli as judged by light microscope observation.

III. Nuclear Purification in Sucrose Gradients

Certain analytical approaches require a nuclear purity beyond that obtainable with the procedures described above. For this purpose discontinuous sucrose gradients are employed to produce highly purified nuclei (Elgin and Boyd, 1975).

1. Thoroughly suspend the nuclear pellet obtained in the last step of the aqueous isolation procedures in isolation buffer containing 1.67 *M* sucrose (15 ml per gram of glands).
2. Layer 2 ml of isolation buffer containing 2.3 *M* sucrose under 10-ml aliquots of the nuclear suspension in polyallomer test tubes.
3. Mix the interface between the two sucrose layers slightly and centrifuge the tubes in the SW40 rotor of a Beckman L2-65B for 42 minutes at 40,000 rpm and 4°C.
4. Suspend the nuclear pellet by vigorous agitation and free the remaining nuclei by scraping the bottom of the tube with a spatula.

IV. Cytological Analysis of Isolated Nuclei

The following method was developed by Dr. W. Beermann for preparing chromosome squashes from isolated nuclei. The entire operation is monitored continuously with a binocular microscope.

1. Allow nuclei contained in a drop of isolation buffer to settle to the surface of a clear microscope slide over a period of about 5 minutes.
2. Carefully draw off the excess fluid from one side of the drop with the aid of filter paper. The flow rate should be slow enough to permit the majority of the nuclei to remain in place.
3. Fix the damp nuclei to the slide by dropping a mixture of ethanol and acetic acid (3:1) directly on them from a height of about 2 cm. Alternatively, if the nuclei float away with this procedure, they can be caused to clump and stick to the slide by quickly surrounding the aqueous drop with a ring of fixing solution.
4. Remove the excess fixing solution with a piece of filter paper, and replace it with acetocarmine (Reflux 1 gm of carmine in 100 ml of 50% acetic acid for 5 hours).
5. After 5 minutes replace the stain with a drop of 45% acetic acid and gently lower a siliconized cover slip onto the preparation. Excess fluid is removed by gently stroking a clean piece of filter paper that has been placed on the preparation.
6. Sharply tap the cover slip over individual nuclei with the tip of a scalpel while holding the cover slip rigidly in place. Tapping is continued over a single nucleus until the chromosome arms are well spread. Since the chromosomes are more difficult to spread from isolated nuclei than from salivary glands, great care must be taken to leave the proper amount of solution under the cover slip. Excess fluid permits the nuclei to float away, and with too little fluid the chromosomes shatter.
7. Finally, flatten the chromosomes into one plane for observation with phase-contrast optics by applying pressure to the preparation through a piece of filter paper.

V. Maniputation and Incubation of Polytene Nuclei

A. Manipulation

Nuclei isolated by aqueous procedures have been shown to retain the normal chromosome banding pattern and the acidic protein associated with chromosomal puffs (Berendes and Boyd, 1969). They also demonstrate an apparently normal capacity for RNA synthesis (Berendes and Boyd, 1969) and DNA synthesis (Boyd and Presley, 1973). These properties can be exploited in metabolic studies by first suspending the pellet obtained from the aqueous isolation procedure in a small volume of incubation buffer. Gentle, repeated pipetting with an Eppendorf pipette, whose tip has been been cut back to provide a larger orifice, produces a uniform suspension which can be quantitatively distributed for incubation. During subsequent incubation the nuclei are resuspended periodically, because they settle rapidly from solution.

The most serious problem encountered in manipulating polytene nuclei isolated with nonionic detergents is their tendency to clump and stick. The problem can only be partially reduced by employing buffers containing detergent, bovine serum albumin, glycerin (Boyd and Presley, 1973), or low pH (Lezzi and Robert, 1972). The most satisfactory solution is to retain the nuclei in relatively small volumes of buffer after the glands have been disrupted. This limitation makes it difficult to obtain completely pure preparations without the use of sucrose gradients, because extensive dilution results in clumping and sticking of the nuclei to the glassware. For this reason resuspension and centrifugation of the original nuclear pellet in large volumes of buffer should be avoided if the isolated nuclei are to be freely suspended. Nuclei exposed to ionic detergents are not as subject to sticking (Ristow and Arends, 1968; Cohen and Gotchel, 1971; Hennig, 1972). Fortunately, the type of contamination associated with standard nuclear preparations is not likely to interfere with metabolic studies.

B. RNA Synthesis in Isolated Polytene Nuclei

For studies of RNA synthesis incubation conditions have been defined that both optimize certain forms of RNA synthesis and permit the retention of normal chromosome morphology (Fig. 1b). Incorporation of RNA precursors into this system is inhibited by actinomycin D and by the absence of unlabeled precursors in the incubation medium (Berendes and Boyd, 1969). The relative amounts of RNA produced at temperature-induced puffs *in vitro* is considerably higher than that observed *in vivo* (Fig. 2). Contrary to an earlier report (Berendes and Boyd, 1969), exogenous RNA polymerase is not required for the production of this specific labeling pattern. RNA synthesis performed in the presence of a buffer designed to optimize DNA synthesis (see Section V, C) results in a more normal distribution of RNA synthesis, with extensive synthesis occurring in the nucleolus. The apparent capacity to suppress the majority of RNA synthesis not associated with temperature-induced puffs makes this system a source of RNA produced primarily at five

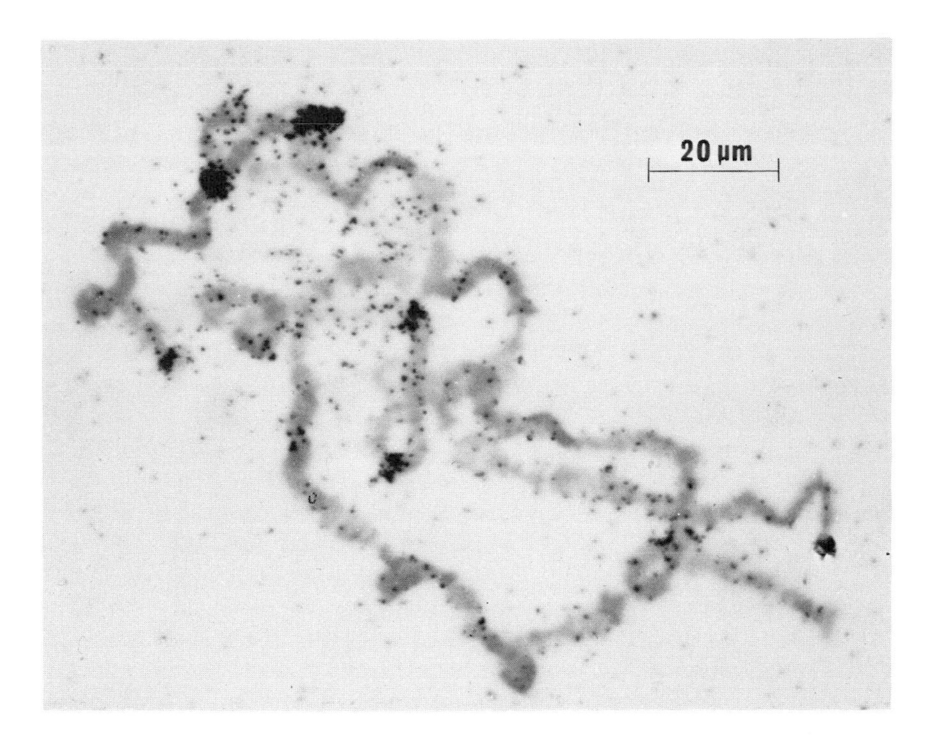

FIG. 2. RNA synthesis in isolated polytene nuclei. Autoradiographs were prepared from nuclei incubated as described in Section V, B. Cytological preparations were dipped in liquid emulsion and exposed for 2 weeks. For the method, see Boyd and Presley (1973).

chromosomal loci. RNA with a specific activity in excess of 10^3 cpm/μg has been recovered from 5×10^3 nuclei incubated under these conditions (Presley *et al.*, 1975).

C. DNA Synthesis in Isolated Polytene Nuclei

Autoradiographic studies of DNA synthesis in *Drosophila* salivary glands have demonstrated the existence of three nuclear labeling patterns: (1) intensively labeled nuclei with continuously labeled chromosomes, (2) moderately labeled nuclei with discontinuously labeled chromosomes, and (3) unlabeled nuclei (Plaut, 1969). This same distribution of label is exhibited by isolated polytene nuclei incubated *in vitro* in the presence of DNA precursors (Fig. 3a). The heavily labeled chromosomes in such a preparation have been shown to be continuously labeled, and many of the moderately labeled nuclei exhibit a discontinuous labeling pattern. In addition, the

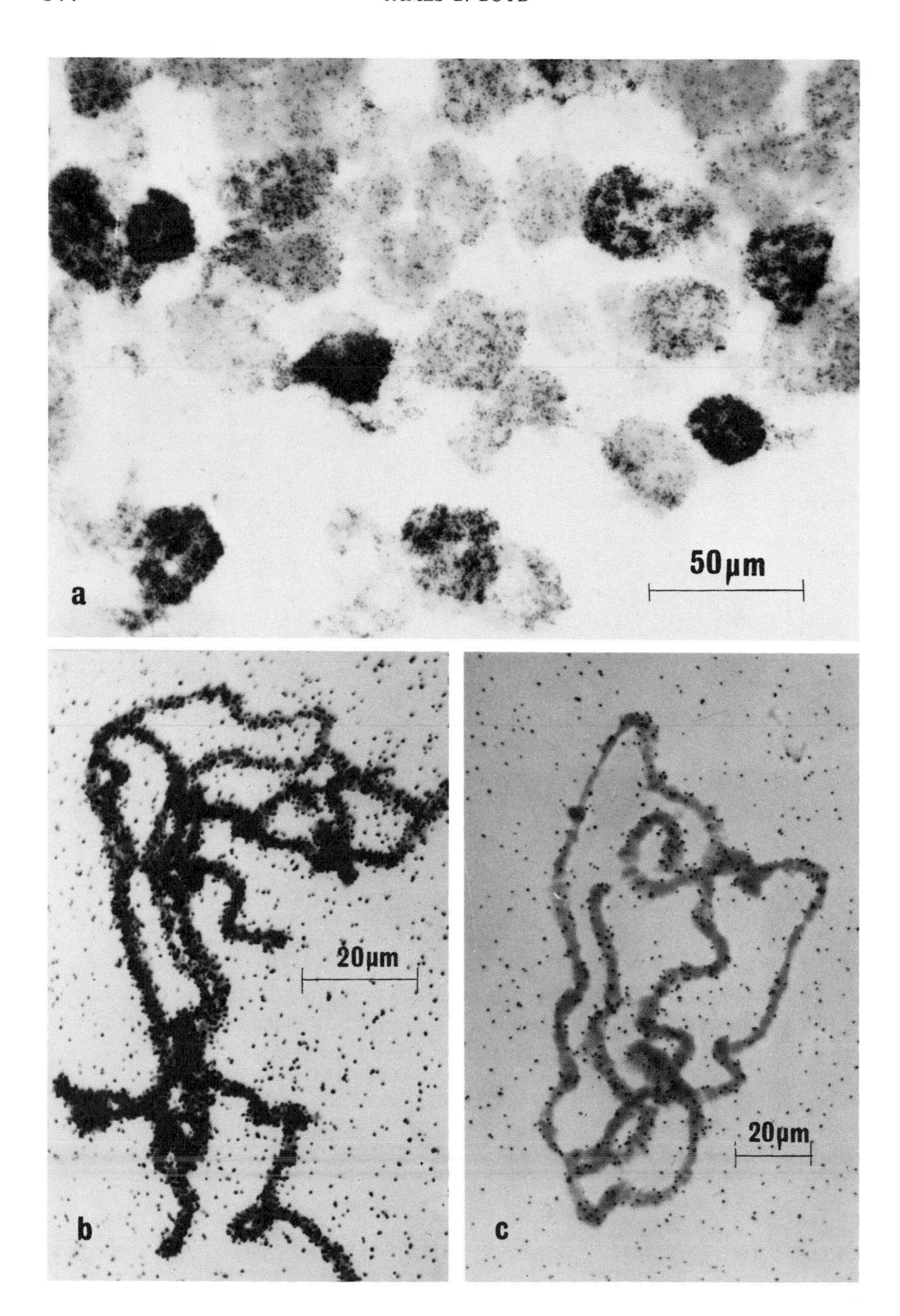
50μm
a
20μm
b
20μm
c

isolated nuclei possess a reserve capacity for DNA synthesis. What is presumed to be a form of repair synthesis is detected in isolated nuclei when the glands have been previously irradiated (Fig. 3b and c). Incorporation of both RNA and DNA precursors is not detected in about 20% of the isolated nuclei. Whether the same nuclei are inactive with respect to both functions is not known.

VI. Related Studies

Several recent studies involving isolated polytene nuclei in *Drosophila* have been performed in the laboratory of H. D. Berendes (Helmsing, 1970; Alonso, 1972; Helmsing and Berendes, 1971). Isolated polytene nuclei have also been employed in studies of hormone binding proteins (Emmerich, 1972). Detailed studies performed with isolated polytene nuclei and chromosomes of *Chironomus* have made important contributions to this field (Kroeger, 1966; Ristow and Arends, 1968; Lezzi and Robert, 1972).

Acknowledgments

Earlier versions of the procedures described here were developed in collaboration with Dr. H. D. Berendes, Dr. H. Ristow, and Prof. W. Beermann. Current operational procedures were worked out with J. M. Presley and J. Chelseth. Charles Cooper skillfully prepared the figures. Support for these studies was provided by the Max-Planck-Institut für Biologie, The Helen Hay Whitney Foundation, National Institutes of Health Grant GM-16298, and National Science Foundation Grant GB-37637.

References

Alonso, C. (1972). *Develop. Biol.* **28**, 372–381.

Beermann, W. (1967). *In* "Heritage from Mendel" (R. A. Brink, ed.), pp. 179–201. Univ. of Wisconsin Press, Madison.

FIG. 3. DNA synthesis in isolated polytene nuclei. (a) Typical distribution of DNA synthesis following *in vitro* incubation of polytene nuclei as described in Section V, C. Nuclei were incubated 30 minutes at 28°C in the presence of 20 m*M* NaCl, 12 m*M* tris, 0.4 m*M* KCl, 4.2 m*M* $MgCl_2$, 0.4 m*M* $(NH_4)_2SO_4$; 0.2 m*M* mercaptoethanol, 50 μCi/ml TTP-^{3}H; 0.6 mg/ml each of dGTP, dCTP, and dATP. Autoradiographs prepared with Kodak NTB-3 were exposed for 10 months. (b) Induction of unscheduled DNA synthesis in isolated nuclei by prior irradiation of salivary glands with 90 kR of x rays (Boyd and Presley, 1973). Typical nucleus from irradiated glands. Isolated glands were irradiated in Ringer's containing 5 μg/ml actinomycin D and 10^{-5} *M* FUdR (Valencia and Plaut, 1969). Isolated nuclei were incubated in the presence of 44 m*M* NaCl, 14 m*M* tris, 4.4 m*M* $MgCl_2$, 0.5 m*M* KCl, 5% (v/v) glycerin, 0.25 mg/ml BSA, 0.12 mg/ml each of dGTP, dCTP, and dATP, and 66 μCi/ml TTP-^{3}H. Autoradiographs were exposed for 5 weeks. (c) Typical nucleus from unirradiated glands.

Beermann, W. (1972). *In* "Results and Problems in Cell Differentiation" (W. Beermann, J. Reinert, and H. Ursprung, eds.), Vol. 4, pp. 1–33. Springer-Verlag, Berlin and New York.
Berendes H. D. (1965). *Chromosoma* **17**, 35–77.
Berendes, H. D. (1972). *In* "Results and Problems in Cell Differentiation" (W. Beermann, J. Reinert, and H. Ursprung, eds.), vol. 4. pp. 181–207. Springer-Verlag, Berlin and New York.
Berendes, H. D., and Boyd, J. B. (1969). *J. Cell Biol.* **41**, 591–599.
Boyd, J. B. (1975). *In* The Genetics and Biology of *Drosophila*" (T. R. F. Wright and M. Ashburner, eds.), Vol. 2. Academic Press, New York (in press).
Boyd, J. B., and Presley, J. M. (1973). *Biochem. Gene.* **9**, 309–325.
Boyd, J. B., Berendes, H. D., and Boyd, H. (1968). *J. Cell Biol.* **38**, 369–376.
Busch, H. (1967). *In* "Methods in Enzymology" (L. Grossman and K. Moldave, eds.), Vol. 12, Part A, pp. 439–466. Academic Press, New York.
Cohen, L. H., and Gotchel, B. V. (1971). *J. Biol. Chem.* **246**, 1841–1848.
Elgin, S. C. R. and Boyd, J. B. (1975). In preparation.
Emmerich, H. (1972). *Gen. Comp. Endocrinol.* **19**, 543–551.
Goulian, N. (1971). *Annu. Rev. Biochem.* **40**, 855–898.
Helmsing, P. J. (1970). *Biochim. Biophys. Acta* **224**, 579–587.
Helmsing, P. J., and Berendes, H. D. (1971). *J. Cell. Biol.* **50**, 893–896.
Helmsing, P. J., and Van Eupen, O. (1973). *Biochim. Biophys. Acta* **308**, 154–160.
Hennig, W. (1972) *J. Mol. Biol.* **71**, 419–431.
Kroeger, H. (1966). *Methods Cell Physiol.* **2**, 61–92.
Lezzi, M., and Robert, M. (1972). *In* "Results and Problems in Cell Differentiation" (W. Beermann, J. Reinert, and H. Ursprung, eds.), Vol. 4, pp. 35–57, Springer-Verlag, Berlin and New York.
Plaut, W. (1969). *Genetics* **61**, Suppl., 239–244.
Prescott, D. M., Rao, M. V. N., Evenson, D. P., Stone, G. E., and Thrasher, J. D. (1966). *Methods Cell Physiol.* **2**, 131–142.
Presley, J. M., Laird, C. D., and Boyd, J. B. (1975). In preparation.
Ristow, H., and Arends, S. (1968). *Biochim. Biophys. Acta* **157**, 178–186.
Rudkin, G. T. (1972). *In* "Results and Problems in Cell Differentiation" (W. Beermann, J. Reinert, and H. Ursprung, eds.), Vol. 4, pp. 59–85. Springer–Verlag, Berlin and New York.
Valencia, J. I., and Plaut, W. (1969). *J. Cell Biol.* **43**, 151a.
Zweidler A., and Cohen, L. H. (1971). *J. Cell Biol.* **51**, 240–248.

Chapter 8

Methods for Microsurgical Production of Mammalian Somatic Cell Hybrids and Their Analysis and Cloning[1]

ELAINE G. DIACUMAKOS

Laboratory of Biochemical Genetics,
The Rockefeller University,
New York, New York

I. Introduction . . . 147
II. Cell Culture . . . 148
  A. Cell Types . . . 148
  B. Stock and Microcultures . . . 148
III. Cell Fusion . . . 148
  A. Microsurgery . . . 148
  B. Fusion Operation . . . 149
IV. Cell Isolation . . . 151
  A. Bicellular Hybrid Cells of the Same Cell Type . . . 151
  B. Bicellular Hybrid Cells of Different Types or Species . . . 151
V. Analyzing and Cloning Hybrid Cells . . . 152
  A. Analyzing Hybrid Cells . . . 152
  B. Cloning Hybrid Cells . . . 152
VI. Additional Comments . . . 155
  References . . . 156

I. Introduction

Methods are presented for fusing cells by microsurgery (Diacumakos and Tatum, 1972) and for analyzing and cloning the fusion products of mammal-

[1]Work reported has been funded by a Grant from the National Foundation-March of Dimes.

ian somatic cells of the same and different cell types and species (Diacumakos (1973a)). These methods, applied to murine somatic cells, extend the microsurgical methodology evolved for human somatic cells in culture (Diacumakos *et al.*, 1970).

II. Cell Culture

A. Cell Types

Human and murine somatic cells of fibroblastic or epithelioid morphology and of normal or abnormal origin are used.

The human cell types include HeLa clone (HC), human fetal (HF), fibroblasts carrying translocations (T and GM73), Lesch-Nyhan (LN), and diploid (141) cells.

The murine cell types include mouse L (L), fibroblastic (A9 and CL1D), neuroblastoma (N), teratoma (Ter), and Ehrlich ascites (EA) cells.

B. Stock and Microcultures

Stock cultures of the different cell types are maintained in appropriately supplemented medium or without additives in Dulbecco's modification of Eagle's medium supplemented with 10% (v/v) fetal bovine serum (Flow Laboratories, Inc.; Rockville, Md.). The materials and procedures have been described (Diacumakos, 1973b).

Microcultures on cover slips are prepared using no. 0 thinness cover slips 18 mm square. Thinner cover slips than necessary for use with Sykes-Moore chambers (Diacumakos, 1973b) are used to ensure focusing on the cells on cover slips or fragments in plastic petri dishes. Because they are so fragile, it is best to cut them diagonally into two equal triangles before, rather than after, acid washing. Dulbecco's medium supplemented with 10% fetal bovine serum is used for all microcultures.

III. Cell Fusion

A. Microsurgery

The equipment, procedures, and microtools required have been described (Diacumakos, 1973b).

B. Fusion Operation

The basic fusion operation has been reported (Diacumakos and Tatum, 1972). This article describes how it has been modified and extended to more cell types.

1. Cells of the Same Type

Invert the microculture growing on an 18-mm-square cover slip on the microsurgical chamber (Diacumakos, 1973b). Fill the microsurgical chamber with growth medium and seal the four ports with oil (Diacumakos, 1973b). Clean the upper surface of the cover slip. Place the chamber on the stage of the microscope and, following the procedures reported (Diacumakos, 1973b), position the microneedles within the chamber, using the 2.5× objective to view them.

Scribe a circle (Diacumakos, 1973b) on the cover slip using the midpoint of the cover slip as the midpoint of the circle. Inspect it with the 2.5× objective. The perfectly scribed circle will be slightly smaller than the field of the 2.5× objective. Raise the objective and put immersion oil on the cover slip. Lower the oil immersion 100× objective until the cells are in focus and center a cell or its nucleus in the field. Raise the objective.

Bring the 10× objective into position and focus on the cells. Locate the cell that was centered with the 100× objective. Bring the microneedles into position under this cell but do not raise them. Raise the 10× objective, bring the 100× objective into position, and focus on the cells. Locate a pair of cells in telophase and record its position using the microscope stage vernier scales as coordinates.

Locate another pair of cells in telophase. Gently detach this pair by nudging it free of the cover slip or other cells with the shaft, not the tip, of a microneedle. Lower this pair within the chamber and steady it with one or more microneedles. The position of this pair and all microneedles should be below the level of mitotic cells attached to the cover slip (these cells have the largest diameter).

Move the chamber *slowly* to the position recorded for the first telophase pair, centering it in the 100× field. Raise the second telophase pair adjacent to the first and attach it to the cover slip with gentle pressure. Pivot the second so that both pairs lie in a straight line with contact between two cells, i.e., one from each pair.

Raise a clean, abruptly tapered microneedle to touch the surface of one of the contiguous cells lightly. A small amount of hyaline cell matrix (Diacumakos *et al.*, 1972) will emerge at that site and will adhere to the tip of the microneedle. Gently propel the matrix over to the other contiguous cell and far enough into that cell to fuse with its matrix. A minute hyaline bridge will be formed at that moment, the moment of cell fusion.

Observe the continuing fusion of the two cells. Within about 30 minutes, the fusion product is seen with two nuclei and two cytoplasms. This is the bicellular hybrid cell. Observe the two unfused sisters, one on either side of the hybrid cell. Properly done, this operation yields three individual cells—two unfused cells and one bicellular cell hybrid.

To designate this fusion, the cell-type letter (or number) code is used and followed by numerals with a bar above the cells fused and then the date; e.g., LN-$\overline{12}$56 indicates two LN cells, the first and second, fused May 6.

2. Cells of Different Types or Species

The microsurgical chamber should be prepared in a sterile hood. Invert a microculture growing on a triangular cover slip on the microsurgical chamber so that the glass rests on three of the four supports (Diacumakos, 1973b). Fill the covered part of the chamber with growth medium and seal the two ports with oil (Diacumakos, 1973b). Invert the second microculture growing on a triangular cover slip on the other half of the chamber and seal the third and fourth ports with oil after the chamber has been filled with growth medium. Carefully clean the upper surfaces of the two triangles and use sterile immersion oil to seal the diagonal seam between the triangles.

Position the chamber on the microscope stage and center it. The field of the 2.5× objective is large enough to show parts of both microcultures with the gap between them running diagonally in the field. Position the microneedles. Move the chamber so that one of the microcultures is in the field.

Switch to the 100× objective and center a cell as described (Section III,B,1). Then locate a pair of cells in telophase and record its position. Raise the 100× objective and bring the 10× objective into position.

Move the chamber so that the other microculture is brought into view. Raise the 10× objective and lower the 100× objective into position, focusing on the cells. Locate a telophase pair in this population. Gently detach it as described in Section III,B,1 and lower it on one or more microneedles. Raise the 100× objective and lower the 10× objective into position, focusing on the cells.

Move the chamber very slowly back to the recorded position of the first telophase pair. Then check to ensure that the second pair is still on the microneedles, raise the 10× objective, and swing the 100× objective into position, focusing on the first telophase pair.

Raise the second telophase pair into focus and follow the same procedure for fusion (Section II,B,1). Properly done, the operation yields three individual cells—two unfused sisters, one of each cell type or species, and one bicellular hybrid cell made up of two different nuclei and two different types of cytoplasm.

Designate the fusion of two different types or species by using the two cell-type letter or number codes. NTer-1114 designates an N cell, one of a telophase pair attached to the cover slip, and a Ter cell, one of a telophase pair brought to it; the first of these two cell types fused on November 4. The cell-type letter (number) codes always appear with the stationary cell type first, followed by the cell type that was brought to it for fusion. These codes are prefixes for derived cell hybrids, their clones, and derived cell lines.

IV. Cell Isolation

A. Bicellular Hybrid Cells of the Same Cell Type

Bicellular hybrid cells are produced within the scribed circle. Viewing with the 100× objective, isolate the three cells by clearing away all others attached to the cover slip within the circular area and for a short distance beyond (Diacumakos, 1973b). Once this is done, retract the needles from the chamber and transfer the chamber to a sterile hood. Use sterile technique thereafter.

Clean the upper surface of the cover slip with ethanol-moistened and then dry cotton-tipped applicators. Transfer the cover slip cell side up to a petri dish containing Dulbecco's medium supplemented with 20% (v/v) fetal bovine serum. Place the petri dish on the stage of a stereoscopic microscope. With a stainless-steel forceps, anchor the cover slip to the floor of the dish and, with a diamond scriber, draw lines from the edge of the circle to the edges of the cover slip. The circular fragment with the cells to be isolated will break free of the cover slip (Diacumakos, 1973b).

Refilter the same medium through a Swinnex filter (GS, Millipore Corp.; Bedford, Mass.). Transfer the circular cover slip fragment to another petri dish containing 5 ml of this medium. The transfer is facilitated by using two pairs of forceps, one to steady and the other to lift the fragment. Once the fragment is positioned in this dish, use one pair of forceps to press it against the floor of the dish. Place this dish within a larger, dry petri dish and transfer it to a water-jacketed incubator with a 95% air–5% carbon dioxide atmosphere with 100% humidity.

B. Bicellular Hybrid Cells of Different Types or Species

Viewing with the 100× objective, clear an area of the triangular cover slip around the cells to be isolated. Choose two identifiable markers at the cut, diagonal edge of the cover slip. Change to the 10× objective. Identify the markers and the cells at this magnification and form an imaginary triangle using these two markers and the cells as vertices; this facilitates identification during isolation at even lower magnifications. Remove the chamber (Section IV,A) to a sterile hood. Use sterile techniques thereafter.

Clean the upper surface of the triangular cover slip underneath which the cells to be isolated are attached. Place the chamber on the stage of a stereoscopic microscope and relocate the markers and cells before removing the cover slip. Low magnification (18×) of the stereomicroscope will show the cleared area, and then the cells can be located at higher magnifications (72 and 144×). Verify the markers at the edge of the cover slip that were selected at both of the higher magnifications. Remove the triangular cover slip and place it cell side up in a petri dish containing Dulbecco's medium with 20% fetal bovine serum. With a diamond scriber and forceps, cut off a rectangular fragment of the cover slip with the desired cells attached. Follow the procedure as in Section IV,A thereafter.

V. Analyzing and Cloning Hybrid Cells

A. Analyzing Hybrid Cells

Hybrid cells may be isolated along with their unfused sisters or as solitary cells.

Place the three cells apart from all other cells. To separate the hybrid cell from its unfused sisters, either brush one of the unfused sisters off of the glass fragment with a microneedle that is fire-polished but blunt or, while in mitosis, shake it off. Transfer the fragment to another petri dish, following the procedure outlined in Section IV,A or B.

Similarly remove and transfer the other unfused sister to a second petri dish. Then transfer the fragment with the hybrid cell to a third petri dish. The hybrid cell is now in solitary isolation to clone.

Figure 1 shows some variations in hybrid formation that have been encountered during experiments to explore optimal fusion conditions.

B. Cloning Hybrid Cells

Hybrid cells that retain and show the proliferative capacity of either or both starter cells that were fused form clones in 8–9 days. Variations in the hybrid cell's morphology reflect either starter cell's characteristics or combinations of the two. These are illustrated in Fig. 2.

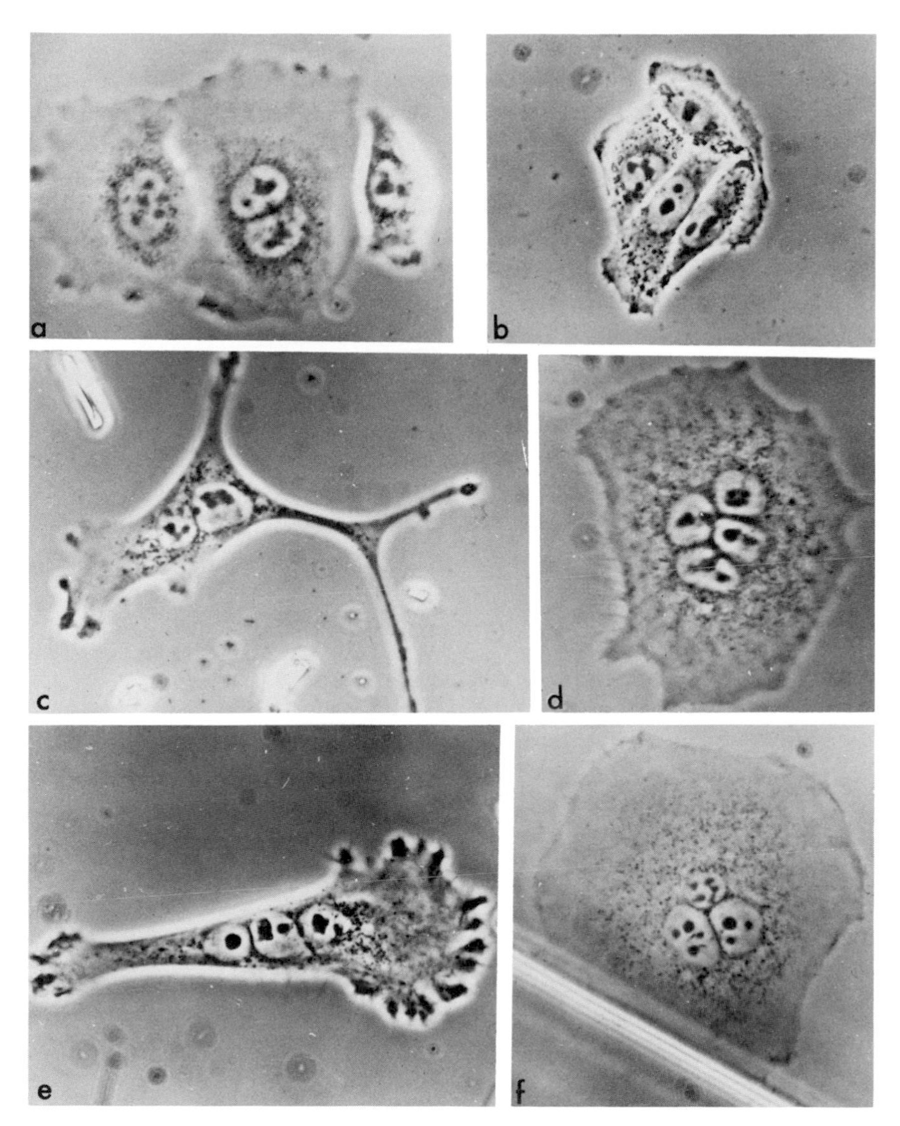

FIG. 1. Some examples of the variations that may occur following microscurgical fusion of cells of the same and different types and species, all ×469. (a) Hybrid HC-$\overline{23}$71272 and the unfused sister cells, one on each side. (b) Four HeLa cells seen from the "fusion" of HC cells 6 and 7 on 7-7-72; the cells did not remain fused. (c) Solitary interspecific hybrid HCL-171472 with one HeLa cell nucleus and one L-cell nucleus with the cytoplasmic complement of the two cells. (d) Hybrid HC-$\overline{1}$24373 in which all four cells have fused. (e) Hybrid LNHC-232273 with two LN nuclei and one HeLa nucleus within the cytoplasmic complement of the three cells. The fibroblastic morphology of the LN cells is predominant. (f) Hybrid HCL-171572 with two HeLa cells and one L cell. The epithelioid morphology of the HeLa cells is predominant. Compare this cell hybrid with (e).

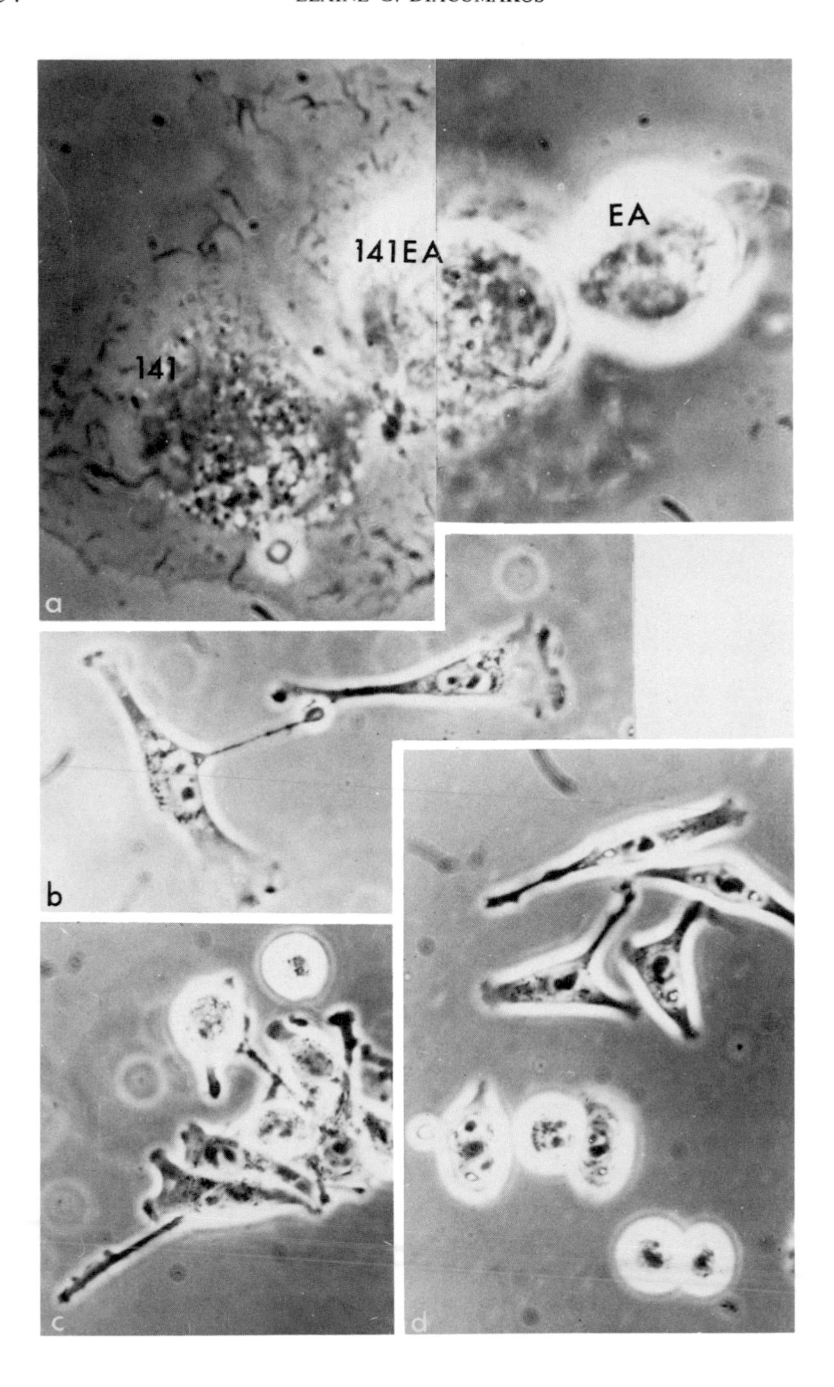
EA
141EA
141
a
b
c
d

Once a clone has formed, i.e., an aggregate of at least 50 cells all derived from solitary hybrid cell, the hybrid cells are transferred to a culture bottle, such as a Falcon flask (no. 3024), with a small amount of Dulbecco's medium supplemented with 10% fetal bovine serum. These cells will, in turn, form small clonal aggregates. Within 2–3 weeks the progeny will be sufficiently numerous to transfer, and a successive transfer schedule can then be initiated.

VI. Additional Comments

It is important to recognize that hybrids produced from preselected starter cells and programmed as described (Diacumakos, 1973a) do not require co-cultivation of different cell populations, nor are selection media necessary.

Each of the starter cells synchronized in telophase represents a reproductive unit of its cell type. Consequently, the variations that occur within the same population of cells derived from fused telophase cells reflect the variable reproduction of cells of that type; likewise, the hybrids that proliferate from that population of cells compound the variation in reproduction of the hybrid cell type.

Acknowledgments

I am grateful to the following investigators for supplying one or more of the cell types used: Drs. R. Cox, J. Jami, L. Ossowski, M. Siniscalco, S. Silverstein, and H. Y. Tan.

I am most grateful to Professor Edward L. Tatum for his helpful discussions.

FIG. 2. The early stages of the evolution of a clone and then a cell line after fusion of an Ehrlich ascites (EA) cell with a normal human fibroblast (141) by microsurgery. Compare (a) and (b) with Fig. 1a and c, respectively. (a) The telophase pair of human (141) cells on the left and the telophase pair of mouse (EA) cells on the right were aligned, and two cells, one from each pair, were fused by a microneedle to form the interspecific somatic cell hybrid (141EA) which had no markers for its selection by other means. The plane of focus is below the level of the cover slip to show the more spherical morphology of the EA cells. $t = 0$. ×1875. (b) The hybrid 141EA gave rise to two mononucleate sister cells at the first cell disjunction (Diacumakos *et al.*, 1972). Both cells were fibroblastic in morphology and attached to the cover slip. $t = 17$ hours ×469. (c) Eight cells have formed in the clonal aggregate; six show fibroblastic and two show the more spherical morphology typical of Ehrlich cells. $t = 66$ hours. × 469. (d) As a result of the two types of attachment, additional cells formed new foci within the dish, and nine of them are seen. $t = 5$ days. ×469. By the sixth day, a clone, i.e., 50 or more cells, all derived from the cell hybrid, 141EA, had formed.

References

Diacumakos, E. G. (1973a). *Proc. Nat. Acad. Sci. U.S.* **70**, 3382.
Diacumakos, E. G. (1973b). *Methods Cell Biol.* **7**, 287.
Diacumakos, E. G., and Tatum, E. L. (1972). *Proc. Nat. Acad. Sci. U.S.* **69**, 2959.
Diacumakos, E. G., Holland, S., and Pecora, P. (1970). *Proc. Nat. Acad. Sci. U.S.* **65**, 911.
Diacumakos, E. G., Holland, S., and Pecora, P. (1972). *Int. Rev. Cytol.* **23**, 27.

Chapter 9

Automated Cell Cycle Analysis

ROBERT R. KLEVECZ

Department of Cell Biology, Division of Biology,
City of Hope National Medical Center,
Duarte, California

I. Introduction 157
Alignment versus Selection Synchrony 158
II. Cell Cycle Analyzer 159
A. Virtues of Automated Synchrony 159
B. Description of the Analyzer 161
C. Operation of the Analyzer 163
III. Optimizing Selection Conditions 164
A. Preparation for Synchrony 164
B. Selection Artifacts 165
C. Yield of Mitotic Cells 165
IV. Calibrating the Cell Cycle 168
V. Limitations and Prospects 171
References 172

I. Introduction

Cells in culture grow randomly, and under the best conditions exponentially. In consequence, all cell products and processes are exponentially expressed, age-averaged values, and even the most obvious events taking place in the individual cell are obscured. In the case in which cells are entering or emerging from confluency, functions are further complicated. Observation and interpretation are simplified when cell cultures of uniform age are used.

Although the results obtained from cell cycle studies are free from the ambiguities of ordinary cell cultures, most work with mammalian cells is still done using random cultures. This must be due in part to the difficulty with which sufficient quantities of adequately synchronized and metabolically unperturbed cells are obtained, and in part to the tedious, often night-long vigils of sample harvesting. This report describes an instrument that automatically stages cells through their cycle and which hopefully can be used to resolve the temporal structure of cells in the same way the electron microscope and the ultracentrifuge have resolved their spatial structure.

A. Alignment versus Selection Synchrony

In establishing synchrony in mammalian cells there are, or should be, two major considerations. One obviously is the purity with respect to age distribution of the starting synchronized cell population. The second deals with perturbations in cellular metabolism emanating from the synchronization process. Selection synchrony, in which cells of a particular age are separated from the remainder of the population on the basis of distinguishing physical or biological characteristics, is superior in both respects to alignment synchrony, in which some process essential to continued cycle traverse is inhibited. The relative virtues and deficiencies of various synchronization protocols have been discussed many times before (Stubblefield, 1968; Mitchison, 1971; Shall, 1973).

In using alignment synchrony one hopes for a single complete block point and a uniform traverse rate to the block point. It is generally agreed that unbalanced growth occurs during collection at the block and that significant perturbations in metabolism can occur on release. Initial synchrony is low, varying upward from 50 to 90%, but cell yield can be quite high. One serious and unresolved problem relates to the cell cycle position of aligned cells. For example, it appears that some S-phase arrest methods tend to align cells in the middle of S phase, not at the beginning (Williams and Ockey, 1970; Comings and Okada, 1973). Included under alignment synchronization are methods that use inhibitors of DNA synthesis (Rueckert and Mueller, 1960), Colcemid-enhanced mitotic selection (Stubblefield and Klevecz, 1965), and isoleucine deprivation (Tobey and Ley, 1970).

Ideally, selection synchrony avoids the problems of unbalanced growth. In practice, mitotic selection, the primary procedure in this class, is often enhanced with a low-calcium medium (Robbins and Marcus, 1964; Peterson *et al.*, 1968), or the collected cells are stored at low temperatures (Sinclair and Morton, 1963). Both of these treatments, which are performed to increase the yield of mitotic cells, can affect subsequent cell cycle events (Lett and Sun, 1970) and reduce the number of traversing cells (Tobey *et al.*, 1972).

Velocity sedimentation separation of cells (Mitchison and Vincent, 1965; Shall and McClelland, 1971) distributes random cells into a continuous array on a gradient and can give high yields. However, an inflection in the volume–

density function of cells with age can put two age groups of cells in the same fraction of the gradient. This is a pitfall that can be avoided if only a single class of cells is selected from the gradient. This method, like all single-point synchronization methods, requires that the investigator be present as the cells progress through the cycle.

II. Cell Cycle Analyzer

A. Virtues of Automated Synchrony

Automated mitotic cell selection attempts to avoid the problems associated with alignment synchrony, and at the same time provide mitotic cells in biochemically usable quantities (Klevecz, 1972). To achieve this end, random roller bottle cultures of mammalian cells are partitioned into a discontinuous series of synchronous populations by repeated automatic selection and subculturing.

The cell cycle analyzer has previously been shown to effectively synchronize V79 Chinese hamster cells (Klevecz, 1972), as well as the human diploid fibroblast WI-38 (Klevecz and Kapp, 1973), but a detailed description of the optimal selection conditions for these cells has not been presented before, nor have other commonly used tissue culture cells been analyzed. In this chapter we have collected the available information regarding selection optima for six different cell lines. Since the original publication describing automated cell cycle analysis, a commercial version of the instrument has been developed (Cell Cycle Analyzer, Talandic Research Corporation, 873 North Holliston Avenue, Pasadena, Calif.). Hopefully, this instrument will make it possible for laboratories with minimal experience with synchronization methods to perform cell cycle experiments. A complete, incubator-contained instrument and a schematic illustrating its operation are shown in Fig. 1.

Several points can be made in support of synchrony studies using the cell cycle analyzer. First, the system operates without the use of inhibitors, an altered medium, or reduced temperatures, any of which can reduce the number of cells traversing the cycle and introduce perturbations into the system. In addition, perfusion conditions can be established prior to the first mitotic selection, ensuring that the selected populations have a similar history. Selection can be optimized in terms of maximum yield of cells consistent with a high mitotic index. Moreover, optimal conditions can be repeatably obtained from selection to selection and from day to day. Most important for laboratories seriously engaged in cell cycle studies is the

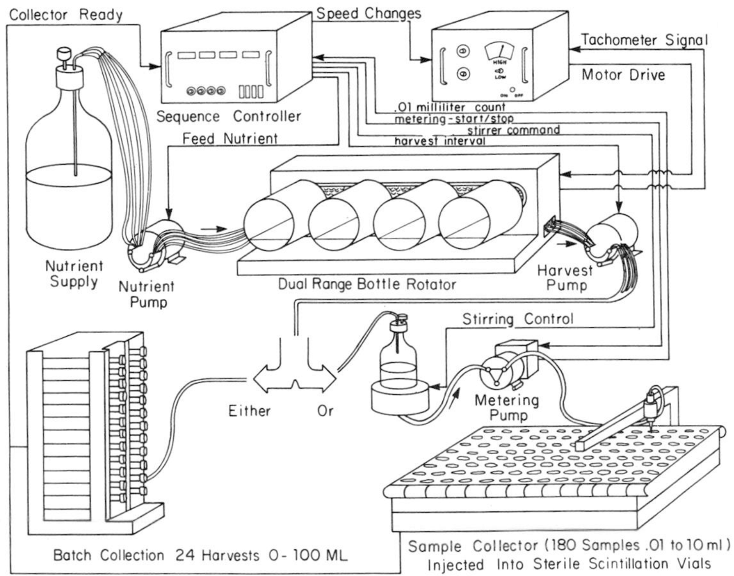

FIG. 1. (a) Complete incubator-contained cell cycle analyzer. (b) Schematic illustration of analyzer components. (Photograph courtesy of Talandic Research Corporation.)

fact that all cultures in the cell cycle analyzer come of age at the same time. This means that there is no need for the investigator to be present through an entire cell cycle. Finally, the sequence controller is sufficiently flexible so that feeding and sampling schedules for experiments not concerned with the cell cycle can be programmed as well.

B. Description of the Analyzer

The system provides for automatic mitotic selection, feeding, and sample collection. The sequence controller coordinates the operation and interaction of all other units. For each of the control functions, numeric values are selected by digit switches (Fig. 2).

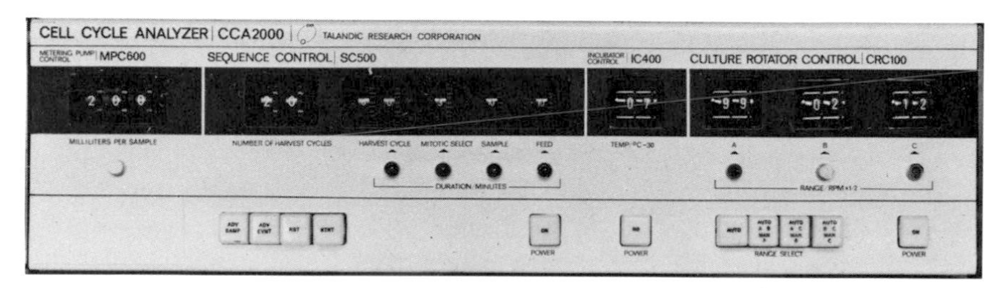

FIG. 2. Cell cycle analyzer sequence control unit. Digit switches on the unit are set for a typical experiment. Terminology is explained below:

1. Sequence control. Coordinates operations listed below.

a. Harvest cycle. Time, in minutes, between high-speed selections of mitotic cells. Within this interval mitotic selection, sampling, and feeding operations take place.

b. Mitotic select. Time, in minutes, during which the culture rotator is at selection speed.

c. Sample. Time, in minutes, during which the selected mitotic cells are pumped from the rotator bottles into growth flasks.

d. Feed. Time, in minutes, during which the fresh medium is pumped from the nutrient bottle into the rotator bottles.

e. Number of harvest cycles. Number of times the complete sequence of operations (a–d above) are performed.

2. Incubator control. Value shown is the temperature, in degrees Celsius minus 30, of the incubator. Display in this form is required by the logic of the circuitry.

3. Culture rotator control. Rotational speeds of the culture bottles are determined for manual and automatic selection.

a. Range select. A = (rpm displayed/1) $\times$ 2 = 200 rpm; B = (rpm displayed/10) $\times$ 2 = 0.4 rpm; C = (rpm displayed/100) $\times$ 2 = 0.24 rpm. If the AUTO and AC MAN B buttons are depressed with the settings shown, the rotator will select mitotic cells at 200 rpm and then return to 0.24 rpm.

4. Metering pump control. Determines the volume of medium distributed to each growth flask or vial.

1. Sequence Controller

The sequence controller controls the system's timing functions. Using the digit switches, the operator selects the number of harvest cycles of an experiment, the duration of the harvest cycle and, within a harvest cycle, the duration of mitotic selection, sampling, and feeding. The unit then sequences the culture rotator speed, the feeding and sampling pumps, and the metering pump for the distributive sample collector, accordingly. Advance sample and advance event buttons permit manual advance of the sampler to the next sample bottle and rapid manual advance of the system to any event in the program sequence. This allows the operator to set up, check out, or modify an experiment before or while it is in progress. A reset button is provided, which returns the control sequence to start and establishes an idle state for the system. A start button is provided, which places the system in automatic operation, thus initiating an experiment.

2. Culture Rotator

The roller apparatus varies significantly from existing models. Bottles are mounted and driven by polyester-filled Teflon derotator caps. Sampling and feeding are performed through stainless-steel tubing mounted within the derotator caps.

Culture Rotator Controller. Speed range is continuously variable from 0.3 to 300 rpm. Speed range selection is made either by the operator when the controller is in the manual mode, or is programmed for mitotic selection or normal speed when the controller is in the automatic mode. By itself the culture rotator can be operated manually to give single or multiple synchronous cultures.

3. Batch Sample Collector

The batch sample collector is a valve system which delivers the total volume from the roller bottles after each selection interval into one of 24 sterile disposable 250-ml flasks.

4. Distributive Sample Collector

The distributive sample collector is an injection system with a rack designed to hold 180 sterile scintillation vials. This system permits high-resolution analysis of life cycle events and/or multiple samples for replicate or parallel analyses. Selected mitotic cells are first pumped to the homogenator where the cells from each of the four roller cultures are mixed to assure uniform cell numbers per volume of medium. Cells are then pumped in equal volumes by the metering pump into the individual scientillation vials. The volume to be pumped into each vial is selected by the metering pump controller.

C. Operation of the Analyzer

Programming the Controller

The operator programs the control unit for the following inputs:
1. Volume, in milliliters, of sample distributed to each sample vial.
2. Number of harvest cycles (a harvest cycle is the time between high-speed mitotic selections) in the complete experiment, after which the system returns to the idle mode.
3. Duration, in minutes, of the harvest cycle.
4. Duration, in minutes, of the mitotic selection process.
5. Duration, in minutes, of the sampling period.
6. Duration, in minutes, of feeding period.
7. Temperature, in tenths of a degree, of the system incubator.
8. Speed, in revolutions per minute, of the culture bottles during normal and mitotic selection times.

To begin a synchrony experiment, the medium transfer apparatus is cleaned and autoclaved. The medium transfer apparatus is then set up with the nutrient supply bottle, pumps, culture rotator, and homogenizing bottle, if the distributive sample collector is used. All control functions are tested by depressing the advance event button and observing that the system is cycling properly.

The operator depresses the start button and the experiment begins. The sequence of events is as follows:
1. The system rotates the culture bottles at normal speed for a full harvest cycle period to allow cell growth.
2. A new harvest cycle is initiated and mitotic selection begins. During this period the culture bottles rotate at high speed.
3. When mitotic selection ends, the sampling period begins. The sample is pumped out of the culture bottles into either a batch sample collector bottle or a homogenator bottle, when the distributive sample collector is used.
4. When sampling is complete, feeding begins. Nutrient is pumped into the culture bottles during the feeding period. If a distributive sample collector is used, the sample vials begin to fill with the homogenator bottle contents.
5. When feeding is complete, and for the remainder of the harvest cycle period, cell growth at low rotational speed is permitted.
6. When the current harvest cycle is complete, the system returns to step 2, provided the selected number of harvest cycles has not been completed.
7. When the selected number of harvest cycles is complete, the system resets itself and remains in the idle mode with the culture bottles rotating at normal slow speed.

III. Optimizing Selection Conditions

A. Preparation for Synchrony

In preparation for synchrony studies, cells are generally subcultured into roller bottles treated with 0.1 *M* Na_3CO_3 and thoroughly rinsed before sterilization. Preparation of the roller bottles and culture conditions during attachment are critical phases of the synchronization procedure. Cells are inoculated at concentrations from 1 to 5×10^7 cells per roller in 100 ml of medium. Bottle rotational speed during attachment should be 0.2 to 0.3 rpm; higher rotational speeds impede rapid attachment to the glass and result instead in cell-to-cell attachment in solution, leading to clumping. Roller cultures are allowed to grow 18 (V79) to 36 (WI-38) hours before the first selection sequence is initiated.

The following parameters have been shown to affect optimum selection conditions: (1) growth phase; (2) selection speed; (3) interval between selections; (4) selection duration, and (5) medium volume during selection. Obviously, certain of these parameters are greatly affected by the cell type used, others less so. Growth phase is especially important, and the timing varies from cell to cell. Optimal selection speed and the optimal interval between selections also vary with cell type. These parameters are discussed in detail in Section III,B. Duration of a given selection and the volume of

TABLE I
YIELD OF MITOTIC CELLS WITH INCREASING SELECTION TIME[a]

Selection interval (minutes at 200 rpm)	Cell number
0	5×10^3
0.5	2.4×10^6
1	3.8×10^6
2	4.4×10^6
3	4.8×10^6
4	4.6×10^6
5	4.9×10^6

[a] A single roller bottle culture was innoculated with 5×10^7 CHO cells 36 hours prior to the first selection as described in the text. For each selection culture medium was removed and 35 ml of fresh medium was added to the roller. Selection for increasing intervals was performed at 30-minute intervals, beginning with the shortest interval.

medium in the roller bottles during selection show minimal variation from one cell type to another. Selection durations of 3 minutes or greater give the maximum yield of cells (Table I). Initially, it was anticipated that long-duration selection intervals (up to 1 hour) would give considerably greater yields than short-interval selections (Klevecz, 1972). Contrarily, it has been observed that three 3-minute selections at 20-minute intervals give more cells with less risk of trauma than a single 1-hour-long selection. Medium volume, within reasonable limits, does not appear to affect the detachment of mitotic cells directly, but rather exerts its influence on the quality and yield of synchronous cells as a consequence of efficiency of medium removal. If the total volume of medium during selection is small, the carried-over volume (medium not removed by the pump), which is nearly constant under all conditions, is relatively a greater proportion of the total volume. Consequently, a considerable number of the selected cells are left behind, reducing the yield and contaminating subsequent selections.

B. Selection Artifacts

Although it operates without the use of inhibitors or an altered medium and maintains a constant temperature, automated mitotic selection introduces a new potential variable. The cell cycle analyzer performs multiple mitotic selections at predetermined intervals, hence each synchronous culture has a slightly different history. If, for example, the analyzer performs 12 selections at hourly intervals, at the end of this time the first cells selected will be 12 hours into the cycle but will have only been exposed to a single mitotic selection, whereas the last mitotic cells selected will have been exposed to 11 selections during the time they were in interphase. The argument can be made that these selections are traumatic and that such cells are less capable of normal cell cycle traverse. Investigations of this problem, which is unique to the analyzer, indicate that the selective procedure is sufficiently gentle to eliminate this concern (Klevecz and Kapp, 1973), since viability, plating efficiency, and clone-forming ability are high, and equal to that of control cultures of V79 and WI-38 cells.

C. Yield of Mitotic Cells

A second, more subtle, problem is related to the growth phase of the cultures prior to selection. This problem is present in all mitotic selection synchrony studies and, in fact, is typical of mammalian cell cycle studies generally. It is made more apparent and more amenable to treatment by the cell cycle analyzer. A question can be posed regarding the effects of treatment in one cell cycle on the behavior in the next: Are cells selected from late exponential cultures the same or different from cells selected from a population that has recently been subcultured? Phase of growth of the cultures and cell cycle position of cells at the time of the first mitotic selection can play a major role in the yield of mitotic cells and may influence the results obtained from the ensuing synchronous cell cycle. Although automated synchronization using the apparatus described here can be performed adequately in the absence of such information, it is desirable that the kinetics of growth in roller bottles be determined for different cells in different media.

Two approaches to this problem can be considered. It may be turned to advantage by exploiting the parasynchronous growth that results from the simple act of subculturing. Mitotic selections can be performed during the parasynchronous mitotic burst. Likewise, one can exploit the serum-stimulated overgrowth (Wiebel and Baserga, 1969) phenomenon to obtain a greater-than-average number of cells in the mitotic phases. This procedure is best limited to experiments requiring a single mitotic selection. It probably is not suitable for a closely spaced series of analyses of an entire cycle. In Table II an experiment in which WI-38 cells were subcultured into roller bottles and allowed to grow for 72 hours in the same medium is presented. The selection program was initiated after 72 hours of growth, and fresh medium was pumped into the rollers. After the first selection, which is generally discarded because of contaminating interphase cells, the yield of mitotic cells remains relatively constant for 15 hours. There is then an

TABLE II

MITOTIC CELL YIELD OF WI-38 AT INTERVALS AFTER THE ADDITION OF FRESH MEDIUM[a]

Selection number	Time after first selection (hours)	Cell number ($\times 10^{-5}$)
1	0	12.1
2	1	5.2
3	2	4.4
4	3	4.9
5	4	4.6
10	9	4.7
15	14	4.5
16	15	13.5
20	19	12.1

[a] WI-38 cells were innoculated into four roller bottles at 1 to 2 $\times$ 10^7 cells each 72 hours prior to first selection. Mitotic cells were selected at hourly intervals at 200 rpm.

abrupt 2- to 4-fold increase in the number of mitotic cells selected. This increase in yield of cells often lasts for 4–6 hours. It appears as though WI-38 cells cultured for a reasonably long period in the same medium are sensitive to localized contact inhibition, even though the culture is not confluent. When fresh medium is added, these cells move rapidly toward mitosis. In this laboratory WI-38 cells have a 19- to 20-hour generation time and a combined S + G_2 of 14–16 hours. The conclusion to be drawn from this is that cells that cease to cycle behave as though they are arrested no closer to M than the G_1/S boundary. The fact that the increase in yield of mitosis lasts beyond the intial burst may indicate that arrested cells are distributed back from the G_1/S boundary through G_1. Cultures of WI-38 subcultured and grown for shorter periods prior to the first mitotic selection do not show this discontinuity in yield of mitotic cells (Klevecz and Kapp, 1973).

To assess the problem of growth phase effects one can exploit the capacity of the instrument to simulate perfusion conditions. If, as appears to be the case, cells are differentially sensitive to growth-stimulating factors in serum, it is possible that various metabolic and macromolecular functions also show differential sensitivities. Conceivably, the expression of an enzyme or the pattern of DNA synthesis in one cell cycle may be altered, depending on where in the pervious cell cycle selected cells were at the time of the first medium addition. In a series of experiments to be reported elsewhere (R. R. Klevecz, unpublished data), the effects on V79 cells of a single exposure to fresh medium and subsequent mitotic selection after 24 hours of growth in the original culture medium were compared with those of perfusionlike culture conditions. Perfusion conditions were attained by removing 10% of the culture medium at hourly intervals and replacing it with fresh medium for 24 hours prior to the first mitotic selection. The pattern of DNA synthesis was the same under both conditions. Whether other, more easily perturbed functions would also be unaffected is problematical.

Optimum selection speeds were determined for several commonly used cell lines. For many cells the fastest selection speeds available were still not beyond the optimum (Table III). Chinese hamster cell lines appear to show the greatest differential between mitotic and interphase cells with respect to detachment. Increasing yields of V79 and CHO cells can be obtained at speeds up to 200 rpm with no decrease in mitotic index. HeLa and L929 cells show the greatest yield of mitotic cells with the fewest interphase cells at speeds near 100 rpm. WI-38 human diploid and Don Chinese hamster cells appear to be quite resistant to detachment. Even at the highest selection speeds microscope observation confirmed the fact that considerable numbers of mitotic cells are still attached. Accordingly, newer versions of the instrument have been provided with a 300-rpm capacity.

TABLE III

YIELD AND INITIAL SYNCHRONY OF MITOTIC CELLS FROM COMMONLY USED CELL LINES[a]

Cell type	Optimum selection speed (rpm)[b]	Mitotic cells per roller detached	Total population (%)	Mitotic index (%)
CHO	200	3.5×10^6	10.9	99
V_{79}	200	6.3×10^5	—	98
Don	100–150	3.5×10^5	0.4	95
HeLa	75–150	1.1×10^6	2.4	94
W138[c]	200	1.2×10^5	0.2	89
L929	100	8.3×10^5	—	88

[a]Cells were innoculated into roller bottles as described in the footnote to Table I. Selections were performed using 40 ml of McCoy's 5a medium per selection. CHO, V_{79}, Don, W138, and HeLa were grown as described in the text. Bottles were innoculated 24 hours prior to use and purged twice by rotation at 200 rpm, 30 minutes before mitotic selection. Collections were made by pumping cells into vessels containing 0.5 ml Colcemid (6 μg/ml). Rotational speed was increased in 25- or 50-rpm increments up to 200 rpm, and then decreased in 50-rpm increments from 175 to 75 rpm. L929 cells were supplied to us by Dr. Igor Tamm of Rockefeller University.

[b]Yield of mitotic cells is from a single experiment. Selection speeds represent best estimates from a series of experiments.

[c]Only speed used.

IV. Calibrating the Cell Cycle

The distribution of mitotic stages from prophase through telophase and early G_1 was scored in selected cultures fixed immediately on detachment and in cultures allowed to incubate for a time after detachment (Table IV). Initially, detached cells were distributed throughout all division stages, with the greatest frequency in telophase. After 10 minutes at 37°C, few undivided cells remained and the vast majority were in telophase or early G_1. This can be confirmed by noting with phase-contrast microscopy that nearly all cells in the field exist as small doublets after 30 minutes of incubation. The degree of initial synchrony and the distribution of freshly detached mitotic cells can also be judged using neutral formalin fixation, fluorescent Feulgen staining, and flow microfluorometric (FMF) analysis as described by Kraemer *et al.* (1973).

FMF analysis can be used to resolve heterogeneous populations of cells into their constituent subpopulations, in this case, according to the DNA-

TABLE IV

DISTRIBUTION OF MITOTIC STAGES AT SELECTION AND IMMEDIATELY THEREAFTER[a]

Selection speed	Mitotic stage[b]						
	P	PM	M	A	T	$G_1(C+)$	I
100 rpm, fixed immediately	2	3	28	32	129	23	3
100 rpm, incubated 10 minutes at 37°C	1	1	6	11	55	104	4
100 rpm, 30 minutes at 37°C	1	3	1	1	4	77	22

[a] CHO cells were prepared for synchrony as described in the footnote to Table I.

[b] P, Prophase; PM, prometaphase; M, metaphase; A, anaphase T, telophase; $G_1(C+)$, condensed chromatin positive posttelophase; I, interphase or dead cells.

specific fluorescence expressed. It can also be used to demonstrate the presence of asynchronous contaminants in a putatively homogeneous population such as selected metaphase cells. Using this approach, one can examine the distribution through the cycle of individual cells, hence determine more precisely the onset and termination of cycle substages. Synchrony seems to decay rather rapidly in mammalian systems, and consequently it can be difficult to obtain an accurate value for the S/G_2 boundary. Since FMF analysis provides a frequency distribution of cells with differing DNA contents, the modal DNA content can be easily determined. It seems quite reasonable to assign values for the duration of cell cycle substages using the time interval during which the major mode of cells had the appropriate DNA content. In Fig. 3 an FMF analysis of the first 3 hours of the CHO cell cycle is shown. If cells are fixed immediately after selection, or if 30 minutes are allowed to elapse before fixation, a bimodal distribution in DNA content is seen (Fig. 3). The greater mode, centering in this instance around channel 38, may represent undivided cells and possibly freshly divided doublets. After 1 hour of incubation, this population is no longer detectable. One interesting observation relates to the apparent DNA content of the G_1 cells between hours 0 and 2. It has been consistently observed that the DNA-specific fluorescence of the detached cells increases for 2 hours following selection. Autoradiography reveals no acid-precipitable tritiated thymidine incorporation during this time. Since the DNA in chromosomes at division and in early G_1 is more condensed, it may be less accessible to the Feulgen staining reaction. As the chromatin becomes more diffuse, the fluorescence per cell increases. There is no change in the modal channel number between hours 2 and 5, and an absence of subpopulations with DNA contents greater than that of G_1. The synchronous entrance of cells into

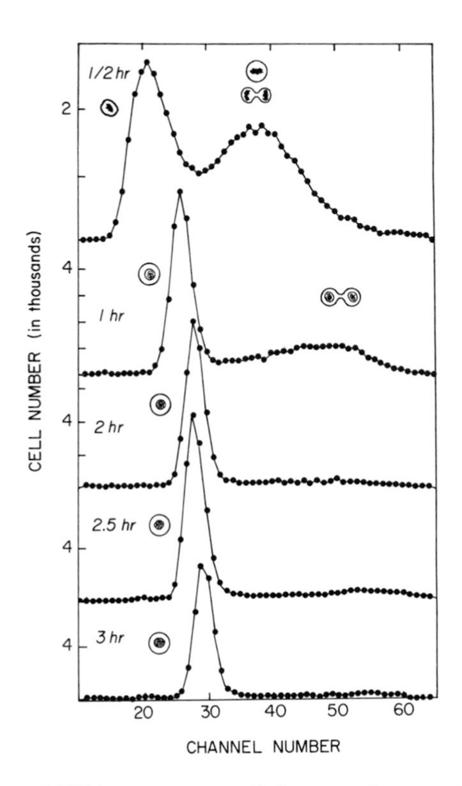

FIG. 3. Distribution of DNA content per cell from synchronous CHO cells. CHO cells were synchronized as described in the text and stained by the acriflavine-Feulgen method of Kraemer *et al.* (1972). Only the first 3 hours of the cycle are shown. Freshly detached cells are either in mitosis or are densely staining early G_1 cells [G_1(C+)]. The change in condensation states of chromatin and the division stage cells is represented in the figure. (From R. R. Klevecz and L. L. Deaven, unpublished.)

S phase can best be measured autoradiographically. The changing DNA content through S can be followed by Feulgen fluorescence in FMF analysis. In Fig. 4 mean and modal DNA contents in Chinese hamster CHO synchronous cells are compared with autoradiographic analysis of the beginning of S phase. The ability of autoradiography to sensitively measure low levels of DNA synthesis is revealed by comparing percent labeled nuclei with modal FMF values for DNA content.

One conclusion to be drawn from this figure is that estimates of S-phase duration vary depending on whether they are made by total radiochemical incorporation, spectrophotometry, or autoradiography. In the first two cases, lower values are reported than if the estimates were made autoradiographically. Bostock and Prescott (1971) observed an 8-hour S phase in rabbit endometrium by measuring total TdR-^{3}H incorporation, while noting

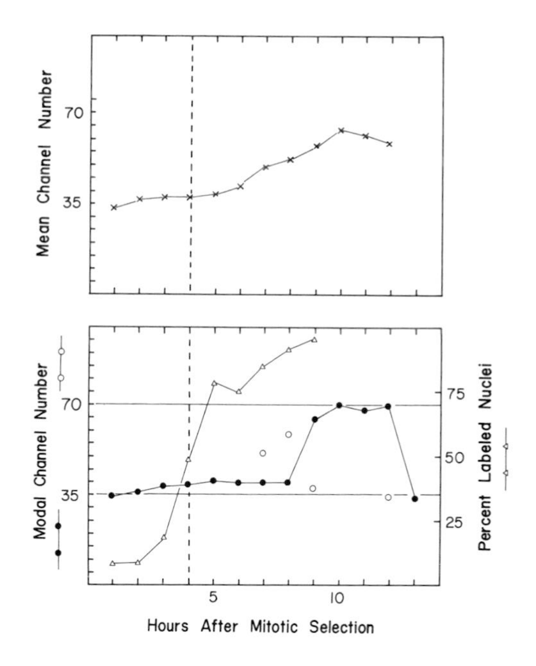

FIG. 4. Mean and modal channel number (DNA content) from a FMF analysis of synchronous CHO cells. Mitotic CHO cells were selectively detached and fixed for FMF analysis at hourly intervals throughout the cell cycle. Major modes are indicated by filled circles, and minor modes by open circles. The beginning of S phase was determined by autoradiographic analysis of the percent labeled nuclei (triangles) and is indicated by the vertical broken line. Mean DNA content was determined using the equation $\bar{c} = \Sigma(fc)/\Sigma f$, where $\bar{c}$ is the mean channel number or center of mass, c is any of 100 channels, and f is the number of cells in that channel. (From R. R. Klevecz and L. L. Deaven, unpublished.)

low levels of incorporation over a considerable portion of the cycle. The labeled cells were detectable autoradiographically and, if scored for percent of labeled nuclei, gave a value for S phase of 18 hours out of a 27-hour cycle. In reporting this finding, these investigators suggested that it might be a peculiarity of rabbit cells, since a similar observation had been made by Painter and Schaeffer (1969) who also used rabbit cells. Our findings suggest that low net synthesis in early S phase is a general occurrence.

V. Limitations and Prospects

The upper limit on the yield of mitotic cells is determined by the volume of medium required and the logistical problems involved in selecting cells from large numbers of roller bottles. For most experiments greater yields are unnecessary. By using four roller bottles with an initial innoculation of 5×10^7 CHO cells each, 2×10^7 cells can be selected every hour. Decreasing the interval between selections from 1 hour to 20 minutes results in only a small decrement in yield per selection. Consequently, if one is willing to pay a slight penalty in sharpness of synchrony, three separate selections can be pooled to give 5×10^7 cells per hour. Increasing the number of roller bottles is possible, but the volumes of medium that result exceed the capacity of the recipient flasks. While this rules out automated synchrony, manual collection from large numbers of rollers may be feasible. The logic of the sequence controller is such that many methods for making cells synchronous can probably be coupled to this sample-handling system. Hopefully, the availability of the cell cycle analyzer will make it possible for all interested laboratories to exploit synchrony as a technique.

ACKNOWLEDGMENT

The work described in this chapter was supported by grant HD-04699 from the National Institutes of Health.

REFERENCES

Bostock, C. J., and Prescott, D. M. (1971). *J. Mol. Biol.* **60**, 151–162.
Comings, D. E., and Okada, T. A. (1973). *J. Mol. Biol.* **75**, 609–618.
Klevecz, R. R. (1972). *Anal. Biochem.* **49**, 407–415.
Klevecz, R. R., and Kapp, L. N. (1973). *J. Cell Biol.* **58**, 564–573.
Kraemer, P. M., Deaven, L. L., Crissman, H. A., and VanDilla, M. A. (1972). *Advan. Cell Mol. Biol.* **2**, 47.
Kraemer, P. M., Deaven, L. L., Crissman, H. A., Steinkamp, J. A., and Petersen, D. F. (1973). *Cold Spring Harbor Symp. Quant. Biol.* **38**, 133–144.
Lett, J. T., and Sun, C. (1970). *Radiat. Res.* **44**, 771–777.
Mitchison, J. M. (1971). "The Biology of the Cell Cycle." Cambridge Univ. Press, London and New York.
Mitchison, J. M., and Vincent, W. S. (1965). *Nature* (*London*) **205**, 987–989.
Painter, R. B., and Schaeffer, A. W. (1969). *J. Mol. Biol.* **45**, 467–479.
Peterson, D. F., Anderson, E. C., and Tobey, R. A. (1968). *Methods Cell Physiol.* **3**, 347–370.
Robbins, E., and Marcus, P. I. (1964). *Science* **144**, 1152–1154.
Rueckert, R. R., and Mueller, G. C. (1960). *Cancer Res.* **20**, 1584–1591.
Shall, S. (1973). *Methods Cell Biol.* **7**, 269–285.
Shall, S., and McClelland, A. J. (1971). *Nature* (*London*) **229**, 59–61.
Sinclair, W. K., and Morton, R. A. (1963). *Nature* (*London*) **199**, 1158–1160.
Stubblefield, E. (1968). *Methods Cell Physiol.* **3**, 25–43.
Stubblefield, E., and Klevecz, R. R. (1965). *Exp. Cell Res.* **40**, 660–664.
Tobey, R. A., and Ley, K. D. (1970). *J. Cell Biol.* **46**, 151–157.
Tobey, R. A., Crissman, H. A., and Kraemer, P. M. (1972). *J. Cell Biol.* **54**, 638–642.
Wiebel, F., and Baserga, R. (1969). *J. Cell. Physiol.* **74**, 191–203.
Williams, C. A., and Ockey, C. H. (1970). *Exp. Cell Res.* **63**, 365–372.

Chapter 10

Selection of Synchronous Cell Populations from Ehrlich Ascites Tumor Cells by Zonal Centrifugation

HANS PROBST AND JÜRGEN MAISENBACHER[1]

Physiologisch-chemisches,
Institut der Universität,
Tübingen, West Germany

I. Introduction 173
II. Materials 174
  A. Ascites Cells 174
  B. Apparatus 176
III. Zonal Centrifugation Run 177
  A. Preparation 177
  B. Zonal Run 178
IV. Results 179
  A. Separation by Zonal Centrifugation 179
  B. Further *in Vitro* Growth of Cells from Different Fractions 181
V. Possible Applications 183
  References 184

I. Introduction

Synchronous cell populations can be obtained from asynchronously growing cells either by induction methods which use in most cases chemical agents to cause reversible blocks of certain cell cycle stages, or by selection

[1] *Present address*: E. Merck, Abteilung klinische Forschung, Darmstadt, West Germany.

methods which separate cells belonging to a distinct phase of the cell cycle from the others. Selection methods largely avoid disturbances of the course of the cell cycle, which occur necessarily when induction methods are used. However, when selection methods are used, the higher the degree of synchrony the smaller, necessarily, is the size of the cell population obtained relative to the asynchronous starting material. Thus a practicable selection method should have a large capacity, and suitable starting cells should be available in larger amounts.

Ascites tumors which usually grow at very high cell concentrations in the abdominal cavity of common laboratory animals can provide large amounts of rapidly growing cells without requiring expensive equipment. Furthermore, ascites tumors consist of suspended cells and are therefore suitable for selection methods based on the different sedimentation properties of cells belonging to different phases of the cell cycle. The preparation of synchronous cell populations by sucrose density gradient centrifugation was first described by Mitchison and Vincent (1965). It can be carried out on a large scale by using large-volume zonal rotors (Warmsley and Pasternak, 1970; Probst and Maisenbacher, 1973).

II. Materials

A. Ascites Cells

In principle all ascites tumor cells strains should be suitable that fulfill the following conditions. (1) The ascites tumor must have a genuine logarithmic growth phase with a proliferative index of nearly 1. (2) The cells of the logarithmically growing tumor must be very uniform in size; size differences between individual cells should mainly reflect their position in the cell cycle. (3) The cells should retain their viability under the conditions occurring during zonal centrifugation. (4) The cells should be able to continue their growth *in vitro*; otherwise, the success of the separation procedure cannot be demonstrated without difficulty. Besides these prerequisites several other properties of the cells could be critical: the (buoyant) density, the resuspendibility after being pelleted, and the broadness of cell cycle time distribution. It cannot be excluded that, for preparing synchronous populations by zonal centrifugation from other ascites tumor cell strains completely other conditions must be chosen than those used by us for Ehrlich ascites cells. Furthermore, it is conceivable that some Ehrlich ascites cell lines exist that are completely unsuitable.

The Ehrlich ascites tumor cells used by us were originally obtained from Dr. K. Karzel, Pharmakologisches Institut der Universität Bonn, Germany, who first succeeded in growing them *in vitro* (Karzel, 1965). They can be grown in mice and in permanent suspension culture and can be transplanted from mouse to culture, and vice versa, without too much difficulty. For transplantation from mouse to mouse, 0.2 ml of ascites fluid (7 days after prior inoculation, containing approximately 3×10^7 cells) is injected into female NMRI/HAN mice weighing 18–22 gm. In order to ensure *in vitro* viability of the cells transplanted *in vivo*, we grew them *in vitro* after every 40 *in vivo* passages for at least 50 passages and again retransplanted them into mice. The general technique for the growth *in vitro* and the medium used are described by Karzel (1965). The detailed procedure for establishing asynchronous primary cultures from *in vivo* tumors is described by Probst and Maisenbacher (1973). Cells from mice inoculated 4 days before were exclusively used for zonal centrifugation experiments. (The logarithmic growth phase of the *in vivo* tumor extends from the second to the fifth day after inoculation.)

Table I summarizes the cell kinetic parameters of a 4-day-old ascites tumor and of cultures derived from a 4-day-old tumor during the first two *in vitro* passages. The data indicate that explantation of *in vivo* cells to *in vitro* cultures causes a remarkable acceleration of cell growth, which occurs stepwise after the explantation itself and after transplantation to the second

TABLE I

KINETIC PARAMETERS OF ASYNCHRONOUS POPULATIONS OF THE EHRLICH ASCITES CELLS USED[a]

Cell cycle parameter	*In vivo* 4-day-old ascites tumor	First *in vitro* passage	Second *in vitro* passage
t_{G_1} (mean ± S.D.)	—	8.3 ± 4.0 hours	4.7 ± 1.9 hours
t_S (mean ± S.D.)	17.8 ± 4.1 hours	11.0 ± 4.1 hours	7.5 ± 1.9 hours
t_{G_2} (mean ± S.D.)	8.4 ± 4.1 hours	3.4 ± 1.0 hours	2.6 ± 1.0 hours
t_C (median)	—	22.0 hours	14.7 hours
Proliferative fraction	—	>0.95	>0.98
Thymidine-^{3}H labeling index (mean ± S.D.)	0.510 ± 0.018	0.486 ± 0.018	0.458 ± 0.018
Cell number doubling time (mean ± S.D.)	34 ± 7 hours	21.1 ± 2.8 hours	14.1 ± 1.6 hours

[a]Data from Probst and Maisenbacher, 1973.

in vitro passage. Consequently, a shortening of all cell cycle stages is observed. During the third and fourth *in vitro* passages, the cells exhibit essentially the same growth characteristics as in the second passage. Usually, between the fifth and the twelfth *in vitro* passages a crisis of cell growth occurs, which is characterized by high cell loss and which persists normally over three to four passages. After this crisis the cells appear to be completely adapted to the *in vitro* conditions (perhaps by selection). Henceforth, they show the growth properties also reported by Karzel and Schmid (1968) for permanent cultures of this cell strain, which are essentially characterized by a reincrease in the generation time to about 22 hours.

B. Apparatus

Figure 1 shows a scheme of the equipment. We use a MSE Type-A zonal rotor in a MSE Mistral 6L centrifuge. For exact control of rotor speed, the centrifuge is connected to an electronic digital revolution counter. The gradient-forming device is placed in an ice bath. It consists of a closed mixing vessel (magnetically stirred; volume, 1.2 liters) connected with an open reservoir and a peristaltic pump (Supra-Schlauchpumpe driven by a Multifix-Constant-Motor, both manufactured by A. Schwinnherr, Schwä-

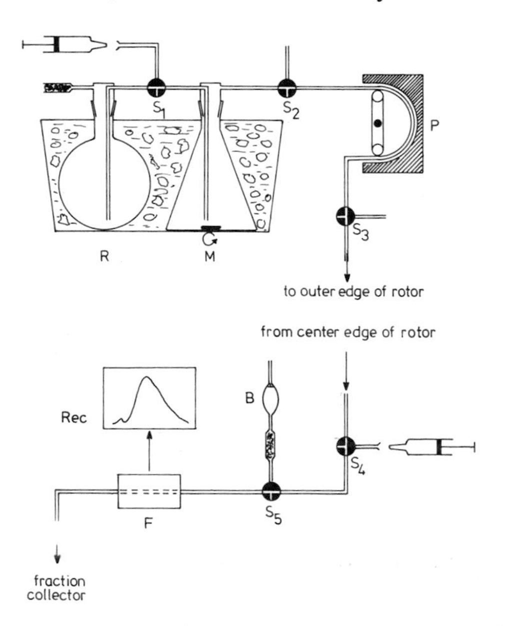

FIG. 1. Scheme of the zonal centrifugation equipment. R, Reservoir; M, mixing vessel; P, peristaltic pump; B, sphygmomanometer ball; F, flow-through cell; Rec, recorder; S_1–S_5, three-way stopcocks.

bisch-Gmünd, West Germany), which leads to the outer edge of the rotor. The effluent from the center edge of the rotor passes a flow-through cell for monitoring the $A_{578}^{1\,cm}$ and then is delivered to a fraction collector. The three-way stopcocks, S_1 to S_5, serve for filling, sterilizing, sample application, and emptying the device. Their particular function is described below. S_1 and S_4 have facilities to accommodate a syringe. The air inlet into the reservoir R is stoppered by a cotton plug. Another cotton plug is placed between the stopcock S_5 and the sphygmomanometer ball B. The fraction collector is advantageously placed in a cold cabinet. All connections are made by 6 × 4 mm silicone tubing.

III. Zonal Centrifugation Run

A. Preparation

1. Sterilization

If it is planned to grow separated cell fractions in culture after the run, sterile conditions must be maintained throughout. When the run is performed to yield synchronous cell populations for biochemical analyses, the sterilization procedure can be omitted.

All solutions required are sterilized by filtration through asbestos filters (Seitz Asbestwerke, Bad Kreuznach, West Germany) in our laboratory. All syringes used are sterile and disposable.

The tubing connections are installed as completely as possible, and the resulting device (without rotor, sphygmomanometer ball, and flow cell) is sterilized by autoclaving. Afterward, the residual connections are installed, and the matching tubing intercept is placed in the peristaltic pump. The rotor is then accelerated to approximately 400 rpm, and a 0.3% solution of diethyl pyrocarbonate (Baycovin, Bayer-Leverkusen, Germany) in water is pumped into the rotor until approximately 500 ml has emerged from the drop outlet on the fraction collector. The rotor is then stopped, and the Baycovin solution–containing assembly is allowed to stand overnight at room temperature while the inlet at S_2 and the outlet on the fraction collector remain immersed in Baycovin solution. Note that some additional liquid emerges from the outlet during this period, because carbon dioxide is produced by hydrolysis of Baycovin. Approximately 3 hours before the planned start of the run, about 3 liters of sterile, ice-cold water is pumped through the rotor in the same way as before the Baycovin solution while the rotor is spinning at 300–400 rpm. The cooling device of the centrifuge is set to

3°–4°C. The water is then displaced from the spinning rotor by gentle air pressure generated by the sphygmomanometer ball at S_5. It leaves the system through S_3. The empty rotor is allowed to spin at minimal possible speed for a further 2–3 hours in order to allow equalization of temperature.

2. Loading of Gradient

The reservoir R is filled with 1.5 liters of Hanks' solution without NaCl but containing 31% (w/v) sucrose, and the mixing vessel is completely filled with Hanks' solution which contains only 7.6 gm NaCl per liter and 2% sucrose. Remaining air is completely displaced from the mixing vessel through S_2 by injecting low-density gradient solution through S_1 by means of a syringe. Air residues in the tubing between S_1 and R are aspirated with the same syringe. S_3 is then opened in outward from P, the stirrer is started, and the pump is operated until gradient solution without air bubbles emerges from S_3. S_3 is then reversed to its normal position, and the gradient is pumped into the spinning rotor at about 50 ml per minute. About 50 ml of the gradient is allowed to flow out from the outlet on the fraction collector before the pump is stopped.

3. Preparation of the Ascites Cells

The desired number of tumor-bearing mice is killed by cervical dislocation. The abdominal musculature is denuded by everting the skin without opening the peritoneal cavity by means of two tissue forceps. Subsequently, 3–5 ml of ice-cold Hanks' solution containing 5 USP units of heparin (Liquemin, Roche) are injected. The diluted ascites is aspirated and transferred to ice-cold Hanks' solution (50 ml per mouse). When it is desired to harvest nearly all the ascites cells, the peritoneal cavity is irrigated by two or three further injections and reaspirations of cold Hanks' solution without heparin. The cells are collected by centrifugation (500 *g*, 5 minutes), suspended in a final volume of 20 ml of an ice-cold 1:1 mixture of Hanks' solution and the less dense gradient solution, and afterward aspirated into a syringe.

B. Zonal Run

After adjusting the rotor speed to 280–300 rpm, the peristaltic pump is restarted for 1–2 seconds in order to ensure complete filling of the system. Then the stopcock S_3 is turned so that the outer edge of the rotor opens outward. The cell suspension is injected within approximately 90 seconds through S_4 into the center edge of the rotor. Subsequently, 30 ml of cold Hanks' solution is injected at the same speed, and S_3 is brought to its normal position. The rotor is then accelerated within 40–60 seconds to 900 rpm.

This rotor speed is maintained for 8 minutes, starting from the onset of acceleration. Digital rotor speed measurements are permanently made at 6-second intervals and, if necessary, readjustments are performed. Meanwhile, the pump is connected by S_2 with a vessel containing ice-cold dense gradient solution. At 8 minutes the revolutions-per-minute knob of the centrifuge is turned to the chock at minimal speed. At the same time the pump, the fraction collector, and the recorder feed are started. The exact pump speed of 25 ml/30 seconds is controlled on the basis of the filling height of the fraction collection tubes which are changed at 30-second intervals. The centrifuge decelerates during 3 minutes to about 280 rpm. (The speed maintained when the revolutions-per-minute knob is turned to the chock can be regulated within certain limits by adjusting a potentiometer on the back of the control panel of the centrifuge.)

When the gradient is completely displaced, another 100 ml of underlay solution is pumped through the rotor in order to spill out any larger cell lumps sedimented to the boundary layer between gradient and underlay.

When aliquots for cell number determination, etc., are to be taken from the collected fractions, the tubes must be inverted several times in order to mix the contents homogenously. As soon as possible, the cells should be freed from sucrose by washing them with cold Hanks' solution or culture medium. When kept at 0°–2°C in Hanks' solution, the cells remain viable for at least 2–3 hours and can be further grown in culture thereafter as described below.

The dense sucrose solution is washed out of the rotor by pumping water through it via S_4 ⟶center edge ⟶outer edge⟶pump⟶S_3. Finally, the water is displaced by gentle air pressure as described above. When a further zonal run is planned on the same day, the outwashing of sucrose is performed with cold Hanks' solution and, if necessary, sterile conditions are maintained. When no trouble, such as obliteration of tubing and channels, occurs, several runs can be performed successively without disassembling the rotor and the other devices. Between these runs the rotor is maintained spinning at minimum speed and at 3°–4°C.

IV. Results

A. Separation by Zonal Centrifugation

Figure 2 shows a sedimentation pattern obtained by sedimenting the ascites cells from three mice. In this case DNA-synthesizing cells were prelabeled by intraperitoneal injection of 25 μCi of thymidine-^{3}H 10 minutes

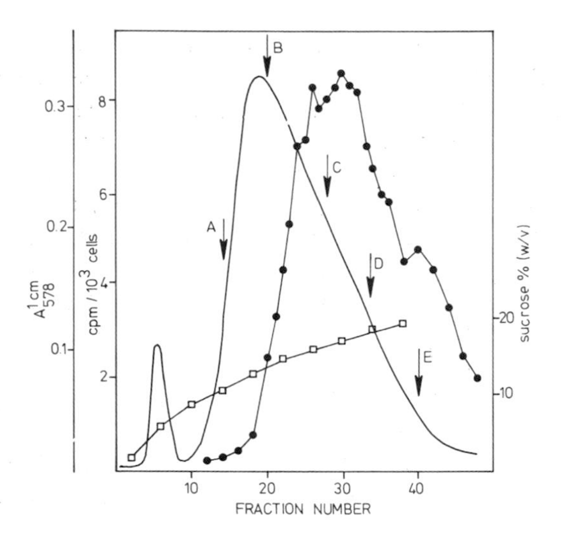

FIG. 2. Zonal centrifugation of prelabeled cells. Three mice bearing 4-day-old ascites tumors were injected intraperitoneally with 25 μCi thymidine-^{3}H in 0.5 ml of sterile Ringer solution. Ten minutes after the injection, the ascites cells were removed as described in Section III,A. Preparation of the cells and zonal centrifugation were carried out as described in the text. Solid line, $A_{578}^{1\,cm}$; circles, counts per minute per 10^3 cells; squares, sucrose concentration (grams per 100 ml) determined refractometrically; arrows, position of the fractions providing the cells for the *in vitro* cultures mentioned in the text.

before withdrawal of the ascites fluid. The first small A_{578} peak is caused by red blood cells which are completely separated from the ascites cells. Since a 4-day-old ascites tumor of the cell strain used is not usually hemorrhagic and this peak does not always occur, we assume that it is caused by bleeding during the ascites withdrawal. The broad second peak represents the ascites cells which are separated according to their different sedimentation properties.

As indicated by the ^{3}H radioactivity incorporated into the cells, the first five fractions of this peak contain cells that did not synthesize DNA shortly before being withdrawn from the animals. As outlined below, these cells belong to the G_1 phase of the cell cycle. In the following 9 to 10 fractions, a sharp increase in thymidine-^{3}H incorporation is observed, which reaches a plateaulike maximum when the cell content of the fractions is already decreasing. The decrease in the thymidine incorporation curve is less sharp than the increase, and the zero level is not reached again.

The sedimentation pattern of the cells and the distribution of radioactivity along the gradient depicted in Fig. 2 are almost completely reproducible when the technical conditions (gradient, revolutions per minute, centrifugation time, temperature, and fractionation speed), as well as the biological properties of the cells used (especially the age of the *in vivo* tumor), are held constant. The deviation of the position of several specific points within the gradient (onset and maximum of the A_{578} peak, onset of the radioactivity peak, and the beginning of the plateau) seldom exceeded one fraction in the experiments of the type shown in Fig. 1 performed in the past $1\frac{1}{2}$ years. Optimal resolution is obtained when the number of cells loaded onto the gradient does not exceed about 2×10^9, which is the content of four to five mice. It is, however, possible to apply more than twice as many cells without obtaining substantially unsatisfactory results. For example, Fig. 4 shows the $A_{578}^{1\,cm}$ pattern obtained in a run with about 5×10^9 cells. Complete separation of the erythrocyte peak was not achieved in this case, but the overall pattern is the same as that presented in Fig. 2. The degree of synchrony of the cells in single fractions of such an "overloaded" run is only unessentially diminished as judged on the basis of further growth in culture (see next paragraph).

The $A_{578}^{1\,cm}$ pattern serves usually as a guide in locating fractions belonging to distinct cell cycle stages according to the rules outlined below. It is, however, no measure of cell density in the fractions obtained. The ratio cell density/A_{578} decreases by about a factor of 2 from the beginning to the end of the ascites cell peak.

B. Further *in Vitro* Growth of Cells from Different Fractions

The cells from all fractions of the gradient can be grown *in vitro* after being washed only once with culture medium to free them of sucrose. In such cultures from any selected fractions, we followed the cell number, the mitotic index, and the thymidine-^{3}H incorporation rate for periods up to 30 hours. These fractions were: (A) region of the ascending limb of the A_{578} curve, which was found to be free of radioactivity if the cells were prelabeled as described above; (B) maximum of the A_{578} peak; (C) second fraction of the radioactivity plateau, which corresponds approximately to fraction 28 of the gradient; (D) the beginning of the decrease in the radioactivity curve (fractions 33 to 34); (E) one-sixth of the maximum height of the A_{578} curve in the descending limb (about fraction 40). The position of the corresponding fractions is indicated in Fig. 2 by arrows. As an example, Fig. 3 shows the results obtained by observation of an *in vitro* culture derived from fraction-D cells. The rate of DNA synthesis, the mitotic index, and the growth rate undergo rhythmic variations which are in an ordered relation to one another. Each maximum of DNA synthesis corresponds to a low mitotic index and to a stagnancy of the cell number. And, vice versa, the mitotic index and the growth rate are found to be elevated when DNA synthesis is low. The first peak of mitotic cells occurs 6 hours after explanation. A

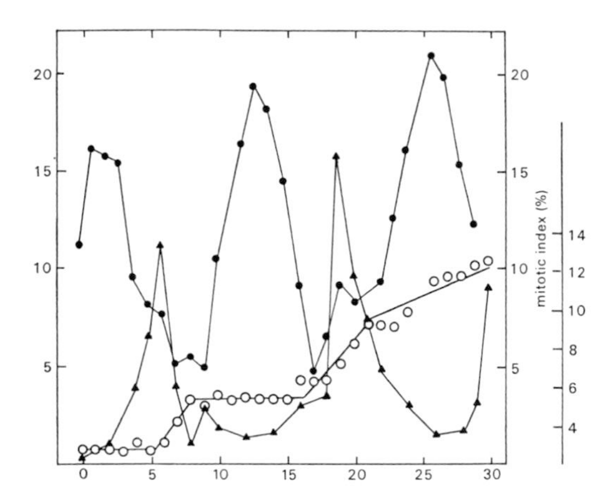

FIG. 3. *In vitro* growth of cells derived from fraction D. Abscissa: time (hours); ordinate (left) ^{3}H-cpm/10^3 cells; (right, outer) cells/ml $\times 10^{-5}$. Cells from the zonal centrifugation fraction D (see Fig. 2) were washed once with medium, suspended in 100 ml of prewarmed medium, and held in an incubator at 37°C. At the times indicated on the abscissa, two milliliter samples were removed from the culture. One milliliter of each sample was incubated for 20 minutes at 37°C with 2.5 μCi thymidine-^{3}H and analyzed for radioactivity incorporated. The rest of the sample was used for the determination of the cell number and the mitotic index. Closed circles, ^{3}H (counts per minute per 10^3 cells); triangles, mitotic index (%); open circles, cells per milliliter. Probst and Maisenbacher (1973).

second peak follows 13 hours later. The distance between the second and the third peak of DNA synthesis is also 13 hours.

These data indicate that the cells of fraction D represent a synchronous population growing with a cell cycle duration of about 13 hours. Examination of the other fractions indicated in Fig. 2 shows that they also contain synchronous populations growing with essentially the same cell kinetics. If we arrange the cultures derived from these fractions in order corresponding to the time after explantation at which mitotic activity begins to rise significantly, we arrive at the following sequence: E, 3 hours; D, 4 hours; C, 6 hours; B, 12 hours; A, 14 hours. This means that, the faster the cells sediment, the closer they are to mitosis. When grown *in vitro* after separation by zonal centrifugation, the cells from fraction A behave as relatively pure G_1 cells; those from fraction B as a mixture of G_1 cells and cells from early S phase; those from fraction C as cells from early S; those from fraction D as cells from middle to late S; and those from fraction E cells as a mixture of G_2 cells and cells of late S phase. A fraction containing pure G_2 cells or mitotic cells is not found in any position of the gradient. In general, the degree of synchrony is relatively low in the last fractions of the gradient because they contain varying amount of aggregated cells from earlier stages of the cell cycle.

The cells of fractions C to E appear to continue cycling immediately after explanation, whereas the cells of fractions A and B show a lag period. We presume that this lag period is caused by exposure of the cells to a serum-free environment during the zonal centrifugation process. These facts are discussed in detail by Probst and Maisenbacher (1973).

V. Possible Applications

There are two principally different types of application: (1) studies on synchronously growing cell cultures derived from distinct zonal centrifugation fractions, and (2) direct examination of the different fractions obtained.

For application (1) cells from the middle third of the descending limb of the A_{578} curve appear to be the most suitable. From this region pure S-phase cells are recovered, and it is possible to choose, within certain limits, the starting point of the culture within the S phase. Such cell populations continue to cycle immediately after being suspended in a culture medium at 37°C. In contrast, G_1 cells found in the preceding fractions are less suitable, because of the lag period mentioned above. Cultures of cells from the last fractions of the gradient have an unsatisfactory degree of synchrony.

Application (2) is perhaps the more interesting one, because it offers the possibility of measuring biochemical parameters, e.g., enzyme activities or metabolite concentrations in cells belonging to different cell cycle stages of a real *in vivo* tumor. The interesting fractions of a "normal" zonal run (with cells from five to six mice) contain 6–60 $\times$ 10^6 cells. This is sufficient starting material for a wide range of biochemical analyses. When such studies include the preparation of distinct cell compartments, e.g., nuclei or mitochondria, relatively large amounts of starting material are frequently necessary. In these cases we mostly resigned the high degree of synchrony that can be achieved by collecting small fractions and pooled neighboring fractions of the gradient as depicted in Fig. 4. On the basis of the *in vitro* growth of cells from the borders of these "large" fractions, we defined their contents as follows. Fraction 1 contains exclusively cells of the G_1 phase; they do not as yet synthesize DNA. Fraction 2 consists of cells of the late G_1 phase and of the early S phase. Fraction 3 contains cells of the middle and the late S phase. Fraction 4 is a mixture of cells from the late S phase and the G_2 phase and, as mentioned above, it contains additionally varying amounts of aggregated cells from earlier stages of the cell cycle. Pure G_2 cells or mitotic cells cannot be obtained by this method. The correctness of the above classification is confirmed by the protein and the DNA content per cell

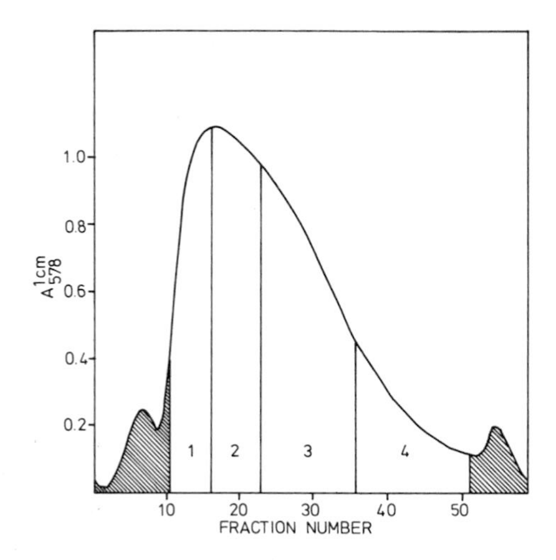

FIG. 4. A_{578} pattern of a zonal run performed with the cells of 10 mice (about 5×10^9). The vertical lines limit the "large" fractions mentioned in the text. The cells of the hatched areas were discarded. The small peak appearing at the end of the gradient is caused by lumped cells which accumulated at the boundary between the gradient and the underlay. This peak is frequently observed when the passage of the boundary is monitored in experiments with large amounts of cells.

number in these fractions. The cells of fraction 4 contain nearly twice as much protein and DNA compared with fraction-1 cells. The protein content increases stepwise from fraction 1 to 4, while the steepest increase in DNA content is between fractions 2 and 3.

ACKNOWLEDGMENT

The work described in this chapter was supported by a grant from the Deutsche Forschungsgemeinschaft (Pr 95/3).

REFERENCES

Karzel, K. (1965). *Med. Pharmacol. Exp.* **12**, 137–144.
Karzel, K., and Schmid, J. (1968). *Arzneim. Forsch.* **18**, 1500–1504.
Mitchison, J. M., and Vincent, W. S. (1965). *Nature (London)* **205**, 987–989.
Probst, H., and Maisenbacher, J. (1973). *Exp. Cell Res.* **78**, 335–344.
Warmsley, A. M. H., and Pasternak, C. A. (1970). *Biochem. J.* **119**, 493–499.

Chapter 11

Methods with Insect Cells in Suspension Culture *I. Aedes albopictus*

ALLAN SPRADLING, ROBERT H. SINGER,[1]
JUDITH LENGYEL, AND SHELDON PENMAN

Department of Biology,
Massachusetts Institute of Technology,
Cambridge, Massachusetts

I. Introduction 185
II. Adaptation of *A. albopictus* Cells to Desired Growth Conditions 186
III. Characteristics of the Cell Line 187
IV. Cell Fractionation and Other Procedures 189
V. Conclusion. 193
References. 194

I. Introduction

Techniques for growing cells and tissues of insects in culture have improved dramatically in the last 10 years. There now exist established lines derived from numerous species, particularly among the Diptera and Lepidoptera, and the number is rapidly increasing as methods favorable to establishment are refined (reviewed by Brooks and Kurtti, 1971). Eventually, it may be possible to obtain cell lines with known genetic alterations by the choice of an appropriate mutant strain as starting material.

The conditions under which these lines must be grown, however, are not well suited to investigations of insect cell biology at the molecular level. It is not always possible, for example, to grow the cells in conventional suspension culture and thereby obtain the quantities of macromolecules neces-

[1] *Present address*: Department of Anatomy, University of Massachusetts Medical School, Worcester, Massachusetts.

sary for many types of analysis. Even when this can be done, the media commonly used for propagating the cells are expensive and extremely complex, often containing high concentrations of precursor molecules which render radioactive labeling inefficient. In addition, techniques for cellular fractionation analogous to those developed for cultured mammalian cells cannot in general be used for insect lines.

Because it was clear that the new insect cell lines would provide interesting systems for basic studies of cell physiology, we undertook several years ago the adaptation of one such line to growth conditions more suitable to this purpose. Over a period of about a year, cells of the *Aedes albopictus* line established by Singh (1967) were adapted to growth in suspension culture in Eagle's medium supplemented only with nonessential amino acids and serum. Exponentially growing cells of this subline may consequently be labeled with radioactive precursor molecules as readily as cultured mammalian cells. The development of fractionation procedures for the mosquito cells proved to be a considerably more difficult problem, especially for RNA, and in some cases, such as the subfraction of nuclei, the problem is still not satisfactorily resolved.

We describe here the methods utilized for adapting *Aedes* cells and for their fractionation, as well as some of the basic cellular parameters as measured by these techniques. More recently, a similar approach has made possible the adaptation of the Schneider *Drosophila* line no. 2 (Schneider, 1972) to similar conditions of growth, as described in Chapter 12.

II. Adaptation of *A. albopictus* Cells to Desired Growth Conditions

The optimum composition of the medium used to propagate a particular line of insect cells has remained a matter of some controversy on both theoretical and practical grounds (Schneider, 1971; Stanley, 1972). Commonly, it has been suggested that parameters of the medium should closely parallel values found in the hemolymph of the species for which it is used. However, the well-known variability of such hemolymph parameters as ion balance (Florkin and Jeuniaux, 1964), in addition to questions of its physiological significance for the unknown cell types in culture, at present leave this question unresolved. Empirically, many insect lines have been found to grow well in a variety of media, some of which differ considerably from hemolymph (Brooks and Kurtti, 1971; Schneider, 1972). In undertaking this project we attempted to utilize, within limits imposed by a constant karyotype and undiminished growth rate, this apparent plasticity of many insect cells.

The *A. albopictus* line was established using the medium of Mitsuhashi and Maramorosch (1964), originally designed for the culture of leafhopper cells. It contains large quantities of yeast extract and lactalbumin hydrolysate, and is supplemented with fetal bovine serum (FBS) for use with mosquito cells. Since conditions of growth were desired that would permit optimum labeling with nucleotides, a minimum change was the elimination of the requirement for yeast extract. Eagle's minimal essential medium (MEM) (Joklik-modified) was used as a basic medium, since the salts and osmotic pressure are similar to M. + M. media and its additional vitamins might help eliminate the yeast extract requirement. Amino acid concentrations were increased by supplementing with 1% nonessential amino acids and 1% lactalbumin hydrolysate. When *Aedes* cells were seeded into 250-ml Falcon plastic tissue culture flasks and fed with the supplemented Eagle's medium plus 10% FBS, growth continued after a brief delay. The cells were allowed to grow under these conditions for several months. Half the medium was replaced every 3 days. New cultures were seeded by briskly shaking a flask to release loosely attached cells into the medium, which was then added to a new flask.

After it was clear that a subline capable of growth in the new media had been obtained, adaptation to suspension culture was undertaken. Because conditions in a suspension culture are physically rigorous, it did not seem advisable to begin this step until a well-adapted line had been obtained in monolayer culture. Indeed, several additional months of passaging between flask and suspension cultures followed before a subline was obtained that grew readily in suspension.

At this point it was determined that supplementation with lactalbumin hydrolysate was no longer necessary. It thus became possible to label cell protein with radioactive amino acids (e.g., leucine), either directly or by preparing the corresponding amino acid–deficient medium. Elimination of the nonessential amino acid supplement caused the cells eventually to stop growing, which suggests that at least one of the seven amino acids not required for growth by mammalian cells is a required amino acid for mosquito cells. Elimination of the serum also resulted in a slow decline in the growth rate of the culture. The nutritional requirements of the cells were not examined further.

III. Characteristics of the Cell Line

The *A. albopictus* subline obtained as a result of these procedures seems to have undergone few changes in its measurable properties. Morphologically, the cells from suspension culture are of two basic types. Spherical cells about 10 μm in diameter predominate over a similar class of cells which contain one or more spinelike projections, commonly referred to as spindle-shaped cells. Each contains a single large nucleus, usually with one prominent nucleolus. The cells display a strong tendency to attach to glass or plastic surfaces and to each other. Thus, in suspension cultures, clumps form, some of which contain many hundreds of cells.

If suspension-grown cells are seeded at a relatively low density into a plastic tissue culture flask, virtually all the cells rapidly attach to the surface. As the culture grows, dense, often multilayered arrays of cells are produced. In addition, a much wider variety of cell morphology is present under these conditions. Cytoplasmic bridges connecting widely spaced cells, multinucleate syncytia, and other complex forms are often common in such a culture.

The growth rate of the cells at 25°C is exponential, with a doubling time of approximately 21 hours. However, as might be expected for cells from a poikilothermic organism, exponential growth is observed over a considerable range of temperature. *Aedes* cells maintained at 32°C, the highest temperature investigated, multiply more rapidly than at the lower temperature, doubling about every 15 hours. Suspension cultures have also been maintained at 20°C with good growth.

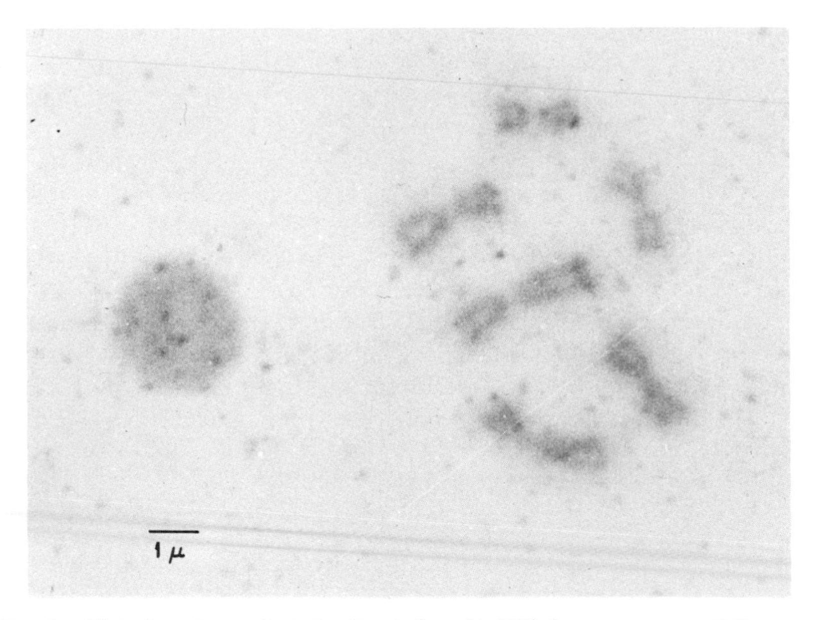

FIG. 1. Metaphase from adapted cells. *Aedes* cells (10^7) from an exponentially growing suspension culture were arrested overnight with 1 μg/ml vinblastine sulfate. The cells were hypotonically treated and then fixed in 3:1 ethanol–acetic acid. Slides were prepared by the air-drying technique and stained with Giemsa. ×6400.

Karyotypes as determined from metaphase-arrested cells passaged about 1 year in suspension culture indicate that the population consists of approximately 70% diploid (six chromosomes), 30% tetraploid, and a small number of cells of higher ploidy. This is similar to the karyotypic frequencies reported by other workers (Stevens, 1970; Bianchi *et al.*, 1971). Even though the number of chromosomes appears to be normal, considerable alteration of the genome can occur as a result of extensive chromosomal translocation and rearrangement. Indeed, some studies of the original Singh line indicate that changes of this nature are detected with high frequency (Bianchi *et al.*, 1971). Although we have not made an extensive analysis, translocations in the adapted cells are easily detected. A typical metaphase is shown in Fig. 1.

Knowledge of the karyotype makes feasible a measurement of the genome size for this species. Diphenylamine assays yield a value, corrected for the fraction of polyploid cells, of 5.7×10^{11} daltons DNA per haploid genome. Since the relative lengths of the stages in the cell cycle have not been determined, however, some error would be introduced should they differ significantly from CHO cells used for comparison. The value obtained is in approximate agreement with a kinetic complexity of 4.1×10^{11} daltons determined by renaturation kinetics of DNA from the cell line (Spradling *et al.*, 1974).

IV. Cell Fractionation and Other Procedures

In many respects the extraction of macromolecules from the *Aedes* cell line proved to be routine. Total nuclear DNA, for example, can be prepared in good yield from isolated nuclei by standard sodium dodecyl sulfate (SDS)–high salt procedures (Marmur, 1961). However, it became clear very early that virtually any procedure that ruptured the cell membrane simultaneously released a RNase activity that rendered recovery of intact molecules of some types of RNA very difficult.

Two tests were employed to guide our efforts at overcoming this problem. First, direct lysis in SDS of whole cells labeled briefly with uridine indicated that *Aedes* nuclei contain RNA molecules of very large size (greater than 50 S), hence any procedure that did not result in a similar recovery of large molecules was clearly inadequate. Second, direct tests of RNase activity were made by coextracting ^{3}H-labeled *Aedes* cells and ^{14}C-labeled HeLa cells. A decrease in the size of the appropriate ^{14}C HeLa fraction compared to a similar preparation of HeLa RNA extracted alone was taken as evidence of degradative activity.

By utilizing this approach, several procedures have been developed for the fractionation of *Aedes* cells based on standard methods used for mammalian cells (Penman, 1966). One of two methods is employed to lyse the cell membrane. Treatment of cells in an isotonic buffer [100 m*M* NaCl, 10 m*M* $MgSO_4$, 30 m*M* tris (pH 8.3)] with 0.5% of the detergent NP40 yields 100% cell lysis, leaving nuclei with virtually no cytoplasmic contamination visible by phase microscopy. Alternatively, when detergents must be avoided, as in the preparation of a mitochondrial fraction, a hypotonic medium is used [20 m*M* NaCl, 5 m*M* $MgSO_4$, 10 m*M* tris (pH 8.3)]. *Aedes* cells are very unstable at this tonicity, and a few strokes with a Dounce homogenizer results in nearly complete lysis. Nuclei are pelleted by centrifugation at 800 *g* for 5 minutes. If the resulting supernatant is centrifuged an additional 10 minutes at 8000 *g*, a crude mitochondrial fraction is obtained.

Early experiments in which poly(A)-containing mRNA was prepared by passing RNA purified from the cytoplasmic fraction over oligo-dT cellulose consistently yielded preparations with a mean sedimentation coefficient of about 15 S compared to 18 S for similar preparations from mammalian cells. Coextraction of *Aedes* and HeLa cells revealed that the RNA was not, however, being extracted intact, despite this uniformity of results. Extensive experimentation suggests that such degradation is totally eliminated by the

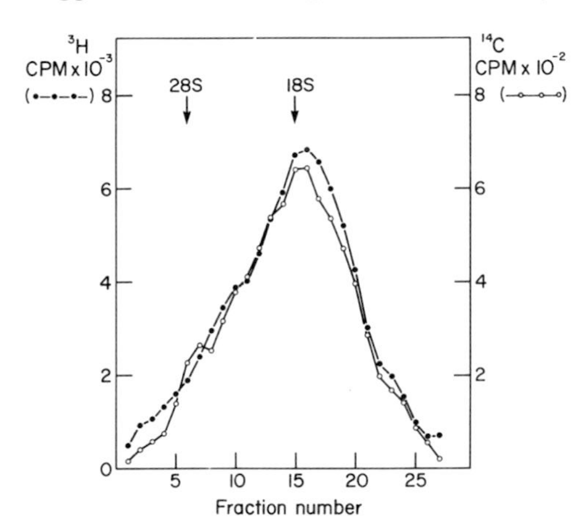

FIG. 2. Coextraction of *Aedes* and HeLa mRNA. *Aedes* cells (2.5×10^7) concentrated 5× to a density of 5×10^6 cells/ml were labeled 2 hours with 10 μCi/ml adenosine-^{3}H (New England Nuclear, 26 Ci/mmole). HeLa cells (1.2×10^7) were concentrated to 2×10^6 cells/ml and labeled for 2 hours with 0.5 μCi/ml adenosine-^{14}C (New England Nuclear, 50 mCi/mmole). The two cultures were then mixed, and a cytoplasmic extract prepared as described for *Aedes* cells. mRNA was purified by phenol extraction and oligo-dT cellulose chromatography. Closed circles, mRNA-^{3}H eluted from oligo-dT cellulose; open circles, mRNA-^{14}C eluted from oligo-dT cellulose.

addition of polyvinyl sulfate (PVS) (25 μg/ml) and diethylpyrocarbonate (0.5%) to the lysis medium just prior to use. Spermidine (35 μg/ml) is added as well, since this polycation counteracts the destabilizing effect of PVS on cell nuclei.

By using these procedures, mRNA from *Aedes* cells labeled for 1 hour with uridine-^{3}H is found to sediment with a mean at about 18 S, very much like mRNA from mammalian cells. This result is probably not due to aggregation of RNA, since the sedimentation distribution is unchanged relative to rRNA when sedimentation is carried out under denaturing conditions in 99% dimethyl sulfoxide (DMSO). The steady-state size distribution of *Aedes* mRNA is somewhat smaller than that of newly labeled material, however (Spradling and Penman, 1974). Figure 2 shows by coextraction that *Aedes* mRNA prepared by these techniques is probably totally intact. mRNA extracted from polysomes (Fig. 3) has a sedimentation distribution identical to that prepared from whole cytoplasm. Treatment of the cytoplasmic extract with EDTA prior to centrifugation causes more than 90% of the polysomal O.D. and the associated poly(A)-containing RNA to sediment near the top of the gradient.

Two properties of *Aedes* rRNA are worthy of note. Figure 4A shows a typical preparation of cytoplasmic RNA coelectrophoresed with HeLa rRNA. Both the *Aedes* species are seen to differ in apparent size from HeLa rRNA, a conclusion also supported by cosedimentation in sucrose gradients. Furthermore, as indicated in Fig. 4A, the relative amounts of large and small rRNA are generally not found in the expected 2:1 ratio. The origin of

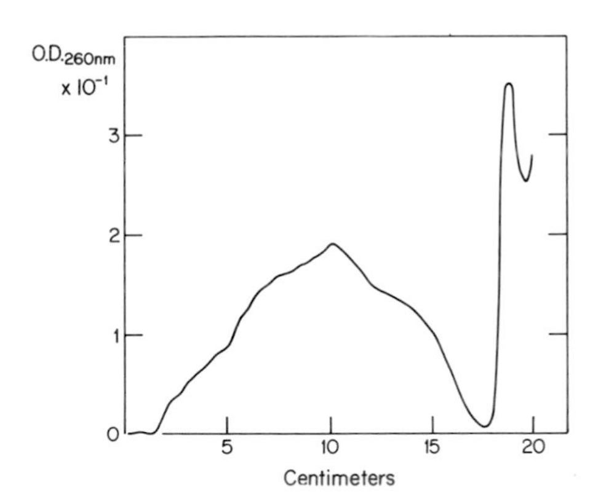

FIG. 3. *Aedes* polysomes. A cytoplasmic extract was prepared from 2 × 10 10^7 *Aedes* cells in 1 ml as described in the text. It was layered over a 16.5-ml, 15–30% (w/w) sucrose gradient in lysis medium and centrifuged 2 hours at 25,000 rpm in a Spinco SW27. Absorbance was monitored by pumping the gradient from the bottom through a continuous-flow spectrophotometer.

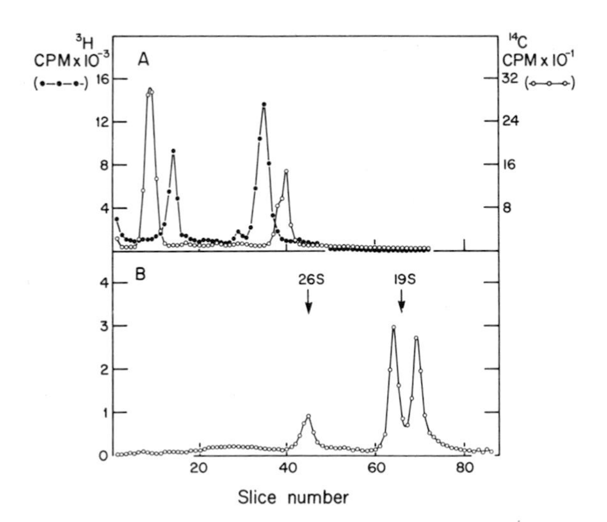

FIG. 4. *Aedes* rRNA. (A) HeLa uridine-^{3}H-labeled whole cytoplasmic RNA (approximately 10^5 cpm) was combined with about 5×10^3 cpm of a similar preparation of *Aedes* RNA labeled with uridine-^{14}C. The sample was electrophoresed as described in Hirsch *et al.* (1974) on a 3.5% polyacrylamide gel for 7 hours. (B) Approximately 2×10^4 cpm of uridine-^{3}H-labeled 26 S *Aedes* rRNA prepared by pooling the appropriate fractions from a sucrose gradient was combined with 10^3 cpm of uridine-^{14}C-labeled *Aedes* cytoplasmic RNA as a marker. The sample was electrophoresed as above, except that the acrylamide concentration was 2.8%.

this anomaly is illustrated in Fig. 4B. 26 S rRNA isolated from a sucrose gradient was heated at 70°C for 5 minutes and then electrophoresed on a 2.8% acrylamide gel. The bulk of the 26 S species dissociated into two component molecules, one of which migrates slightly faster and one somewhat slower than 19 S rRNA. Some breakdown occurs during cell fractionation, and under most conditions the product species comigrate with the small rRNA thus giving rise to the anomalous ratio. The instability of the large rRNA of many invertebrate species has been described previously (Applebaum *et al.*, 1966; Shine and Dalgarno, 1973).

Appropriately labeled whole *Aedes* cells or isolated nuclei, when lysed directly with an SDS-containing buffer, are found to contain large heterogeneous nuclear RNA as well as rRNA precursors. So far, however, it has not been possible to subfractionate the nuclei into nucleolar and nucleoplasmic fractions consistently while maintaining the full size of the HnRNA.

When working with an unfamiliar cell line, often the most elementary procedures can cause problems. For example, early measurements of the growth characteristics of the *Aedes* cells yielded erratic results. Eventually, it became clear that this was due to the tendency of the cells to clump together while growing. Big clumps tend to be lost from a sample, since they settle rapidly. In addition, unless the cells are stirred at rapid rate, a portion of them will attach to the culture bottle.

These problems were overcome by a combination of care in maintaining constant conditions of culture agitation and by lysing aliquots of cells with NP40 and counting cell nuclei rather than cells in the hemocytometer. The detergent plus a brief vortexing disperses the clumps and yields accurate counts.

Stocks of *Aedes* cells may be stored frozen at −80°C (or preferably −170°C) if 10% glycerol or DMSO is added as a cyroprotectant. Even when cells are simply allowed to stand in a flask at room temperature without feeding, they remain viable for very long periods of time (greater than 6 months).

V. Conclusion

The development of an adapted line of *A. albopictus* that can be manipulated in much the same manner as cultured mammalian cells should significantly increase the utility of this line in many areas of cell biology. The RNA metabolism of togaviruses during infection of insect cells, for example, has not been well characterized, at least in part because of a lack of the methods described here. Yet such studies might lead to a better understanding of the mechanisms that result in lytic infection of vertebrate hosts by togaviruses which establish a persistent infection in the cells of their insect vectors (Stollar *et al.*, 1972). Another use of the line is in making comparisons between cells from organisms that are widely divergent on an evolutionary scale. Several aspects of RNA metabolism have already been examined in this regard (Hirsch *et al.*, 1974; Spradling *et al.*, 1974). The fact that we have been able to adapt two lines (see Chapter 12) to favorable growth conditions suggests that many insect lines could be similarly propagated. The availability of a variety of such cell lines would undoubtedly be a stimulus for increasing the contributions of work on cultured insect cells to our understanding of eukaryotic molecular biology.

Acknowledgments

This work was supported by grants from the National Institutes of Health (NIH 5 R01 CA08416; NIH CA12174) and from the National Science Foundation (NSF GB37709X). J.A.L. is the recipient of an NIH postdoctoral research fellowship (GM54218–02).

References

Applebaum, S. W., Ebstein, R. P., and Wyatt, G. R. (1966). *J. Mol. Biol.* **21**, 29–39.
Bianchi, N. O., Sweet, B. H., and Ayres, J. (1971). *Exp. Cell Res.* **69**, 236–239.
Brooks, M. A., and Kurtti, T. J. (1971). *Annu. Rev. Entomol.* **16**, 27–52.
Florkin, M., and Jeuniaux, C. (1964). *In* "The Physiology of Insecta" (M. Rockstein, ed.), 1st ed., Vol. 3, pp. 109–152. Academic Press, New York.
Hirsch, M., Spradling, A., and Penman, S. (1974). *Cell* **1**, 31–35.
Marmur, J. (1961). *J. Mol. Biol.* **3**, 208–218.
Mitsuhashi, J., and Maramorosch, K. (1964). *Contrib. Boyce Thompson Inst.* **22**, 435–460.
Penman, S. (1966). *J. Mol. Biol.* **17**, 117–136.
Schneider, I. (1971). *Curr. Top. Microbiol. Immunol.* **55**, 1–12.
Schneider, I. (1972). *J. Embryol. Exp. Morphol.* **27**, 353–365.
Shine, J., and Dalgarno, L. (1973). *J. Mol. Biol.* **75**, 57–72.
Singh, K. R. P. (1967). *Curr. Sci.* **36**, 506–508.
Spradling, A., and Penman, S. (1974). In preparation.
Spradling, A., Penman, S., Campo, M. S., and Bishop, J. O. (1974). *Cell* **3**, 23–30.
Stanley, M. S. M. (1972). *In* "Growth, Nutrition, and Metabolism of Cells in Culture" (G. H. Rothblat and V. J. Cristofalo, eds.), Vol. 2, pp. 327–370. Academic Press, New York.
Stevens, T. M. (1970). *Proc. Soc. Exp. Biol. Med.* **134**, 356–361.
Stollar, V., Shenk, T. E., and Stollar, B. D. (1972). *Virology* **47**, 122–132.

Chapter 12

Methods with Insect Cells in Suspension Culture II. Drosophila melanogaster

JUDITH LENGYEL, ALLAN SPRADLING AND SHELDON PENMAN

Department of Biology,
Massachusetts Institute of Technology,
Cambridge, Massachusetts

I. Introduction 195
II. Adaptation of the Line to a New Medium and to Suspension Growth 197
A. Composition of the New Medium 197
B. Adaptation to Suspension Culture 198
III. Characteristics of the Cells 200
A. Morphology 200
B. Karyotype 200
C. Macromolecular Components 201
D. Endogenous Virus 202
IV. Cell Fractionation 203
A. Procedures 203
B. Size of Nuclear RNA, mRNA, and rRNA—Absence of Degradation . . 204
V. Summary and Conclusions 207
References 208

I. Introduction

The fruit fly *Drosophila melanogaster* has for many years been the subject of intensive genetic analysis and has provided much of our present understanding of the organization of the eukaryotic chromosome. It is in many ways an ideal organism for studying, at the molecular level, mechanisms by which expression of the eukaryotic genome is controlled. In addition to the fact that it is amenable to genetic analysis, *Drosophila* has several other useful features. It has one of the smallest metazoan genomes known (Shapiro, 1970), containing approximately 10×10^{10} daltons of DNA (Rasch *et al.*, 1971; Laird, 1971), of which at least 80% is a unique sequence (McConaughy *et al.*, 1969; Wu *et al.*, 1972). Also, the chromosomes of a number of larval tissues are polytenized and display a highly reproducible banding pattern. This makes possible the correlation of genetic and cytological maps, and also the localization on the chromosome of sequences complementary to given RNAs or DNAs by cytological hybridization *in situ* (Pardue *et al.*, 1970).

The recent establishment of *Drosophila* cell lines in culture (Kakpakov *et al.*, 1969; Echalier and Ohanessian, 1970; Schneider, 1972) has increased the advantages of using *Drosophila* for studying gene transcription. With a cell line it is possible to ask detailed biochemical questions of a homogeneous population of cells, and to manipulate the cells in ways that are not possible using the whole organism.

We became interested in studying RNA metabolism in the Schneider *Drosophila* line as a first step toward understanding controls involved in *Drosophila* gene transcription. Some of the growth characteristics of the Schneider line, however, were unsuitable for the types of analysis we wished to undertake. Some cells grew attached to the plate, while others were free in the medium; thus there did not appear to be a homogeneous population. Furthermore, although Schneider's medium supports excellent cell growth, it does not permit efficient labeling of nucleic acids to high specific activity, since it contains yeast extract and therefore large amounts of nucleic acid precursors.

In order to analyze unstable RNA species such as HnRNA and mRNA under physiological conditions, it is essential to be able to label RNA to high specific activity in the same medium in which the cells are growing (preferably in a suspension culture). The advantage of suspension culture in providing a homogeneous, logarithmically growing population of cells in large quantity has been recognized by many investigators. Several workers have in fact adapted the Schneider *Drosophila* line to grow in suspension in Schneider's medium or a modification thereof (Hanson and Hearst, 1973; E. Weinberg and R. Sederoff, personal communication). However, nucleic acids in these cells can be labeled to high specific activity only if the cells are transferred to a medium lacking yeast extract; such a "shiftdown" may induce alterations in RNA metabolism. Furthermore, such a protocol allows for brief labeling only, and does not allow labeling to high specific activity over a long period of cell growth.

We therefore adapted the Schneider *Drosophila* line 2 to grow in suspension in a modified Dulbecco's medium. While the adaptation to growth in flasks in the new medium paralleled that of *Aedes albopictus* cells (described in

Chapter 11), adaptation to growth in spinner culture proved more difficult. These techniques are therefore given in some detail, as they may prove useful in the adaptation of other *Drosophila* lines, such as those derived from various mutants (see, for example, Parádi, 1972).

After the prolonged adaptation period, most of the cells were euploid, (approximately half were diploid) and contained normal-appearing *Drosophila* chromosomes. The cells contain the expected amounts of DNA and RNA. We have developed techniques for fractionating the cells into cytoplasm and nucleus, and have characterized the size distribution of nuclear RNA, rRNA, and mRNA from the cell line.

II. Adaptation of the Line to a New Medium and to Suspension Growth

A. Composition of the New Medium

For the reasons outlined above, we sought to adapt the Schneider *Drosophila* line to a medium better fitting our experimental requirements. A simple medium, lacking nucleic acid precursors, which would be inexpensive and commercially available, was desired. The only media satisfying these criteria were several used for mammalian cell culture. These media differ drastically from Schneider's medium with respect to pH, ion balance, and content of organic ions, since the former are patterned on mammalian serum while the latter is based on *Drosophila* larval hemolymph. Schneider's medium, like *Drosophila* hemolymph, has a lower pH (6.7–6.8), a lower Na^+/K^+ ratio, lower monovalent/divalent cation ratios, and a much higher amino acid content (contributing 40% of the total tonicity) than mammalian media, and also contains the unusual sugar trehalose. The attempt to adapt *Drosophila* cells growing in Schneider's medium to a mammalian culture medium might therefore appear to be a dubious undertaking. However, the necessity of using a medium based on hemolymph for the growth of insect cells has been debated (Jones, 1966; Brooks and Kurtii, 1971; Stanley, 1972). Insect cells are capable of growth in a wide variety of media; particularly, the ability of the Schneider *Drosophila* line to grow in several different media has been demonstrated (Schneider, 1972).

Dulbecco's modified Eagle's medium was chosen as the basis of the new medium. Dulbecco's modified Eagle's medium is relatively simple and inexpensive, and contains a fairly high concentration of amino acids and vitamins, making it more similar to Schneider's medium than, for example, Joklik modified Eagle's medium, F-10, or F-12. To further increase the amino acid content of the medium, Bactopeptone [0.5% (w/v)] and 1 × Gibco (Grand Island, New York) MEM nonessential amino acids were added. The medium was supplemented with 10% fetal bovine serum; penicillin (100 units/ml) and streptomycin sulfate (100 μg/ml) were also included.

Schneider *Drosophila* line 2 cells, growing in a flask culture, were adapted to this medium by gradual replacement of the Schneider's medium. All cell cultures described here and subsequently were maintained at 25°C. After a period of months, Bactopeptone was gradually eliminated from the medium. The adapted cells did not adhere firmly to the bottom of the flask; rather, many were floating freely in the medium, and many more could be dislodged from the flask by vigorous shaking. It therefore appeared that the cells might have the capacity to grow completely in suspension if the appropriate growth conditions and adaptation conditions were provided.

B. Adaptation to Suspension Culture

After several fruitless attempts were made to adapt the cells in Dulbecco's medium to suspension growth, it seemed likely that the medium was inadequate to support growth under the additional stress imposed by adaptation to a suspension culture (cells had been growing very slowly in this medium in a stationary culture). Therefore supplements were added back to the medium. Bactopeptone and lactalbumen hydrolysate (Gibco) were added at concentrations of 0.5% (w/v) and 0.1% (w/v), respectively. Biotin (0.15 μg/ml) was also added, since it is a fairly common vitamin which insect cells may require (Vaughn, 1971). Finally, hypoxanthine (4 μg/ml) was added, since it is a nucleic acid precursor which might be expected to have some of the stimulatory effect of the yeast extract present in Schneider's medium, yet does not interfere with adenosine or uridine labeling.

The first suspension cultures were begun in roller bottles, since it was felt that this might be less damaging to the cells than spinner culture. However, the selection of containers in which to grow the cells presented a problem. Attempts to grow the cells in glassware that had been washed and autoclaved were not successful. Since washing procedures can introduce residues into glassware which are toxic to cultured cells (Parker, 1961), cultures were started in unwashed, empty fetal bovine serum bottles.

Cells suspensions (150 ml, 2–5 × 10^5 cells/ml) from flasks were inoculated into 500-ml bottles and rolled at 2 rpm. The growth of the cells was very erratic; they usually went through several divisions during the first few days of rolling suspension culture, and then ceased to grow for periods of up to a week. They were fed at least once every 5 days, with replacement of at least 25% of the medium. This infrequent feeding was adequate to sustain the cells, and was preferable to diluting nongrowing cells by repeated feeding. When a bottle of cells remained at a low density for as long as a week, it was removed from the roller and kept stationary until the cells started to grow again (usually about 2 weeks), at which time feeding and rolling were resumed. Using this procedure three roller bottle cultures of cells were developed, in which the cells grew to densities of $\sim 6 \times 10^5$/ml with a generation time of approximately 48 hours. As adaptation of these cells to growth in the roller bottles was continued, further alterations in their growth characteristics occurred. The cells grew to higher densities ($\sim 3 \times 10^6$/ml), with a shorter generation time (30–36 hours), and began to attach to the walls of the bottle; roller speed was then increased to 3 rpm. The medium the cells were growing in became more acidic, probably because of increased cell density.

Once stably growing roller bottle cultures had been established, adaptation to growth in spinner bottles was undertaken in order to provide larger volumes of cells. Because Bactopeptone interferes with uridine incorporation (probably because of a low level of bacterial nucleic acids), it was eliminated from the medium, and the lactalbumen hydrolysate concentration increased to 0.5%. For the first spinner bottle cultures, it was necessary to alternate periods of spinning with periods of stationary growth, over a period of a month, in order to adapt the cells to the bottle. It was then found that this period of adaptation to the spinner bottle could be reduced to several days if the bottle were partially filled with distilled water and then autoclaved. This apparently removes inhibitory residues. Autoclaving with distilled water is no longer necessary when making subcultures from a spinner bottle. Spinner bottle cells now grow exponentially at densities of $1\text{–}4 \times 10^6$/ml, with a generation time of 30 hours. Densities as high as 1×10^7/ml have been reached.

We have now eliminated hypoxanthine and biotin from the medium without observing any deleterious effects on growth; the present medium consists of Dulbecco's modified Eagle's medium supplemented with 10% fetal bovine serum, 0.5% lactalbumen hydrolysate, and 1× Gibco MEM nonessential amino acids, and contains penicillin and streptomycin sulfate. Presumably, it is possible to adapt the cells to a medium lacking lactalbumen hydrolysate, as has been described for *A. albopictus* in Chapter 11.

The cells can be maintained in stationary culture, from which logarithmically growing spinner cultures can be started within 5 days, or they can be kept frozen at −70°C in a culture medium containing 10% glycerol and 10% additional serum.

Although the effect of different culture conditions on cell growth has not been investigated systematically, continued observation of cell growth over a period of 6 months allows several conclusions to be reached regarding conditions most important to the successful growth of these cells. Maintenance of a low pH (below 7) is probably most important. Cells growing in a flask culture in which the pH becomes too high detach from the surface and lyse. The culture medium is therefore adjusted to pH 6.9, and cultures gassed with a 5% CO_2–air mixture. Cell density is also important; cells grow more rapidly when kept at a fairly high density (at least 1×10^6/ml). This effect of density on growth rate is presumably due to conditioning of the medium by the cells; part of this conditioning may consist of acidification of the medium.

Finally, it may be of interest that *Drosophila* cells also grow well in flasks in F-10 or F-12 medium supplemented with nonessential amino acids and lactalbumen hydrolysate. Thus it is probably possible to adapt the cells to growth in suspension in these media as well. Clearly, the cells are rather plastic in terms of the different media in which they can grow.

III. Characteristics of the Cells

A. Morphology

Even after the prolonged period of adaptation, the cells remain somewhat variable in morphology. Both spherical cells and spindle-shaped cells, some of which bear long rhizoidal processes, are present; many cells have a halo of short, filamentous, hairlike protrusions. The cells contain a well-defined nucleus with a single nucleolus.

During adaptation to a spinner culture, the cells became significantly smaller than they were when growing in a flask culture, or as originally described by Schneider (1972). The diameter of the adapted cells ranges from 5 to 14 μm, with an average of 7.5 μm, whereas the original cell line consisted predominantly of cells 5–11 μm in width and 11–35 μm in length (Schneider, 1972).

B. Karyotype

For genetically relevant work to be done with these cells, it is important to show that, after adaptation to a new medium and to growth in suspension culture, the chromosome complement is not dramatically different from the original predominantly diploid XX state. A number of metaphase figures, both from normally growing cells and from cells arrested for 16 hours with vinblastine sulfate (1 μg/ml) were examined. The chromosomes were enumerated, without counting the fourth chromosome, since it is quite small and thus easily obscured. The majority of metaphases contained at least one-

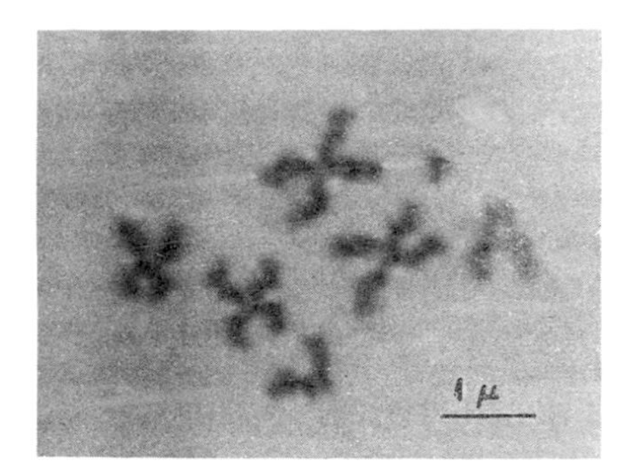

FIG. 1. Metaphase figure from the adapted cell line. Cells from a spinner culture were inoculated into a flask and incubated for 16 hours in the presence of 1 μg/ml vinblastine sulfate. An aliquot was then centrifuged through hypotonic medium [0.01 *M* tris (pH 7.4), 0.01 *M* NaCl, and 0.0015 *M* $MgCl_2$] onto a cover slip, fixed in acid alcohol, and stained with Giemsa (Hanson and Hearst, 1973). Two pairs of large autosomes, two V-shaped X chromosomes, and a single fourth chromosome are seen. ×2600.

fourth chromosome, however, and many contained two. Of 75 metaphases analyzed, 50% were diploid, 25% were tetraploid, 15% were probably tetraploid, and the remaining 10% were aneuploid. (Although some metaphase figures were too condensed to allow a determination of the exact number of chromosomes, they were included to avoid skewing of the data toward a lower average chromosome number). A typical metaphase containing a haplo-4 but otherwise normal appearing *Drosophila* chromosome complement is shown in Fig. 1a.

C. Macromolecular Components

The DNA content of the cells is approximately that which would be expected from the karyotype analysis. Diphenylamine assays (Burton, 1956) give a value of 0.51 pg of DNA per cell. If the ploidy of the cells is taken into account (see Section III,B) and it is assumed that 25% of the cells have replicated their DNA but have not divided (Puck *et al.*, 1964), the haploid DNA content of the cells is calculated to be 0.15 pg. This is in good agreement with the size of the *Drosophila* genome as determined by hybridization kinetics and by Feulgen staining of sperm nuclei (Laird, 1971; Rasch *et al.*, 1971).

The ribosomal content of the cells was determined by sedimenting the RNA of a known aliquot of cells on a sucrose gradient and measuring the area under the OD_{260} peaks of the 26 and 19 S rRNA. The amount of rRNA per cell determined in this way is 2.8 pg, which is a 5.6 times the DNA content of the cell.

D. Endogenous Virus

A striking feature of these cells, which has been reported by other investigators (Williamson and Kernaghan, 1972), is the number of viruslike particles, approximately 40 nm in diameter, seen in both nucleus and cytoplasm (Fig. 2). Particles with relatively lightly staining centers, which may be

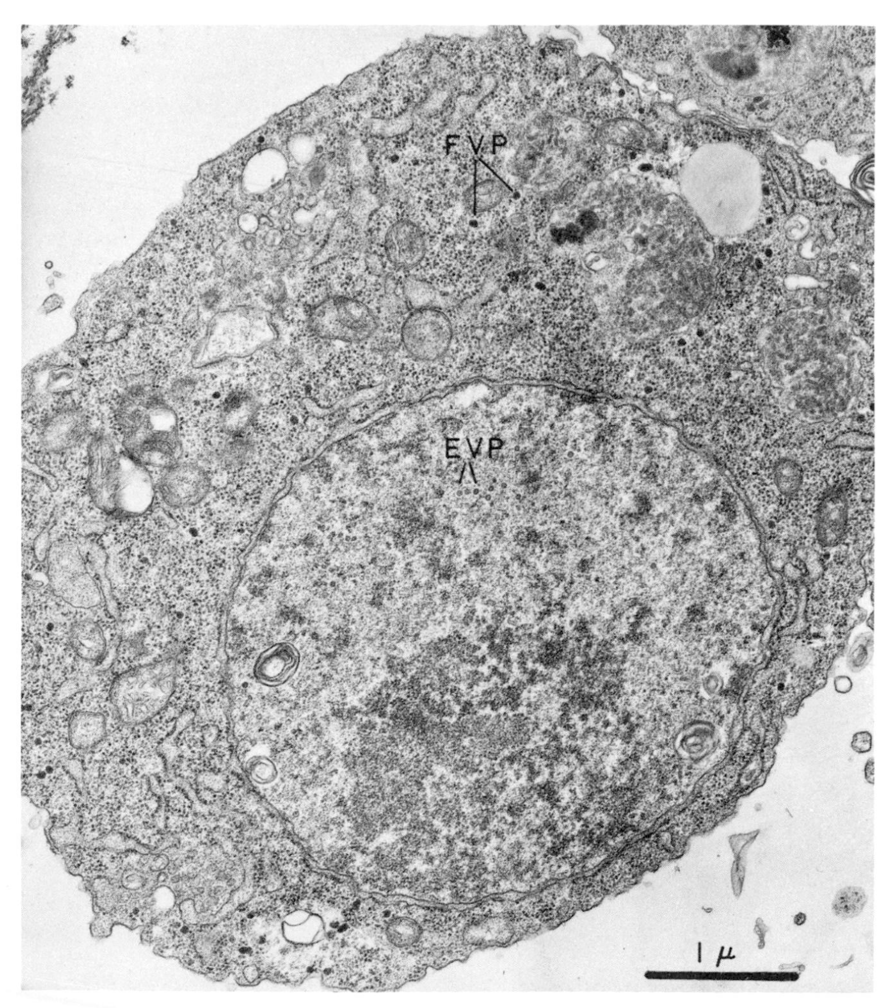

FIG. 2. Electron micrograph of a typical cell from a suspension culture. A cell pellet was collected by centrifugation at 800 rpm for 10 minutes and prepared for electron microscopy by standard techniques, including fixation with glutaraldehyde and osmium and staining with lead citrate and uranyl acetate (Lenk and Penman, 1971). Note the darkly staining (full?) viruslike particles (FVP) in the cytoplasm and the empty-appearing viruslike particles in the nucleus (EVP). ×24,000.

empty capsids, are found mainly in the nucleus, while darkly staining particles, which may be filled capsids, are found only in the cytoplasm. The particles we observed appear identical to those seen by other workers in both *Drosophila* tissue culture cells (Williamson and Kernaghan, 1972) and in tissues of adult flies (Kernaghan *et al*., 1964; Rae and Green, 1967; Akai *et al*., 1967).

There is some evidence that this virus has an RNA genome. When DNA-dependent RNA synthesis is inhibited with 10 μg/ml of actinomycin D, the cells still synthesize a small amount of RNA. Some of this RNA binds to oligo-dT cellulose and has a sedimentation coefficient of 14 S. This 14 S RNA may be transcripts of the viral genome or the virus genome itself.

IV. Cell Fractionation

A. Procedures

Methods for fractionation of the *Drosophila* cells into nuclei and cytoplasm were developed from procedures used for fractionation of mammalian cells. The most significant modifications, essential for nuclear stability and RNA integrity, are lysis in an isotonic buffer (since the nuclei are unstable in a hypotonic buffer) and use of nuclease inhibitors to inactivate the endogenous nucleases of the *Drosophila* cells. Either a nonionic detergent or shear can be used to break the cells.

For lysis with a detergent, an aliquot of 10^7 to 10^8 cells is collected by centrifugation, washed with isotonic buffer [0.03 *M* tris (pH 8.3), 0.1 *M* NaCl, and 0.01 *M* $MgCl_2$], and resuspended in 1–2 ml of the same buffer containing polyvinyl sulfate (25 μg/ml) and diethyl pyrocarbonate (0.25%) as nuclease inhibitors, and spermine (10^{-3} *M*) to stabilize the nuclei in the presence of polyvinyl sulfate. The nonionic detergent NP40 is added to 0.5% to lyse the cytoplasmic membrane; nuclei are then removed by centrifugation at 2000 rpm for 4 minutes.

Breakage of the cells by this procedure is virtually quantitative, while nuclei remain intact. Very little large RNA is found in the cytoplasmic fraction after a 15-minute labeling period, indicating that no significant leakage of RNA from the nuclei occurs. Contamination of nuclei with cytoplasmic material was estimated by determining the amount of adherent 19 rRNA, since this species exists rapidly from the nucleus, at least in mammalian cells. By this criterion the nuclei carry approximately 5% of the cytoplasmic material. Half of this can be removed by resuspending the nuclei in 2 ml of the same isotonic buffer plus inhibitors, containing 0.15 ml of a mixed detergent solution [2 parts 10% Tween 40 and 1 part 10% sodium deoxycholate (Penman, 1969)].

Cells can also be broken without the use of detergents, if the preparation of mitochondrial or membrane fractions is desired. The cells are washed in the same isotonic buffer, and then resuspended in a hypotonic buffer [0.03 *M* tris (pH 8.3), 0.01 *M* $MgCl_2$, and 0.01 *M* NaCl] plus inhibitors, omitting diethyl pyrocarbonate (which tends to fix the outer cytoplasmic membrane and thus prevent disruption by shear) and including the alkylating agent *N*-ethylmaleimide (5 m*M*). The cells are allowed to swell in this buffer for 10 minutes at 0°C, and are then broken by four to eight strokes with a Potter homogenizer. Nuclei and unbroken cells are removed by centrifugation at 2000 rpm for 4 minutes. The disadvantage of this procedure is that breakage of the cells is not quantitative.

B. Size of Nuclear RNA, mRNA, and rRNA—Absence of Degradation

Pulse-labeled RNA extracted from fractionated *Drosophila* nuclei is found to be rather small (Fig. 3). This result was at first thought to be due to

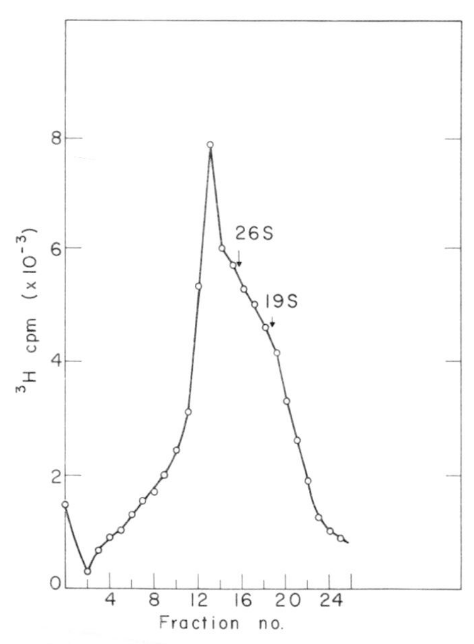

FIG. 3. RNA extracted from nuclei of pulse-labeled *Drosophila* cells. Suspension culture cells were concentrated 10× and labeled for 15 minutes with uridine-^{3}H (50 μCi/ml, 24 Ci/mmole). Cells were fractionated into nuclei and cytoplasm by lysis with NP40, as described in the text. Nuclei were resuspended in SDS buffer and phenol-extracted, and an aliquot equivalent to 5 ml of unconcentrated cells centrifuged on a 15–30% SDS sucrose gradient in an SW41 rotor for 10 hours at 25,000 rpm.

degradation during the fractionation procedure; however, the same size distribution is found if pulse-labeled cells are lysed directly with sodium dodecyl sulfate (SDS) buffer containing nuclease inhibitors. Further evidence that the small size of *Drosophila* nuclear RNA is not a degradation artifact comes from coextraction experiments. When ^{14}C-labeled *Aedes* cells and ^{3}H-labeled *Drosophila* cells are mixed and then lysed with SDS buffer, the size distribution of the *Drosophila* RNA is the same as that from fractionated nuclei, while a significant proportion of the *Aedes* RNA is present as high-molecular-weight material (Fig. 4). The same respective RNA patterns are seen when the cells are lysed separately. These results indicate that the small size observed for *Drosophila* nuclear RNA probably reflects the *in vivo* situation, and that the fractionation procedures do not reduce the size of this RNA.

Drosophila cytoplasmic mRNA, defined by its capacity to bind to oligo-dT cellulose (and thus possession of a poly A segment) has the same size distribution as that from HeLa cells. This is demonstrated by a coextraction experiment in which ^{14}C-labeled HeLa cells are mixed with ^{3}H-labeled *Drosophila*

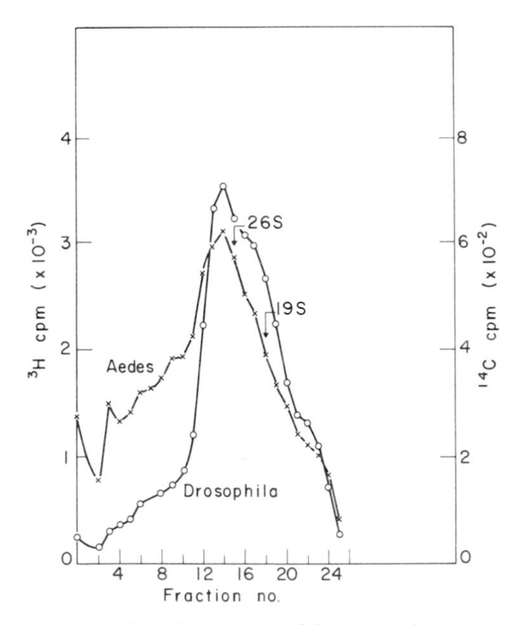

FIG. 4. Coextraction of pulse-labeled *Drosophila* and *Aedes* RNA. Five milliliters each of 5× concentrated *Drosophila* and *Aedes* cells (Spradling *et al.*, this volume) were labeled for 15 minutes with uridine-^{3}H (5 μCi/ml, 24 Ci/mmole) and uridine-^{14}C (5 μCi/ml, 57 mCi/mmole), respectively. Half of each incubation mix was then pipetted into the same tube of iced Earle's buffer. The cell pellet was collected by centrifugation, resuspended in SDS buffer, phenol-extracted, and centrifuged on a 15–30% SDS sucrose gradient as described in the legend for Fig. 3. ○, ^{3}H-labeled *Drosophila* RNA; x, ^{14}C-labeled *Aedes* RNA.

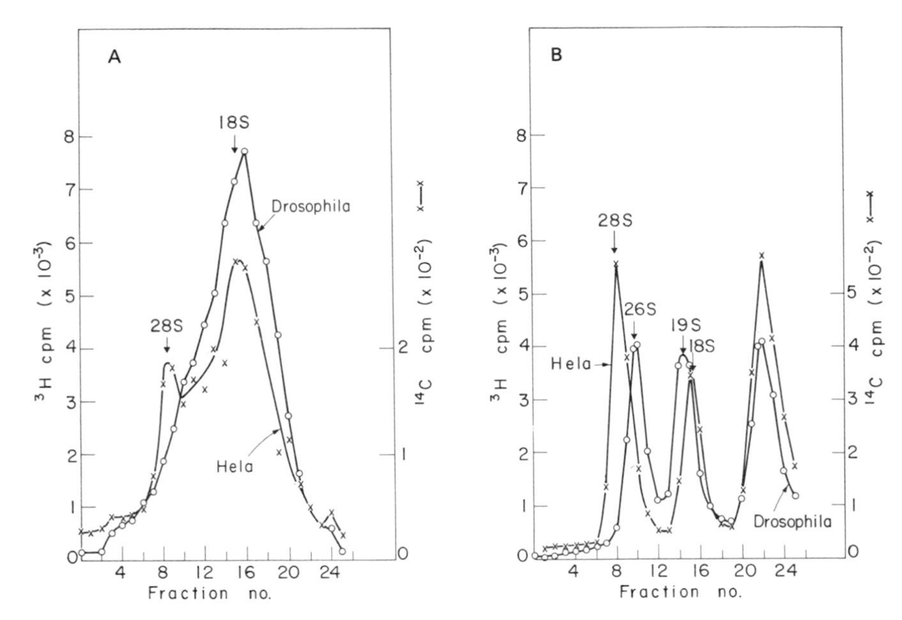

FIG. 5. Coextraction of *Drosophila* and HeLa cell cytoplasmic RNA. Five milliliters each of 5× concentrated *Drosophila* and HeLa cells were labeled for 3 hours with uridine-^{3}H (10 μCi/ml, 24 Ci/mmole) and uridine-^{14}C, (0.2 μCi/ml, 57 mCi/mmole), respectively. Half of each incubation mix was then pipetted into the same tube of iced Earle's buffer, and the cell pellet collected by centrifugation. Cells were fractionated into nuclei and cytoplasm by NP40 lysis, as described in the text. The cytoplasmic fraction was exhaustively extracted with phenol. The RNA thus obtained was passed over oligo-dT cellulose to select poly-A-containing RNA (Singer and Penman, 1973); both bound and unbound fractions were centrifuged on 15–30% SDS sucrose gradients in an SW41 rotor for 16 hours at 25,000 rpm. (A) Material bound to oligo-dT cellulose (poly A$^+$). (B) Material not bound to oligo-dT cellulose (poly A$^-$). ○, ^{3}H-labeled *Drosophila* RNA; x, ^{14}C-labeled HeLa RNA.

cells, cytoplasm prepared by NP40 lysis, and poly-A-containing cytoplasmic RNA obtained as material which binds to oligo-dT cellulose (Fig. 5a).

One of the principal reasons for adapting the cells to a new medium was to have the capacity to label RNA to high specific activity in the same culture medium in which the cells are growing. It is thus of interest that we routinely obtain mRNA labeled to a specific activity of 3×10^6 dpm/μg from cells labeled for 4 hours with uridine-^{3}H (50 μCi/ml, 24 Ci/mmole) in modified Dulbecco's culture medium.

The rRNA of *Drosophila* differs from mammalian rRNA in several respects. The two *Drosophila* ribosomal species have sedimentation coefficients of 26 and 19 S, determined by cosedimentation with the 28 and 18 S ribosomal species of HeLa cells (Fig. 5b). These size relationships are the same as those found for *Aedes* which have also been confirmed by electrophoretic analysis (see Chapter 11). Another interesting property of *Drosophila* rRNA, and of insect RNA in general, is that the 26 S species is dissociable by heat into two species which cosediment with the 19 S rRNA (Shine and Dalgarno, 1973). Some 26 S dissociation occasionally occurs during the extraction procedure, as seen in Fig. 5b in which the ratio of the *Dorsophila* rRNA peaks is 1:1 instead of the expected 2:1.

The 32 S peak seen in both *Drosophila* and *Aedes* pulse-labeled cells (Figs. 3 and 4) has been demonstrated by methylation experiments to be a ribosomal precursor. Even after labeling periods as short as 2 minutes, however, a separate peak of 37–38 S, which is believed to be the size of the *Drosophila* rRNA precursor (Greenberg, 1969), is not seen. This may be due to very rapid processing of this species in the cultured cells, or to a difference in estimating the S value of the precursor.

V. Summary and Conclusions

Drosophila melanogaster cultured cells have been adapted to grow in suspension in a modified mammalian culture medium. This allows labeling of nucleic acids to high specific activity, as well as production of large quantities of relatively homogeneous, logarithmically growing cells. The fact that such an adaptation was possible and that the adapted cells retain a normal *Drosophila* chromosome complement (although tetraploidy is increased) is a rather dramatic demonstration that close adherence of culture medium to the composition of insect hemolymph is probably not necessary once the cell line has been established.

Procedures have been developed for fractionation of *Drosophila* cultured cells into nucleus and cytoplasm in the presence of nuclease inhibitors. Thus it is possible, despite the high level of endogenous nucleases in these cells, to obtain what appears to be undegraded nuclear and cytoplasmic mRNA. These techniques should be useful for studies of this cell line from the point of view of molecular biology, and hopefully will contribute to the utility of *Drosophila* as a model system for eukaryotes.

Acknowledgments

We thank Elaine Lenk for the preparation and examination of samples in the electron microscope, and Jonathan King for electron microscope facilities. J.L. is the recipient of an NIH postdoctoral research fellowship (GM 54218–02). This work was supported by grants from the National Institutes of Health (NIH 5 ROI CAO8416; NIH CA 12174) and from the National Science Foundation (NSF GB 37709X).

References

Akai, H., Gateff, E., Davis, L. E., and Schneiderman, H. A. (1967). *Science* **157**, 810–813.

Brooks, M. A., and Kurtti, J. J. (1971). *Annu. Rev. Entomol.* **16**, 27–52.

Burton, K. (1956). *Biochem. J.* **62**, 315.

Echalier, G., and Ohanessian, A. (1970). *In Vitro* **6**, 162–172.

Greenberg, J. R. (1969). *J. Mol. Biol.* **46**, 85–98.

Hanson, C. V., and Hearst, J. E. (1973). *Cold Spring Harbor Symp. Quant. Biol.* **38**, 341–346.

Jones, B. M. (1966). *In* "Cells and Tissues in Culture" (E. N. Willmer, ed.), Vol. 3, pp. 397–457. Academic Press, New York.

Kakpakov, V. T., Gvosdev, V. A., Platova, T. P., and Polukarova, L. C. (1969). *Genetika* **5**, 67–75.

Kernaghan, R. P., Bonneville, M. A., and Pappass, G. D. (1964). *Genetics* **50**, 262.

Laird, C. D. (1971). *Chromosoma* **32**, 378–406.

Lenk, R., and Penman, S. (1971). *J. Cell. Biol.* **49**, 541–546.

McConaughy, B., Laird, C., and McCarthy, B. (1969). *Biochemistry* **8**, 3289.

Parādi, E. (1972). *Drosophila Inform. Serv.* **49**, 53.

Pardue, M. L., Gerbi, S. A., Eckhardt, R. A., and Gall, J. G. (1970). *Chromosoma* **29**, 268–290.

Parker, R. C. (1961). "Methods of Tissue Culture." Harper (Hoeber), New York.

Penman, S. (1969). *In* "Fundamental Techniques in Virology" (K. Habel and N. P. Salzman, eds.), Vol. 1, pp. 36–48. Academic Press, New York.

Puck, T. T., Sanders, P., and Peterson, D. (1964). *Biophys. J.* **4**, 441.

Rae, P. M. M., and Green, M. M. (1967). *Virology* **34**, 187–189.

Rasch, E. M., Barr, H. J., and Rasch, R. W. (1971). *Chromosoma* **33**, 1–18.

Schneider, I. (1972). *J. Embryol. Exp. Morphol.* **27**, 353–365.

Shapiro, H. S. (1970). *In* "Handbook of Biochemistry" (H. A. Sober, ed.), 2nd ed., pp. 104–116. Chem. Rubber Publ. Co., Cleveland, Ohio.

Shine, J., and Dalgarno, L. (1973). *J. Mol. Biol.* **75**, 57–72.

Singer, R., and Penman, S. (1973). *J. Mol. Biol.* **78**, 321–334.

Stanley, M. S. M. (1972). *In* "Growth, Nutrition, and Metabolism of Cells in Culture" (G. H. Rothblat and V. J. Cristofalo, eds.), Vol. 2, pp. 327–370. Academic Press, New York.

Vaughn, J. L. (1971). *In* "Invertebrate Tissue Culture" (C. Vago, ed.), Vol. 1, pp. 4–40. Academic Press, New York.

Williamson, D. L., and Kernaghan, R. P. (1972). *Drosophila Inform. Serv.* **48**, 58.

Wu, J.-R., Hurn, J., and Bonner, J. (1972). *J. Mol. Biol.* **64**, 211–219.

Chapter 13

Mutagenesis in Cultured Mammalian Cells

N. I. SHAPIRO AND N. B. VARSHAVER

Biological Department,
Kurchatov Institute of Atomic Energy,
Moscow, U.S.S.R.

I. Introduction . . . 209
II. Conditions for Experiments on Mutagenesis . . . 210
A. Cell lines . . . 210
B. General Material and Methods . . . 211
C. Selective Conditions . . . 212
D. The Effect of Cell Population Density on Mutant Recovery (Concentration Effect) . . . 212
E. Forward Mutations . . . 215
F. Reverse Mutations . . . 219
III. Spontaneous Mutagenesis . . . 220
A. Cell Plating in a Nonselective Medium . . . 220
B. Cell Plating in a Selective Medium . . . 222
C. Isolation and Testing of Colonies . . . 223
D. Proof of Spontaneous Mutagenesis . . . 224
E. Methods of Spontaneous Mutation Rate Determination . . . 226
IV. Induced Mutagenesis . . . 227
A. Dependence of Cell Survival on Mutagen Dose . . . 227
B. Time Interval between Mutagen Treatment and Cell Transfer to a Selective Medium . . . 230
C. Description of Experiment on Gene Mutation Induction . . . 232
References . . . 233

I. Introduction

It has always been a complicated problem for geneticists to study mutagenesis experimentally in mammals. This difficulty arises from the necessity to use a large number of animals in any experiment, as well as from the length of time needed. Besides, there are certain methodological difficulties connected, for instance, with the necessity to distinguish between gene mutations and chromosome aberrations.

The availability of gene mutations in experiments with somatic mammalian cells *in vitro* opened up new possibilities for studies of mutagenesis in this class of animals. The main difficulty mentioned above was successfully overcome.

The study of mutagenesis in somatic cells *in vitro* became possible as a result of the use of proven methods of microbial genetics. The possibility of obtaining progeny of a single cell in culture, together with the creation of selective conditions, has led to a dramatic increase in the resolving power of genetic analysis of the mutation process. It was found possible, in comparatively simple and short experiments, to score events, the probabilities of which equal 10^{-7}–10^{-8}. The utilization of microbiological methods made it possible to concentrate not so much on morphological peculiarities of cells as on their biochemical characteristics. All this permitted strictly quantitative investigations of spontaneous as well as induced mutagenesis in cultured mammalian cells. At present, almost all main types of gene mutations are involved in such studies: (1) those controlling sensitivity and resistance to various external agents, (2) those altering growth requirements, and (3) conditionally lethal temperature-sensitive mutations. In this article we concentrate on methods of investigation of spontaneous and induced mutagenesis; mutations of resistance to purine base analogs provide the main example.

II. Conditions for Experiments on Mutagenesis

A. Cell Lines

Certain factors should be taken into consideration when choosing cell lines for experiments on mutagenesis. Thus, first, one should decide whether a diploid or an aneuploid cell line should be used. It seems that, for the modeling of *in vitro* processes in an organism, diploid cells would be preferred. However, methodological difficulties arising when working with diploid cells (low plating efficiency, the possibility of culturing only for a limited number of cell generations, etc.) interfere with their extensive use.

It is easier to work with established aneuploid lines. Besides, there is as yet no evidence that much difference between these two cell types would be observed in mutagenesis studies. Thus special experiments on a comparison of the spontaneous mutability of diploid cells of embryonic human lung and aneuploid cells of the same origin gave similar values (Marshak and Varshaver, 1970). In this experiment mutations of resistance to purine base analogs were studied. This similarity is also revealed when comparing data on spontaneous mutation rates of diploid and aneuploid mammalian cells obtained by various investigators (DeMars and Held, 1972; Harris, 1971; Morrow, 1971; Rappaport and DeMars, 1973).

In studying induced mutagenesis there is no basis whatever for the assumption of distinctive differences between these cell types. The probability of dominant mutation recovery is the same in both cases. However, the possibility of the existence in aneuploid lines of monosomic regions, besides the X chromosome, makes them preferable in investigation of recessive mutations. To recover an autosome recessive mutation of particular gene in diploid cells, a cell line heterozygous for this gene should be employed. At present this is rather difficult (Clive *et al.*, 1972, 1973), and sometimes quite impossible.

Another cell characteristic to be taken into consideration when choosing lines is their chromosome set. Investigation of gene mutations is often accompanied by chromosome analysis. In this case a cell line with a relatively small number of well-identifiable chromosomes is preferred.

Last, lines with a short cell cycle have considerable advantage as regards the duration of the experiment. The control or absence of PPLO contamination in the chosen cell line is also necessary.

B. General Material and Methods

Practical examples given in this article to illustrate experimental investigations of spontaneous and induced mutagenesis are taken from work in our laboratory. These investigations were carried out on a hypodiploid clone (237_1) of Chinese hamster cells and its subclones, as well as on a quasi-diploid clone 431 (initial line BIId-ii-FAF28). Cells were cultivated in Eagle's medium with 10% bovine serum. When single colonies were grown, 30% serum was added to the medium. When glutamine-dependent auxotrophs were isolated, serum dialyzed for 24 hours was added to a glutamine-deficient medium. For isolation of cells able to proliferate in a medium with a low glucose concentration (so-called glucose-independent mutants), Eagle's medium without glucose and dialyzed serum with trace amounts of glucose were used. The final concentration of glucose was 15–20 μg/ml.

In experiments cells were grown in petri dishes 60 mm in diameter, in 5 ml of medium, at pH 7.1–7.2. To count colonies they were stained vitally with methylene blue. Only colonies with 100 or more cells were scored.

C. Selective Conditions

The main requirement for quantitative investigation of mutagenesis in cultured mammalian cells, as well as in microorganisms, is to create conditions for selective survival of mutants and simultaneous death of nonmutant cells. It is essential to choose conditions under which nonmutant cells very soon cease to divide, i.e., residual growth is absent. Residual growth complicates the estimation of rates of spontaneous mutations and mutations induced by various agents. The frequency of mutants detected in a selective medium should then be related not to the number of plated cells, but to the number of cells present at the moment of termination of cell proliferation under selective conditions. When choosing selective conditions, the presence of residual growth should be verified by counting cells at regular time intervals, e.g., after 24, 48, and 72 hours, until there is no further increase in the number of cells.

D. The Effect of Cell Population Density on Mutant Recovery (Concentration Effect)

It has been found in various microorganisms (Horowitz and Leupold, 1951; Newcombe, 1948) and mammalian cells (Bridges and Huckle, 1970; Khalizev *et al.*, 1966, 1969) that under selective conditions the number of detected mutants is not always proportional to that of initial cells per dish. The disproportion may be differently directed, i.e., a decrease as well as an increase in mutant yield may take place with an increase in cell population density.

It seems that different mechanisms of concentration effect may exist. The phenomenon of metabolic cooperation of cells, sensitive and resistant to purine base analogs, is well known (Cox *et al.*, 1970; Dancis *et al.*, 1969; Subak-Sharpe *et al.*, 1966, 1969). At high cell population densities, when cells contact, some molecular species necessary for the incorporation of the analog is transferred from sensitive to resistant cells. By acquiring the ability to incorporate the analog, cells die in a selective medium. This distorts the proportion between mutant yield and number of plated cells. If the number of plated cells is very high ($\sim$1 million per dish), total death of mutants may occur (Khalizev *et al.*, 1966). An analogous decrease in mutant survival may take place as well at medium exhaustion, as a result of residual growth of wild-type cells (Khalizev *et al.*, 1969).

The reverse, i.e., a nonproportional increase in mutant yield, may be observed in the case of cross-feeding as a result of "conditioning" of the medium by nonmutant cells (Chu *et al.*, 1972; Eagle and Piez, 1962; Varshaver *et al.*, 1971).

The optimal conditions for mutant recovery for cells plated in a selective medium can be determined in two ways: by detection of background mutants with increasing numbers of plated cells, and by survival of a given number of cells of some mutant line plated in artificial mixtures with various numbers of wild-type cells (reconstruction experiments).

In the first case cells are plated in a selective medium over a range of arbitrarily chosen numbers. For instance, between 10^4 and 10^6 cells per dish. Multiple ratios are preferable, but fractional numbers are also possible. At good coincidence of results, two experiments with five dishes for each variant are enough. In Table I is presented the survival of background mutants resistant to 15 μg/ml 6-mercaptopurine (6MP) in relation to the number of cells plated in a selective medium. The data indicate that a decrease in mutant yield is observed at 10^5 cells per dish. However, this threshold number may greatly vary in different lines, therefore it should be determined for each type of cell and each selective condition.

This method allows one to evaluate the relative plating efficiency of mutants at various cell population densities in a selective medium. However, it does not help in determining what fraction of preexisting mutant cells remained unrecovered. The second method, i.e., the plating of artificial mixtures, is more satisfactory.

Two series of dishes are plated. In a control series, as in the first case, background mutants are recovered. The selection of cell numbers and other procedures are identical to those described for the first method. In parallel

TABLE I

SURVIVAL OF BACKGROUND MUTANTS, RESISTANT TO 15 μG/ML 6MP, IN RELATION TO THE NUMBER OF CELLS PLATED IN A SELECTIVE MEDIUM[a,b]

Number of cells per dish ($\times 10^3$)	Mean number of colonies per dish	Mutant frequency ($\times 10^{-4}$)
5	0.6	1.20
10	1.4	1.40
50	7.2	1.44
100	9.1	0.91
200	12.8	0.64
400	6.9	0.17
1200	3.8	0.03

[a] From Khalizev *et al.* (1966).

[b] Results obtained with clone-237_I Chinese hamster cells.

with this series are plated artificial mixtures which consist of a given small number of mutant cells available and nonmutant cells in the same numbers as chosen for the first series. After scoring the colonies the mutants' plating efficiency in a mixture with various numbers of wild-type cells may be determined by the difference between the number of colonies in experimental and control series.

In Table II is presented the plating efficiency of glutamine-independent Chinese hamster cells mixed with various numbers of glutamine-dependent cells. Results of the experiments show that in this case a disproportionate increase in mutant survival occurs with increased population density when plating takes place in a minimal medium. The number of 4×10^5 cells per dish was found optimal. At greater plating density the scoring of colonies becomes difficult because of the residual growth of nonmutant cells which in some cases remain on the surface of the dish.

TABLE II

SURVIVAL OF GLUTAMINE PROTOTROPHS WHEN PLATED IN A MINIMAL MEDIUM IN ARTIFICIAL MIXTURES WITH VARIOUS NUMBERS OF GLUTAMINE *ts* AUXOTROPHS[a,b]

Number of cells per dish				
Auxotrophic cells ($\times 10^3$)	Prototrophic cells	Mean number of colonies per dish	Frequency of background mutants ($\times 10^{-6}$)	Plating efficiency of prototrophs, (% difference)
15	0	0	0	—
15	300	0	—	0
25	0	0	0	—
25	300	0	—	0
50	0	0	0	—
50	300	0.6	—	0.2
100	0	0	0	—
100	300	16.5	—	5.5
200	0	0	0	—
200	300	31.2	—	10.4
400	0	0.8	0.2	—
400	300	135.0	—	44.7
600[c]	0	4.0	6.7	—
600	300	260.0	—	85.2

[a] M. I. Marshak and N. B. Varshaver (unpublished data).

[b] Results obtained with clone-237_I Chinese hamster cells incubated at 40°C.

[c] In some experiments, when 6×10^5 cells were plated per dish, degeneration of auxotrophic cells was not complete, which made it difficult to score colonies at this number of cells.

E. Forward Mutations

1. Recovery of Drug-Resistant Mutants

The diversity of various types of resistance to chemicals may be extremely great in theory. However, at present only mutations for resistance to purine base analogs are widely studied. Somewhat less common are mutations for resistance to the pyrimidine analog 5-bromodeoxyuridine (BUdR). Mutants resistant to these substances were obtained in various cell lines (Chu and Ho, 1970; Chu, 1971; DeMars and Held, 1972; Hsu and Somers, 1962; Roosa *et al.*, 1962; Shapiro *et al.*, 1966; Szybalski and Smith, 1959). It has been shown that resistance to purine base analogs as a rule is determined by the complete absence or decrease in activity of hypoxanthine-guanine-phosphoribosyl transferase (HGPRT) (Brockman *et al.*, 1962; Lieberman and Ove, 1960; Littlefield, 1964a; Szybalski *et al.*, 1961). So far as resistance to BUdR is concerned, it is connected in most cases with the absence of thymidine kinase (TK) (Kit *et al.*, 1963). However, some cases have been described in which resistance was determined by a defect in BUdR transport through the cell membrane (Breslow and Goldsby, 1969). Resistance to purine base analogs may as a rule be obtained by single-step selection (DeMars and Held, 1972; Shapiro *et al.*, 1966; Szybalski and Smith, 1959). On the contrary, clones resistant to BUdR have been in most cases isolated by multistep selection in the medium with successively increasing analog concentrations (Chu and Ho, 1970; Littlefield, 1965).

It is essential, when preparing for experiments on mutagenesis, to choose the selective agent concentration by determining dose–survival curves. If resistance to some substance, which was never investigated earlier, is to be studied, a wide range of concentrations should be first tested. If resistance to a substance has already been studied, it is easier, according to the data available, to choose concentration limits, although the survival of selected cells should still be determined since various lines can differ greatly in this characteristic.

When choosing the selective agent concentration for experiments on mutagenesis, it is important that the frequency of background mutants in the population be low enough. It is particularly essential when evaluating the effect of weak mutagens. A high background of mutants allows the detection of their effect only if the experiment is carried out on a very large scale.

When investigating mutations for resistance to purine and pyrimidine base analogs, the population may be cleared of preexisting resistant cells. For this, before beginning experiments on mutagenesis, cells of an initially sensitive line are plated in HAT medium (Section II,F,1), in which resistant forms die. However, this procedure does not necessarily screen out all resistant cells (Albertini and DeMars, 1973). Besides, the transfer from HAT medium to a normal one entails certain difficulties (Section II,F,1).

If high analog concentrations are chosen, not only is a low level of background ensured but also a higher degree of resistance of recovered mutants. Besides, the probability of colony formation due to residual growth of non-mutant cells is decreased, i.e., the number of registered colonies will better represent the true number of mutants. However, one should not forget that the frequency of mutations scored may be in inverse proportion to the level of resistance (Bridges and Huckle, 1970; Littlefield, 1965).

2. Detection of Auxotrophic Mutants

At present one can obtain auxotrophs requiring various metabolites: nonessential amino acids, vitamins, and other substances (Jones and Puck, 1973; Kao and Puck, 1967, 1972; Varshaver *et al.*, 1971). To avoid background mutants one has only to cultivate initial prototrophic cells in a medium minimal in this component. To remove the substance from serum, the latter is intensively dialyzed or a standard macromolecular serum fraction is used.

The main procedure for auxotroph detection from mammalian cell populations, developed by Puck and Kao (1967), consists of creating conditions for the selective death of prototrophs, proliferating in the minimal medium, and the simultaneous survival of auxotrophs. To this end the sensitization of cells, which had incorporated BUdR into DNA, to the effect of near-visible light was used. By this method clones with various auxotrophic markers may be recovered. An attempt was even made to evaluate quantitatively the production of mutations to auxotrophy due to various mutagens (Kao and Puck, 1969).

The procedure for auxotroph detection in our laboratory amounts to the following (an example of glutamine-dependent mutants is given). Glutamine-independent Chinese hamster cells (clone 237_1-glu$^+$) were taken as the parental line. Cells were cultivated in glutamine-deficient Eagle's medium with the addition of 10% dialyzed serum. Auxotrophs isolated after mutagen treatment were used, as well as auxotrophs that arose spontaneously. To accumulate spontaneous mutants, an initially prototrophic population was cultivated for several passages in complete medium.

The scheme of the experiment is presented in Fig. 1. Two series of dishes (1.5×10^5 cells per dish) were plated in minimal medium, without BUdR and with BUdR at 15 μg/ml. After 48 hours the cells were twice thoroughly washed with Earle's saline, trypsinized, and counted. During this period usually two to three cell generations have taken place. As a rule, no difference was observed in the number of generations in controls and after

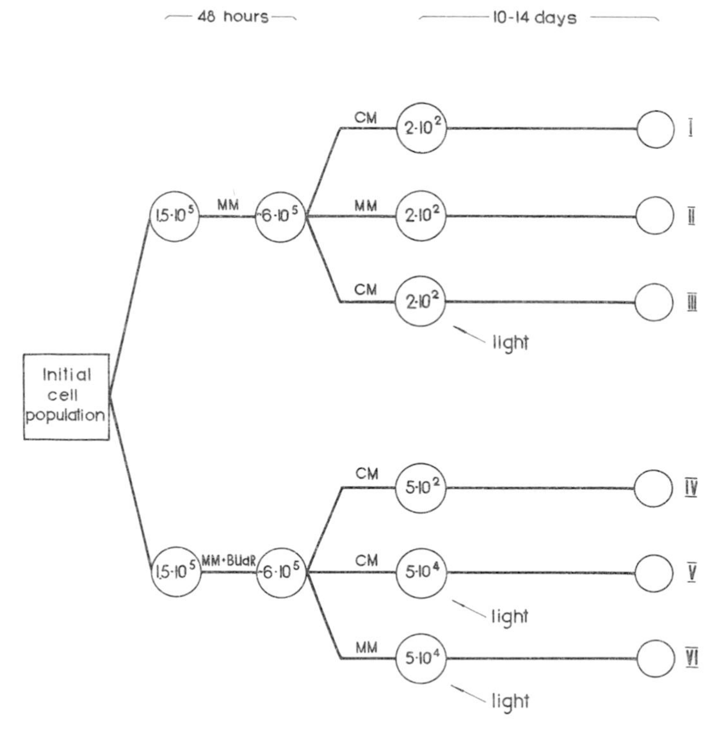

FIG. 1. Schematic diagram of procedures used for detection of auxotrophic mutants. CM, Complete medium; MM, minimal medium. In the circles in Figs. 1, 2, and 4 are given the number of cells per petri dish in 5 ml of medium.

incubation with BUdR. Next the cells were replated according to the standard scheme. Respective variants were exposed to near-visible light. The experimental scheme made it possible to evaluate: (1) wild-type cell survival in a complete and in a minimal medium (I and II); (2) the lethal effect of irradiation (III) and BUdR (IV) separately; (3) the degree of sensitization of cells to light after BUdR incorporation, as well as the frequency of supposed auxotrophic mutants (V); (4) the fraction of prototrophs surviving after treatment with BUdR and light, i.e., the degree of enrichment with auxotrophs (VI). When visible colonies were formed, they were stained and scored.

A number of dishes from (V) was left for isolation and further testing of the colonies after treatment with BUdR and light. This is essential not only to obtain auxotrophic clones, but also to ascertain if all grown colonies are mutant. After proliferation of isolated colonies in a complete medium, they were tested in parallel in complete and minimal media. The plating in two types of media is made from one cell suspension. The initial prototrophic clone serves as control; it is also plated in both media. The selection of the number of cells per dish (Section II,D) is very important. It is shown that sometimes cells not growing under selective conditions in small numbers are capable of proliferating in great numbers (Meiss and Basilico, 1972). Therefore it is advisable to plate at two chosen numbers of cells, e.g., at 5×10^2 and $10–20 \times 10^3$ cells per dish. Another way is possible, however, if one has to try a large number of clones, i.e., in the order of hundreds. After a short period one should replate the cells in two dishes with minimal and complete media, respectively, at approximately $5–10 \times 10^3$ per dish. This method of rough evaluation of cell behavior under selective conditions makes it possible to screen nonmutant clones and clones with weakly expressed auxotrophy. Mutant clones are further tested. When isolating and investigating various nutritional mutants, one should not forget that often they display considerable leakiness. This preliminary test helps to detect them in a shorter time.

3. Detection of Temperature-Sensitive (*ts*) Mutants

The main principle of *ts* mutant detection is analogous to that of auxotroph detection, i.e., conditions are created for the selective death of temperature-resistant cells, proliferating at a nonpermissive temperature, and survival of *ts* cells. At present three modifications of the procedure are used: incorporation of BUdR with further irradiation (Naha, 1969, 1970; Scheffler and Buttin, 1973), the lethal effect of thymidine-H^3 (Thompson *et al.*, 1970, 1971), and killing with 5-fluorodeoxyuridine (Meiss and Basilico, 1972). Replica plating (Smith and Chu, 1973) is also possible. However, because of technical difficulties it is not yet widely used for mammalian cells.

Before beginning experiments on *ts* clone isolation, it is necessary to determine the temperature limits between which the initial population of cells can proliferate. Since cell populations may be heterogeneous and consist of cells with various levels of temperature sensitivity, it is essential to obtain temperature-resistant clones which will be later employed in experiments on mutagenesis. For this, from a series of clones obtained from the parental line, one should be selected that has the same plating efficiency at normal and nonpermissive temperatures. Nonpermissive temperatures of 39°C (Meiss and Basilico, 1972) to 41°C (Smith and Chu, 1973) have been used for various cell lines. The permissive temperature can be lowered from 37°C to 33–35°C or even to 31°C (Roscoe *et al.*, 1973) to increase the interval between the two temperatures. It has practically no effect on cell survival.

To isolate *ts* mutants we use the same scheme as for auxotrophs (Fig. 1). Selective conditions are created by incubation with BUdR at 40°–40.5°C. The period of incubation is the same as shown in Fig. 1. Cells in all variants are plated in a complete medium. Cells in II and VI are cultivated at a nonpermissive temperature, and the rest at 36°C. Isolated colonies are tested in parallel at the two temperatures, and with two numbers of cells, 5×10^2 and $10–20 \times 10^3$ cells per dish.

Our experience with auxotroph and *ts* clone isolation shows that, among colonies isolated after one cycle of selection with the help of BUdR and irradiation with near-visible light, a considerable number of clones is present that do not exhibit a mutant phenotype or hardly expresses it (Varshaver *et al.*, 1971, 1972). To overcome this unwanted phenomenon, one has to repeat the selection. For this, 4–5 days later, after the first irradiation the medium is removed from the dishes and medium with BUdR is added. Cells are incubated with BUdR for 24 hours. In controls the medium is renewed, but without BUdR. The medium is minimal for auxotroph isolation; it is complete for *ts* mutant isolation, but cells are placed in a thermostat at 40°C. Twenty-four hours later all dishes are twice washed with Earle's saline, a medium without BUdR is added, and the dishes are irradiated. In this case an increase in the fraction of mutants among growing colonies is observed.

F. Reverse Mutations

1. Detection of Drug-Sensitive Revertants

Selection of cells sensitive to purine and pyrimidine base analogs is made in HAT medium (Szybalski *et al.*, 1962). The medium includes hypoxanthine, aminopterin, and thymidine. The quantitative ratio of ingredients may vary somewhat according to various investigators. Some workers add glycine. In this medium purine and pyrimidine synthesis *de novo* is blocked by aminopterin, therefore cells remain viable only if they can use exogenous precursors. Sensitive cells with a normal level of HGPRT and TK activity incorporate hypoxanthine and thymidine during synthesis of nucleic acids and survive in HAT medium. Mutants resistant to purine or pyrimidine base analogs with HGPRT and TK activity absent (or greatly reduced) die under these conditions. The modified medium of Littlefield (1964b) composed of 13.6 μg/ml hypoxanthine, 0.18 μg/ml aminopterin, 12.5 μg/ml thymidine, and 7.5 μg/ml glycine is found to be reliable enough for screening resistant forms (Volkova and Kakpakova, 1972). When studying reverse mutations, from resistance to purine base analogs to sensitivity, the number of plated cells does not influence mutant yield. This is to be expected from the biochemical mechanism determining this type of resistance.

Difficulties arise when transferring cells from HAT medium to a standard one. To maintain their viability cells should be first replated in a medium containing hypoxanthine, thymidine, and glycine in the same concentrations. After one passage in this medium, cells begin to proliferate well in standard media.

2. Detection of Prototrophic Revertants

Auxotrophic cells are plated in a minimal medium. When choosing the number of cells, it is essential to take into consideration the possibility of cross-feeding at high numbers of cells per plate (Section II,D). In the absence of cross-feeding, only prototrophs grow.

3. Detection of Temperature-Resistant Revertants

Temperature-sensitive cells are incubated at a nonpermissive temperature. Determination of the optimal number of cells is as essential as in the detection of prototrophs.

III. Spontaneous Mutagenesis

To evaluate spontaneous mutation rates, it is good practice to use the fluctuation test scheme (Luria and Delbrück, 1943). This ascertains not only the very fact of spontaneous mutagenesis but also makes it possible to determine the spontaneous mutation rate by utilizing various statistical methods. Other experimental schemes are also possible (Demerec, 1946). We concern ourselves only with the fluctuation test in mammalian cells. The standard experiment performed in our laboratory to estimate the spontaneous mutation rate (Marshak and Varshaver, 1970; Shapiro *et al.*, 1966, 1972a; Varshaver *et al.*, 1969) is described below. The scheme of the experiments is presented in Fig. 2. The experiment includes three stages: cell growth in a nonselective medium, cell transfer and growth in a selective medium, and isolation and testing of colonies.

A. Cell Plating in a Nonselective Medium

The experiment should begin with the selection of the size of initial inoculum for parallel culture plating. This value is determined by the mutant frequency in the parental cell population (mutant background). An approximate background level is usually defined by estimation of the dependence of cell survival on selective agent concentration. Since background mutant frequency may vary greatly at different times for unknown reasons, the estimation should be made just before the experiments on determination of spontaneous mutation rate. When choosing the number of cells per dish

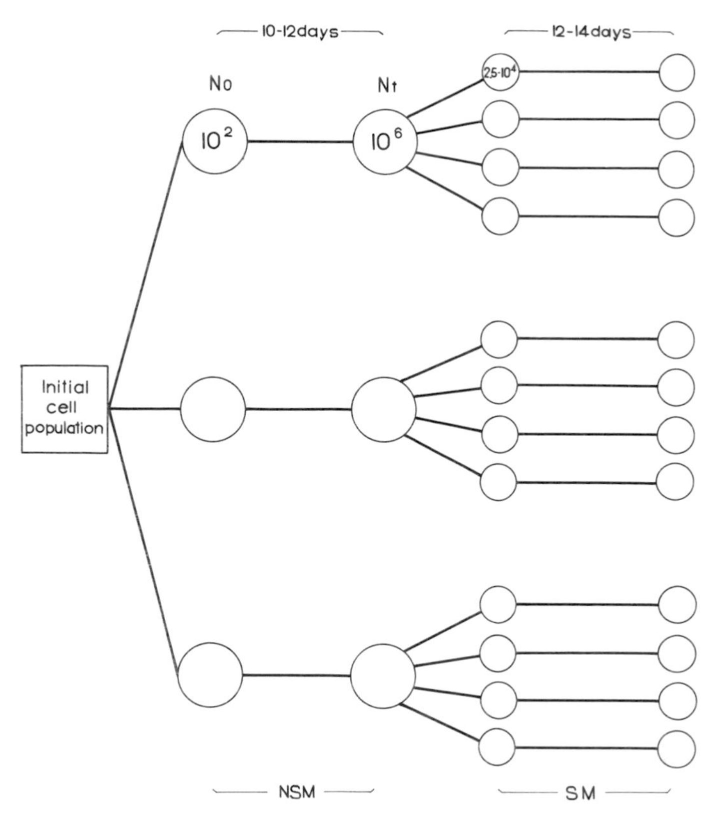

FIG. 2. Schematic diagram of the experiment used for determination of the spontaneous mutation rate (fluctuation test). N_0, Initial number of cells per culture; N_t, number of cells per culture at the time of transfer to selective medium; NSM, nonselective medium; SM, selective medium.

to be plated in a selective medium, the concentration effect should be assessed in all cases.

To estimate the frequency of background mutants resistant to 15μg/ml 6MP in a Chinese hamster cell population (clone 237_1), 2.5×10^4 cells were plated per dish (10 to 15 dishes) in a selective medium. The medium was changed every 3–4 days. In parallel, survival of the same cells in a medium without 6MP has to be determined. 2×10^2 cells per dish (five dishes) were plated in a nonselective medium; 8–10 days later, when colonies were visible to the eye, they were stained and scored. To avoid introduction of pre-existing mutant cells when plating parallel cultures, 10^2 cells per culture were chosen for the initial inoculum. Judging from the background level detected in our experiments (10^{-4} to 10^{-5}), the probability of mutant cell introduction was sufficiently small for fluctuation test conditions.

The plating of 10 to 15 parallel cultures gives quite reliable results. The period of parallel culture growth depends on the rate of cell proliferation and supposed mutation rate. It is advisable to proceed on the basis of a mutation rate in the order of 10^{-6}. The cell number in parallel cultures should be brought up to a level that ensures that mutations have arisen in most cultures. It is important to avoid the formation of necrotic regions in the centers of growing colonies, which would introduce errors in the estimation of the final number of cells because of cell death. If, however, to reach the required number, cultivation has to be prolonged, each culture should be replated separately.

Clone 237_{I} cells were propagated for 10 days. During this period the number of cells per dish increased from 10^2 to $1\text{–}2 \times 10^6$. The number of cell generations was estimated with the formula

$$g = \frac{\ln N_2 - \ln N_1}{\ln 2}$$

(Luria and Delbrück, 1943), where N_1 and N_2 are respectively, the number of cells per dish at the beginning and end of the experiment.

B. Cell Plating in a Selective Medium

When cells are transferred to a selective medium, the plating from each culture is done separately. If the final number of cells is not large, all the cells from each culture may be plated. This makes possible mutation rate estimation by an additional method (Table V, method 4). If the number of cells is large enough and the concentration effect is already perceptible at a relatively small number of cells per dish, samples should be plated.

When estimating the final number of cells per culture, cells may be counted in a sample of cultures (four to five dishes), and, if variation among different dishes is not great, one may proceed from the mean number of cells per dish when plating in a selective medium. If the number of cells in parallel cultures varies considerably, it is advisable to count cells for each culture. This ensures an accurate estimate of the number of cells per culture and sample.

When plating, a number of cells per dish is chosen that does not cause a concentration effect. This number was 2.5×10^4 for clone 237_{I} with 15 μg/ml 6MP in the medium. To obtain a large enough total sample, one should plate several dishes from each culture, e.g., eight dishes when plating 2.5×10^4 cells per dish. The summed sample value from each culture amounts to 2×10^5 cells in this case. This procedure is important not only for the increase in the number of cells per sample, but also because it permits comparison of the distribution of the number of mutants in samples taken from one culture and in those taken from parallel cultures. The type of distribution in both cases is a criterion for determination of the presence of spontaneous mutagenesis.

The duration of incubation in a selective medium is determined by two factors: nonmutant cells degeneration rate and growth rate of mutants. It is essential for accurate scoring of colonies not to leave initial cells on the surfaces of dishes. After 12–14 days of growth in a selective medium, colonies were stained and counted.

It is very important to determine the viability of initial and mutant cells. Cell survival in parallel cultures may be easily estimated if cells from parallel cultures, when plated in a selective medium, are simultaneously seeded in a standard medium in numbers allowing assessment of their plating efficiency (100 to 200 cells per dish). If, as is usual, the plating efficiency approaches 100%, the sample taken for the experiment corresponds to the number of viable cells. If, however, the plating efficiency is 50% or less (as it is for diploid cells), the actual number of viable cells is less than the number of cells in samples, and a corresponding correction should be made.

It remains to be determined whether survival of mutants corresponds to that of wild-type cells. This may be estimated in the following way. To control mutant cell survival, artificial mixtures of mutant (100 to 200 cells) and nonmutant cells in numbers chosen for this experiment are plated in a selective medium in parallel with the plating of cells from independent cultures (Section II,D). The scoring of colonies at the end of the experiment makes possible the determination of mutants plating efficiency from the difference between the number of colonies in dishes containing mixtures and corresponding dishes to which no mutants were added. It permits the introduction of corrections and determination of the true number of mutants per culture. However, this correction is only of relative value, since viability of cells of an established mutant line and that of newly formed mutants may be different.

C. Isolation and Testing of Colonies

One of the methods for establishing the hereditary nature of a character is to study its stability when it is cultured under nonselective conditions. Therefore it is important not only to score grown colonies but also to retain a certain number of dishes so that colonies can be isolated and the stability of a mutant phenotype tested.

Ways to transfer colonies are various (Ham, 1972). The simplest one is transfer with a Pasteur pipette. The medium is decanted from the dishes, then trypsin is added and immediately removed. This procedure ensures some loosening of cell attachment to the dish. A colony at a distance from all other colonies is chosen. It may be transferred to a vial containing a small amount of medium, so that it will be convenient to pipette the cell suspension. After pipetting, the suspension is transferred into a dish, 20–30 mm in diameter, with 3–4 ml of medium. The colony may be transferred from the first dish into a drop of trypsin. This facilitates single-cell suspension production on pipetting. Colonies should be transferred and propagated in a medium not containing the analog. It is essential because (1) when colonies are transferred, not all of them will proliferate, because of unknown causes; if, when transferred to a selective medium, the colony does not grow, the possibility that cells of this colony are not true mutants and are incapable of long-term proliferation in the presence of a selective agent cannot be easily ruled out; (2) in addition, cultivation in a nonselective medium is necessary for further testing of character stability.

After a sufficient number of cell generations (not less than 15 to 20), the clone should be tested in a selective medium. For this, 100 to 200 cells per dish are plated in parallel from one suspension in selective and nonselective media. After the formation of visible colonies, they should be stained and counted. A comparison of plating efficiency in the two media allows one to establish whether one is dealing with a hereditary character and, in this case, what the degree of resistance of the cells is.

D. Proof of Spontaneous Mutagenesis

As previously indicated, the experimental scheme presented in Fig. 2 makes it possible to establish whether we are dealing with spontaneous mutagenesis or whether colony-forming ability in a selective medium is a result of adaptation. To establish the fact of spontaneous mutagenesis, two criteria may be utilized: (1) comparison of the distribution of the number of mutants in a series of independent cultures and in samples taken from one culture, and (2) comparison of experimental and theoretical variances calculated for parallel culture series (Luria and Delbrück, 1943).

In Table III are presented examples of the distribution obtained when investigating mutations controlling the ability of Chinese hamster cells (clone 431) to proliferate in a medium with a low glucose concentration (15–20 μg/ml). The variation in the number of mutants in samples from one culture is determined by chance error of sampling. It follows the pattern of normal or Poisson distribution in which the variance must not exceed the mean. This equality was indeed observed in examples presented in Table III for samples from one culture. If spontaneous mutations arise during the period from parallel culture plating to the moment of cell replating in a selective medium, the mutant number distribution in samples taken

TABLE III

DISTRIBUTION OF THE NUMBER OF "GLUCOSE-INDEPENDENT" MUTANTS IN SAMPLES FROM ONE CULTURE[a,b]

	N of culture			
N of sample	1	2	3	4
1	2	1	5	1
2	2	3	8	0
3	3	2	4	0
4	3	1	2	1
5	1	5	4	1
6	1	2	2	0
7	5	5	4	0
8	2	4	3	0
Mean	2.4	2.9	4.0	0.4
Variance (experimental)	1.7	2.9	4.3	0.3

[a] From Varshaver *et al.* (1969).
[b] 12.5×10^4 clone-431 Chinese hamster cells per sample.

from parallel cultures depends on the time of mutation appearance. If a mutation arises at early stages of culture growth, the mutant cell will have time to produce a large colony with many mutants. If a mutation arises late the colony will contain a small number of cells. As a result, after replating in a selective medium, the number of mutant colonies differs markedly in various parallel cultures. This leads to considerable increase in the variance over the mean. A comparison of means and variances for the distribution of the number of mutants in samples from parallel cultures, presented in Table IV, indicates that we are observing spontaneous mutagenesis.

The presence of spontaneous mutagenesis is also demonstrated by the comparison of experimental and theoretical variances for parallel culture series. Theoretical variance is calculated with the formula $\sigma^2_{\text{theor.}} = Ca^2n_t + r_s$, where C is the parallel culture number, a is the mutation rate (Section III,E), n_t is the number of cells per sample, and r_s is the mean number of mutants per sample (the formula allows for the error in sample taking) (Luria and Delbrück, 1943). In spontaneous mutagenesis experimental variance for parallel cultures must be equal to or exceed the theoretically expected variance. In examples presented in Table IV, experimental variances exceeded theoretical ones many times, which also confirms the presence of spontaneous mutagenesis.

TABLE IV
DISTRIBUTION OF THE NUMBER OF MUTANTS IN SAMPLES FROM PARALLEL CULTURES[a,b]

N of culture	*N* of experiment			
	1	2	3	4
1	1	3	5	0
2	0	0	56	16
3	97	0	26	2
4	3	1	7	1
5	1	108	1	0
6	2	0	14	0
7	28	0	1	0
8	0	5	0	1
9	0	0	5	0
10	1	0	7	0
11	0	—	77	—
12	0	—	—	—
13	0	—	—	—
14	0	—	—	—
15	3	—	—	—
Mean per sample	9.1	11.7	18.1	2.0
Variance, experimental	641.8	1146.0	648.7	24.5
Variance, theoretical	33.2	33.5	89.7	3.3

[a] From Varshaver *et al.* (1969).
[b] 10^5 clone-431 Chinese hamster cells per sample, 12.5×10^4 cells per dish.

E. Methods of Spontaneous Mutation Rate Determination

At present there is a series of methods for spontaneous mutation rate estimation, which makes possible determination of the number of mutational events from the number of mutants produced as a result of mutation in a wild-type cell population, as well as due to proliferation of mutants. Data from the aforementioned fluctuation test serve as material for these calculations. In this case assessment of the spontaneous mutation rate may be made with formulas suggested by Luria and Delbrück (1943), Lea and Coulson (1949), and Newcombe (1948).

Along with the fluctuation test, factual material may be obtained in another experiment which makes it possible to establish the number of produced mutations from the number of mutant colonies (Table V, method 6) (Demerec, 1946). The spontaneous mutation rate may also be evaluated from the increase in the fraction of mutants during cell proliferation in a nonselective medium. In these experiments the initial number of cells per culture should be large enough so that in the first generation in a culture series an average of one mutation per culture can be obtained (Table V, method 5).

The advisability one method of assessment of spontaneous rate or another is determined by the specificity of the object under study and of the experiment (initial presence of mutants in the cell population, experiment duration, etc.). Correct calculations naturally demand acquaintance with cited papers. Here we confine ourselves only to presenting a table that demonstrates general characteristics of the main methods for calculating mutation rates.

In some cases it seems wise to utilize several methods of calculation of spontaneous mutation rate when handling data from an experiment.

IV. Induced Mutagenesis

The possibility of gene mutation induction in somatic mammalian cells *in vitro* was established quite recently. It was demonstrated simultaneously and independently in 1968 in three laboratories (Chu and Malling, 1968; Kao and Puck, 1968; Shapiro *et al.*, 1968). The results of many investigations on the induction of various mutations have been published. In these studies physical mutagens (ionizing radiation, ultraviolet rays), as well as chemical ones (BUdR, ethyl methanesulfonate, *N*-nitrosomethylurea (NMU), *N*-methyl-*N'*-nitro-*N*-nitrosoguanidine, etc.) were employed. These experiments are interesting in many respects, first, for understanding the mechanism of mutation production in mammals. In addition, the method of estimation of induced mutagenesis in somatic cells may provide a basis for the estimation of genetic danger to man from various environmental factors.

A. Dependence of Cell Survival on Mutagen Dose

Before investigating the mutagenic effect of any agent, it is essential to establish the quantitative dependence of cell death on mutagen dose. The dose–effect curve is utilized to evaluate the number of cells retaining viability after mutagen treatment at various doses. Proceeding from these data, one chooses a dose range (or a single dose) for work on mutation induction and determines the number of cells necessary for mutagen treatment. The dose–effect curve does not eliminate the necessity for controlling in each experiment the mutagen's lethal effect at the moment of mutation

TABLE V

CHARACTERIZATION OF DIFFERENT METHODS OF SPONTANEOUS MUTATION RATE ESTIMATION[a,b]

			Determination of mutation rate is influenced by:			
Method	Main formulas	Limiting conditions	Differences in generation time and probability of elimination of mutant and nonmutant cells	Phenotypic lag	Early mutations	Reference
1	$r = m \ln (Cm)$ $a = \frac{m \ln 2}{N_t - N_0}$	$R_1 = 0$	$a \lessgtr a_{act}$ (−)	$a < a_{act}$ (+)	$a > a_{act}$ (+)	Luria and Delbrück (1943)
2	$r_0/m - \ln m = 1.24$	$0 = R_1 \ll 1$;	$a \lessgtr a_{act}$ (−)	$a > a_{act}$ (−)	No influence	Lea and Coulson (1949)
	$a = \frac{m \ln 2}{N_t - N_0}$	$r_0 > 1$				
3	$a = \frac{(h - r^1) \ln 2}{C/N_t - N_0(h - r^1)}$	$R_1 = 0$	$a \lessgtr a_{act}$ (−)	No influence	—	Newcombe (1948)

4	$a = \dfrac{-(\ln P_0)\ln 2}{N_t - N_0}$	$0.9 > P_0 > 0.01$ $0 = R_1 \ll 1$ Not samples, but all cells from each culture should be plated	No influence	$a < a_{act}$	No influence	Luria and Delbrück (1943)
5	$a = \dfrac{2(\ln 2)(R_2/N_2) - (R_1/N_1)}{g}$	$N_1 \gg \dfrac{\ln 2}{a}$	$a \lessgtr a_{act}$	No influence	No influence	Luria and Delbrück (1943)
6	$a = \dfrac{(M_2 - M_1)\ln 2}{N_2 - N_1}$	t = One passage; cells are cultured on solid substrate	No influence	$a < a_{act}$	No influence	Demerec (1946)

[a] From Khalizev (1969).

[b] a = Mutation rate per cell per generation as determined experimentally; a_{act} = actual mutation rate per cell per generation; C = number of parallel cultures; g = number of cell generations during growth under nonselective conditions; h = maximal number of mutants in one of the parallel cultures; m = mean number of mutants per culture; M_1 and M_2 = mean number of mutants per culture at the beginning and at the end of the experiment; N_0 = initial number of cells per culture; N_1 = mean number of cells per culture at the beginning of the experiment; N_2 and N_t = mean number of cells per culture at the end of the experiment; P_0 = fraction of cultures containing no mutants at the end of the experiment; r = mean number of mutants per culture; r_0 = median number of mutants in a series of parallel cultures; r^1 = mean number of mutants per culture in a series, except the one with the maximal number of mutants (h); R_1 and R_2 = mean number of mutants per culture at the beginning and at the end of the experiment; t = duration of the experiment, i.e., duration of growth in a nonselective medium; (+) = decrease in error with increase in t; (–) = increase in error with increase in t.

induction. This control is important because of the great variability in cell survival after treatment with the same mutagen dose. A precise definition of cell survival after mutagen treatment is also essential because, for comparative assessment of the efficiency of various mutagenic agents, doses similar in their effect on cell survival have to be compared.

When choosing mutagen doses, it should be kept in mind that, although large mutagen doses lead to great mutation yield, they also lead to the death of a considerable number of treated cells. The latter may enormously limit the resolving power of the experiment. If the employed mutagen is toxic to cells, we recommend choosing mutagen doses leading to the death of not less than 50% of the treated cells.

To characterize the mutagen under study, it is advisable not to limit the investigation to one mutagen dose, but to obtain dose–dependence of mutation yield. In some cases this dependence helps to define the nature of the mutagen effect. Again, with dose–effect curves, a comparison of various mutagen effects may be correctly made.

B. Time Interval between Mutagen Treatment and Cell Transfer to a Selective Medium

It is essential, when determining the time necessary for expression of newly produced mutations, to define the kinetics of mutation formation. If the same mechanisms operate for mammalian cells as for microorganisms (Freese, 1971), data on phenotypic lag will make it possible to assess, although indirectly, the molecular basis of induced mutations. It is also necessary in order to determine the time at which is expedient to transfer cells to selective conditions so that maximal numbers of induced mutations can be defined.

In Fig. 3 is presented the frequency of recovered mutants resistant to 6MP, depending on the number of cell generations between NMU treatment and transfer to a selective medium (Shapiro *et al.*, 1970). Experiments were carried out according to the scheme presented in Fig. 4 (for details see Section IV,C). Replating in a selective medium took place directly after mutagen treatment and also every day for 8–10 days. Cell counts at the moment of replating made possible determination of the number of cell generations. For the same time period, this number varies from experiment to experiment. Therefore it may be that, according to the number of cell generations, experiments made during various periods of time belong to one category. The data presented here show that the maximal number of mutants is detected if plating in a selective medium takes place two to four generations after NMU treatment. Thereafter, the number of detected mutants decreases. The possibility of this decrease should always be kept in mind,

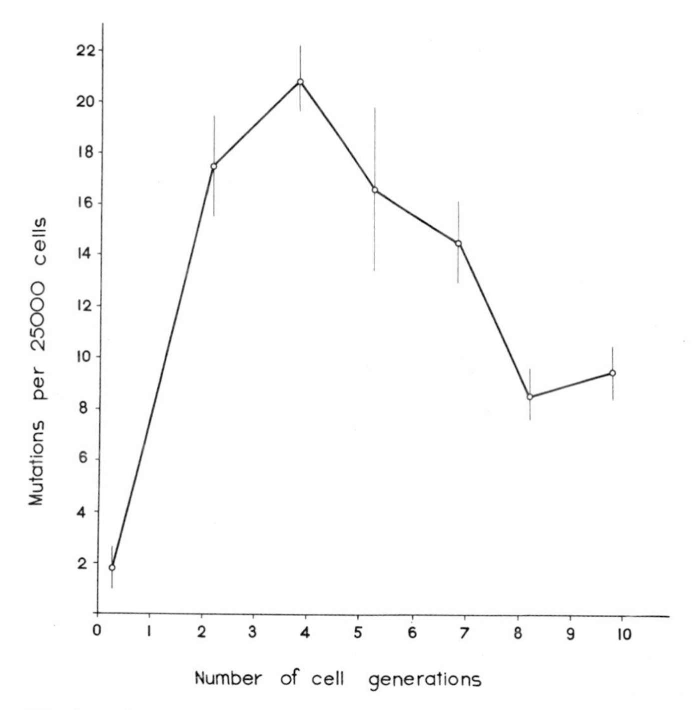

FIG. 3. Kinetics of expression of 6MP resistance mutations induced by NMU. (From Shapiro *et al.*, 1970).

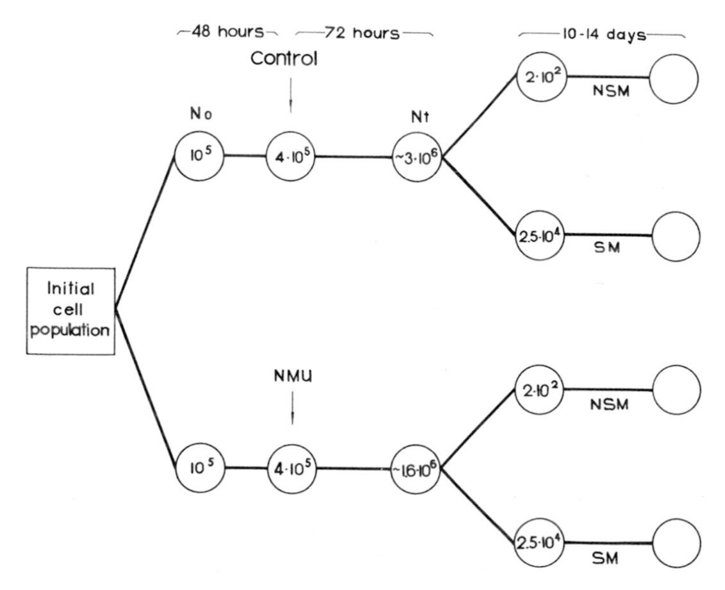

FIG. 4. Schematic diagram of an experiment on induction by NMU of mutation to 6MP-resistance. N_0, initial number of cells per dish; N_t, number of cells at the time of transfer to selective medium; NSM, nonselective medium; SM, selective medium.

since the phenomenon is not rare (Arlett and Harcourt, 1972; Duncan and Brooks, 1973). Without thoroughly considering its causes, it may be assumed that elimination of mutants, because of the selective advantage of wild-type cells, may be one of the factors involved.

C. Description of Experiment on Gene Mutation Induction

In Fig. 4 is presented the scheme of the experiment on induction of mutations for resistance to 6MP in Chinese hamster cells treated with NMU. Cells were plated at 10^5 per dish. After the culture entered the logarithmic growth stage, i.e., about 2 days later, cells were treated with the mutagen. Proceeding from the survival curve determination, a mutagen concentration of 0.1 mg/ml and a treatment duration of 2.5 hours were chosen. The medium was removed from dishes and replaced by fresh Eagle's medium containing the mutagen with 10% serum. Cultures were placed in an incubator at 37°C and pH 7.1–7.2. After 2.5 hours the medium was decanted, and the cells were washed twice with Earle's saline. The same was done in controls without the mutagen. To determine survival cells were removed from two control dishes and mutagenized cultures and plated at 200 cells per dish. Survival was not more than 50% of the control. However, variations in the experiments were great (Shapiro *et al.*, 1972a).

Other dishes, after washing, were filled with fresh medium and incubated for 48–72 hours. In parallel, daily counts of cells in control and NMU-treated cells were made (two dishes from each variant). This made possible the determination of the number of cell generations that occurred after treatment and the selection of a time sufficient for phenotypic expression of mutations. Cells were transferred to a selective medium after two to four generations following mutagenic treatment. Cells in control and mutagenized cultures were plated in two variants: 200 cells per dish in a medium without 6MP to determine survival (five dishes) and 2.5×10^4 cells per dish in selective medium with 6MP (40 dishes). The selective medium was replaced every 3 days. Colonies were scored to determine survival after 10 days, following incubation in a selective medium after 12–14 days. The isolation and testing of mutants is also necessary (Section III,C).

The experiment performed according to this schme makes it possible to assess the mutagenic effect of the agent under study by the "induction value." By this is meant the frequency of induced mutants in relation to the number of viable cells plated in a selective medium. To determine this frequency, corrections for survival should be made for experimental and control cells. For instance, if the average number of colonies per dish were 0.5 and the plating efficiency 20% of the number of plated cells, the true number of resistant colonies at 100% survival must evidently be 0.5×5, i.e., 2.5 colonies per 2.5×10^4 cells per dish. After correction in control and mutagenized cells, the mean number of mutants per dish and their frequency are determined by the difference in the number of colonies in treated and untreated dishes. It is shown at present that the value of induction may vary for different mutagens from one to several orders of magnitude as compared to the spontaneous mutation rate.

REFERENCES

Albertini, R. J., and DeMars, R. (1973). *Mutat. Res.* **18**, 199–224.

Arlett, C. F., and Harcourt, S. A. (1972). *Mutat. Res.* **16**, 301–306.

Breslow, R. E., and Goldsby, R. A. (1969). *Exp. Cell Res.* **55**, 339–346.

Bridges, B. A., and Huckle, J. (1970). *Mutat. Res.* **10**, 141–151.

Brockman, R. W., Roosa, R. A., Law, L. W., and Stutts, L. J. (1962). *J. Cell. Comp. Physiol.* **60**, 65–84.

Chu, E. H. Y. (1971). *Mutat. Res.* **II**, 23–24.

Chu, E. H. Y., and Ho, T. (1970). *Mammalian Chromosome Newslett.* **11**, 58–59.

Chu, E. H. Y., and Malling, H. V. (1968). *Proc. Nat. Acad. Sci. U.S.* **61**, 1306–1312.

Chu, E. H. Y., Sun, N. C., and Chang, C. C. (1972). *Proc. Nat. Acad. Sci. U.S.* **69**, 3459–3463.

Clive, D., Flamm, W. G. Machesko, M. R., and Bernheim, N. J. (1972). *Mutat. Res.* **16**, 77–87.

Clive, D., Flamm, W. G., and Patterson, J. B. (1973). *In* "Chemical Mutagens. Principles and Methods for their Detection" (A. Hollaender, ed.), Vol. 3, pp. 79–103. Plenum, New York.

Cox, R. P., Krauss, M. R., Balis, M. E., and Dancis, J. (1970). *Proc. Nat. Acad. Sci. U.S.* **67**, 1573–1579.

Dancis, J., Cox, R. P., Berman, P. H., Jansen, V., and Balis, M. E. (1969). *Biochem. Genet.* **3**, 609–616.

DeMars, R., and Held, K. R. (1972). *Humangenetik* **16**, 87–110.

Demerec, M. (1946). *Proc. Nat. Acad. Sci. U.S.* **32**, 36.

Duncan, M. E., and Brookes, P. (1973). *Mutat. Res.* **21**, 107–118.

Eagle, H., and Piez, K. (1962). *J. Exp. Med.* **116**, 29–43.

Freese, E. (1971). *In* "Chemical Mutagens. Principles and Methods for their Detection" (A. Hollaender, ed.), Vol. I, pp. 1–56. Plenum, New York.

Ham, R. G. (1972). *In* "Methods in Cell Physiology" (D. M. Prescott, ed.) Vol. 5, pp. 37. Academic Press, New York.

Harris, M. (1971). *J. Cell. Physiol.* **78**, 177–184.

Horowitz, N. H., and Leupold, V. (1951). *Cold Spring Harbor Symp. Quant. Biol.* **16**, 65–72.

Hsu, T. C., and Somers, C. E. (1962). *Exp. Cell Res.* **26**, 404–410.

Jones, C., and Puck, T. T. (1973). *J. Cell. Physiol.* **81**, 299–303.

Kao, F.-T., and Puck, T. T. (1967). *Genetics* **55**, 513–524.

Kao, F.-T., and Puck, T. T. (1968). *Proc. Nat. Acad. Sci. U.S.* **61**, 1275–1281.

Kao, F.-T., and Puck, T. T. (1969). *J. Cell. Physiol.* **74**, 245–257.

Kao, F.-T., and Puck, T. T. (1972). *J. Cell. Physiol.* **80**, 41–50.

Khalizev, A. E. (1969). *Genetika* **5**, No. 3, 157–168.

Khalizev, A. E., Petrova, O. N., and Shapiro, N. I. (1966). *Genetika* **2**, No. 12, 18–24.

Khalizev, A. E., Petrova, O. N., Luss, E. V., Varshaver, N. B., and Shapiro, N. I. (1969) *Genetika* **5**, No. 1, 58–65.

Kit, S., Dubbs, D. R., Piekarski, L. J., and Hsu, T. C. (1963). *Exp. Cell Res.* **31**, 297–312.

Lea, D. E., and Coulson, C. A. (1949). *J. Genet.*, **49**, 264–285.

Lieberman, J., and Ove, P. (1960). *J. Biol. Chem.* **235**, 1765–1768.
Littlefield, J. W. (1964a). *Nature (London)* **203**, 1142–1144.
Littlefield, J. W. (1964b). *Science* **145**, 709–710.
Littlefield, J. W. (1965). *Biochim. Biophys. Acta* **95**, 14–22.
Luria, S. E., and Delbrück, M. (1943). *Genetics* **28**, 491–511.
Marshak, M. I., and Varshaver, N. B. (1970). *Genetika* **6**, No. 2, 130–138.
Meiss, H. K., and Basilico, C. (1972). *Nature (London), New Biol.* **239**, 66–68.
Morrow, J. (1971). *J. Cell. Physiol.* **77**, 423–426.
Naha, P. M. (1969). *Nature (London)* **223**, 1380–1381.
Naha, P. M. (1970). *Nature (London)* **228**, 166–168.
Newcombe, H. B. (1948). *Genetics* **33**, 447–476.
Puck, T. T., and Kao, F.-T. (1967). *Proc. Nat. Acad. Sci. U.S.* **58**, 1227–1234.
Rappaport, H., and DeMars, R. (1973). *Genetics* **75**, 335–345.
Rossa, R. A., Bradley, T. K., Law, L. W., and Herzenberg, L. A. (1962). *J. Cell. Comp. Physiol.* **60**, 109–126.
Roscoe, D. H., Read, M., and Robinson, H. (1973). *J. Cell. Physiol.* **82**, 325–332.
Scheffler, J. E., and Buttin, G. (1973). *J. Cell. Physiol.* **81**, 199–216.
Shapiro, N. I., Khalizev, A. E., Luss, E. V., Marshak, M. I., Petrova, O. N., and Varshaver, N. B. (1972a). *Mutat. Res.* **15**, 203–214.
Shapiro, N. I., Khalizev, A. E., Luss, E. V., Manuilova, E. S., Petrova, O. N., and Varshaver, N. B. (1972b). *Mutat. Res.* **16**, 89–101.
Shapiro, N. I., Petrova, O. N., and Khalizev, A. E. (1966). *Genetika* **2**, No. 12, 5–17.
Shapiro, N. I., Petrova, O. N., and Khalizev, A. E. (1968). Preprint No. 1782. Kurchatov Inst. At. Energy, Moscow (Russian with English summary).
Shapiro, N. I., Petrova, O. N., and Khalizev, A. E. (1970). *Genetika* **6**, 138–140.
Smith, D. B., and Chu, E. H. Y. (1973). *Mutat. Res.* **17**, 113–120.
Subak-Sharpe, H., Bürk, R. R., and Pitts, J. D. (1966). *Heredity* **21**, Part 2, 342–343.
Subak-Sharpe, H., Bürk, R. R., and Pitts, J. D. (1969). *J. Cell Sci.* **4**, 353–367.
Szybalski, W., and Smith, M. J. (1959). *Proc. Soc. Exp. Biol. Med.* **101**, 662–666.
Szybalski, W., Szybalska, E. H., and Brockman, R. W. (1961). *Proc. Amer. Ass. Cancer Res.* **3**, 272.
Szybalski, W., Szybalska, E. H., and Ragni, G. (1962). *Nat. Cancer Inst. Monogr.* **7**, 75–88 and 88–89 (discuss.)
Thompson, L. H., Mankowitz, R., Baker, R. M., Till, J. E., Siminovitch, L., and Whitmore, G. F. (1970). *Proc. Nat. Acad, Sci. U.S.* **66**, 377–384.
Thompson, L. H., Mankovit, R., Baker, R. M., Wright, J. A., Till, J. E., Siminovitch, L., and Whitmore, G. F. (1971). *J. Cell. Physiol.* **78**, 431–440.
Varshaver, N. B., Luss, E. V., and Shapiro, N. I. (1969). *Genetika* **5**, No. 11, 67–78.
Varshaver, N. B., Reznik, L. G., Bagrova, A. M., and Chernikov, V. G. (1971). *Genetika* **7**, No. 3, 89–94.
Varshaver, N. B., Chernikov, V. G., Marshak, M. I., and Shapiro, N. I. (1972). *Genetika* **8**, No. 9, 54–60.
Volkova, L. V., and Kakpakova, E. S. (1972). *Genetika* **8**, No. 12, 120–127.

Chapter 14

Measurement of Protein Turnover in Animal Cells[1]

DARRELL DOYLE AND JOHN TWETO

Department of Molecular Biology,
Roswell Park Memorial Institute,
Buffalo, New York

I. Introduction 235
II. Methods for the Measurement of k_s 238
Radioisotopic Methods 238
III. Measurement of k_d 243
A. Single-Isotope Administration Methods for Measuring k_d 244
B. Double-Isotope Administration Method for Measuring k_d 249
C. Continuous-Administration Methods for Measuring k_d 255
D. Kinetic Methods for Measuring k_d 257
IV. Conclusions 258
References 259

I. Introduction

Over 30 years ago Schoenheimer (1942) used the phrase "dynamic state of body constituents" to emphasize that the proteins within animal tissues are not static but are continually being degraded and replaced by new synthesis. While the acceptance of this idea did not come easily (Hogness *et al.*, 1955), the concept was correct, and now it is well established that essentially all proteins in animal cells turn over (Schimke, 1970). Several recent excellent reviews discuss in detail the importance of the interplay between synthesis and catabolism in the overall regulation of the intracellular level of many specific proteins and enzymes in animal tissues (Schimke and Doyle, 1970;

[1] Research from the authors' laboratory is supported by the National Institutes of Health through grants HD 18410 and GM 19521.

Pine, 1972; Goldberg *et al.*, 1974; Goldberg and Dice, 1974; Schimke, 1969, 1973; Rechcigl, 1971). Thus it is not necessary to stress again that the level of a protein in an animal cell is controlled by both the rate of synthesis and the rate of degradation, and that a change in protein level can be the result of a change in either or both rates.

Within the last several years there has developed a renewed interest in the whole process of intracellular protein turnover and particularly in the role of protein degradation in this process. One reason for this interest has been the realization that degradation, on balance at least, is just as important as synthesis in regulating the concentration of a protein in an animal cell. But, in comparison to what is known about the cellular events involved in polypeptide synthesis, little is known about the cellular mechanisms involved in protein degradation.

Contributing to the renewed interest in protein turnover has been the development in recent years of better techniques to measure both the synthesis and the degradation of animal protein. It seems particularly appropriate at this time to review these methods, pointing out the advantages and limitations of each. Some of the methods, either because they are easier to handle experimentally or because they give a better estimate of the rate of protein synthesis or degradation, are used more frequently than others. We concentrate in this article on the methods we think are the most reliable for measuring protein turnover. We hope to present sufficient experimental detail that a turnover experiment can be done with minimal reference to other sources.

Before proceeding to a discussion and critique of methodology, it is helpful to present some terminology and to develop a mathematical formulation which is useful in defining the relative roles of protein synthesis and degradation in the overall regulation of protein level in animal cells (Price *et al.*, 1962; Segal and Kim, 1965; Berlin and Schimke, 1965). Turnover is defined here as the overall process of renewal of tissue proteins, and includes both their synthesis and degradation. In the steady-state condition, *i.e.*, when the amount of specific protein is not changing, the rates of synthesis and degradation of the protein must be equal. Any change in the level of an enzyme or any protein can be described by

$$dE/dt = k_s - k_d E \qquad (1)$$

where E is the tissue concentration of the specific protein, k_s is a zero-order rate constant of synthesis (usually expressed as units or mass of specific protein synthesized per time per weight of tissue), and k_d is a first-order rate constant of degradation (time^{-1}). The rate of degradation is expressed as a first-order constant simply because degradation with few exceptions (most notably the erythrocyte and its hemoglobin, Shemin and Rittenberg, 1946) conforms to first-order kinetics while protein synthesis conforms to zero-order kinetics. Protein degradation is usually expressed as a half-life which is related to the rate constant of degradation by the expression

$$t_{\frac{1}{2}} = \ln \frac{2}{k_{\mathrm{d}}}$$

In the steady state, when $dE/dT = 0$,

$$E = \frac{k_{\mathrm{s}}}{k_{\mathrm{d}}} \tag{2}$$

Hence in the steady state the level of a specific protein is determined by both its rate of synthesis and its rate of degradation. However, the steady state should be considered a limited case of the more general condition in which tissue levels of specific proteins are changing continuously in response to various alterations in the local environment. In these cases the level of a protein can be altered by changing k_{s} to a new rate of synthesis k'_{s}, or by changing k_{d} to a new rate constant of degradation k'_{d}, or any combination of the two. An equation then can be derived describing the time course of change from one steady-state level to a new level of protein.

$$\frac{E_t}{E_0} = \frac{k'_{\mathrm{s}}}{k'_{\mathrm{d}}E_0} - \frac{k'_{\mathrm{s}}}{k'_{\mathrm{d}}E_0 - 1}\, e - k'_{\mathrm{d}}t \tag{3}$$

In Eq. (3), E_0 is the original concentration of enzyme, and E_t is the concentration of enzyme at some time t after the change to the new k'_s or k'_d. E_t/E_0 describes the fold change in enzyme protein such as might occur in a hormone induction experiment. The equation shows that the new steady-state level of enzyme protein is a function of the new rate constant of synthesis and/or rate constant of degradation, but that the time required to reach the new steady-state level is a function only of the rate constant of degradation k'_{d}.

It should be obvious that both the synthesis and the degradation of a protein are complex cellular processes involving many different steps, and that the mathematical formulation is a very simplified treatment of these processes. Indeed, the steps involved in protein degradation for the most part are not even known; more of course is known about the pathway of protein biosynthesis. However, in a turnover experiment synthesis is usually measured simply as the appearance of "new" protein molecules in the tissue, while degradation is usually measured as the loss of a protein from an intracellular pool of like molecules.

Several important considerations for the study of protein turnover are apparent from Eq. (3): First, proteins with short half-lives respond to a physiological effector such as a hormone or a change in nutritional status much more rapidly than proteins with long half-lives. That is, E_t/E_0 or the fold change will be faster and more apparent if the protein turns over rapidly; and second, it is necessary to measure k_s, k'_s, k_d, and k'_d in order to define the mechanism responsible for a change in the level of a specific protein. The methods for measuring these rates are the subject of the rest of this article.

The choice of method for any particular experiment depends on the system being studied. Systems vary from the intact animal to isolated cells in continuous culture and also include perfused organs, organ explants, tissue slices, and even cell-free systems (Li and Knox, 1972a,b; Ganschow and Chung, 1974). Since liver has been the organ of choice for most turnover studies, we use this tissue as an example of the different methods for studying turnover in the intact animal, mentioning where appropriate variations that must be applied to study protein turnover in other organs. Perfused systems, organ cultures, and tissue culture systems all represent attempts to isolate the cell of interest from the influences of other body constituents and to obtain a system more amenable to experimental manipulation. As an example of these systems, we use mainly cells in continuous culture. The methods developed for cultured cells should be applicable with only slight modification to other isolated systems.

II. Methods for the Measurement of k_s

Radioisotopic Methods

The most rigorous isotopic method for measuring the rate of synthesis of a protein is to determine over an interval of time the change in specific radioactivity of the protein as a function of the change in specific radioactivity of the free intracellular pool of the precursor. This approach requires the constant administration of a labeled precursor, usually an amino acid. Experimentally, the procedure is complicated by the difficulty of measuring the specific radioactivity of the immediate precursor for protein synthesis, the aminoacylated tRNA. Instead, the specific radioactivity of the free amino acid in the intracellular pool usually is measured. Loftfield and Harris (1956) used this approach to measure the rate of rat liver ferritin synthesis after the administration of iron.

By using a labeled precursor that equilibrates rapidly between the tissue and the extracellular compartment, it should be possible to measure an *absolute* rate of protein synthesis, even in the intact animal. Examples of such precursors are carbonate ^{14}C (Swick *et al.*, 1968; McFarlane, 1963) which labels primarily the guanidino group of arginine, and arginine-*guanidino*-^{14}C itself. At least in liver there is a rapid equilibrium between the guanidino group of arginine and urinary urea. The latter conversion is due to the high concentration of the enzyme arginase within the hepatocyte. It should be possible to measure the free amino acid pool of these precursors, and a knowledge of the amino acid composition of the specific protein being studied should allow calculation of the absolute rate of biosynthesis.

In most protein turnover studies, it usually is not necessary to measure an absolute rate of protein synthesis, because the question usually being asked is whether an alteration in enzyme synthesis has occurred; i.e., whether k_s has been changed to k'_s. The most common method of measuring a relative rate of protein biosynthesis in all systems, from the intact animal to cells in culture, is to measure the amount of radioactivity incorporated into the protein of interest after a single administration of a labeled amino acid. An example illustrating the use of this method for measuring a relative rate of protein biosynthesis is presented in Table I. A single gene in inbred mice controls the tissue activity of δ-aminolevulinate dehydratase, the second enzyme in the pathway of heme biosynthesis (Russell and Coleman, 1963). Mice of the inbred strain DBA/2J have about three times more activity of this enzyme in liver than mice of the inbred strain C57BL/6J. Titration with a specific antibody showed that the difference in enzymatic activity between the two strains was due to a difference in the quantity of enzyme protein (Doyle and Schimke, 1969). The question asked was whether the difference in amount of dehydratase protein was due to a difference in the rate of dehydratase synthesis.

TABLE I

RELATIVE RATES OF SYNTHESIS OF HEPATIC δ-AMINOLEVULINATE DEHYDRATASE OF INBRED AND HYBRID MOUSE STRAINS[a]

Mouse strain	δ-Aminolevulinate dehydratase concentration (mg/gm liver)	(1) Radioactivity in δ-aminolevulinate dehydratase (cpm/gm liver)	(2) Radioactivity in total cell protein (cpm/gm liver)	(1)/(2) × 100
DBA/2J	0.41	400	360,000	0.110
C57BI/6J	0.18	200	373,000	0.052
F_1 (DBA/2J × C57BL/6J) (DBA/2J × C57BL/6J)	0.26	260	376,000	0.070

[a] From Doyle and Schimke (1969).

Five microcuries of leucine-4-5-^{3}H in a volume of 0.1 ml was administered by intraperitioneal injection to each of nine mice of each strain. Each mouse weighed about 20 gm. Two hours later the animals were killed, and the livers were perfused via the splenic vein with cold 0.85% NaCl. A 25% (w/v) homogenate of liver was prepared, and a high-speed supernatant fraction of this homogenate was obtained by centrifugation at 105,000 *g* for 1 hour. This fraction was then heated at 67°C for 10 minutes. The latter step increased the specific activity of δ-aminolevulinate dehydratase about 5-fold over that in the homogenate with little reduction in yield. Sufficient monospecific antiserum was added to precipitate all the enzymatic activity (the amount of antiserum required was determined by previous titration). The reaction mixtures were incubated at 37°C for 30 minutes. The immune precipitates were then collected by centrifugation and washed three times with cold 0.85% NaCl. The washed immune precipitates were dissolved in an organic base solubilizer and were counted in a liquid scintillation spectrometer. Preimmune control serum precipitated less than 5% of the radioactivity precipitated by immune serum. Radioactivity incorporated into total liver protein was determined by precipitating an aliquot of the liver homogenate with 10% trichloroacetic acid (final concentration) directly onto glass-fiber filter papers. The filter papers were washed according to standard procedures (Siekevitz, 1952) and then placed in 1.0 ml of an organic base solubilizer for 8 hours at 55°C. Scintillation fluid was added, and radioactivity in protein was determined in a liquid scintillation spectrometer.

The ratio of radioactivity in δ-aminolevulinate dehydratase to that in total protein, shown in the last column in Table I, is used to correct for possible differences in injection technique, body weight, and pool sizes. The assumption involved in using this correction factor is that the enzyme is synthesized from the same free precursor pool as the total cell protein.

It is apparent from the data presented in Table I that the difference in hepatic δ-aminolevulinate dehydratase concentration between the two inbred mouse strains is due to a lower rate of enzyme biosynthesis in the C57BL/6J strain relative to the DBA/2J strain. F_1 progeny of the two strains have intermediate rates of biosynthesis. Thus the genetic locus controlling the tissue level of this enzyme acts at the level of δ-aminolevulinate dehydratase synthesis.

The experiment in Table I is presented in detail, because essentially the same protocol can be used to measure the incorporation of a labelled amino acid into any isolatable protein in any animal system; indeed it has been used many times for many different systems (see Schimke and Doyle, 1970, for review). In this type of experiment, it is imperative that the time of labeled precursor incorporation be short relative to the half-life of the protein whose rate of biosynthesis is being measured. Otherwise, some of the incorporated radioactivity is lost through degradation. Hepatic δ-aminolevulinate dehydratase turns over with a half-life of 5–6 days in the two mouse strains used in the study in Table I. Hence a 2-hour incorporation period is short relative to the half-life of this protein. Other liver proteins turn over with half-lives ranging from less than 1 hour to as long as 20 days (Schimke, 1973; Goldberg and Dice, 1974). A 2-hour pulse of precursor would not give an accurate measure of the synthesis of a protein that turns over with a half-life of 1 hour. Thus some information about the rate of degradation of the protein is needed in order to measure biosynthesis.

Radioactively labeled leucine was used as the precursor for protein biosynthesis in the experiment in Table I, because there is little conversion in liver of leucine to other metabolites (Arias *et al.*, 1969). However, other amino acids can also be used. For example, tyrosine may be a good choice for studying muscle protein biosynthesis or degradation. It has been shown that this amino acid is not metabolized significantly in muscle, and tyrosine equilibrates readily with the intracellular precursor pool in this tissue (Li *et al.*, 1973). For the same reasons phenylalanine might be the precursor of choice for studies on the biosynthesis and turnover of heart muscle proteins (Morgan *et al.*, 1971), and valine for perfused liver systems (Mortimer and Mondon, 1970).

In the experiment in Table I, each mouse was injected with 5 μCi of labeled leucine (25 μCi/100 gm body weight, specific activity 300 mCi/mole). This amount of precursor was sufficient to label δ-aminolevulinate dehydratase to the extent of 200–400 cpm in the enzyme from a gram of liver. The dehydratase accounts for about 0.1% of the total protein synthesized in the liver of a DBA/2J mouse. However, in mice and probably also in other animals, the amount of labeled precursor incorporated into protein is proportional to the amount administered. That is, had each mouse been injected with 200 μCi of labeled leucine instead of 5 μCi, the amount of isotope incorporated into both total protein and δ-aminolevulinate dehydratase would have been about 40-fold higher. Thus it is possible to label proteins present in quite small concentrations by increasing the dose of labeled precursor (R. E. Ganschow, 1974, personal communication, 1974).

In order to study the synthesis and also the degradation of a specific protein using isotopic methods, a method must be available for isolating the protein in homogeneous form from crude tissue extracts. A specific antiserum prepared to the protein in question has been used most frequently for this purpose, and was used in the experiment in Table I. Occasionally, specific proteins or enzymes have been isolated from labeled extracts by classic methods of purification. But this approach is restricted primarily to proteins that are present in relatively high concentrations or which can be purified easily. Furthermore, the approach is rather tedious if repeated isolations are required when measuring rates of synthesis or degradation over some time interval.

The use of an antibody to isolate a specific protein from an extract, while easier, is not without problems. The protein must be purified at least once and in sufficient quantity to provoke an immune response in a rabbit, goat, or other species, sometimes not an easy task. Once a monospecific antiserum to the protein is available, the main problem involved in its use is the nonspecific entrapment of other labeled tissue proteins in the immune complex. The use of an IgG fraction, rather than whole serum, may eliminate some nonspecific coprecipitation (Kabat and Mayer, 1961). But the best way to decrease the amount of coprecipitation is to purify the enzyme partially, maintaining yield, before adding the specific antiserum or IgG fraction. For some proteins and enzymes, such as δ-aminolevulinate dehydratase, a heat step gives good purification and prevents much of the nonspecific coprecipitation. Other proteins of course are not stable to heat and require a different purification procedure, while still others cannot be purified easily without marked reduction in yield. Conditions must be worked out for each enzyme and system.

Sometimes it is possible to remove much of the adventitious protein from the immune precipitate by a more vigorous washing procedure than with cold saline. For example, the immune precipitate can be centrifuged through a 1.0 *M* sucrose solution that is also 0.85% in saline, 0.5% in deoxycholate, and 0.5% in Triton X-100 (Palmiter *et al.*, 1972; Rhoads *et al.*, 1973). Another possibility is to wash the immune precipitate in concentrations of urea that do not break the immune complex but remove loosely bound proteins (R. E. Ganschow, personal communication, 1974).

Finally, the immune precipitate can be dissociated in sodium dodecyl sulfate (SDS) in the presence of a sulfhydryl reductant, and the component polypeptides separated by SDS–polyacrylamide gel electrophoresis (Weber and Osborn, 1969; Laemmli, 1970). This procedure should eliminate all nonspecific contamination, except that from nonspecific polypeptides of the same size as the subunit(s) of the protein of interest. SDS–polyacrylamide gel electrophoresis as a final step increases significantly the specificity of immunochemical isolations and is used routinely in our laboratory.

A control formerly used to correct for the presence of adventitious protein in the immune precipitate was first to remove all the labeled enzyme (or other protein) from the extract with monospecific antiserum. Then an amount of unlabeled ensyme, equivalent to that removed in the first precipitation, was added to the original extract. The amount of radioactivity precipitated by an additional amount of specific antiserum was then subtracted from the radioactivity in the original immune precipitate. It now is known that the second immune precipitation may not be equivalent to the first (Schimke, 1973). Thus this procedure in most cases is not a rigorous control for nonspecific coprecipitation, and it is better, if possible, to use SDS–polyacrylamide gel analysis as the final step in demonstrating specificity.

Finally, possibly the simplest way to estimate the rate of synthesis of a protein, at least in the steady-state condition is to solve Eq. (2). However, the solution requires a knowledge of both the rate constant of degradation and the steady-state level of the protein. Often this information is not available. However, this method has been used to calculate the steady-state rate of synthesis of several enzymes (Schimke, 1969; Fritz *et al.*, 1969; Rechcigl, 1971).

III. Measurement of k_d

It already has been pointed out that the term *rate of protein degradation* is not used synonymously here with the term *protein turnover*. To avoid any confusion about the concepts under consideration in this section, we emphasize again that the term *turnover* means the overall process of protein renewal comprising both synthesis and degradation.

At the outset we wish to consider in more detail the theoretical basis for the measurement of protein degradation. It has been consistently observed that the degradation of most proteins is characterized by first-order kinetics. This has been taken to mean that, once a protein has been synthesized, it has the same chance of being degraded as any other like molecule within the cell. Hence the degradation reaction can be written:

$$\mathrm{E} \xrightarrow{k_d} \mathrm{P}$$

where E is the native enzyme, P is some unspecified product of an irreversible rate-limiting step in the degradation of E, and k_d is the rate constant for the reaction. The rate equation for this reaction is then:

$$dE(t)/dt = -k_\mathrm{d}E(t) \tag{4}$$

or on integration,

$$E(t) = E(0)e^{-k_\mathrm{d}t} \tag{5}$$

or

$$\ln E(0)/E(t) = k_\mathrm{d}t \tag{6}$$

where E(0) is the initial number of protein molecules present. E(t) is the number of molecules remaining after time t has elapsed, and k_d is the rate constant of degradation. The half-life of a protein is the time required for one-half of the protein molecules present initially in the pool to be lost. The half-life, as mentioned, is related to k_d by the expression

$$t_{\frac{1}{2}} = \frac{\ln 2}{k_d}$$

and is a useful parameter when comparing rates of degradation among different proteins. Furthermore, the use of the term half-life avoids the ambiguity involved in describing rates of protein degradation when what is actually meant is rate constants of protein degradation.

A. Single-Isotope Administration Methods for Measuring k_d.

The introduction of amino acids containing heavy nuclides in the 1940s provided a great impetus to the study of animal cell protein turnover (Schoenheimer, 1942). Today use of labeled amino acids is still the most prevalent means for measuring protein degradation, except that radioactive nuclides have for the most part replaced heavy nuclides. In its simplest form the procedure for measuring a rate constant of degradation involves following the loss of label from a protein or a cell fraction with time after the single administration of a radioactivity labeled amino acid. A plot of the logarithm of $E(t)$ or the amount of radioactivity remaining in the protein after an interval of time t has elapsed against time should yield a straight line with a slope equal to $-k_d$. Under steady-state conditions the

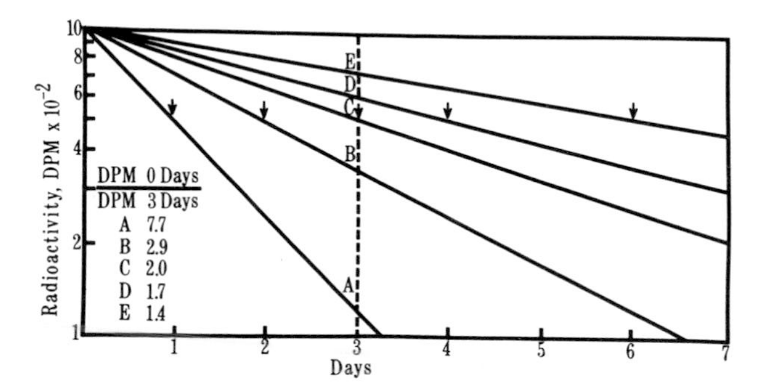

FIG. 1. The loss of radioactivity expected for five hypothetical proteins having different half-lives and undergoing first-order decay. Half-lives for the five proteins, A to E, are indicated by arrows. The numerical values indicate the ratios of the initial radioactivity to the radioactivity remaining in the protein after 3 days of decay. From Glass and Doyle (1972).

loss of specific radioactivity is usually followed, while in nonsteady state conditions in which the level of protein is changing the loss of total radioactivity in the protein is measured.

The type of result expected for five hypothetical proteins having different half-lives and undergoing first-order decay is shown in Fig. 1. This method for obtaining a rate constant of degradation or a half-life for a protein assumes that the radioactive precursor is administered as a short pulse which disappears rapidly from the free intracellular pool. This condition is difficult to obtain experimentally, because of the problem of extensive precursor reutilization in most animal cells. In rats 50% or more of the free intracellular amino acid pool of the liver may come from protein catabolism, even under normal dietary conditions (Gan and Jeffay, 1967). Significant reutilization also occurs in tissue culture cells (Eagle *et al.*, 1959). Thus it is almost impossible to administer labeled amino acids as a pulse.

The problem is illustrated in Fig. 2 which shows the loss of labeled leucine from the free liver pool after a single administration to rats. There is a fairly rapid decay during the first day after injection, but then a rather

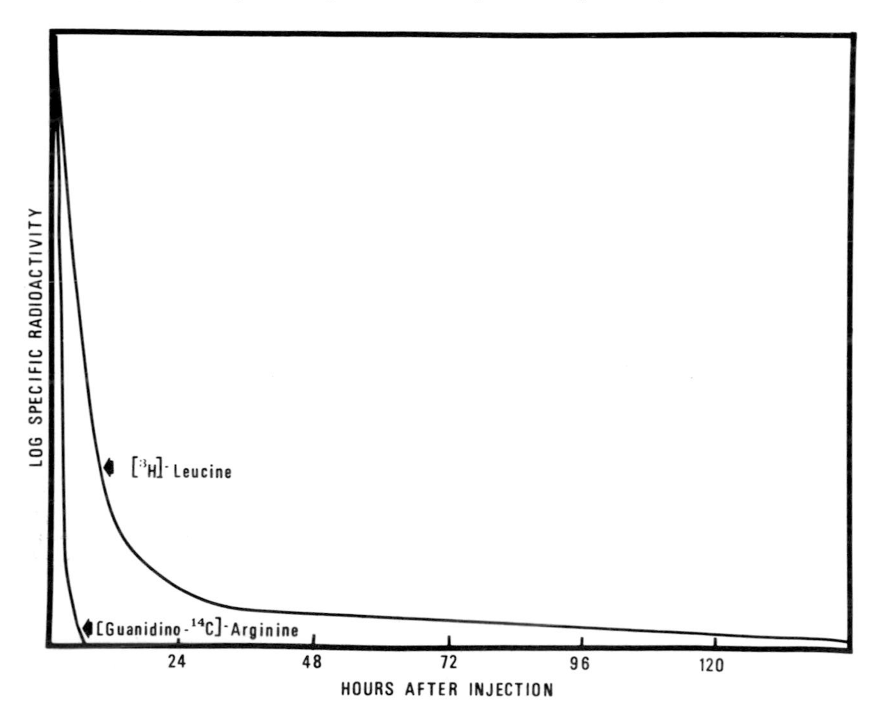

FIG. 2. Schematic of the time course of the free intracellular levels of leucine-^{3}H and arginine-*guanidino*^{14}C following a single administration of the labeled amino acids to rats. After Schimke (1964) and Poole (1971).

slow decay, such that even after 9 days there is still some labeled precursor in the liver cells (Poole, 1971). When leucine is used as the precursor, as it commonly is in turnover studies, label is still entering the protein during the time the degradation measurements are being made. Thus the use of precursors subject to extensive reutilization gives values for the half-life that are greater than the actual value, the degree of overestimation being a function of the extent to which the amino acid is subject to reutilization. This problem is most serious in estimating half-lives of cell proteins turning over rapidly (Koch, 1962). A variety of experimental approaches have been devised to minimize precursor reutilization. Swick (Swick and Handa, 1956), for example, introduced the use of arginine labeled with ^{14}C in the guanidino group as a precursor to label liver proteins. The high concentration of arginase in liver rapidly hydrolyzes the guanidino carbon to urea and, as shown in Fig. 2, this precursor is rapidly lost from the free intracellular pool.

Table II shows a comparison of the half-lives of several rat liver proteins as measured by the rate of loss of radioactivity following a single injection of either arginine-*guanidino*-^{14}C or leucine-^{3}H. It is apparent from the table that guanidino-labeled arginine gives shorter values for the half-lives than does leucine. However, even guanidino-labeled arginine is reutilized to some extent in liver, and the degree of reutilization is a function of the physiological state of the animal (McFarlane, 1963). Experiments by Poole *et al.* (1969) illustrate this point. These investigators measured the half-life of rat liver catalase by several different methods. The single administration of leucine gave a value for the half-life of about 3.5 days, while the rate of loss of guanidino-labeled arginine gave a value of about 2.5 days. But the loss of radioactivity from the heme prosthetic group of catalase labeled *in vivo*

TABLE II

COMPARISON OF THE HALF-LIVES OF SEVERAL RAT LIVER PROTEINS AS DETERMINED BY THE RATE OF LOSS OF RADIOACTIVITY FOLLOWING A SINGLE INJECTION OF EITHER ARGININE-*GUANIDINO*-^{14}C OR LEUCINE-^{3}H[a]

	Half-life (days)	
Protein	Arginine-*guanidino*-^{14}C	Leucine-^{3}H
Ferritin	1.3	3.6
δ-Aminoleuvulinate dehydratase	5.5	6.6
Catalase	2.5	3.5

[a] From Glass and Doyle (1972) and Poole *et al.* (1969).

with aminolevulinic-^{14}C acid gave a half-life of 1.8 days. The latter method is not complicated by reutilization, but it measures the half-life of the heme prosthetic group and not the catalase protein. However, the 1.8-day half-life for the heme group of catalase was comparable to the half-life obtained for the catalase protein when the catalytic activity was inhibited irreversibly with the drug aminotriazole (Price *et al*, 1962; Ganschow and Schimke, 1969). The latter method for measuring a half-life also is not complicated by reutilization (see below). Thus it may be that in this case the turnover of the prosthetic group reflects the turnover of the catalase protein. If so, guanidino-labeled arginine overestimates the half-life of catalase by about 1 day.

A factor contributing to the reutilization of guanidino-labeled arginine by the liver is that after injection the precursor is taken up by the extrahepatic tissues in which it also is incorporated into protein and then released again through degradation. Most tissues, in contrast to liver, are poor in arginase and can act as a trap for the precursor, slowly releasing it to be used again for protein biosynthesis. Although the labeled arginine is diluted by unlabeled amino acid, it may well persist in the animal for some time (Swick and Ip, 1974).

Obviously, there is no advantage in using guanidino-labeled arginine to measure half-lives of proteins in tissues, such as muscle, that do not have arginase activity. Even in tissues that have a urea cycle, there are problems other than reutilization associated with the use of guanidino-labeled arginine to measure protein half-lives. The most serious of these is that the protein synthetic machinery of the cell must compete with the urea cycle for the precursor. Consequently, most of the precursor is hydrolyzed to urea, and only a small amount is incorporated into liver cell protein. Thus the use of guanidino-labeled arginine is restricted to the study of proteins that account for a significant proportion of the total protein synthesized in the liver.

Swick and his collaborators (Swick and Handa, 1956; Swick *et al.*, 1968; Swick and Ip, 1974) used a modification of the guanidino-arginine labeling method which may prove to be the most useful of the single-administration techniques for the measurement of liver protein half-lives. In this method carbonate-^{14}C is used as the precursor for protein biosynthesis. Since only liver has an appreciable urea cycle, it is only in this tissue that the carbonate becomes incorporated into the guanidino group of arginine. There then can be little transfer of guanidino-labeled arginine to the liver from extrahepatic tissues. However, carbonate-^{14}C also becomes incorporated into the carboxyl group of arginine and several other amino acids. Consequently, it was thought necessary to isolate and degrade the arginine residues at each time point on the decay curve of the protein—needless to say, a laborious procedure. This would not be required if all the amino acids labeled with carbonate had a low probability of reutilization. This indeed may be the case. Swick and Ip (1974) measured the rate of loss of both total radioactivity and arginine radioactivity from albumin and from total liver protein after the administration of carbonate-^{14}C to rats. The resulting decay curves were similar, suggesting that it is not necessary to isolate the arginine residues from the protein when using labeled carbonate to measure degradation. Furthermore, the rate at which radioactivity was lost from albumin was 30% slower with guanidino-labeled arginine than with labeled carbonate, again suggesting that extrahepatic tissues release guanidino-labeled arginine to the liver.

This effect is quite striking in regenerating liver in which there appears to be no protein degradation as measured with guanidino-labeled arginine, but almost normal degradation as measured with labeled carbonate. The experiments of Swick described above indicate that carbonate-^{14}C (which is inexpensive and available with high specific activity) is the precursor of choice in studying the turnover of liver proteins. It may also be useful in the measurement of protein turnover in other tissues that lack a urea cycle, such as muscle. In these tissues the carbonate is incorporated into the carboxyl group of such amino acids as glutamate and aspartate. As shown, the carboxyl groups of these amino acids, at least in liver, also have a low probability of reutilization, possibly because the label becomes diluted with tissue carbonate. However, in muscle, it may be necessary to isolate the labeled glutamate and aspartate residues because some of the carbonate becomes incorporated into other amino acids which are subject to reutilization (Millward, 1970; Swick and Song, 1974).

In summary, of the single-administration techniques, carbonate-^{14}C is the best precursor for measuring rates of protein degradation, at least for liver proteins. It has significant advantages as a precursor over arginine-*guanidino*-^{14}C but reasonable half-lives for liver proteins can be obtained with the latter precursor. Other amino acids, such as leucine, significantly overestimate the "true" value of the half-life. Indeed, as shown by Poole (1971), all liver proteins with half-lives of less than 2 days show half-lives of 3–4 days with leucine-^{3}H, if the first time point is taken at 1 day after injection.

Precursors other than amino acids have been used to measure degradation. We have already mentioned that labeling the heme group of catalase with aminolevulinic acid gives a half-life of 1.8 days, which is similar to the half-life of the catalase protein.

Flavin covalently bound to succinic dehydrogenase is also lost at about the same rate as succinic dehydrogenase activity in glucose-repressed yeast

(Grossman *et al.*, 1973). In both these cases the prosthetic group seems to be turning over at the same rate as the protein. If this were always the case, the use of such labels would be a technical advantage, since only a small number of proteins, in some cases maybe only one, would be labeled. However, this is not always the case. For example, the prosthetic group of rat liver fatty acid synthetase, 4-phosphopantetheine, is exchanged many times during the *in vivo* lifetime of a fatty acid synthetase molecule (Tweto *et al.*, 1971). Hence it is not valid to assume that the rate of loss of a prosthetic group reflects the rate of loss of the protein backbone.

B. Double-Isotope Administration Method for Measuring k_d

Arias *et al.* (1969) introduced a modification of the single-administration technique for the measurement of protein degradation, known as the double-isotope method.

The essentials of the double-isotope technique are depicted schematically in Fig. 1 which shows the loss of radioactivity expected for five hypothetical proteins with different first-order rate constants of degradation. It is assumed that the proteins in Fig. 1 were labeled by a short exposure to a precursor which is not reutilized. The loss of radioactivity from protein A with a half-life of 1 day is much faster than that from protein E with a half-life of 6 days. Proteins B, C, and D, with half-lives of 2, 3, and 4 days, respectively, have first-order decay slopes intermediate between those of proteins A and E. In the double-isotope technique, one form of an amino acid, leucine-^{14}C, is administered initially and is allowed to decay a specified length of time. Then a second isotopic form of the same amino acid, leucine-^{3}H, is given, and the animal is killed a short time thereafter. The ^{3}H counts represent the initial time point on the decay curve for any one

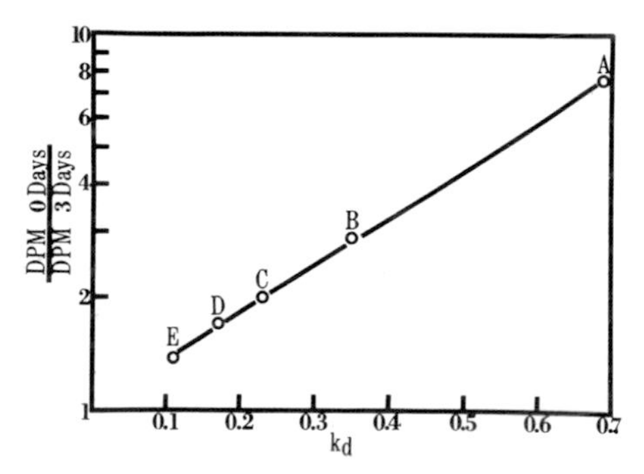

FIG. 3. Dependence of the rate constant of degradation on the logarithm of the ratio of initial radioactivity to radioactivity remaining after 3 days for the five proteins shown in Fig. 1. From Glass and Doyle (1972).

of the five hypothetical proteins, while the ^{14}C counts represent the amount of radioactivity remaining in the protein after the specified interval between injections. The ratio of the initial radioactivity to that remaining after 3 days is shown in Fig. 1 for the five hypothetical proteins. This ratio is equivalent to the $^{3}H/^{14}C$ ratio for a 3-day interval between injections of precursor. The logarithm of this ratio is a function of the first-order decay of the protein, and when it is plotted against the rate constant of degradation k_d a smooth curve is obtained (Fig. 3).

The dependence of the rate constant of degradation k_d on the logarithm of the $^{3}H/^{14}C$ ratio, or the ratio of the initial radioactivity to the radioactivity remaining in a protein after some specified interval, can be seen from the equation describing first-order decay [Eq. (6)]. The $^{3}H/^{14}C$ ratio is equivalent to $E(0)/E(t)$. t is constant for any given experiment and re-

TABLE III

$^{3}H/^{14}C$ RATIOS OF RAT LIVER CELL FRACTIONS AND PROTEINS WITH 3-DAY INTERVALS BETWEEN INJECTIONS[a,b]

Fraction	^{3}H (dpm/mg protein)	^{14}C (dpm/mg protein)	$^{3}H/^{14}C$ ratio
Homogenate	11,400	2,300	5.0
Mitochondria	8,000	1,900	4.2
Microsomes	15,600	2,500	6.3
Supernatant	10,600	2,400	4.3
Cytochrome b_5	1,200	360	3.3
NADPH-cytochrome c reductase	5,200	900	5.9
Catalase	8,140	1,200	6.7
Ferritin	15,300	720	20
Lactate dehydrogenase	8,200	3,200	2.5
δ-Aminolevulinate dehydratase	5,200	1,760	2.9

[a] From Glass and Doyle (1972).

[b] Each of eight rats received an intraperitoneal injection of 25 μCi of leucine-^{14}C. Three days later each rat received 75 μCi of leucine-^{3}H. Four hours after the last injection, the animals were killed and the cell fractions and proteins isolated. Lactate dehydrogenase was purified from about 30 gm of liver by classic purification procedures (Hsieh and Vestling, 1966). Catalase, ferritin, and δ-aminolevulinate dehydratase were isolated from about 2 gm of liver, with their specific antisera. The radioactivities in the purified proteins are expressed per milligram of protein, a value that was estimated from the specific enzymatic activity of the pure enyyme and the activity of the enzyme in crude extracts.

presents the interval between the first and second injections of precursor. The logarithm of $E(0)/E(t)$ or the logarithm of $^3H/^{14}C$ when plotted against k_d should yield a straight line.

In using the double-isotope technique to measure rate constants of degradation, the choice of the proper interval between first and second administrations of isotope is important. As shown in Figs. 1 and 3, a 3-day interval between injections would easily distinguish protein A from protein B, which have half-lives of 1 and 2 days, respectively, but may not differentiate among proteins C, D, and E. By extending the interval to 7 days or longer, the half-lives of proteins, C, D, and E can be measured, but the half-life of protein A cannot. Results obtained from a typical double-isotope experiment illustrate the latter point (Table III).

Lactate dehydrogenase and δ-aminolevulinate dehydratase had ratios of 2.5 and 2.9 in this experiment with a 3-day interval between injections. But Table IV shows a control experiment in which rats were given leucine-^3H and leucine-^{14}C simultaneously. The results presented in Table IV indicate the degree of error inherent in the double-isotope method, with fractions varying in ratio from 2.9 to 3.1, and also gives the isotope ratio to be expected without the effect of degradation, about 3. By comparing the ratios for lactate dehydrogenase and δ-aminolevulinate dehydratase from Table III to the ratios obtained in the control experiment, it can be seen that the two proteins underwent little degradation during the 3-day interval between the first and second injections of isotope.

By extending the interval between injections to 10 days, the relative degradation of lactate dehydrogenase and δ-aminolevulinate dehydratase could be measured as shown in Table V.

TABLE IV
$^3H/^{14}C$ RATIOS OF RAT LIVER CELL FRACTIONS WITH SIMULTANEOUS INJECTION[a,b]

Fraction	^3H	^{14}C (dpm/mg protein)	$^3H/^{14}C$
Homogenate	12,700	4,000	3.1
Mitochondria	5,700	1,800	3.1
Microsomes	14,400	5,000	2.9
Supernatant	9,700	3,400	2.9

[a] After Glass and Doyle (1972).

[b] Two rats each received a simultaneous intraperitoneal injection of 25 μCi of leucine-^{14}C and 75 μCi of leucine-^3H. Four hours later the animals were killed, and the cell fractions were isolated and counted.

TABLE V

$^3H/^{14}C$ RATIOS OF RAT LIVER CELL FRACTIONS AND PROTEINS WITH A 10-DAY INTERVAL BETWEEN INJECTIONS[a]

Fraction	^3H (dpm/mg protein)	^{14}C (dpm/mg protein)	$^3H/^{14}C$ ratio
Homogenate	17,400	1,600	11.1
Mitchondria	9,000	1,400	6.2
Microsomes	13,100	800	16.7
Supernatant	10,543	1,059	10.0
Ferritin	14,190	304	50.0
Lactate dehydrogenase	7,450	1,240	6.2
δ-Aminolevulinate dehydratase	4,400	659	6.7

[a]Experimental details are exactly the same as given in the footnotes to Tables III and IV.

For proteins with half-lives of less than 1 day, it may be better to use arginine-*guanidino*-^{14}C or carbonate-^{14}C as the first isotope and arginine-^3H as the second. The use of a precursor subject to less reutilization than leucine for the first injection would probably give better estimates for the relative rate of degradation of proteins with short-half-lives, because the interval between injections would necessarily have to be short for these proteins.

The double-isotope technique has definite advantages for the determination of relative rates of protein degradation. It is particularly useful for comparing the relative degradation of several different proteins or groups of proteins arranged as intracellular organelles (Dehlinger and Schimke, 1970, 1971; Dice and Schimke, 1972, 1973; Glass and Doyle, 1972). The method is rapid and easy to use, in that a protein need be isolated homogeneously only once. Both the rate of synthesis (from the second isotope administration) and the relative rate of degradation are obtained at the same time. Furthermore, the method is quite reproducible. Since the ratio is obtained from the specific ^3H and ^{14}C radioactivities in a protein isolated only once from the same pool of tissue, there is very little scatter in the data compared to most other methods for measuring protein degradation. The double-isotope technique can also be used to compare the relative degradation of a protein from two different groups of animals. In this case it is better to correct for possible differences in administration technique by relating the $^3H/^{14}C$ ratio in the specific protein to the $^3H/^{14}C$ ratio in the total protein, obtaining a "turnover index" for the specific protein (Ganschow and Schimke, 1969; Ciarnello and Axelrod, 1973).

The double-isotope method as presented thus far does not give the rate constant of degradation or the half-life of a protein, but a ratio of ^3H to ^{14}C radioactivity in the protein. This ratio is sufficient to answer the questions usually asked in most turnover studies, namely, has the rate of degradation of a protein been altered by some effector such as a hormone, a change in diet, or a gene mutation, or is there heterogeneity in the rates of synthesis and degradation of a group of proteins in an animal cell?

However, because the double-isotope method yields a ratio of radioactivity and not a rate constant of degradation, it has been difficult to assess the accuracy and reliability of this method compared to other methods for measuring degradation. But since the ratio $^3H/^{14}C$ is equivalent to $E(0)/E(t)$, it is possible to determine how well the ratio reflects the rate constant of degradation. A plot of the logarithm of $^3H/^{14}C$ against k_d should theoretically give a straight line by reference to Fig. 1 and 3. Glass and Doyle (1972) did this for the proteins and cell fractions listed in Tables III and V. Rate constants of degradation for the specific liver proteins and cell fractions were obtained from the rate of loss of radioactivity after a single administration of arginine-*guanidino*-^{14}C to rats. The results shown in Figs. 4 and 5 were obtained. It is obvious that the ratio is an adequate reflection of the rate constant of degradation.

By constructing standard curves such as those shown in Figs. 4 and 5, it is possible to obtain an estimate of k_d with the double-isotope technique.

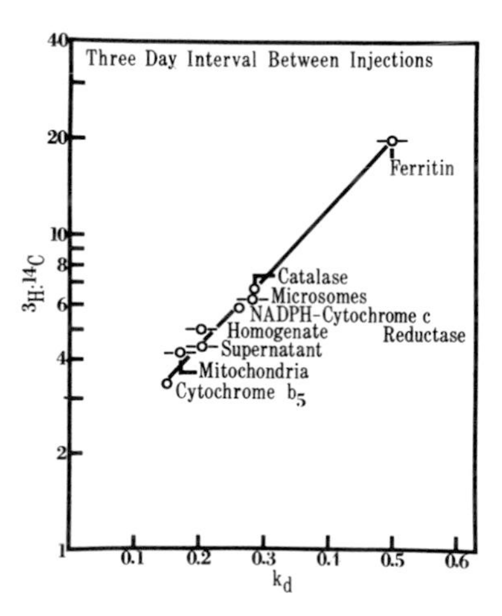

FIG. 4. A plot of the logarithm of the $^3H/^{14}C$ ratio against k_d as determined by arginine-*guanidino*-^{14}C for cell fractions and proteins of rat liver; 3-day interval between injections of isotope. From Glass and Doyle (1972).

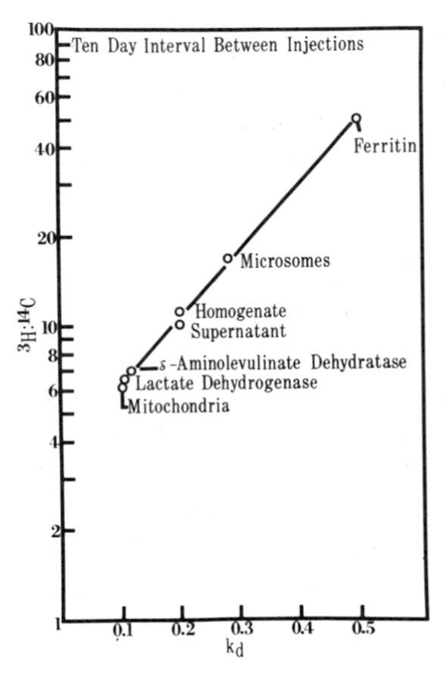

FIG. 5. A plot of k_d against the logarithm of the $^3H/^{14}C$ ratio for cell fractions and proteins of rat liver; 10-day interval between injections. The ratios of ferritin, homogenate, microsomal, mitochondrial, and supernatant fractions were plotted against k_d of these fractions as determined by arginine-*guanidino*-^{14}C. Then the ratios for lactate dehydrogenase and δ-aminolevulinate dehydratase were placed on the curve to estimate their rates of degradation. From Glass and Doyle (1972).

The rate constants of degradation so obtained are equivalent in accuracy to those used to construct the standard curve. The standard curves in Figs. 4 and 5 were constructed from proteins or cell fractions whose half-lives were determined from arginine-*guanidino*-^{14}C decay. Thus the values for the rate constant of degradation are corrected for reutilization, at least to the extent that arginine-*guanidino*-^{14}C is reutilized for protein synthesis. As mentioned previously, this precursor is subject to some reutilization.

The assumptions inherent in the use of the double-isotope technique have been described in detail (Arias *et al.*, 1969; Glass and Doyle, 1972). Briefly, they are: (1) The isotope is not metabolized into other products which are incorporated into the protein or cell fraction; (2) the proteins follow exponential decay kinetics; (3) the rates of synthesis of the proteins of the liver are the same at the time the first and second injections of isotope are given; and (4) at the time the animal is killed, the labeled proteins are undergoing isotopic decay. The first three assumptions in fact have been shown to be valid. Leucine is not metabolized in any significant amount to products incorporated into rat liver proteins (Arias *et al.*, 1969). Most cell fractions and proteins decay with apparent first-order kinetics; there are, however, some exceptions (Shemin and Rittenberg, 1946). The amount of leucine-^{3}H incorporated into the liver cell fractions is about the same whether the isotope is administered simultaneously with, 3 days after, or 10 days after leucine-^{14}C. The fourth assumption is not strictly valid for reasons already discussed. Namely, most amino acids including leucine cannot be administered as a pulse. As shown in Fig. 2, some leucine-^{3}H is still present in the free pool as long as 10 days after administration. However, the double-isotope technique obviates many of the problems associated with a single administration of labeled leucine, and the ratio obtained adequately reflects the "true" rate constant of protein degradation.

The most important advantage of this method is the ability to compare the rates of degradation of many proteins in a single experiment. From such experiments insights into the mechanism of protein degradation have been obtained (Dehlinger and Schimke, 1971; Dice and Schimke, 1973; Glass and Doyle, 1972). Furthermore, it may be possible to develop other dual-labeling methods to study the turnover of specific cell proteins and organelles. For example, the turnover of plasma membrane proteins could be studied independently of the rest of the cell's protein by labeling the membrane with ^{125}I and ^{131}I with lactoperoxidase (Phillips and Morrison, 1971; Hubbard and Cohn, 1972; Tweto and Doyle, 1974). Similarly, the turnover of membrane glycoprotein could be studied independently of the rest of the membrane protein by a dual-labeling technique specific for carbohydrates (Steck and Dawson, 1974).

Poole and Wibo (1973) used a dual-labeling technique to follow the kinetics of degradation of cellular proteins of both slow and fast turnover in cultured fibroblasts. In their method cells are grown in leucine-^{14}C for an extended period of time (20 hours), followed by a 1-hour period of labelling with leucine-^{3}H. Release of ^{3}H radioactivity reflects primarily degradation of the protein population with a fast turnover, while the ^{14}C radioactivity reflects the degradation of proteins with slow turnover. The degradation of both classes of proteins is measured in the same experiment.

C. Continuous-Administration Methods for Measuring k_d

When a labeled amino acid is given to an animal over an extended interval, the time required for the specific radioactivity of cell protein to approach the specific radioactivity of the amino acid is a function of the rate constant of degradation [Eq. (3)]. The labeled amino acid is usually given in the diet or by infusion. In practice the specific radioactivity of the protein never reaches the exact specific radioactivity of the diet. Therefore the time required for the protein to reach constant specific activity is usually taken as an approximation for the complete replacement in the protein of unlabeled amino acid with labeled amino acid. From a plot of $\ln(E^*_{max} - E^*_t)$ against time, k_d can be calculated. E^*_{max} in this case is the maximum specific radioactivity attained by the protein, and E^*_t is the specific radioactivity at some time t after the start of isotopic labeling.

Continuous-administration methods have the advantage that precursor reutilization is less of a problem than with single-administration methods for measuring k_d. However, to obtain reliable results from the continuous-administration method, it is essential that the specific radioactivity of the amino acid precursor reach a maximum *in vivo* quickly relative to the half-life of the protein being studied and then remain constant. In rats there is a lag in the time required for the specific radioactivity of the free pool to reach a maximum. Consequently, continuous-administration methods have been used primarily to measure rates of degradation of long-lived proteins under steady-state conditions. [See Fritz *et al.* (1974) for a discussion of the measurement of k_d under nonsteady-state conditions.]

Arginase (Schimke, 1964) and the isoenzymes of lactate dehydrogenase (Fritz *et al.*, 1969, 1973) are examples of long-lived proteins whose rates of degradation were measured using the continuous-administration method. The value obtained for the k_d of arginase from the continuous administration of lysine-^{14}C of constant specific radioactivity was similar to that obtained from the rate of loss of guanidino-labeled arginine, about 4–5 days. By the twentieth day of isotope administration, about 90% of the lysine in arginase was replaced from the diet.

Measurement of the turnover of the isoenzymes of lactate dehydrogenase in different tissues by the continuous-administration method gave rise to the important concept that *in vivo* exchange of subunits may contribute to the characteristic tissue lactate dehydrogenase isoenzyme pattern (Fritz *et al.*, 1971). However, heterogeneity of cell types in animal tissues has also been invoked to explain the apparent nonrandom distribution of the two dissimilar subunits among the different lactate dehydrogenase forms (Lebherz, 1974). Thus while most turnover studies have assumed homogeneity of cell type, all animal tissues are in fact composed of several different cell types and such heterogeneity can potentially complicate the interpretation of any turnover experiment.

An interesting variation of the continuous-administration method was used by M. Pine (personal communication, 1974) to measure total protein degradation in cultured cells. When cells in culture are grown in the presence of tritiated water, the tritium rapidly exchanges with amino acid hydrogen because of transamination reactions. However, once the amino acid is in a peptide bond, significant exchange no longer occurs. Therefore protein degradation can be followed from the rate of loss of radioactivity in protein. Whether this approach is applicable to other systems remains to be seen.

D. Kinetic Methods for Measuring k_d.

Equation (3) shows that the time required for an enzyme to go from one steady-state level to another is strictly a function of the rate constant of degradation. Consequently, it is possible to obtain an estimate of k_d of a protein whose steady-state level in the cell can be changed, such as by a hormone or a change in diet. Experimentally, an animal or a cell in culture is exposed to some "inducer" for a sufficient time for the enzyme to reach the new steady state. By measuring the change in activity as a function of time, and plotting ln $(E_{max} - E_t)$ against time, k_d can be calculated. E_{max} denotes the enzymatic activity at the new steady state, and E_t is the enzymatic activity at some time t after addition of inducer. This method of measuring k_d has the advantage that the enzyme is not purified; only its catalytic activity is measured. It is assumed that the change in catalytic activity reflects a change in enzyme protein. The other assumptions of this method are the same as the assumptions used in the development of Eq. (3). They are:

1. The change in the rate of enzyme synthesis is instantaneous and is then maintained constant.
2. The inducer does not affect k_d. That is, k_d is not changed to k'_d.
3. Finally, it is important in using this approach to establish that the new steady state actually is reached, which is sometimes difficult to do.

In view of the limitations of this method, it is surprising how well the "time required to approach or decay from the new steady state" actually reflects the "true" k_d for many enzymes (Reel and Kenny, 1968; Auricchio *et al.*, 1969).

Rate constants of degradation of proteins or enzymes have also been obtained by measuring the rate of loss of catalytic activity after the inhibition of protein synthesis by such drugs as chloramphenicol, cycloheximide, and puromycin. This method for estimating k_d is easy to apply, but unfortunately the results are not easy to interpret because the drug itself may affect the rate of degradation of the enzyme being studied (Barker *et al.*, 1971). A more promising drug-based approach to the measurement of enzyme degradation is the use of drugs that specifically and irreversibly inhibit the catalytic activity of the ensyme in question.

We have already mentioned that aminotriazole was used to measure the rate constant of degradation of catalase. This drug, when given to intact animals, combines irreversibly and rather specifically with the catalase tetramer, inhibiting catalytic activity. The return of catalase activity, again a function of k_d indicates a half-life for the catalase molecule of about 1.5 days (Price *et al.*, 1962; Ganschow and Schimke, 1969). There are other rather specific enzyme inhibitors such as methotrexate for folate reductase (Hakala, 1973). Unfortunately, however, the number of such inhibitors is not large. Furthermore, the inhibitors must be able to reach the specific enzyme *in vivo* and be relatively nontoxic to the animal. Consequently, this approach has not been used very often.

IV. Conclusions

One might conclude from the previous discussion that each of the methods for measuring a rate constant of degradation seems to have serious limitations. The question is then whether such a rate can be measured at all and, if so, what the best method for doing it is. Indeed, all the methods have one or more limitations. Some of the limitations, however, are more serious than others. Still there is no one best method for measuring a rate constant of degradation.

Often the method of choice depends on the system and the question being asked. When determining the relative turnover of a group of proteins, the double-isotope method might be the one of choice. It is easy to use for this type of study, and the rate of synthesis and the relative rate of degradation can be obtained from the same set of data.

The single-carbonate-administration method might be the best choice if one is interested in a value for the half-life of a liver protein that best reflects the true half-life. Arginine-*guanidino*-^{14}C might be the precursor of choice in studying the degradation of a rapidly turning over protein. A protein that turns over rapidly is synthesized rapidly. Guanidino-labeled arginine would be incorporated extensively into such a protein, and reutilization in this case would not be a serious problem. For proteins synthesized at a very low rate, a continuous administration technique may be the only practicable way of labeling the protein sufficiently to study its turnover. A single administration with labeled leucine, despite reutilization, may be sufficiently sensitive to distinguish a change in the rate of protein degradation. In cell culture studies in which the medium can be manipulated readily, this precursor or another reutilizable amino acid may be the one of choice.

As a final conclusion, it is noted that while the method depends on the study it is always better to use a second or even a third method to confirm the results of the first. Then one can have some confidence in the value obtained for the rate constant of degradation.

REFERENCES

Arias, I. M., Doyle, D., and Schimke, R. T. (1969), *J. Biol. Chem.* **244**, 3303.

Auricchio, F., Martin, D., Jr., and Tomkins, G. (1969). *Nature (London)* **224**, 806.

Barker, K. L., Lee, K. L., and Kenney, F. T. (1971). *Biochem. Biophys. Res. Commun.* **43**, 1132.

Berlin, C. M., and Schimke, R. T. (1965). *J. Mol. Pharmacol.* **1**, 149.

Ciarnello, R. D., and Axelrod, J. (1973). *J. Biol. Chem.* **248**, 5616.

Dehlinger, P. J., and Schimke, R. T. (1971). *J. Biol. Chem.* **246**, 2574.

Dehlinger, P. J., and Schimke, R. T. (1970). *Biochem. Biophys. Res. Commun.* **40**, 1973.

Dice, J. F., and Schimke, R. T. (1972). *J. Biol. Chem.* **247**, 98.

Dice, J. F., and Schimke, R. T. (1973). *Arch. Biochem. Biophys.* **158**, 97.

Doyle, D., and Schimke, R. T. (1969). *J. Biol. Chem.* **244**, 5449.

Eagle, H. Plef, K. A., Fleischman, R., and Oyama, V. I. (1959). *J. Biol. Chem.* **234**, 592.

Fritz, P. J. Vesell, E. S., White, E. L., and Pruitt, K. M. (1969). *Proc. Nat. Acad. Sci. U. S.* **62**, 558.

Fritz, P. J., White, E. L., Vesell, E. S., and Pruitt, K. M. (1971). *Nature (London), New Biol.* **230**, 119.

Fritz, P. J., White, E. L., Pruitt, K. M., and Vesell, E. S. (1973). *Biochemistry* **12**, 4034.

Fritz, P. J., White, E. L., Osterman, J., and Pruitt, K. M. (1974). *Proc. Joint U.S.-Jap. Symp. Protein Turnover, 1975* (in press).

Gan, J. C., and Jeffay, H. (1967). *Biochim. Biophys. Acta* **148**, 448.

Ganschow, R. E., and Schimke, R. T. (1969). *J. Biol. Chem.* **244**, 4649.

Ganschow, R. E., and Chung, A. C. (1974). *Fed. Proc., Fed. Amer. Soc. Exp. Biol.* **33**, 1534.

Glass, R. D., and Doyle, D. (1972). *J. Biol. Chem.* **247**, 5234.

Goldberg, A. L., and Dice, J. F. (1974). *Annu. Rev. Biochem.* **43**, 835.

Goldberg, A. L., Howell, E. M., Martel, S. B., Li, J. B., and Prouty, W. F. (1974). *Fed. Proc., Fed. Amer. Soc. Exp. Biol.* **33**, 1112.

Grossman, S., Obley, J., Hogue, P. K., *et al.* (1973). *Arch. Biochem. Biophys.* **158**, 744.

Hakala, M. (1973). *In* "Drug Resistance and Selectivity" (E. Mihich, ed.), p. 263. Academic Press, New York.

Hogness, D. S., Cohn, M., and Monod, J. (1955). *Biochim. Biophys. Acta* **16**, 99.

Hsieh, W. T., and Vestling, C. S. (1966). *Biochem. Prep.* **11**, 69.

Hubbard, A. L., and Cohn, Z. A. (1972). *J. Cell. Biol.* **55**, 390.

Kabat, E. A., and Mayer, M. M. (1961). "Experimental Immunochemistry." Thomas, Springfield, Illinois.

Koch, A. L. (1962). *J. Theor. Biol.* **3**, 283.

Laemmli, U. K. (1970). *Nature* (*London*) **227**, 680.

Lebherz, H. (1974). *Experientia*, **30**, 655.

Li, J. B., and Knox, W. E. (1972a). *J. Biol. Chem.* **247**, 7546.

Li, J. B., and Knox, W. E. (1972b). *J. Biol. Chem.* **247**, 7550.

Li, J. B., Falks, R. M., and Goldberg, A. L. (1973). *J. Biol. Chem.* **248**, 7272.

Loftfield, R. B., and Harris, A. (1956). *J. Biol. Chem.* **219**, 151.

McFarlane, A. S. (1963). *Biochem. J.* **89**, 277.

Millward, D. J. (1970). *Clin. Sci.* **39**, 577.

Morgan, H. E., Earl, D. C. N., Broadus, A., Wolpert, E. B., Giger, K. E., and Jefferson, L. S. (1971). *J. Biol. Chem.* **246**, 2152.

Mortimore, G. E., and Mondon, C. E. (1970). *J. Biol. Chem.* **245**, 2375.

Palmiter, R. D., Palacios, R., and Schimke, R. T. (1972). *J. Biol. Chem.* **247**, 3296.

Phillips, D. R., and Morrison, M. (1971). *Biochemistry* **10**, 1766.

Pine, M. J. (1972). *Annu. Rev. Microbiol.* **26**, 103.

Poole, B. (1971). *J. Biol. Chem.* **246**, 6587.
Poole, B., and Wibo, M. (1973). *J. Biol. Chem.* **248**, 6221.
Poole, B., Leighton, F., and DeDuve, C. (1969). *J. Cell Biol.* **41**, 536.
Price, V. E., Sterling, W. R., Tarantola, V. A., Hartley, R. W., Jr., and Rechcigl, M., Jr. (1962). *J. Biol. Chem.* **237**, 3468.
Rechcigl, M., Jr. (1971). "Enzyme Synthesis and Degradation in Mammalian Systems," (E. M. Rechcigl, ed.), p. 237. Univ. Park Press, Baltimore, Maryland.
Reel, J. R., and Kenny, F. T. (1968). *Proc. Nat. Acad. Sci. U.S.*, **61**, 200.
Rhoads, R. E., McKnight, G. S., and Schimke, R. T. (1973). *J. Biol. Chem.* **248**, 2031.
Russel, R. L., and Coleman, D. L. (1963). *Genetics* **48**, 1033.
Schimke, R. T. (1964). *J. Biol. Chem.* **239**, 3808.
Schimke, R. T. (1969). *Curr. Top. Cell Regul.* **6**, 77.
Schimke, R. T. (1970). *Mammalian Protein Metab.* **4**, 177.
Schimke, R. T. (1973). *Advan. Enzymol.* **37**, 135.
Schimke, R. T., and Doyle, D. (1970). *Annu. Rev. Biochem.* **39**, 1929.
Schoenheimer, R. (1942). "Dynamic State of Body Constituents." Harvard Univ. Press, Cambridge, Massachusetts.
Segal, H. L., and Kim, U. S. (1965). *J. Cell. Comp. Physiol.* **66**, Suppl. 1, 11.
Shemin, D., and Rittenberg, D. (1946). *J. Biol. Chem.* **166**, 627.
Siekevitz, P. (1952). *J. Biol. Chem.* **195**, 549.
Steck, T. L., and Dawson, G. 1974. *J. Biol. Chem.* **249**, 2135.
Swick, R., and Handa, D. T. (1956). *J. Biol. Chem.* **218**, 557.
Swick, R. W., and Song, H. (1974). *J. Anim. Sci.* **38**, 1150.
Swick, R. W., and Ip, M. M. (1974). *J. Biol. Chem.* **249**, 6836.
Swick, R. W., Rexroth, A. K., and Stange, J. L. (1968). *J. Biol. Chem.* **243**, 3581.
Tweto, J., and Doyle, D. (1974). *Proc. Joint U.S.-Jap. Symp. Protein Turnover, 1975* (in press).
Tweto, J., Liberati, M., and Larrabee, A. R. (1971). *J. Biol. Chem.* **246**, 2468.
Weber, K., and Osborn, M. J. (1969). *J. Biol. Chem.* **244**, 4406.

Chapter 15

Detection of Mycoplasma Contamination in Cultured Cells: Comparison of Biochemical, Morphological, and Microbiological Techniques

EDWARD L. SCHNEIDER

Gerontology Research Center,
N.I.A., N.I.H.,
Baltimore, Maryland

I. Introduction 261
II. Microbiological Culture 262
III. Morphological Techniques 263
  A. Acetoorcein Staining (Light Microscopy) 263
  B. Electron Microscopy 263
  C. Autoradiography 263
IV. Biochemical Techniques 265
  A. Polyacrylamide Gel Electrophoresis (PAGE) 265
  B. Differential Incorporation of Exogenous Uridine and Uracil into Cellular RNA 268
  C. Measurement of Arginine Content in Cell Culture Media 268
V. Comparative Studies of Biochemical, Morphological, and Microbiological Techniques for the Detection of Mycoplasma Contamination 268
  A. Screening of Human Skin Fibroblast Cultures 268
  B. Screening of Amniotic Fluid Cell Cultures 270
VI. Sensitivities of Biochemical, Morphological, and Microbiological Techniques for Mycoplasma Detection. 272
VII. Discussion 274
  References 275

I. Introduction

The diverse cellular metabolic abnormalities induced by mycoplasma contamination are well documented in several recent reviews (Stanbridge,

1971; Maniloff and Morowitz, 1972; Barile, 1973; Schneider and Stanbridge, 1974). In brief, these include altered nucleic acid metabolism, cytopathology, diminished viral yields, impaired drug responses, and increased chromosomal breakage. Mycoplasma contamination of cultured cells has become a serious problem not only for the cell biologist but also for the clinician. In fact, recent studies indicate that the karyological and metabolic effects of mycoplasma may pose a potential hazard to successful pre- and postnatal diagnosis of inherited chromosomal and biochemical disorders (Schneider *et al.*, 1974b; E. L. Schneider, J. Tallman, and E. J. Stanbridge, personal communication, 1975; E. J. Stanbridge, J. Tischfield, and E. L. Schneider, personal communication, 1975). The extent of mycoplasma contamination is illustrated by a survey of cell strains, which revealed that up to 60% were infected with these microorganisms (Hayflick, 1965).

Since mycoplasma-infected cells often grow well and appear normal by light microscopy, contamination by these organisms is notoriously difficult to detect. Currently, the most widely used technique for mycoplasma detection involves growth of these organisms on specially prepared solid agar and broth (Hayflick, 1965). However, several recent studies employing biochemical, serological, and morphological techniques have demonstrated mycoplasma contamination in cultures that were negative by this microbiological method (Markov *et al.*, 1969; Harley *et al.*, 1970; Todaro *et al.*, 1971; Levine, 1972; Perez *et al.*, 1972; Hopps *et al.*, 1973; Schneider *et al.*, 1973, 1974b; Brown *et al.*, 1974). Unfortunately, there have been few studies in which microbiological techniques for mycoplasma detection are compared with histological and biochemical techniques (Schneider *et al.*, 1973, 1974b). This chapter is devoted to a description of the techniques employed in these studies and to a comparison of their success in screening human cell cultures for mycoplasma contamination. Special attention is paid to two new biochemical techniques: identification of labeled mycoplasma RNAs by polyacrylamide gel electrophoresis, and measurement of the incorporation of exogenous bases and nucleotides into RNA.

II. Microbiological Culture

Because of their fastidious nutrient requirements, specialized media have been developed for the growth of mycoplasma. The most commonly employed medium (Hayflick, 1965) is composed of Difco beef heart for infusion, Difco Bacto Peptone and NaCl supplemented with yeast extract and horse agammaglobulin serum. For maximal detection both cells and cell culture media are inoculated into broth and transferred to Difco Bacto-PPLO agar (supplemented as above) twice weekly for 3 weeks. Additional samples are streaked directly onto agar. All cultures are incubated aerobically and in a 5% CO_2–95% N_2 environment. The agar plates are examined microscopically for mycoplasma colonies. Cultures are considered negative if no colonies are observed after 3 weeks.

III. Morphological Techniques

A. Acetoorcein Staining (Light Microscopy)

Cells are grown on slide-flaskettes (Lab-Tek), suspended in a hypotonic buffer, fixed with acetic acid–methanol (1:3), air-dried, and stained with orcein by the method of Fogh and Fogh (1964). By phase microscopy, mycoplasmas can be detected in the swollen cell cytoplasm and in the intracellular spaces. Care is taken to avoid confusing cell organelles or cell debris with mycoplasma.

B. Electron Microscopy

Cell pellets are fixed in 3% glutaraldehyde in cacodylate buffer (pH 7.2), postfixed in 2% osmium tetroxide, dehydrated in graded alcohol, embedded in a mixture of Epon and Araldite, and stained with uranyl acetate (Schneider *et al.*, 1973). Mycoplasmas, when present, were observed along the cell membrane (Fig. 1A).

C. Autoradiography

Cell cultures, grown on slide-flaskettes, are incubated for 1 hour with thymidine-^{3}H (specific radioactivity 17 Ci/mmole) or 20 minutes with uridine-^{3}H (specific radioactivity 28 Ci/mmole) at a concentration of 5 μCi/ml. The slides are then detached, rinsed briefly with 5% perchloric acid, fixed with acetic acid–methanol (1:3), air-dried, dipped in Kodak NTB-2 emulsion, developed after a 1-week exposure, and stained with Giemsa. Uncontaminated cells should have grains limited to the nuclear site of RNA and DNA synthesis. In mycoplasma-infected cells, such as the one seen in Fig. 1B, grains are found over the cell cytoplasm, particularly near the cell periphery.

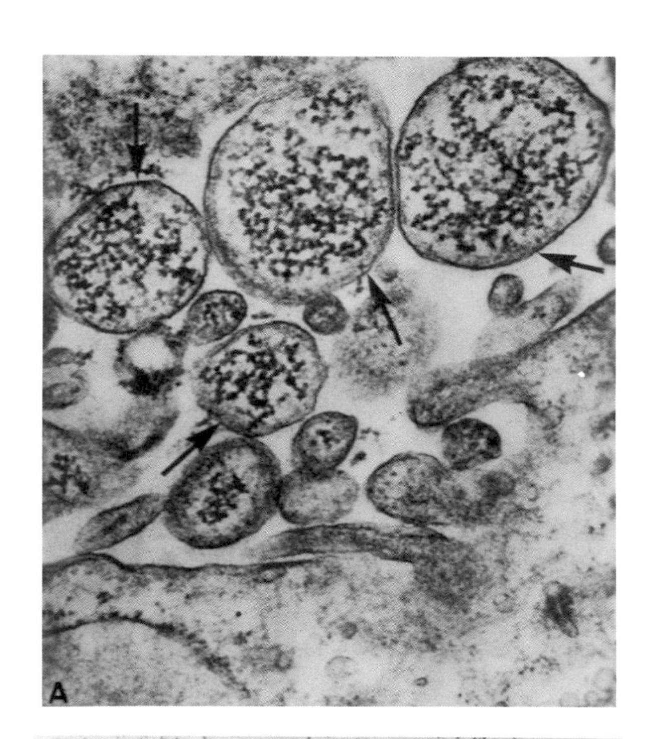
A

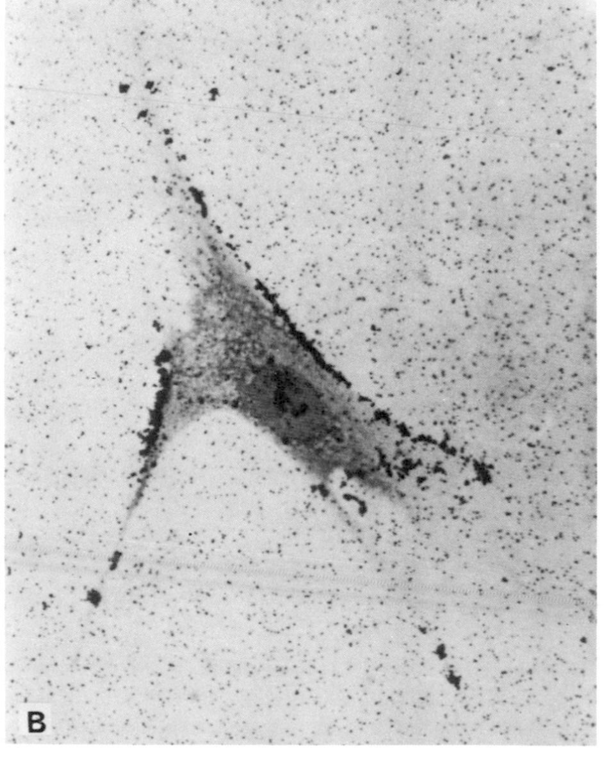
B

IV. Biochemical Techniques

A. Polyacrylamide Gel Electrophoresis (PAGE)

1. Background

A variety of biochemical techniques has been introduced for mycoplasma detection. Most have involved measurement of specific mycoplasma enzymes (Barile and Schimke, 1963; Horoszewicz and Grace, 1964; Levine, 1974). Another biochemical approach takes advantage of the size difference between mycoplasma and mammalian cell RNAs. Levine *et al.* (1968) first demonstrated that mycoplasma RNAs could be distinguished from eukaryotic 28 and 18 S RNAs. The introduction of polyacrylamide gel electrophoresis has led to clear separation of mycoplasma 23 S_E (Svedberg constant by electrophoresis) and 16 S_E RNAs from host cell RNA species.

2. Description of PAGE Technique

Cells (1×10^6) from cultures to be tested are inoculated into 75-ml plastic flasks (Falcon). When the cell monolayers are near confluency (usually 3 days), they are labeled overnight with 5 μCi/ml uridine-^{3}H (specific radioactivity 28 Ci/mmole). The labeled cells are then removed from the flask with 0.1% pronase, washed with ice-cold, phosphate-buffered saline (Dulbecco's), and suspended in 0.01 *M* Na acetate and 0.1 *M* NaCl buffer (pH 5.1) with 1% sodium dodecyl sulfate. Cellular protein is then removed by three extractions with redistilled phenol. The resultant ^{3}H-labeled RNA (10-μg aliquots) is electrophoresed on 100-mm 2.5% acrylamide–bisacrylamide gels in Plexiglas tubes for 2 hours at 5 mA per tube. Replicate gels are either fixed (1% acetic acid for 15 minutes), stained (methylene blue in 0.4 *M* Na acetate and 0.4 N acetic acid buffer) for 1 hour, destained in distilled water, and photographed, or sliced into 1-mm fractions and the radioactivity of these fractions measured (see Chapter 16, Section III,C).

3. Separation of Mammalian and Mycoplasma RNAs

Mycoplasmas, as well as other prokaryotic microorganisms, have ribosomal subunits and RNAs that are considerably smaller than their mammalian (eukaryotic) ribosomal counterparts. Stained polyacrylamide gels of RNA extracted from both mycoplasma-contaminated and uncontam-

---

FIG. 1. (A) Electron micrograph of a cell from a human fibroblast culture that demonstrated mycoplasma 23 and 16 S_E RNAs on PAGE. Note the pleomorphic mycoplasma (arrow) located along the cell membrane. ×35,000. (B) Autoradiograph of a cell from the same culture as in (A) labeled with 5 μCi/ml of uridine-^{3}H for 20 minutes. Note the localization of grains to the cell periphery characteristic of mycoplasma-contaminated cells. × 680.

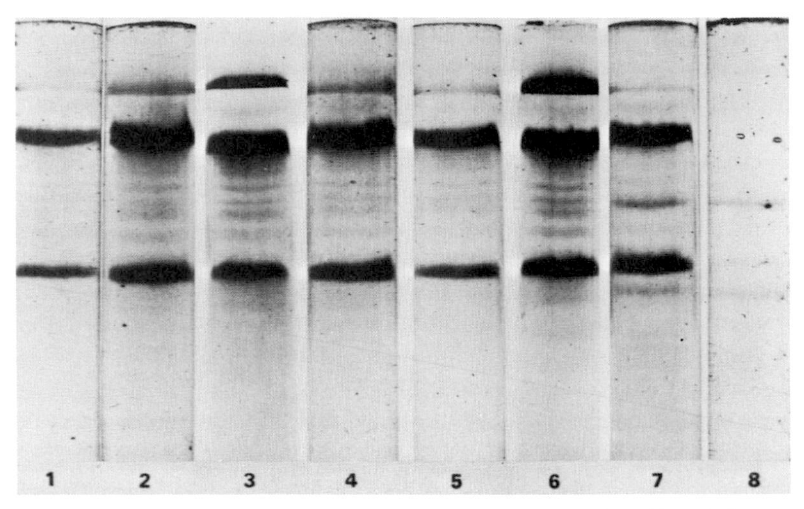

FIG. 2. Photograph of polyacrylamide gels stained with methylene blue. Gels 1 to 6 are cellular RNAs extracted from uncontaminated (1) and mycoplasma-contaminated (2 to 6) human fibroblast cell cultures. Several light-staining bands are seen between the dark major 28 and 18 S human rRNAs. However, there was insufficient mycoplasma RNA present to produce the distinctive 23 and 16 S_E RNA bands observed in gel 7, a composite gel of purified mycoplasma RNA electrophoresed together with uncontaminated whole-cell RNA. Gel 8 indicates the mobility of the purified mycoplasma RNA.

inated cultures are seen in Fig. 2 (gels 1 to 6). Although there are several intermediate bands visible between the darkly stained eukaryotic 28 and 18 S RNAs in these gels, it is only when purified mycoplasma RNA is added to cellular RNA (gel 7) that distinct 23 and 16 S_E bands can be distinguished. Fortunately, the rapid incorporation of uridine-^{3}H into mycoplasma RNA compensates for the small quantities of RNA present in contaminated cells. The radioactive profiles of uridine-^{3}H-labeled RNA from uncontaminated and mycoplasma-contaminated cell cultures clearly demonstrate the presence of ^{3}H-labeled 23 and 16 S_E mycoplasma RNAs in the contaminated cells (Fig. 3).

4. SEPARATION OF MYCOPLASMA AND BACTERIAL RNAS

The same gel electrophoretic techniques described above can distinguish the larger 23 S_E mycoplasma RNA species from its 22 S_E bacterial counterpart (Fig. 4). This separation may be useful in identifying organisms that cannot be detected by standard microbiological culture assay.

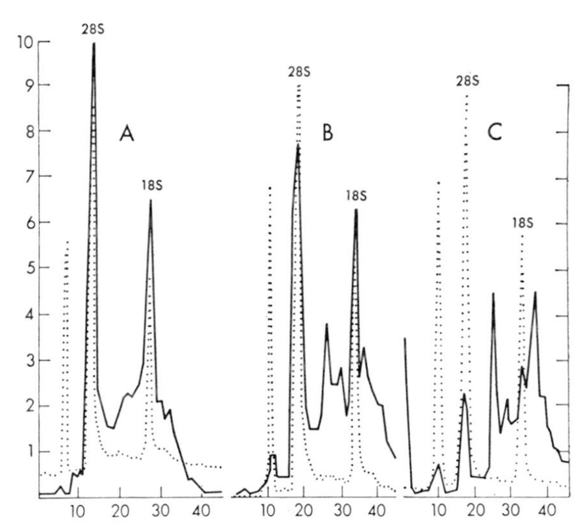

FIG. 3. Abscissa: gel slice number; ordinate: solid line, radioactivity per fraction (percent total counts per minute); broken line, relative optical density. PAGE profiles of uridine-^{3}H labeled cellular RNA extracted from cell strain UC 99 from group I (A), and UC 69 (B) and UC 15 (C) from group II. The 28 and 18 S eukaryotic rRNA peaks are indicated. In the uncontaminated cell strain seen in (A) the ^{3}H-labeled RNA profile parallels the optical density distribution of cellular 28 and 18 S ribosomal species. In cell strains that were positive by standard microbiological testing, seen in (B) and (C), prominent ^{3}H-labeled mycoplasma 23 and 16 S_E peaks are observed.

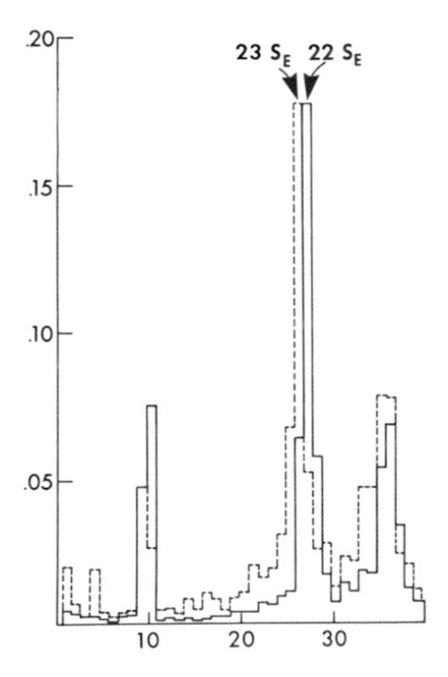

FIG. 4. Abscissa: gel slice number; ordinate: fraction of total radioactivity. Uridine-^{3}H-labeled *M. hyorhinis* RNA (broken line) was coelectrophoresed with uridine-^{14}C-labeled *E. coli* RNA (solid line) on 2.5% polyacrylamide gels. Examination of their radioactive profiles reveals that the 23 S_E mycoplasma rRNA can be distinguished from its bacterial 22 S_E counterpart.

B. Differential Incorporation of Exogenous Uridine and Uracil into Cellular RNA

This new, simple biochemical test is discussed in great detail in the next chapter (Schneider and Stanbridge, Chapter 16).

C. Measurement of Arginine Content in Cell Culture Media

Many mycoplasma species utilize arginine as their chief energy source, leading to selective depletion of this amino acid in media from mycoplasma-contaminated cell cultures. Media from cultures suspected of contamination are deproteinized with 20% sulfosalicyclic acid, and then aliquots of the media supernatants are run on column B of an amino acid analyzer and arginine levels determined. Unfortunately, this test, as well as enzymatic tests which involve measurement of the mycoplasma enzyme responsible for the depletion (Barile and Schimke, 1963), is limited because not all mycoplasma species utilize arginine as their energy source.

V. Comparative Studies of Biochemical, Morphological, and Microbiological Techniques for the Detection of Mycoplasma Contamination

A. Screening of Human Skin Fibroblast Cultures

In our initial study (Schneider *et al.*, 1973), 18 human skin fibroblast cultures were screened by the following previously described techniques:

1. Microbiological culture (Hayflick, 1965).
2. PAGE.
3. Electron microscopy.
4. Autoradiography.
5. Media arginine determination.

The results based on the first two techniques permitted us to divide the 18 cultures into three distinct groups (Table I). There was agreement between these two techniques in seven cell strains with either concurrently negative (group I) or positive (group II) results. The PAGE profile of uridine-^{3}H-labeled RNA from a representative uncontaminated cell strain, UC 99 from group I (Fig. 3A), parallels the distribution of the cellular 28 and 18 S RNAs. By contrast, mycoplasma-contaminated strains UC 68 (Fig. 3B) and UC 15 (Fig. 3C) revealed prominent ^{3}H-labeled 23 and 16 S_E noneukaryotic RNA peaks as well as a reduction in the size of the eukaryotic ^{3}H-labeled 28 and 18 S RNA peaks.

TABLE I

COMPARISON OF STANDARD MICROBIOLOGICAL CULTURE AND PAGE ANALYSIS OF URIDINE-^{3}H LABELED CELLULAR RNA FOR THE DETECTION OF MYCOPLASMA CONTAMINATION

Cell strain	Standard mycoplasma culture result (number of colonies)	^{3}H-labeled 23 and 16 S peaks on PAGE	Specific activity of extracted RNA $\left(\frac{\text{cpm} \times 10^{-6}}{\text{mg RNA}}\right)$
Group I			
UC99	–	–	52.85
UC14	–	–	63.69
UC84	–	–	84.03
UC 8	–	–	61.53
Group II			
UC12	+ (>10^5)	+	7.44
UC15	+ (>10^6)	+	1.74
UC69	+ (<10^2)	+	3.27
Group III			
UC48	–	+	2.16
UC47	–	+	2.35
UC83	–	+	1.79
UC31	–	+	1.67
UC37	–	+	3.24
UC57	–	+	2.14
UC56	–	+	1.60
UC34	–	+	2.18
UC71	–	+	4.51
UC75	–	+	1.04
UC70	–	+	1.84

Mean specific activity for group I cell strains = 65.53 ±6.60 (S.E.).
Mean specific activity for group II cell strains = 4.15 ±1.70 (S.E.).
Mean specific activity for group III cell strains = 2.29 ±0.28 (S.E.).
Mean specific activity for group II and III cell strains = 2.64 ±0.44 (S.E.).

Most of the cell strains tested (11 of 18) fell into group III with positive PAGE findings despite negative microbiological cultures. Figure 5 is a composite of PAGE profiles from three representative cell strains in this group.

Another indication of the presence of mycoplasma is the marked decrease in the specific radioactivity of the cellular RNA extracted from contaminated cells (Table I). There was no significant difference in mean specific radioactivities between group II (4.15 ±1.7 S.E.) and III (2.29 ±0.28 S.E.) strains. However, the difference between the mean specific radioactivities of the

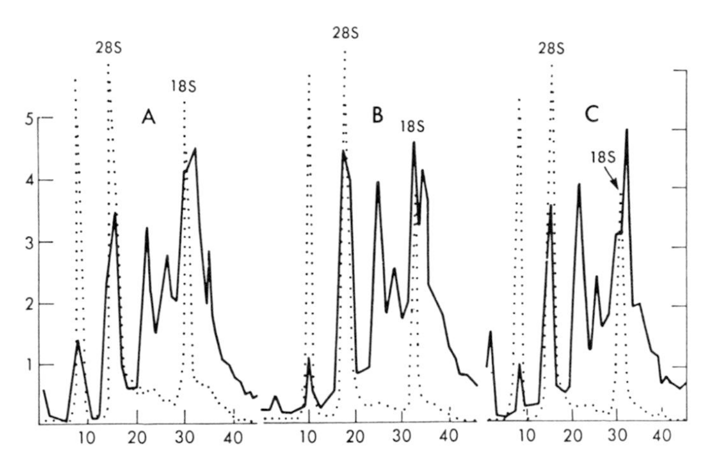

FIG. 5. Abscissa: gel slice, number; ordinate: solid line, radioactivity per fraction (percent of total counts per minute); broken line, relative optical density. PAGE, as in Fig, 3, from cell strain UC 37 (A), and UC 57 (B) and UC 34 (C) from group III. Note the presence of ^{3}H-labeled 23 and 16 S_E RNAs in these cell cultures.

combined group II and III cell strains and the group I cell strains (65.53 ± 6.6 S.E.) was highly significant, $p \ll 0.001$. Of equal importance was the lack of overlap between PAGE-positive and PAGE-negative strains (Table I).

The three remaining techniques were utilized to demonstrate that the source of the ^{3}H-labeled noneukaryotic RNAs observed in group III cell cultures were mycoplasma. An autoradiograph of a uridine-^{3}H-pulse-labeled cell from group III is seen in Fig. 1B. Instead of the usual nuclear concentration of grains, numerous grains were observed along the cell periphery, a characteristic location for mycoplasmic contamination.

Electron microscope survey of cells from group III revealed numerous pleomorphic 0.1- to 0.6-μm mycoplasmas (Fig. 1A) located along the cell membrane. No organisms were seen in cells from group I strains.

Arginine determinations revealed a selective depletion of this amino acid in culture media from several cell strains in group III.

Based on the above findings and the subsequent demonstration of culturable mycoplasma in group III strains, it was concluded that mycoplasma was indeed the source of the noneukaryotic RNA peaks observed in these cultures.

B. Screening of Amniotic Fluid Cell Cultures

Because of the potential hazard to prenatal chromosomal diagnosis posed by mycoplasma, a series of amniotic fluid cell cultures was screened for the presence of these organisms (Schneider *et al.*, 1974b). To ensure against contamination by the cell culture facility, all amniotic fluid specimens were divided into two parallel cultures which were kept in separate incubators, subcultured on alternate days, and grown in different batches of media. In addition to these precautions, the parallel cultures were supplemented with either penicillin (10 units/ml) and streptomycin (100 μg/ml) or with specific antimycoplasma drugs, gentamycin (Schering, 50 μg/ml) or aureomycin (Lederle, 50 μg/ml).

The following screening techniques were utilized:

1. Standard microbiological culture (Hayflick, 1965).
2. PAGE.
3. Urd/U determinations.
4. Acetoorcein staining.

A flow diagram of the cell culture screening, shown in Fig. 6, emphasizes that all testing was performed on parallel replicate cultures. The results are shown in Table II. Slightly over half (11 of 20) the cultures examined revealed prominent 23 and 16 S_E prokaryotic RNA peaks in addition to the human 28 and 18 S rRNA peaks on PAGE. Coelectrophoresis of contaminated cellular RNA with both mycoplasma and bacterial RNA indicated that these prokaryotic 23 S_E RNA peaks were mycoplasmic in origin (Fig. 7). Urd/U results agreed with PAGE analysis in all cultures examined, with Urd/U ratios above 374.6 in cultures with only ^{3}H-labeled eukaryotic RNA species. Acetoorcein staining was concordant with PAGE results in 17 of the 20 cultures. However, microbiological cultures for mycoplasma, as well as for bacteria and fungi, were consistently negative. Once again, bio-

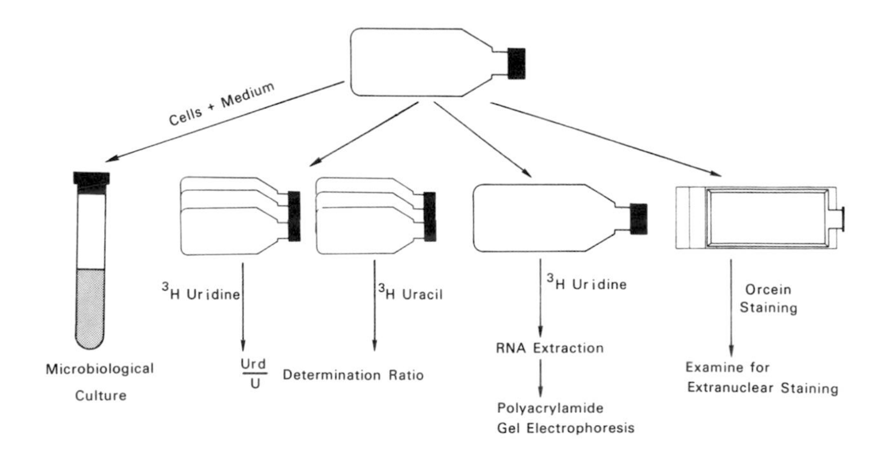

FIG. 6. Flow diagram for the screening of cultured amniotic fluid cells for mycoplasma contamination.

TABLE II

RESULTS OF BIOCHEMICAL, MORPHOLOGICAL, AND MICROBIOLOGICAL SCREENING OF CULTURED AMNIOTIC FLUID CELLS[a]

Cell culture	Antibiotic	23 and 16 S_E RNAs on PAGE	Urd/U	Acetoorcein staining	Microbiological culture
91	P	−	840.0	+	−
96	P	−	699.0	−	−
98	P	−	408.0	+	−
98	A	−	374.6	−	−
99	P	−	492.5	−	−
99	A	−	681.0	−	−
107	P	−	482.4	−	−
107	A	−	778.5	−	−
115	P	−	850.0	−	−
90	P	+	176.5	+	−
90	G	+	1.4	+	−
91	G	+	4.3	+	−
95	P	+	4.6	+	−
95	G	+	10.4	+	−
96	G	+	8.7	+	−
97	P	+	215.0	+	−
97	G	+	178.0	+	−
110	P	+	62.7	−	−
110	A	+	82.7	+	−
115	A	+	57.9	+	−

[a]Cultures were grown in media supplemented with penicillin and streptomycin (P), gentamycin (G), or aureomycin (A). From Schneider *et al.* (1974b). Copyright 1974 by the American Association for the Advancement of Science.

chemical and histological techniques demonstrated contamination, while microbiological methods were negative.

VI. Sensitivities of Biochemical, Morphological, and Microbiological Techniques for Mycoplasma Detection

Theoretically, no biochemical or morphological screening technique can be as quantitatively sensitive as microbiological testing which can detect a single contaminating organism. However, as the above studies have shown, as well as reports of other investigators, there have been numerous examples

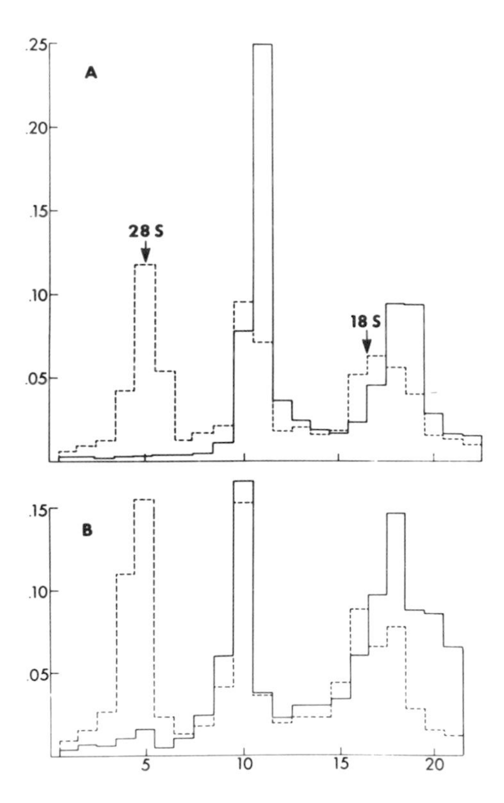

FIG. 7. PAGE of uridine-^{3}H-labeled contaminated cellular RNA (broken line) together with either (A) uridine-^{14}C-labeled *E. coli* RNA or (B) uridine-^{14}C-labeled *M. hyorhinis* RNA (solid line) Abscissa: gel slice number; ordinate: fraction of total radioactivity. The 28 and 18 S human RNAs are indicated. The larger mycoplasma RNA migrated to a position identical to the contaminant RNA species, while the larger *E. coli* RNA migrated to a distinct position.

of mycoplasma contamination of cell cultures that were negative by standard microbiological culture. By contrast, of the hundreds of cell cultures screened by PAGE, no cultures were found to produce negative PAGE results with positive microbiological culture of mycoplasma. Urd/U testing has also been equally free of false negative results.

Unfortunately, comparative quantitative studies of these biochemical and morphological detection techniques have been limited. Nevertheless preliminary results of the biochemical and histological techniques are most encouraging. For example, acetoorcein staining has been reported to detect mycoplasma contamination at levels of 3×10^5 colony-forming units (CFU)/ml (Fogh, 1973). In contrast, initial findings by E. J. Stanbridge and E. L. Schneider (personal communication, 1975) indicate that both

Urd/U and PAGE techniques can detect mycoplasma contamination at levels as low as 10^3–10^4 CFU/ml.

VII. Discussion

These comparative studies emphasize the difficulty of screening cell cultures for mycoplasma contamination by microbiological methods. Although there are other cultural techniques (Zgorniak-Nowosielska *et al.*, 1967), which may be more sensitive, the Hayflick (1965) technique was intentionally chosen because of its widespread use. Considering the marked heterogeneity of mycoplasma species and their fastidious nutrient requirements, it is not surprising that certain species may be missed by cultural testing. It can be argued that contaminating organisms detected by biochemical techniques should not be called mycoplasma, since their taxonomic definition requires their growth on agar (Hayflick, 1965). However, the studies described here, as well as those of other laboratories (Markov *et al.*, 1969; Todaro *et al.*, 1971; Levine, 1972; Perez *et al.*, 1972; Hopps *et al.*, 1973; Brown *et al.*, 1974; Zeigel and Clark, 1974), have demonstrated that the contaminating microorganisms possess properties of mycoplasma in every other conceivable way. In two of these studies (Perez *et al.*, 1972; Schneider *et al.*, 1973), culturable mycoplasma was demonstrated on subsequent subcultivation of biochemically "positive," microbiologically "negative" cells. In a third study, Hopps *et al.* (1973) detected *Mycoplasma hyorhinis* serologically in cells that were negative to standard culture techniques. On further subcultivation of these cells, *M. hyorhinis* was detected by culture on agar. In aureomycin-treated cell cultures intentionally infected with mycoplasma, microbiological testing became negative, while biochemical tests (Urd/U and PAGE) remained positive (Schneider *et al.* (1974a). These results may reflect a need to expand the taxonomic definition of these microorganisms.

Although the classic microbiological culture technique may not be as sensitive as biochemical methods in detecting mycoplasma under certain conditions, it is invaluable in the screening of fluids, such as fetal bovine serum, and in situations in which very low levels of contaminatory mycoplasma are suspected. It is therefore recommended that for routine testing of cell cultures both microbiological and biochemical techniques be used. In the latter case it should be noted that PAGE is very sensitive and is free of false positive results to date. However, the Urd/U technique, described in Chapter 16, is recommended because it is equally as sensitive and reproducible, but more rapid and simple.

ACKNOWLEDGMENTS

Parts of this work were reported in two previous publications (Schneider *et al.*, 1973, 1974b).

REFERENCES

Barile, M. F. (1973). *In* "Contamination in Tissue Culture" (J. Fogh, ed.), pp. 132–172. Academic Press, New York.

Barile, M. F., and Schimke, R. T. (1963). *Proc. Soc. Exp. Biol. Med.* **114**, 676–679.

Brown, S., Teplitz, M., and Revel, J. (1974). *Proc. Nat. Acad. Sci. U.S.* **71**, 464–468.

Fogh, J. (1973). *In* "Contamination in Tissue Culture" (J. Fogh, ed.), pp. 66–107. Academic Press, New York.

Fogh, J., and Fogh, H. (1964). *Proc. Soc. Exp. Biol. Med.* **117**, 899–901.

Harley, E. H., Rees, K. R., and Cohen, A. (1970) *Biochim. Biophys. Acta* **213**, 171–182.

Hayflick, L. (1965). *Tex. Rep. Biol. Med.* **23**, Suppl. 1, 285–303.

Hopps, H. E., Meyer, B. C., Barile, M. F., and Del Giudice, R. A. (1973). *Ann. N. Y. Acad. Sci.* **225**, 265–276.

Horoszewicz, J. S., and Grace, J. T. (1964). *Bacteriol. Proc.* **64**, 131.

Levine, E. M. (1972). *Exp. Cell Res.* **74**, 99–109.

Levine, E. M. (1974). *In* "Methods in Cell Biology" (D. M. Prescott, ed.), Vol. 8, pp. 229–248. Academic Press, New York.

Levine, E. M., Thomas, L., McGregor, D., Hayflick, L., and Eagle, H. (1968). *Proc. Nat. Acad. Sci. U.S.* **60**, 583–589.

Maniloff, J., and Morowitz, H. J. (1972). *Bacteriol. Rev.* **36**, 263–290.

Markov, G. G., Bradvarova, I., Mintcheva, A., Petrov, P., Shiskov, N., and Tsanev, R. G. (1969). *Exp. Cell Res.* **57**, 374–384.

Perez, A. G., Kim, J. H., Gelbard, A. S., and Djordjevic, B. (1972). *Exp. Cell Res.* **70**, 301–310.

Schneider, E. L., and Stanbridge, E. J. (1974). *In Vitro* (in press).

Schneider, E. L., Epstein, C. J., Epstein, W. L., Betlach, M., and Abbo-Halbasch, G. (1973). *Exp. Cell Res.* **79**, 343–349.

Schneider, E. L., Stanbridge, E. J., and Epstein, C. J. (1974a). *Exp. Cell Res.* **84**, 311–318.

Schneider, E. L., Stanbridge, E. J., Epstein, C. J., Golbus, M., Abbo-Halbasch, G., and Rodgers, G. (1974b). *Science* **184**, 477–479.

Stanbridge, E. J. (1971). *Bacteriol. Rev.* **35**, 206–277.

Todaro, G. J., Aaronson, S. A., and Rands, E. (1971). *Exp. Cell Res.* **65**, 256–257.

Zeigel, R. F., and Clark, H. F. (1974). *Infec. Immunity* **9**, 430–443.

Zgorniak-Nowosielska, I., Sedwick, W. D., Hummeler, K., and Koprowski, H. (1967). *J. Virol.* **1**, 1227–1237.

Chapter 16

A Simple Biochemical Technique for the Detection of Mycoplasma Contamination of Cultured Cells

EDWARD L. SCHNEIDER AND ERIC J. STANBRIDGE

Gerontology Research Center,
N.I.A., N.I.H.,
Baltimore, Maryland
and Department of Medical Microbiology,
Stanford University School of Medicine,
Stanford, California

I. Introduction . . . 278
II. Alterations in Exogenous Uridine and Uracil Incorporation into RNA in Mycoplasma-Infected Cells . . . 278
III. Measurement of the Ratio of Uridine-^{3}H (Urd) to Uracil-^{3}H (U) Incorporated into Cellular RNA . . . 279
A. Labeling of Cells . . . 279
B. Cell Harvest and Extraction of Nucleic Acids . . . 280
C. Measurement of Uridine-^{3}H and Uracil-^{3}H Incorporated into RNA . . . 280
D. Calculation of Urd/U Ratio . . . 281
IV. Evaluation of Results . . . 282
A. Human Diploid Fibroblasts Intentionally Contaminated with Common Mycoplasma Species . . . 282
B. Screening of Human Cell Cultures . . . 282
V. Effect of Incubation Time, Cell Concentration, Antibiotics, and Serum Batch, Age, and Concentration on Urd/U Ratios . . . 284
A. Incubation Time . . . 284
B. Cell Concentration . . . 285
C. Antibiotics . . . 286
D. Serum . . . 287
VI. Comparison of Parallel Single-Isotope and Double-Isotope Labeling for Urd/U Determinations . . . 287
VII. Discussion . . . 289
References . . . 290

I. Introduction

The problems of detecting mycoplasma by microbiological methods (involving growth on complex acellular media) have been outlined in the preceding chapter (E. L. Schneider, Chapter 15). There is an obvious need for noncultural detection tests that can be performed in the majority of cell culture laboratories. Several biochemical techniques have been described in the literature, but these for the most part require considerable time and expertise, or are relatively insensitive (Barile and Schimke, 1963; Horoszewicz and Grace, 1964; Markov *et al.*, 1969; Todaro *et al.*, 1971; Schneider *et al.*, 1973). We have directed our attention toward a simple biochemical technique which can be performed in any laboratory and can detect low levels of mycoplasma contamination.

II. Alterations in Exogenous Uridine and Uracil Incorporation into RNA in Mycoplasma-Infected Cells

In developing this technique, we have taken advantage of the fact that mammalian cells readily incorporate exogenous nucleosides into their nucleic acids, while exogenous free bases are incorporated to a negligible extent. Mycoplasmas, by contrast, readily incorporate exogenous free bases as well as nucleosides into their nucleic acids (Stanbridge, 1971).

In the gel electrophoretic studies described in the preceding chapter, a 90–98% decrease was observed in the specific radioactivity of uridine-^{3}H-labeled RNA extracted from mycoplasma-infected cells when compared to control uninfected cultures (E. L. Schneider, Chapter 15). This striking decrease in exogenous uridine incorporation has also been reported by several other laboratories (Levine *et al.*, 1967; Harley *et al.*, 1970; Perez *et al.*, 1972). The recent work of Levine (1972) suggests that this observed decrease is related to the cleavage of exogenous uridine to uracil by the mycoplasmal enzyme uridine phosphorylase. The characteristic peripheral localization of these microorganisms places them in an ideal position to prevent exogenous uridine from reaching the nuclear sites of RNA synthesis.

Increased exogenous uracil incorporation is another important finding observed in mycoplasma-infected cell cultures (Levine, 1972; Perez *et al.*, 1972). This increased incorporation is not unexpected and is probably directly related to uptake of the free base by the contaminating microorganisms.

We have devised a technique to measure both these parameters, exogenous uridine and uracil incorporation into nucleic acids. By combining

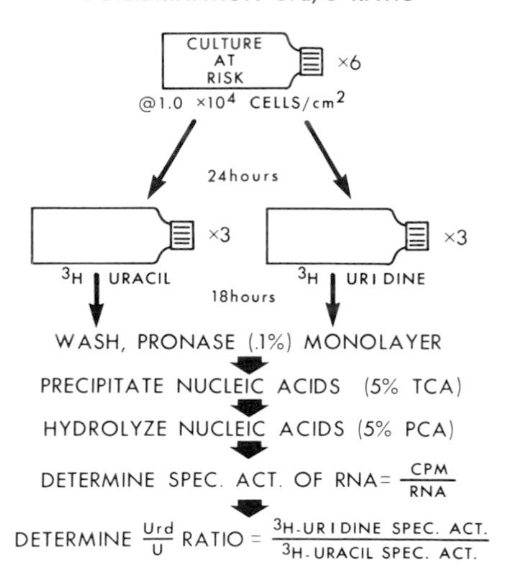

FIG. 1. Flow diagram for the determination of a Urd/U ratio.

these measurements into a ratio, the difference between contaminated and uncontaminated cell cultures is amplified. Exogenous uridine-^{3}H (Urd) and uracil-^{3}H (U) incorporation into RNA are determined in parallel cultures, and the ratio of the specific radioactivity of uridine-^{3}H-labeled RNA to uracil-^{3}H-labeled RNA (Urd/U) is calculated (see Fig. 1). High ratios would be expected in uncontaminated cultures, while decreased ratios should result from decreased uridine incorporation, increased uracil incorporation, or a combination of these mycoplasma-induced alterations.

III. Measurement of the Ratio of Uridine-^{3}H (Urd) to Uracil-^{3}H (U) Incorporated into Cellular RNA

A. Labeling of Cells

Cell cultures are grown in Eagle's minimal essential medium (MEM) supplemented with glutamine and 10% fetal calf serum (FCS). Antibiotic supplements are omitted, particularly those that may suppress mycoplasma growth. Cells to be tested are removed from their substrate by treatment with 0.25% trypsin or 0.1% pronase and resuspended in supplemented MEM. The cell suspension is dispensed at a concentration of 1.0×10^4 cells/cm^2 (2.5×10^5 cells) into each of six 25-ml plastic flasks (Falcon) and incubated overnight (18–24 hours) at 37°C to allow for cell attachment. Flasks are preferred to petri dishes, since they can be gased with 5% CO_2 and sealed, thus preventing cross-contamination among cell cultures. The use of closed flasks also permits cell growth and Urd/U testing in non-CO_2 incubators.

The two parallel sets of three flasks are labeled with uridine-^{3}H and uracil-^{3}H in fresh medium at a concentration of 5 μCi/ml (specific radioactivities of 20 Ci/mmole). Lower isotope concentrations can be utilized for Urd/U measurement. We chose 5 μCi/ml to maximize the efficiency of radioactivity determinations, particularly the incorporation of uracil-^{3}H into uninfected cell cultures. Commercial uracil-^{3}H does not contain appreciable amounts of uridine-^{3}H, and although uridine-^{3}H is contaminated with small amounts of uracil-^{3}H this does not significantly affect Urd/U determinations. Recently, Lavelle has shown that commercial uridine can be contaminated with mycoplasma. We therefore recommend heat inactivation of uridine-^{3}H and uracil-^{3}H before use (Lavelle, 1974).

The labeled cultures are then incubated for a further 18-hour period (overnight). During this time the cells should be in a subconfluent, actively dividing phase of growth.

B. Cell Harvest and Extraction of Nucleic Acids

At the end of the overnight incubation period, the cell monolayers are washed twice with cold phosphate-buffered saline (PBS), and the cells removed with 0.25% trypsin or 0.1% pronase. The resultant cell suspensions are pelleted in conical glass centrifuge tubes at 200 *g* for 10 minutes in a refrigerated centrifuge. The cell pellets are then washed with 1 ml of ice-cold PBS and recentrifuged. The washed pellets are suspended in 1 ml of ice-cold, 5% trichloracetic acid (TCA) for 30 minutes in order to precipitate the cellular nucleic acids, and then centrifuged at 900 *g* for 15 minutes. The TCA pellet is washed with additional ice-cold 5% TCA, recentrifuged, and then hydrolyzed by a 15-minute treatment with 0.6 ml of 5% perchloric acid (PCA) at 90°C (marbles should be placed over the centrifuge tubes to prevent evaporation of PCA). The suspension is cooled on ice after the PCA hydrolysis and centrifuged at 900 *g* for 15 minutes.

C. Measurement of Uridine-^{3}H and Uracil-^{3}H Incorporated into RNA

Aliquots of the PCA supernatants (0.1 ml) are solubilized with 1 ml of NCS (Amersham, Buckinghamshire, England) and mixed with 10 ml of toluene containing a mixture of PPO and POPOP fluors (liquiflor, New

England Nuclear, Boston, Mass.), and the radioactivity measured in a liquid scintillation counter. Additional aliquots (0.4 ml) of the same PCA supernatants are taken for determination of RNA content by the orcinol reaction (Shatkin, 1969).

RNA levels should be the same in the parallel sets of flasks incubated with uridine-^{3}H and uracil-^{3}H if cell inocula are equal. The procedure can therefore be considerably shortened by eliminating the aliquot for RNA measurement. However, if there is a possibility of unequal cell inoculation, either the orcinol reaction or the more simple measurement of optical density at 260 nm is advised to correct for differences in nucleic acid content.

D. Calculation of Urd/U Ratio

The specific radioactivity of the RNA derived from each flask is calculated as counts per minute per milligram of RNA. The Urd/U ratio is then determined by dividing the mean uridine-^{3}H-labeled RNA specific activity by the mean uracil-^{3}H-labeled RNA specific activity (i.e., the mean of the three individual flasks labeled with uridine-^{3}H and uracil-^{3}H respectively). If the orcinol measurements are omitted, Urd/U ratios are determined by dividing the mean uridine-^{3}H counts per minute by the mean uracil-^{3}H counts per minute.

A typical calculation of an Urd/U ratio is presented in Table I. In this example incorporation of exogenous uridine into RNA decreased 20-fold, while uracil incorporation increased 10-fold in the mycoplasma-infected cultures. Together there was a 200-fold difference in Urd/U ratio between mycoplasma-infected and control cultures.

TABLE I

DETERMINATION OF THE RATIO OF URIDINE-^{3}H TO URACIL-^{3}H INCORPORATION INTO RNA (URD/U) IN UNCONTAMINATED AND DELIBERATELY INFECTED WI-38 CULTURES

	Uridine-^{3}H specific radioactivity $\left(\frac{\text{cpm} \times 10^{-6}}{\text{mg RNA}}\right)$	Uracil-^{3}H specific radioactivity $\left(\frac{\text{cpm} \times 10^{-6}}{\text{mg RNA}}\right)$	Urd/U
Uncontaminated	112.0	0.141	
WI-38 cells	141.0	0.136	830.0
	97.0	0.148	
WI-38 cells	5.88	1.64	
infected with	5.22	1.62	3.5
A. granularum	5.60	1.50	

IV. Evaluation of Results

A. Human Diploid Fibroblasts Intentionally Contaminated with Common Mycoplasma Species

Human diploid fibroblast cell strain WI-38 was infected with four common mycoplasma contaminants of cell cultures: *Acholeplasma laidlawii*, *Acholeplasma granularum*, *Mycoplasma hyorhinis*, and *Mycoplasma arginini*. Cell cultures were tested between 3 days and 2 weeks after infection. The specific radioactivities of the uridine-^{3}H and uracil-^{3}H-labeled RNAs, as well as the Urd/U ratios of control and mycoplasma-infected cell cultures are shown in Table II. Control Urd/U ratios were over 400, while contaminated cultures had Urd/U values ranging from 1.6 to 6.3.

B. Screening of Human Cell Cultures

Over the last 2 years, we have used the Urd/U technique to screen over 100 human cell cultures for mycoplasma contamination. Representative results from uncontaminated and contaminated cultures are listed in Tables III and IV. We define contamination by the presence of 23 and $16S_E$ noneukaryotic rRNAs on polyacrylamide gel electrophoretic (PAGE)

TABLE II
URD/U DETERMINATIONS ON CULTURES INTENTIONALLY INFECTED WITH FOUR COMMON MYCOPLASMA CONTAMINANTS[a]

	Uridine-^{3}H specific radioactivity $\left(\frac{\text{cpm} \times 10^{-6}}{\text{mg RNA}}\right)$	Uracil-^{3}H specific radioactivity $\left(\frac{\text{cpm} \times 10^{-6}}{\text{mg RNA}}\right)$	Urd/U
M. hyorhinis	2.94	1.75	1.7
A. granularum	5.57	1.59	3.5
M. arginini	16.50	10.50	1.6
A. laidlawaii	—	—	6.3
Control 1	116.7	0.141	830.0
Control 2	103.3	0.169	598.0
Control 3	130.3	0.264	493.0

[a]WI-38 cells were infected with four common mycoplasma species and tested from 3 days to 4 weeks after infection. Three uninfected WI-38 cell cultures were tested as controls.

TABLE III

URD/U RATIOS IN UNCONTAMINATED CELL CULTURES[a]

Cell culture[b]	Uridine-^{3}H specific radioactivity $\left(\frac{\text{cpm} \times 10^{-6}}{\text{mg RNA}}\right)$	Uracil-^{3}H specific radioactivity $\left(\frac{\text{cpm} \times 10^{-6}}{\text{mg RNA}}\right)$	Urd/U
Skin fibroblast cultures			
UC 153 P	54.0	0.069	777.9
UC 154 P	64.3	0.072	893.0
UC 183 A	193.8	0.208	929.0
UC 184 A	219.2	0.186	1178.4
Amiotic fluid cell cultures			
A 91 P	86.4	0.103	840.0
A 96 P	43.2	0.064	669.0
A 99 A	31.2	0.046	681.0
A 115 P	73.3	0.082	892.0

[a]Cell cultures that contained only human rRNA species on PAGE of their uridine-^{3}H labeled whole-cell RNA.

[b]The letter following the culture number designates the antibiotic supplement. P, penicillin plus streptomycin; G, gentamycin; A, aureomycin.

TABLE IV

URD/U RATIOS IN CONTAMINATED CELL CULTURES[a]

Cell culture[b]	Uridine-^{3}H specific radioactivity $\left(\frac{\text{cpm} \times 10^{-6}}{\text{mg RNA}}\right)$	Uracil-^{3}H specific radioactivity $\left(\frac{\text{cpm} \times 10^{-6}}{\text{mg RNA}}\right)$	Urd/U
Skin fibroblast cultures			
UC 34 A	7.3	3.7	1.9
UC 34 G	8.8	4.5	2.0
UC 69 P	5.3	0.116	45.6
UC 70 P	8.4	1.14	7.3
Amniotic fluid cell cultures			
A 90 G	3.6	2.6	1.4
A 95 G	5.7	0.53	10.4
A 96 G	13.3	1.50	8.8
A 115 A	11.7	0.203	57.9

[a]Cell cultures in which 23 and 16 S_E mycoplasma RNA species were observed on PAGE of uridine-^{3}H labeled whole-cell RNA.

[b]The letter following the culture number designates the antibiotic supplement. P, penicillin plus streptomycin; G, gentamycin; A, aureomycin.

analysis of uridine-^{3}H-labeled cellular RNAs. Uncontaminated cultures contain only eukaryotic 28 and 18 S rRNAs. The rationale for this definition is discussed in the preceding chapter (E. L. Schneider, Chapter 15).

In Table III a considerable variation in the specific radioactivity of both uridine-^{3}H- and uracil-^{3}H-labeled RNAs is seen. This is probably related to the normal variation in cell growth observed in diploid cells in tissue culture. However, despite this variation in specific radioactivity the Urd/U ratio was consistently over 400 in uncontaminated cultures. This reinforces our choice of dual parameters (Urd/U) over measurement of a single parameter (either uracil-^{3}H or uridine-^{3}H incorporation).

To date microbiological testing of cultures with Urd/U ratios over 400 has always been negative for mycoplasma.

The decrease in Urd/U ratios observed in mycoplasma-contaminated cultures (Table IV) was the result of decreased uridine specific radioactivity combined with increased uracil specific radioactivity. Often one of these parameters was more severely altered than the other, and thus the frequency of successful detection of mycoplasma is improved by monitoring both parameters. In mycoplasma-contaminated cell cultures, Urd/U ratios were usually below 100. Occasionally, Urd/U values as high as 215.0 have been observed in cultures which reveal 23 and 16 S_E mycoplasma RNAs by PAGE. However, on retesting, Urd/U values usually fell to below 100. We therefore suggest repeat testing of cell cultures that give Urd/U ratios of between 100 and 400. If the ratio remains in this intermediate range, the more complex PAGE technique described in Chapter 15 should be utilized to determine if the cell culture is contaminated.

V. Effect of Incubation Time, Cell Concentration, Antibiotics, and Serum Batch, Age, and Concentration on Urd/U Ratios

A. Incubation Time

The lack of effect of incubation time on the determination of Urd/U ratios is illustrated in Table V. Control and mycoplasma-infected cultures were incubated with uracil-^{3}H and uridine-^{3}H for 1–24 hours. Both uridine-^{3}H and uracil-^{3}H incorporation into RNA increased 30-fold in the uncontaminated cells, the Urd/U ratio remaining well over 400. At each time period uridine-^{3}H incorporation decreased, and uracil-^{3}H incorporation increased in the mycoplasma-contaminated cells, the resultant low Urd/U ratios remaining well below 100. Although a 1-hour incubation time is sufficient for Urd/U determination, overnight radioisotope incubation is recommended to produce the best results.

TABLE V

EFFECT OF INCUBATION TIME ON THE DETERMINATION OF THE URD/U RATIO

Incubation time (minutes)	Uridine-^{3}H specific radioactivity $\left(\frac{\text{cpm} \times 10^{-6}}{\text{mg RNA}}\right)$	Uracil-^{3}H specific radioactivity $\left(\frac{\text{cpm} \times 10^{-6}}{\text{mg RNA}}\right)$	Urd/U
Uncontaminated WI-38 cells			
60	12.6	0.012	1050
120	24.6	0.037	665
240	66.3	0.047	1412
1440	391.9	0.452	865
WI-38 cells infected with *M. hyorhinis*			
90	0.615	0.064	9.6
180	1.64	0.723	2.2
360	6.45	0.745	8.7
960	6.85	1.115	6.1

B. Cell Concentration

WI-38 cells intentionally infected with *M. hyorhinis*, and uninfected WI-38 cells, were inoculated at concentrations of 0.25, 0.5, 1.0, and 2.0 × 10^4 cells/cm^2 into six small 25-ml flasks each. The results of Urd/U determinations seen in Table VI show no appreciable difference in Urd/U ratios in the uninfected cultures. An increase in Urd/U values was observed at the lowest cell concentration in the mycoplasma-infected cultures. This may reflect decreased growth of mycoplasmas in the presence of very low numbers of cells (Stanbridge *et al.*, 1969). For routine Urd/U testing we therefore advise a cell concentration of 1.0 × 10^4 cells/cm^2. Urd/U testing should not be performed on confluent cells since contact inhibition will result in decreased uridine-^{3}H incorporation.

TABLE VI

EFFECT OF CELL CONCENTRATION ON THE DETERMINATION OF URD/U IN UNCONTAMINATED AND CONTAMINATED WI-38 CELLS

Cell concentration (10^{-4}/cm^2)	Urd/U uncontaminated WI-38 cells	Urd/U WI-38 cells contaminated with *M. hyorhinis*
2.0	596.6	3.9
1.0	1218.7	5.7
0.5	844.3	13.3
0.25	1275.6	28.9

C. Antibiotics

Control WI-38 cells and WI-38 cells contaminated with *M. hyorhinis* were grown from 2 weeks to 2 months in media supplemented with 100 units/ml penicillin, or 50 μg/ml aureomycin (Lederle), or 50 μg/ml gentamycin (Schering), or 500 μg/ml lincomycin, or 60 μg/ml tylosin (Gibco, Grand Island, New York). In addition to Urd/U determinations, PAGE of cellular RNA and microbiological testing were performed on all the cultures.

Uncontaminated cultures grown for several weeks in media supplemented with the antibiotics listed above showed no appreciable differences in Urd/U ratios or PAGE profiles of uridine-^{3}H-labeled cellular RNAs. Mycoplasma could be isolated from all the antibiotic-treated infected cultures (with the exception of the aureomycin-treated cells), and Urd/U ratios were in the infected range (Table VII). The Urd/U ratio of the aureomycin-treated cultures was border-line (287.0) at 2 weeks, and in the contaminated range (92.8) at 5 weeks. PAGE profiles of the uridine-^{3}H-labeled RNA from these cultures showed a similar pattern; eukaryotic RNA species only were observed at 2 weeks, while 23 and 16 S_E mycoplasma RNAs were seen at 5 weeks. Mycoplasma could not be isolated by microbiological culture from the aureomycin-treated cells. In order to optimize conditions for both biochemical and microbiological testing, we therefore suggest that the cells under study be carried in antibiotic-free medium if possible.

TABLE VII

EFFECT OF ANTIBIOTIC THERAPY ON MYCOPLASMA-INFECTED WI-38 CELLS: COMPARISON OF URD/U, PAGE, AND STANDARD MICROBIOLOGICAL CULTURE[a]

Antibiotic (dosage)	Urd/U	PAGE presence of 23 and 16 S_E RNA peaks	Standard microbiological culture results
Penicillin (100 units/ml)	5.7	+	+
Penicillin (100 units/ml) plus streptomycin (200 μg/ml)	6.0	Not done	+
Tylosin (60 μg/ml)	12.0	+	+
Gentamycin (50 μg/ml)	6.0	+	+
Lincomycin (500 μg/ml)	143.4	+	+
Aureomycin (50 μg/ml)	287.0	−	−
Aureomycin (50 μg/ml)	92.8	+	−

[a] All cultures tested by Urd/U, PAGE, and microbiological culture at 2 weeks, except for aureomycin which was reexamined 5 weeks after the onset of antibiotic therapy.

TABLE VIII
EFFECT OF DIFFERENT FCS BATCHES ON THE DETERMINATION OF URD/U IN UNCONTAMINATED WI-38 CELLS[a]

FCS batch	Uridine-^{3}H specific radioactivity $\left(\frac{\text{cpm} \times 10^{-6}}{\text{mg RNA}}\right)$	Uracil-^{3}H specific radioactivity $\left(\frac{\text{cpm} \times 10^{-6}}{\text{mg RNA}}\right)$	Urd/U
A	141.7	0.192	736
B	148.0	0.207	716
C	181.3	0.213	854
D	145.1	0.190	763

[a]FCS samples (100 ml) from four separate serum batches supplied by the Grand Island Biological Co., San Jose, Calif.

D. Serum

We examined the possibility that calf serum supplements may be a source of exogenous nucleosides and bases which would interfere with the uptake and incorporation of uridine-^{3}H and uracil-^{3}H. Several different batches of FCS were tested. Urd/U determinations of WI-38 cells grown in the presence of these individual serum batches for at least 2 weeks before testing revealed no significant differences (Table VIII). Storage of serum for 6 months at $-20°$C did not alter Urd/U ratios. When dialyzed FCS was used in place of undialyzed FCS as the growth supplement, a decreased Urd/U ratio was observed in the control WI-38 cells. This was primarily due to increased uracil-^{3}H incorporation. In WI-38 cells infected with *M. hyorhinis*, both uridine-^{3}H and uracil-^{3}H incorporation were increased with dialyzed FCS, the Urd/U ratio remaining relatively unchanged. Based on these results, the use of dialyzed FCS with Urd/U determinations is unnecessary as well as undesirable.

VI. Comparison of Parallel Single-Isotope and Double-Isotope Labeling for Urd/U Determinations

In the early stage of the development of the Urd/U technique, both single- and double-isotope labeling were employed. Since uridine is more rapidly incorporated into RNA, cells to be tested by double-isotope labeling were incubated with low levels (0.01 μCi/ml) of uridine-^{14}C and the standard concentration (5.0 μCi/ml) of uracil-^{3}H. Results of both routine Urd/U determinations by parallel uridine-^{3}H and uracil-^{3}H incubation, and by double-isotope incubation, are seen in Table IX. When the Urd/U values obtained by double labeling are corrected for differences in isotope concentration (500-fold), radioisotope counting efficiencies (52% for ^{14}C versus 28% for ^{3}H), and spillage of ^{14}C counts into ^{3}H radioactivity counting channels, they approximate values obtained by parallel single-isotope incubation.

The advantages of double-isotope labeling are that no correction for uneven cell inoculation is necessary and fewer flasks are needed for Urd/U measurements. Theoretically, Urd/U ratios can be determined from a single flask of cells. However, in practice at least two replicates are advised. The disadvantages of double-isotope labeling include the need for a dual-channel liquid scintillation counting capability and the more complicated mathematical calculations required. In addition, because of the wide range

TABLE IX

COMPARISON OF PARALLEL SINGLE-ISOTOPE (URIDINE-^{3}H URACIL-H^{3}) AND DOUBLE-ISOTOPE (URIDINE-^{14}C AND URACIL-^{3}H) LABELING FOR URD/U DETERMINATIONS

	Specific radioactivity $\frac{\text{Urd-}^{3}\text{H}}{\text{U-}^{3}\text{H}}$ (cpm)[a]	$\frac{\text{Urd-}^{14}\text{C}}{\text{U-}^{3}\text{H}}$ (dpm)[b]	$\frac{\text{Urd-}^{14}\text{C}}{\text{U-}^{3}\text{H}}$, corrected[c] (dpm)[c]
Uninfected WI-38 cells			
Control 1	830.0	2.8	762.0
Control 2	598.0	2.3	620.0
Control 3	493.0	1.6	431.0
WI-38 cells infected with			
M. hyorhinis	1.7	0.0077	2.1
A. granularum	3.5	0.0270	7.3
M. arginini	1.6	0.0082	2.2

[a]Ratio of uridine-^{3}H specific radioactivities to uracil-^{3}H specific radioactivities (counts per minute 10^{-6} per milligram of RNA) 5 μCi/ml of both isotopes used for parallel cell incubations.

[b]Ratio of uridine-^{14}C disintegrations per minute to uracil-^{3}H disintegrations per minute corrected for background and spillage of ^{14}C disintegrations into ^{3}H counting channels; 0.01 μCi/ml of uridine-^{14}C and 5.0 μCi/ml uracil-^{3}H were used for dual-isotope incubation.

[c]Ratio of uridine-^{14}C disintegrations per minute to uracil-^{3}H disintegrations per minute corrected for differences in radioisotope concentration (500) and counting efficiencies (52% for ^{14}C versus 28% for ^{3}H).

of possible Urd/U values (0.5 to 1500), it is difficult to predetermine the $^{14}C/^{3}H$ ratio necessary to avoid too high spillage of ^{14}C counts into ^{3}H counting channels or too low ^{14}C counts for efficient radioisotope counting.

VII. Discussion

Investigations of the relative sensitivity of Urd/U determinations have indicated that low levels (10^3 to 10^4 colony-forming units per milliliter) of mycoplasma contamination can be detected (Stanbridge and Schneider, 1975). It is obvious that no biochemical test can be quantitatively as sensitive as a microbiological culture in which theoretically a single organism can be detected. However, the preceding chapter (E. L. Schneider, Chapter 15) documents the many instances in which cultures have been shown to be contaminated by Urd/U determinations when standard microbiological cultures were negative. However, we have not yet screened a cell culture that had a Urd/U ratio over 400 and demonstrable mycoplasma by microbiological culture.

Because of our clinical and research interests, the Urd/U technique was developed and applied primarily to human diploid cell cultures. Several permanent cell lines have been examined and found to have the same marked decrease in Urd/U ratio in cultures intentionally infected with mycoplasma. At this time we have not had enough experience with these lines to define the numerical ranges of Urd/U for uninfected and mycoplasma-infected cell cultures. However, there is no indication that the Urd/U technique cannot be successfully applied to permanent cell lines.

Like other biochemical techniques, Urd/U measurements cannot absolutely identify a contaminating organism. In fact, microbiologists define mycoplasma by their ability to produce characteristic colonies on specially prepared agar (Hayflick, 1965). However, to a researcher or clinician, the important question is whether or not the cell culture is contaminated. We therefore recommend that cell cultures be screened by both the Urd/U technique and standard microbiological culture to maximize detection of mycoplasma contamination.

Acknowledgments

The expert technical assistance of Ms. Gail Rodgers was invaluable to the development of this method. In addition, we thank Ms. Gisela Abbo-Halbasch, Ms. Lillian Kwok, and Mr. Robert Spencer for their assistance. Parts of this work reported in previous publications (Schneider *et al.*, 1974a,b), and supported by USPHS research contract NIH 69-2053 within the Special Virus Cancer Program, NCI and NIH training grant AI-00, 082 from the National Institute of Allergy and Infectious Diseases.

References

Barile, M. F., and Schimke, R. T. (1963). *Proc. Soc. Exp. Biol. Med.* **114**, 676–679.

Harley, E. H., Rees, K. R., and Cohen, A. (1970). *Biochim. Biophys. Acta* **213**, 171–182.

Hayflick, L. (1965). *Tex. Rep. Biol. Med.* **23**, Suppl. 1, 285–303.

Horoszewicz, J. S., and Grace, J. T. (1964). *Bacteriol. Proc.* **64**, 131.

Lavelle, G. C. (1974). *Science* **186**, 870.

Levine, E. M. (1972). *Exp. Cell Res.* **74**, 99–109.

Levine, E. M., Burleigh, I. G., Boone, C. W., and Eagle, H. (1967). *Proc. Nat. Acad. Sci. U.S.* **57**, 431–438.

Markov, G. G., Bradvarova, I., Mintcheva, A., Petrov, P., Shishkov, N., and Tsanev, R. G. (1969). *Exp. Cell Res.* **57**, 374–384.

Perez, A. G., Kim, J. H., Gelbard, A. S., and Djordjevic, B. (1972). *Exp. Cell Res.* **70**, 301–310.

Schneider, E. L. (1975). *In* "Methods in Cell Biology" (D. M. Prescott, ed.),

Schneider, E. L., Epstein, C. J., Epstein, W. L., Betlach, M., and Abbo-Halbasch, G. (1973). *Exp. Cell Res.* **79**, 343–349.

Schneider, E. L., Stanbridge, E. J., and Epstein, C. J. (1974a). *Exp. Cell Res.* **84**, 311–318.

Schneider, E. L., Stanbridge, E. J., Epstein, C. J., Golbus, M., Abbo-Halbasch, G., and Rodgers, G. (1974b). *Science* **184**, 477–479.

Shatkin, A. J. (1969). *In* "Fundamental Techniques in Virology" (K. Habel and N. P. Salzman, eds.), Vol. 1, p. 231. Academic Press, New York.

Stanbridge, E. J. (1971). *Bacteriol. Rev.* **35**, 206–227.

Stanbridge, E. J., and Schneider, E. L. (1975). In preparation.

Stanbridge, E. J., Onen, M., Perkins, F. T., and Hayflick, L. (1969). *Exp. Cell Res.* **57**, 397–410.

Todaro, G. J., Aaronson, S. A., and Rands, E. (1971). *Exp. Cell Res.* **65**, 256–257.

Chapter 17

Purity and Stability of Radiochemical Tracers in Autoradiography

E. ANTHONY EVANS

The Radiochemical Centre Limited,
Amersham, Buckinghamshire,
England

I. Introduction 291
II. Design of Tracer Experiments Using Autoradiography 293
A Choice of Radionuclide 293
B. Specific Activity and Chemical Form of the Tracer 295
III. Concepts of Purity 296
A. Types of Purity 296
B. Methods of Analysis 297
C. Effects of Impurities in Tracer Experiments 304
IV. Self-Decomposition of Radiochemicals—Observations and Control . . . 308
A. Control of Decomposition 309
B. Observations 312
V. Importance of the Specificity of Labeling 312
A. Uridine-[5–^{3}H] as a Tracer for RNA 312
B. Biological Hydrogen Transfer Reactions 319
C. In Metabolism 319
D. Specific Enzyme Reactions 319
E. Patterns of Labeling 320
VI. Factors Associated with Autoradiographic Techniques 320
VII. Concluding Remarks 321
References 321

I. Introduction

One of the major facets of the scientific progress that has taken place during the last 20 years is the detailed development of the methodology and applications of high-resolution autoradiography, especially in biological research. High-resolution autoradiographic techniques are of special value in cell biology for locating the site of action of substances in cells, and several detailed texts on this subject have been published (Baserga and Malamud, 1969; Benes, 1967; Evans, 1974; Fischer and Werner, 1971; Gahan, 1972; Gude, 1968; Herz, 1969; Rogers, 1973; Roth and Stumpf, 1969; Salpeter and Bachmann, 1972).

As a technique for measuring radioactivity, autoradiography is used as a means for making visible the results of tracer investigations; it gives a measure of events that have already occurred. It is important therefore to identify and to correct for any problems that are likely to have arisen in the use of the tracer compound *prior* to the application of autoradiographic techniques and the interpretation of the data obtained from the experiment. There are of course necessary precautions to be taken also in the actual autoradiographic techniques themselves in order to obtain correct valid data.

Autoradiography makes it possible to trace substances labeled with a radioisotope in animal or plant organisms macroscopically and down to subcellular structures. The methods employed are normally of three kinds: macroscopic, microscopic, and electron microscopic autoradiography. The technique is used qualitatively more often than quantitatively to ascertain the position of a labeled drug in a biological sample, for example, rather than to estimate or quantify precisely how much of the drug is located in a particular organ. Note especially that the technique of autoradiography does not differentiate between pure and impure radioactive products present in the system under examination. The purity of the tracer compound is therefore a very important parameter to establish in the use of radiochemical tracers in autoradiography.

Measurement of a radionuclide in a tracer investigation, whether by autoradiography or by an other technique such as liquid scintillation counting, follows the fate of labeled atoms put into the experiment. This is an indisputable fact, but it does not indicate either the chemical form introduced into the experiment or the form in which it is present at the end of the experiment and at the moment of autoradiography. One disadvantage therefore of autoradiography, which is indeed common to all techniques based on the detection of radioactivity, is that there is no way of establishing whether chemical changes have occurred in the labeled molecules. In order to determine whether the autoradiography indicates the distribution of the unchanged substance or metabolism (metabolites), recourse must be made to chemical methods, in particular to chromatographic mobility and isotope dilution analysis.

This chapter is intended to highlight some of the basic problems that occur and that can influence interpretation of the facts derived from the use of autoradiography in biological investigations and in cell biology in particular. All research investigators should be fully aware of these problems when using radiochemical tracers.

II. Design of Tracer Experiments Using Autoradiography

In addition to the normal considerations of the appropriate photographic emulsion and the sample preparation techniques to be used for the autoradiographs, the following points need to be carefully considered in the dèsign of a tracer investigation in order to achieve useful valid data.

1. Correct choice of radionuclide.
2. Appropriate molar specific activity and chemical form of the tracer.
3. Checking of the tracer compound for suitable purity.
4. Knowledge of the position of the labeled atoms(s) in the tracer molecule and its stability under the experimental conditions.
5. Establishing the stability of the tracer compound under the experimental conditions.
6. Consideration of the possible errors arising from the limitations of the autoradiographic technique used.

A. Choice of Radionuclide

Radiation affects photographic emulsions, and in practice almost all radionuclides can be detected by autoradiography of course, but those most commonly used for biological research are tritium, carbon-14, sulfur-35, chlorine-36, phosphorus-32, and iodine-125. Some of the essential physical properties of these radionuclides are listed in Table I.

The physical (nuclear) properties of tritium (hydrogen-3), the low maximum beta energy (18 keV) of the radiation, and the high maximum specific activity (29 Ci/mg atom of hydrogen), make tritium the isotope of choice for high-resolution autoradiography. The very short range of the weak beta particles in a photographic emulsion (1 μm) permits the precise localization of tritium compounds, drugs and hormones for example, in biological samples. Such precision is not possible for compounds labeled with other radionuclides, and tritium is unique in this respect (Evans, 1974).

Because of the low energies of the electrons emitted from iodine-125 ($<$4–34 keV), compounds labeled with this radionuclide are sometimes used in autoradiographic studies (Morris and Cramer, 1968; Feinendegen *et al.*,

TABLE I

PHYSICAL PROPERTIES OF SOME RADIONUCLIDES USED IN AUTORADIOGRAPHY

Radionuclide	Half-life	Beta energy, maximum (meV)	Photon energies (meV)	Specific activity: Maximum (mCi/matom)[a]	Specific activity: Common values for compounds (mCi/mmole)	Daughter nuclide (stable)
Tritium	12.35 years	0.018	—	2.92×10^4	10^2–10^4	Helium-3
Carbon-14	5730 years	0.159	—	62.4	1–10^2	Nitrogen-14
Sulfur-35	87.2 days	0.167	—	1.50×10^6	1–10^2	Chlorine-35
Chlorine-36	3×10^5 years	0.714	—	1.2	—	Argon-36
Phosphorus-32	14.3 days	1.71	—	9.2×10^6	10–10^3	Sulfur-32
Iodine-125	60 days	—	0.035	2.18×10^6	10^2–10^4	Tellurium-125

[a] A milliatom is the atomic weight of the element in milligrams.

1966; Ada *et al.*, 1966). However, for histological sections cut at 3–5 μm, iodine-125 is about seven times as effective per disintegration as tritium. This higher efficiency is of course due to the higher-energy electrons emitted by iodine-125, but their greater penetrating power diminishes the value of this isotope in experiments in which precise localization is required (Ada *et al.*, 1966), in common with other radionuclides except tritium.

The weak energy of the radiation from tritium, although excellent for autoradiography, causes difficulties in the measurement and analysis of tritium compounds, and problems of decomposition during the storage of tritium compounds. These aspects are discussed in Sections III and IV, respectively.

B. Specific Activity and Chemical Form of the Tracer

For almost all studies that use autoradiographic techniques to make the results visible, a compound with high molar specific activity is normally required for the investigation. For tritium compounds this specific activity is the range 1–100 Ci/mmole. High specific activity is necessary because biological samples or sections for autoradiography are usually very thin and contain very small amounts of the tracer compound. In order that autoradiographs may be developed in a reasonable time, often several days but sometimes several weeks, the small amount of chemical compound present in the specimen must contain sufficient radioactivity to meet this requirement. Unduly long exposure times for autoradiographs result in high background and a loss in sensitivity, which must be taken into account (Moffatt *et al.*, 1971).

Tritium compounds at specific activities in the range 1–5 Ci/mmole are normally satisfactory if the autoradiographs are to be examined with a light microscope. Autoradiography at the electron microscope level requires the examination of extremely thin sections (Salpeter and Bachmann, 1972) and, for satisfactory autoradiographs to be obtained from a relatively short exposure time, high-specific-activity compounds are required; these are usually above 10 Ci/mmole for the best results.

In some experiments of course only a small fraction of the tracer compound is actually involved in the binding or incorporation into other (macromolecular) structures, and it is essential to use compounds of high specific activity for such studies.

The chemical form of a tracer compound depends of course on the nature of the investigation. Special attention is often necessary in choosing a compound with a clearly defined specificity of labeling, i.e., the tracer atom is in a known position in the molecule, the significance of which is discussed in Section V.

III. Concepts of Purity

Refinements in autoradiographic techniques that have been developed in recent years have permitted high sensitivity in the detection of the tracer. While the importance of using compounds of known radiochemical purity is now widely recognized, it is necessary to take steps to ensure that the data obtained are related to the radiochemical tracer compound and are not a consequence of the presence of radiochemical impurities. This is especially important in autoradiography in which it is difficult to monitor for the likely effects of impurities, or indeed to identify false data due to their presence.

If a research investigator is to know exactly what he has put into his experiment, he must consider radiochemical purity, methods of analysis, and the chemical (or radiochemical) stability of the compound (or label) used.

A. Types of Purity

Three kinds of purity are distinguished as follows.

1. *Radionuclidic* (*radioisotopic*) *purity*. The fraction of radioactivity present as the stated radionuclide. For carbon-14 and tritium compounds, for example, this can be taken as 100% from the methods used for the manufacture of these two radionuclides (Arrol *et al*., 1954; Catch, 1961). Phosphorus-32, however, is often contaminated with a percent or so of phosphorus-33. Contamination of, for example, tritium with a more energetic beta emitter, assuming both radionuclides to be present in the same chemical form, destroys resolution and this should be obvious from the autoradiograph.

2. *Radiochemical purity*. This is the fraction of radioactivity present as the stated form of the compound, and is the most important parameter in the use of radiochemicals. In cases in which the labeling is in a defined position, the presence of the compound containing the label in a different position in the molecule must also be regarded as a radiochemical impurity.

3. *Chemical purity*. This has the usual meaning, i.e., is the chemical identity of the compound the same as that stated on the bottle? It is true that all radiochemicals contain some impurities, by the very nature of their being radioactive. Many research investigators put blind faith in the purity specifications of labeled compounds supplied commercially. While these compounds are normally of very high purity, far higher in fact than many unlabeled chemicals and drugs, investigators sometimes forget that compounds of high molar specific activity can decompose rapidly at a rate quite independent of the half-life of the radionuclide involved. A long physical half-life does not necessarily imply a long shelf-life for the compound at high purity.

B. Methods of Analysis

To modify the statement of Ondetti (1968), "impurities are facts of life in synthesis, and the better one is equipped to identify and get rid of them the greater are the chances of success." Because of the great sensitivity of the radiotracer techniques, it should be a relatively simple matter in detecting impurities. A prudent investigator always checks a radiotracer compound being used for radiochemical impurities immediately before use.

Normal chemical methods of analysis such as the determination of boiling point, melting point, and ultraviolet light absorption, and infrared, nuclear magnetic resonance, and mass spectrometric methods, for example, are usually not sensitive enough to detect radiochemical impurities that may be present only in trace chemical amounts in a compound. It is possible to have a tracer compound that is chemically 99.99% pure and to have all the radioactivity present in the 0.01% impurity. For example, thymidine labeled by exposure to tritium gas (Wilzbach method) after purification to constant specific activity from butanol had a melting point and mixed melting point identical to those of pure thymidine. The ultraviolet light absorption spectrum was also identical to that of pure thymidine. However, chromatographic methods showed that the tracer compound had a radiochemical purity of only 70% (Evans, 1974). Many investigations, particularly in the 1960s, used recrystallization of the tracer compound to constant specific activity as the principal criterion for radiochemical purity. This evidence alone is now known to be unreliable, and additional more sensitive methods are necessary to establish confidence in the radiochemical purity of a compound.

Radiochemical purity is usually determined by chromatographic techniques and by reverse isotope dilution analysis, which are not without their pitfalls (Catch, 1968; Evans, 1974; Sheppard, 1972).

1. Problems in Chromatographic Analysis of Radiochemicals

The measurement of radiochemical purity by chromatography involves the following steps.

1. Selection of an eluant and support capable of separating the products.
2. Application of the sample to the chromatogram.
3. Elution of the chromatogram.
4. Measurement and distribution of the radioactivity along the chromatogram.
5. Interpretation of the data obtained.

Each step can result in artefacts leading to misinterpretation of the data.

a. *Selection of Eluant and Support.* Failure of the chromatography system to resolve the impurities is always a likely problem. Several publications have indicated that the choice of solvent or support is not always easy and that separation of impurities may be achieved only after exploring several solvent systems. For example: tritiated cholic acid appeared pure by thin-layer plate chromatography over silica gel in toluene–acetic acid–water (10:10:1) but grossly impure when chloroform–acetic acid (9:1) was used as solvent (Evans, 1972); cholesterol-[^{3}H] subsequently found to contain about 40% of the radioactivity as cholestanol-[^{3}H] appeared radiochemically pure in as many as four different paper chromatographic systems and three thin-layer systems. Only the use of silver nitrate–impregnated plates of silica gel, and then only one solvent system (chloroform–acetic acid, 98:2), resolved the impurities (Bayly and Evans, 1968; Evans, 1972). Equally difficult are the separations of some amino-[^{14}C] acids such as histidine and lysine (Sala *et al.*, 1966), and some mixtures of steroids and fatty acids such as cholesterol-[^{14}C] and palmitic-[^{14}C] acid (Hansen, 1968).

In order to have reasonable confidence in the radiochemical purity of a tracer compound, the need to use several different solvent systems for chromatographic analysis is apparent.

b. *Application of the Sample to the Chromatogram.* The loading of the sample onto the chromatogram is perhaps the most critical operation and can lead to artefacts if not correctly carried out with care. In general, spots should be air-dried and not heated to avoid decomposition of the compound. Compounds that are sensitive to oxidation may need to be run as "wet" spots in an atmosphere of nitrogen. Examples are methionine and other sulfur-containing amino acids (Sheppard, 1972). Exposure of the spot to light or ultraviolet radiation before elution of the chromatogram may also aggravate decomposition and is a practice to be avoided.

When applied to chromatograms, solutions of compounds at very high molar specific activity may produce artifacts as a result of irreversible adsorption or decomposition of the compound at the origin. This is usually prevented by the addition of a pure inactive carrier compound to the solution used for the analysis, i.e., by effecting an increase in the chemical concentration of the compound. At The Radiochemical Centre the routine practice is to add 4 mg of carrier compound per milliliter of the application solution to samples before chromatography. About 2 μl is then taken for the spot corresponding to 8 μg of the compound. Overloading the chromatogram can cause other problems, namely, poor resolution, streaking, and double peaking. These effects are illustrated in Fig. 1. It is not recommended that known R_f values be relied on for identification of the compound, since they are related not only to the conditions used for the chromatography (solvent, temperature, support, etc.) but also to the loading on the chromatogram. Folic-[2-^{14}C] acid, for example, has been shown to chromatograph at completely different R_f values (ranging from 0.0 to 0.42,

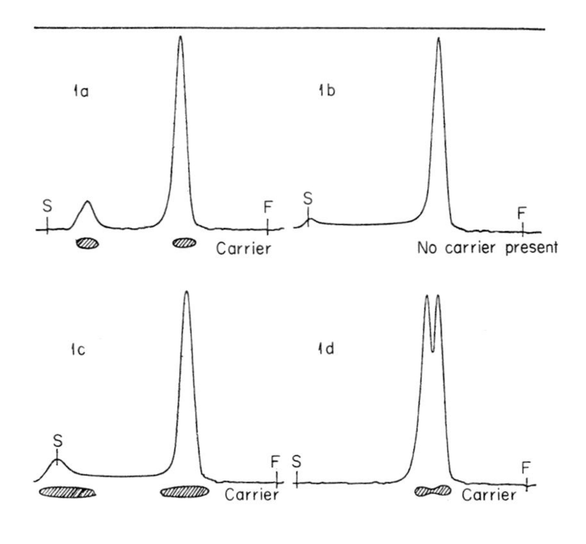

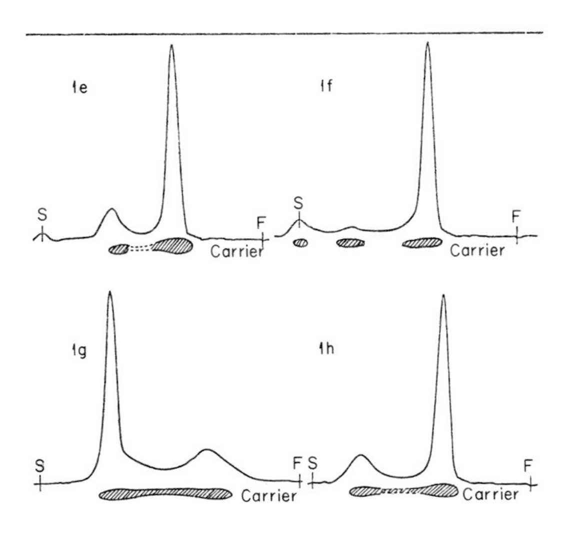

FIG. 1. Possible artifacts in paper and thin-layer chromatography. (a) Decomposition during application. (b) Absorption to support due to inadequate carrier loading. (c) Overloading with carrier. (d) Double peaking due to poor initial application. (e) Double peaking due to the existence of two interequilibrating forms. (f) Overloading and double peaking. (g) Decomposition during elution to faster-running compound. (h) Decomposition during elution to slower-running compound.

compared with carrier R_f 0.0) depending on the loading of the carrier used (Blair and Dransfield, 1969).

Double peaking (Fig. 1e) is frequently observed in the analysis of labeled amines, especially polyamines such as arginine, putrescine, and spermidine, and other molecules that readily form ionic (charged) species. The loading of such compounds on chromatograms is very critical. An example is D-mannosamine-[1-^{14}C] hydrochloride which shows the double-peaking phenomenon in *n*-butanol–ethanol–water (52:33:15) as eluant (Sheppard, 1969). The compound was shown to be pure by reverse isotope dilution analysis.

c. *Elution of the Chromatogram.* It is important of course to choose a solvent system that not only resolves impurities that may be present but also does not cause decomposition of the compound under analysis. The following examples illustrate this point.

1. The use of strongly acidic solvents lead to artefacts in the chromatographic analysis of carbon-14- and tritium-labeled adrenaline and noradrenaline as a result of complex reactions (Forrest *et al.*, 1970; Roberts and Broadley, 1967) and decomposition of tryptophan-[^{14}C] (Lipton *et al.*, 1967; Opienska-Blauth *et al.*, 1962).
2. The elution of cysteine-[^{14}C], especially in ammoniacal eluants, on thin-layer plates in the presence of air, results in extensive oxidation to cystine during chromatography (Sheppard, 1972).
3. The multiple elution of glycine-[^{14}C] on paper chromatograms causes extensive peak broadening and streaking, similar to the results observed on thin-layer plates (Stahl, 1969). Elution of glycine-[^{14}C] in phenolic solvents results in gross decomposition of the compound (Huggins and Moses, 1961).
4. The formation of several spots during the paper chromatography of glutamic-[^{14}C] acid may be due to the formation of pyrrolidone-[^{14}C] carboxylic acid (Nauman and Galster, 1968) or esters (Nauman, 1968).
5. The elution of inulin-[^{3}H] in solvents contaminated with traces of acid results in degradation of the compound to low-molecular-weight fragments (Sheppard, 1972).
6. The elution of [-^{14}C] or nucleoside-[^{3}H] 5′-triphosphates with acid solvents, such as the widely used *n*-butanol–acetone–formic acid–water (7:5:3:5), may lead to hydrolysis of the triphosphate, especially on prolonged elution.
7. Tovey and Roberts (1970) found an artifact in the chromatography of sugar nucleotides using solvents containing ammonium acetate. Paper chromatography of adenosine diphosphoglucose-[^{14}C] in 95% ethanol–1 *M* ammonium acetate at pH 7.5 (5:2), after equilibration of the chromatogram in the solvent vapor, resulted in about 10% of an impurity identified as glucose 1,2-cyclic phosphate. Cyclic phosphates of this type are typical

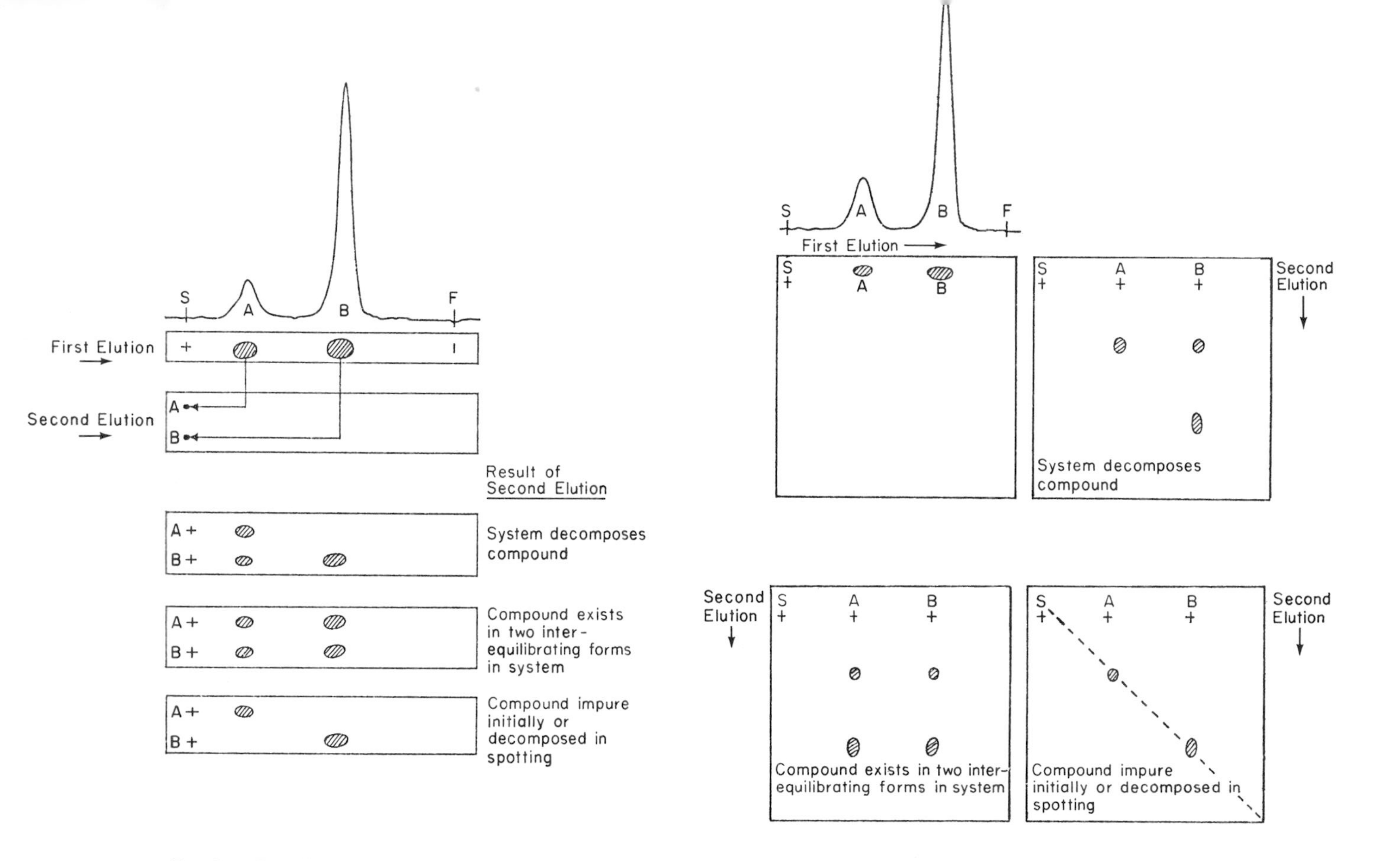

FIG. 2a. Double elution technique.

FIG. 2b. Two-way elution technique.

alkaline degradation products of sugar nucleotides in which the 2-position of the sugar moiety is not blocked. The simple expedient of preequilibrating the chromatography tank with ethanol–water alone was sufficient to overcome the problem in this case.

8. Cis- and trans-isomerization may occur in the solvent, resulting in the formation of an equilibrium mixture during the chromatography and suggesting that the compound is continuously decomposing during the elution. An example is the cis and trans isomers of diethylstilbestrol-[^{14}C] (Sheppard, 1972); each pure isomer separately chromatographed gives rise to two spots.

If decomposition during elution is suspected from the radioactivity scan, two techniques are available to verify it—double elution and two-way elution. These techniques are illustrated in Fig. 2a and 2b and are described in more detail by Sheppard (1972).

Although lowering of the temperature of elution may be appropriate for reducing decomposition during chromatography, it may also have the effect of reducing the resolving power or efficiency of the solvent system, allowing impurities to go undetected. An example is the analysis of 25-hydroxycholecalciferol-[26,27-*methyl*-^{3}H] on thin-layer plates of silver nitrate–impregnated silica gel in the solvent chloroform–acetone (9:1). Impurities resolved at 25°C were not resolved at 2°C (Sheppard, 1972).

It must be remembered that paper and thin-layer plate chromatographic techniques detect only nonvolatile impurities, and false values for radiochemical purity result from either volatility of the tracer compound or of any impurities present. Substances often regarded as nonvolatile are often significantly volatile to give misleading results. An example is the low values obtained for the radiochemical purity of fatty-[^{14}C] acids determined by

TABLE II

THE RADIOCHEMICAL PURITY OF LOWER FATTY [^{14}C] ACIDS DETERMINED BY PAPER CHROMATOGRAPHY AND BY REVERSE ISOTOPE DILUTION ANALYSIS

Compound	Radiochemical purity (%) by	
	Paper chromatography[a]	Dilution analysis
Sodium formate-[^{14}C]	80	87
Sodium acetate-[2-^{14}C]	79	91
Sodium acetate-[U-^{14}C]	90	97[b]
Sodium butyrate-[1-^{14}C]	92	99
Sodium valerate-[1-^{14}C]	91	96

[a]Lowest purity value, eluant *n*-butanol saturated with 10% ammonia.
[b]Radiochemical purity determined by radiogas chromatography was 96%.

paper chromatography, which were found to be due to the fatty acid being lost during the drying of the papers before scanning, whereas nonvolatile impurities were retained. Table II compares the values of the radiochemical purity of lower fatty acids determined by paper chromatography and by reverse isotope dilution analysis (F. G. Stanford, The Radiochemical Centre, Amersham, unpublished results, 1971). All members of the homologous series of monocarboxylic acids up to and including valeric acid are significantly volatile; hexanoic-[^{14}C] acid can be analyzed by thin-layer or paper chromatography without appreciable loss of radioactivity.

d. Measurement of the Radioactivity of the Chromatogram. In analysis autoradiographic methods are used to give a qualitative rather than a quantitative result. An autoradiograph can be quantified by densitometry, but this is not very accurate. In general, scanning techniques are used to establish the distribution pattern of the radioactivity along the chromatogram (Catch, 1961; Evans, 1974; Sheppard, 1972). The development of the 100-channel gas-flow proportional counter has solved the problems of poor counting statistics achieved by single-detector scanning techniques (Sheppard, 1972).

A special word of caution concerning the analysis of tritium compounds should be included. Because of the weak beta radiation from tritium, self-adsorption problems can arise with chromatograms, and care must be taken not only in the methods used to spot the material on the chromatogram but also in the measurement technique. For example, the papers should be air-dried to avoid "transverse" chromatography, and scanning of papers should be 4π not 2π, to avoid differences in the response from opposite sides of the chromatogram (Phillips and Waterfield, 1969).

e. Interpretation of the Data. The factors that determine the accuracy with which a purity figure can be derived from a paper or thin-layer chromatogram of a radiochemical were discussed by Mellish (1970) and by Sheppard (1972).

2. Problems in Reverse Isotope Dilution Analysis

Reverse isotope dilution analysis depends on the ability to remove trace *chemical* amounts of impurity, which may contain a high proportion of the radioactivity, from the compound after dilution with about 1000 times its weight of pure carrier (unlabeled) compound. The main problem with this method is that the principal separation technique used, recrystallization several times to constant specific activity, does not always result in separation of the impurities. An example is the analysis of mixture of cholesterol and cholestanol (Catch, 1968). Suitable derivatives need to be used in such instances. The correct choice of solvent for recrystallization is another important consideration, so that decomposition does not occur during the crystallizing process.

Reverse isotope dilution analysis can sometimes indicate the presence of an impurity, while chromatographic evidence suggests the tracer compound is pure. An example is the finding of 10% methyl-β-glucopyranoside-[^{14}C] in methyl-α-glucopyranoside-[^{14}C], although chromatographic evidence suggested it was pure (Bartlett and Sheppard, 1969).

The result of reverse isotope dilution analysis is influenced by the purity and the chemical form of the carrier used or derivative formed. Disagreements between radiochemical purity values obtained by chromatographic methods and by reverse isotope dilution analysis may sometimes be explained in this way. In spite of the obvious limitations of the technique, it is particularly useful in measuring the content of specific impurities, such as optical enantiomorphs and isomers (Bayly, 1962; Evans, 1974), which are not normally separated by conventional thin-layer or paper chromatographic techniques.

3. Other Methods and Problems

Volatile impurities in radiochemicals are normally detected by radiogas–liquid chromatography. Like all column methods, gas–liquid chromatography has great selectivity and high resolving power, but is best applied to measurement of the content of identifiable specific impurities of known retention times.

Complex biologically active compounds, such as antibiotics, peptides, proteins and hormones, and other macromolecules of high molecular weight, are often the most difficult classes of compounds to measure and to obtain in a state of radiochemical purity adequate for a particular investigation. It is exceedingly difficult to be certain that biological activity and radioactivity are associated together in the tracer compounds. If the labeled compound has been prepared biologically, there is a reasonably good chance that this criterion is satisfied.

Although there are numerous publications describing isotope effects in the analysis of labeled compounds, this has not proved to be a complicating feature in the measurement of radiochemical purity. Such isotope effects are usually quite small and difficult to detect and measure (Evans, 1974).

In order to minimize the problems and artefacts discussed and to be reasonably confident of the radiochemical purity of the tracer compound being used, as many different methods as practical should be used for the analysis of the labeled compound.

C. Effects of Impurities in Tracer Experiments

Since autoradiography is merely a means for making visible events that have already occurred, by identification of the products, the presence of radioactive impurities can present major problems. It is important that the conclusions drawn from the autoradiographs be based on the behavior of the tracer compound and not on a radiochemical impurity.

Whether radioactive impurities in a tracer compound interfere and result in artefacts in high-resolution autoradiography depends largely on the nature and the behavior of these impurities in the experimental system. However, it is important to remember that, in many biological experiments at the cellular level, enough of the tracer compound is used to saturate the system and only a small fraction of the labeled material may be taken up into the cells. *Selective* uptake of radioactive impurities by various tissues could give rise to quite misleading results and an incorrect distribution pattern in various tissues and cellular structures as indicated by autoradiography.

Autoradiographic techniques are used frequently as a means of determining protein and nucleic acid synthesis by following the incorporation of labeled precursors into these macromolecules. Spurious incorporation of radioactivity into these macromolecules can result in three ways:

1. Incorporation of impurities initially present in the precursor.
2. Degradation of the precursor to yield radioactive products which are Incorporated into macromolecules.
3. Incorporation of radioactivity from the precursor into impurities in (or associated with) the macromolecule acceptor.

The controls normally used in such experiments do not always correct adequately for the spurious incorporation of radioactivity, and failure to identify the product, or the true precursor, can lead to false inference of macromolecular syntheses. Such problems in the radiometric assessment of protein and nucleic acid synthesis are discussed in detail by Oldham (1971) and by Monks *et al.* (1971).

There are very few publications that have described or have specifically recognized the significance of radiochemical impurities in a biological experiment using a tracer compound. Unfortunately, usually very little is known about the nature or identity of the radiochemical impurities present in solutions of radiochemicals; consequently, their effects in a particular investigation can be determined only by direct experimentation with the isolated impurities. One such investigation by Wand *et al.* (1967) involved the study of the uptake of impurities, produced by the self-radiolysis of tritiated thymidine in solution, by *Tetrahymena pyriformis* (amicronucleate strain GL).

Self-radiolysis of tritiated thymidine in solution produces several major decomposition products including thymine, thymine glycols and peroxides, and thymidine glycols and peroxides. Because of the difficulties in separating each of these impurities, they were tested collectively after separation from the tritiated thymidine by paper chromatography (Evans, 1972). Radiochemically pure thymidine and the tritiated impurities were separately incubated with synchronized cells (~60%) of *T. pyriformis*, and the uptake into the cells measured first by beta scintillation liquid counting and then by autoradiography. The results indicated that the cells were heavily labeled by both the pure tritiated thymidine sample and by the sample of radiochemical impurities, but only in the former could the label be displaced from the cells on treatment with DNase. A detailed examination of the autoradiographs showed that the impurities failed to label the nucleus of the cells, while the pure tritiated thymidine showed the characteristic heavy labeling of DNA in the nucleus. However, the tritiated impurities showed heavy labeling of the cytoplasm increasing with time. Neither DNase nor RNase removed the label in this case.

These results clearly demonstrate the need to test the cells with DNase and RNase (Baserga and Kisielski, 1963) and, perhaps the most important, the need to test the effects of radiochemical impurities as a control experiment.

Similar labeling of the cytoplasm of cells has been found by Diab and Roth (1970) in their study of the effects of radiochemical impurities in tritiated thymidine on the silver grain density in autoradiographs of mouse crypt cells after injection. The results are summarized in Table III and clearly show that the techniques used for preparing the biological specimens for autoradiography play an important role in determining the observed effects of radiochemical impurities and can even result in a *qualitatively*

TABLE III

SILVER GRAIN DENSITY IN AUTORADIOGRAMS OF INTESTINAL CRYPT CELLS AFTER THYMIDINE-^{3}H INJECTION

System[a]	Total count of 20 cells[b]		Nuclear count		Extranuclear count		Nuclear grains (%)
	Mean	S.D.	Mean	S.D.	Mean	S.D.	
A-I	154	±27	136	±25	18	±4	89
A-II	292	±26	278	±25	14	±4	95
B-I	147	±19	78	±10	69	±11	53
B-II	157	±21	147	±20	10	±3	94

[a]System refers to the source of the autoradiograms according to treatments of animals and processing of tissues as follows: A-I, with pure thymidine, freeze-dried; A-II, with pure thymidine, conventional histological method; B-I, with impure thymidine, freeze-dried; B-II, with impure thymidine, conventional method.

[b]Average silver grain counts were obtained from 10 different areas of 20 labeled cells each. All autoradiograms for counting data were exposed for 7 days.

correct autoradiograph even though an impure radiochemical is used in the investigation. When impure thymidine was used, radioactivity was uniformly distributed in the nucleus, cytoplasm, and extracellular spaces provided autoradiographs were prepared of *freeze-dried* sections, whereas with radiochemically pure thymidine mainly nuclear DNA labeling was observed. Diab and Roth (1970) also showed that in conventional histological methods the solvents used extracted unbound labeled substances and therefore even with impure thymidine mainly nuclear incorporation is observed, the radioactive impurities being washed out by the solvents. Although the impure tritiated thymidine used contained only 38% impurities, the distribution of radioactivity in the cells suggested a selective uptake of these impurities by the cytoplasm.

Observed incorporation of radioactivity into RNA and protein fractions during *in vivo* labeling of rat liver with tritiated thymidine has been ascribed to radiochemical impurities (Castagna, 1967). Nonspecific (i.e., non-DNA) incorporation of tritiated thymidine into *Spirogyra grevilleana* (Meyer, 1966) and into hepatocytes of rat liver (Morley and Kingdon, 1972), and cytoplasmic labeling during long-term labeling of HeLa cells (Girgis and Vieuchange, 1966; Girgis, 1967a,b), are not readily explained. These observations may be due to selective uptake of trace radiochemical impurities or to the incorporation of metabolites (Schneider and Greco, 1971) of tritiated thymidine, produced by enzymatic degradation. Impurities from the self-radiolysis of thymidine were ruled out as being the cause of the incorporation of tritium from tritiated thymidine into proteins of the mouse. The incorporation was thought to be due to metabolites formed by the *in vivo* degradation of the thymidine (Bryant, 1966). Marsh and Perry (1964) recovered from a suspension of human leukemic cells about 1% tritiated thymidine activity in protein and 2% in RNA after 1 hour. Possible uptake of impurities by the tritiated thymidine cannot be ruled out in this case.

Sakai and Kihara (1968) observed the incorporation of tritiated uridine into mouse liver proteins (up to 10%) and into rat kidney proteins, as well as into RNA. This rather unexpected labeling has been ascribed to metabolites of the tritiated uridine rather than to radiochemical impurities in the uridine-^{3}H. Nucleoside phosphorylases, which are present in many tissues and microorganisms, can cause serious degradation of the labeled nucleoside precursors and thus decrease their incorporation into nucleic acids (Oldham, 1971; Cleaver, 1967). These degradation products may then become involved in the synthesis of amino acids and thus become incorporated into macromolecules other than nucleic acids. A straightforward interpretation of the incorporation of uridine-^{3}H into animal tissues, especially with autoradiographic techniques, may lead to erroneous conclusions if such problems are not ruled out.

It is indeed fortunate that many investigations using impure radiochemicals have yielded a qualitatively correct result in that the impurities have either failed to incorporate into the biological specimen or have been removed during subsequent preparation of the autoradiographs. However, one may still wonder just how many results in the published literature are erroneous from this cause? Until work is repeated this will not be known.

IV. Self-Decomposition of Radiochemicals—Observations and Control

Because of the likely problems from the use of impure radiochemicals, it is of course important for any supplier and user of labeled compounds to have them available in a high state of purity for immediate use. Unfortunately, self-decomposition limits the useful shelf life of labeled compounds. The problem is best solved with tritium compounds, because of their very high molar specific activity, necessary for autoradiogrphy, and the weak penetrating power of the low-energy beta particles which results in almost complete absorption of radiation energy by the system.

Half-life and shelf life should not be confused; while the natural decay of tritium, for example, results in a decrease in the specific activity of a compound by about 5% per annum, it normally contributes much less than 5% to any radioactive decomposition products arising during this time. Other factors are the dominant causes of decomposition.

Self-decomposition yielding radioactive impurities arises in several ways; the two major modes are by direct alteration of a labeled molecule hit by a (beta) particle, and by the interaction of reactive species (e.g., radicals), produced by the radiation, with radioactive molecules. It should also be rememberd that radiochemicals are often used at very low chemical concentrations in various solvents, and that many biochemicals are sensitive to chemical decomposition, especially under such conditions. Chemical purity, while usually much less critical than radiochemical purity in tracer investigations, is nevertheless important for some studies, for example, chemical impurities may inhibit enzyme kinetics or the uptake of the tracer by cells. High chemical purity can also be a major factor in prolonging the shelf life of a radiochemical, and therefore factors that affect chemical decomposition, heat, light (photochemical effects), microbiological contamination, etc., must be remembered.

A. Control of Decomposition

Methods for controlling decomposition are by:

1. Reducing the molar specific activity of the compound, i.e., do not use or store compounds at an unnecessarily high specific activity. However, for autoradiography, compounds with high specific activities are normally required (*loc. cit.*).

2. Dispersal of the radioactive molecules to minimize direct interaction with a (beta) particle; this is best done by using a solvent. Suitable solvents include benzene, toluene, ethanol, and aqueous solutions. Aromatic solvents like benzene are excellent, as they are good absorbers of energy, but the actual choice of solvent depends on the type of labeled compound to be stored. It is most important to make sure that the solvents used are chemically pure (Geller and Silberman, 1967, 1969).

3. Using a radical scavenger, such as alcohol, if a compound is stored in an aqueous solution in which the beta radiation produces hydroxyl radicals.

4. Reducing the temperature, thereby raising the activation energy for radical–solute interaction.

5. Storage at liquid-nitrogen temperature (−196°C) if this is convenient, or above the vapor of liquid nitrogen (−140°C).

6. Minimizing the radioactive concentration of the compound in solution. In general, for compounds at high specific activity the rate of decomposition is proportional to the radioactive concentration, while for compounds of lower specific activity (e.g., less than 5 Ci/mmole for tritium compounds) decomposition is less dependent on the radioactive concentration.

A very important point to note is that the rate of decomposition of a radiochemical may accelerate with time of storage (Evans, 1974; Sheppard, 1972; Geller and Silberman, 1967, 1969). Thus it is usually necessary to carry out purity checks at frequent intervals during the shelf life of the radiochemical.

Two important aspects of self-decomposition not always clearly understood by investigators are:

1. *The effect of temperature.* The beneficial effects of reducing the temperature of solutions of radiochemicals for storage are clearly recognized; however, if the solution freezes, molecular clustering of the solute (Fig. 3) can occur, which causes an increase in the effective radiation dose to the compound, i.e., accelerates its rate of decomposition.

Molecular clustering is most pronounced when solutions are slowly frozen. Fast freezing is recommended for any radiochemical that needs to be stored frozen in solution. In general, unless a compound can be stored at −140°C (or below), it is best to store it at the lowest temperature practical

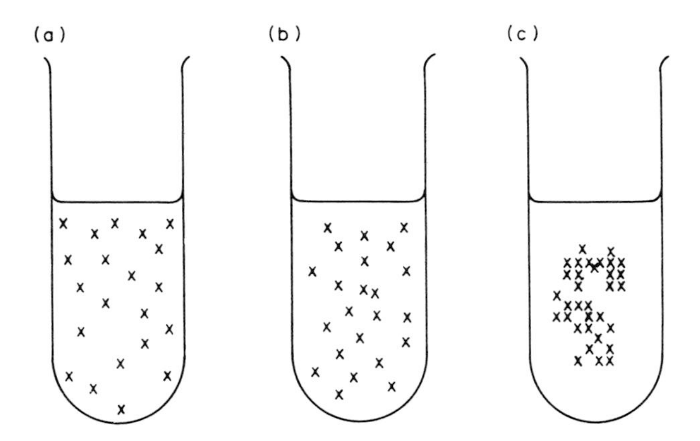

FIG. 3. Distribution of solute molecules (x) in aqueous solutions and in frozen aqueous solutions—molecular clustering effect on slow freezing. (a) +2°C. (b) Frozen at −196°C. (c) Frozen slowly at −20° to −40°C.

without causing the solution to freeze. Repeated freezing and thawing of solutions should also be avoided, as this tends to cause decomposition of the radiochemical. Solutions are therefore best subdivided into aliquots prior to storage.

The weak beta radiation from tritium and the consequent high localized concentration of hydroxyl radicals result in a much more pronounced effect of molecular clustering in the case of tritium compounds, compared with frozen solutions of other radiochemicals.

2. *The effect of radical scavengers.* For radiochemicals stored in aqueous solutions, the most damaging species is normally the hydroxyl radical. Hydroxylated products are found therefore on storage of radiochemicals in aqueous solutions. An example is the formation of tyrosine on storage of solutions of [-^{14}C] or phenylalanine-[^{3}H]. The addition of hydroxyl radical scavengers to solutions of radiochemicals should therefore result in protection of the labeled compound from self-decomposition. This indeed is found to be so in practice, and ethanol (Evans, 1966, 1974), benzyl alcohol (Evans, 1966; Bayly and Evans, 1966, 1967, 1968), sodium formate (Evans, 1966; Bayly and Evans, 1968), glycerol (Apelgot *et al.*, 1964; Apelgot and Ekert, 1965; Apelgot and Frilley, 1965, 1968), cysteamine (Apelgot *et al.*, 1964; Apelgot and Ekert, 1965; Apelgot and Frilley, 1965, 1968; Anet, 1965), and mercaptoethanol (Sheppard, 1972) have all been used with varying degrees of success. To be really effective in affording protection for a radiochemical, the hydroxyl radicals should react faster with the scavenger than with the labeled compound. The radical scavenger must also react preferentially and rapidly with the reactive species, and yet the products of the scavenger–radical reaction must not themselves react with the

labeled compound. For example, Waldeck (1973) has pointed out that the irradiation of aqueous ethanol produces acetaldehyde which can react readily with substrates such as dihydroxyphenylalanine to form tetrahydroisoquinolines, and with 5-hydroxytryptamine to form tetrahydro-β-carbolines, respectively. However, ethanol is widely used for the protection of many different classes of labeled compounds (Evans, 1974); it has the advantage that it can be removed easily by lyophilizing the solution and, at the normally low concentrations (2%) employed, has resulted in very few difficulties in the use of the compounds. Higher concentrations of ethanol (10%) are usually necessary to give effective protection to compounds that react faster than ethanol with hydroxyl radicals.

It must not be assumed that ethanol (or any other free-radical scavenger) provides protection for all labeled compounds in solution, and ethanol is by no means a universal panacea. For example, ethanol affords very little protection for solutions of -[^{14}C] or methionine-[^{3}H] because of the very rapid reaction of methionine with hydroxyl radicals, as can be seen from Table IV.

A compilation by Anbar and Neta (1967) of specific bimolecular rate constants for the reaction of hydroxyl radicals with organic compounds in aqueous solutions gives further data for the selection of other possible scavengers to be used for controlling self-decomposition of radiochemicals.

TABLE IV
REACTION RATES OF OH RADICALS

Compound	pH	Reaction rate, k ($mol^{-1}\ sec^{-1}$)
Hydroquinone	6–7	12.4×10^{9}
Phenol	6–7	10.6×10^{9}
p-Aminobenzoic acid	6–7	9.9×10^{9}
Cysteine	1	7.9×10^{9}
Methionine	7	5.1×10^{9}
Mercaptoethanol	6–7	5.1×10^{9}
Phenylalanine	2	4.4×10^{9}
Phenylalanine	9	2.6×10^{9}
Phenylalanine	6	3.5×10^{9}
Isopropanol	7	3.9×10^{9}
Ethanol	7	1.1×10^{9}
Thymine	1	3.1×10^{9}
Thymidine	2–7.5	2.8×10^{9}
Uracil	5–7	3.1×10^{9}
Glycine	6	1.0×10^{7}

B. Observations

Some examples illustrating the effectiveness of the methods currently used for the control of self-decomposition of radiochemicals are shown in Tables V–VII.

The possible effect on the results of autoradiography arising from a decrease in molar specific activity on storage of some tritium compounds in aqueous solution, due to isotope exchange, should not be forgotten (Waterfield *et al.*, 1968).

There are several comprehensive reviews that discuss the mechanisms, observations, and control of the self-decomposition of radiochemicals for those investigators who wish to delve more into this subject (Bayly and Evans, 1966, 1967, 1968; Evans, 1966, 1974; Pritasil *et al.*, 1969; Sheppard, 1972).

V. Importance of the Specificity of Labeling

In recent years the importance of knowing the specificity of labeling in tracer compounds has become recognized (Evans, 1974). As in many other applications of tritium compounds, tritium is used in autoradiographic studies as a tracer for carbon structures, and in this respect the integrity of the tritium label is important. If the tritium atom is lost (often as tritiated water), the tracer compound goes undetected and therefore the location of metabolites, for example, can be missed.

Labeling with carbon-14 or sulfur-35 by synthesis is usually *completely* specific and unambiguous, in contrast with tritium labeling. Any radioactivity appearing to be in the wrong position is almost certainly the result of impurities, unsound methods of analysis, or wrong interpretation of the experimental data. Unfortunately, one cannot be so forthright about tritium labeling. Unexpected loss of the label or, in cases in which loss of the label is expected, retention of the labeled atom(s), can be very misleading. Some examples will make this point clear.

A. Uridine-[5-^{3}H] as a Tracer for RNA

For several years now uridine-[5-^{3}H] has been used as a specific precursor for RNA labeling (Hayhoe and Quaglino, 1965). However, autoradiographic studies have shown nuclear (DNA) labeling when uridine-[5-^{3}H] was used in cytological investigations, whereas only RNA labeling (cytoplasm or nucleolus) is expected if the pathway to DNA involving deo-

TABLE V

USE OF SOLVENTS TO MINIMIZE SELF-DECOMPOSITION OF RADIOCHEMICALS

Compound	Specific activity (mCi/mmole)	Solvent	Radioactive concentration (mCi/ml)	Temperature (°C)	Storage time (months)	Decomposition (%)[a]
Cholesterol-[4-^{14}C]	55.8	Benzene	—	20	3	N.D.
Cholesterol-[1,2-^{3}H]	46,000	Benzene	1	20	12	<1
Corticosterone-[1,2,6,7-^{3}H]	106,000	Ethanol	1	−20	8	4
Cortisol-[1,2,6,7-^{3}H]	97,000	Benzene–ethanol (9:1)	1	2	3	20
Estradiol-[2,4,6,7-^{3}H]	100,000	Benzene–ethanol (95:5)	1	2	16	15
Guanosine-[8-^{3}H] 5′-triphosphate	11,800	Ethanol–water (1:1) (+1% ammonium bicarbonate)	0.5	−20	14	N.D.
Lactose-[D-*glucose*-1-^{3}H]	3,000	Water	1	2	3	N.D.
Linolenic-[U-^{14}C] acid	139	Benzene	—	20	29	N.D.
Prednisolone-[G-^{3}H]	2,200	Ethanol	1	−20	41	13
Prostaglandin-[5,6-^{3}H] E_1	53,000	Ethanol–water (7:3)	0.2–1	−20	5.75	2
Progesterone-[4-^{14}C]	36.1	Benzene	—	20	24	N.D.
Progesterone-[1,2,6,7-^{3}H]	110,000	Benzene	1	20	5	9
Stearic-[U-^{14}C] acid	92	Benzene	—	20	31	N.D.
Testosterone-[1,2,6,7-^{3}H]	100,000	Benzene	1	20	7	5
Thiosemicarbazide-[^{35}S]	196	Water	2	−30	16	8
Thymidine-[*methyl*-^{3}H] 5′-monophosphate	12,000	Ethanol–water (1:1) (+1% ammonium bicarbonate)	1	−20	12	N.D.

[a] N.D., No detectable decomposition.

TABLE VI

EFFECT OF TEMPERATURE ON THE STORAGE OF RADIOCHEMICALS

Compound	Specific activity (mCi/mmole)	Storage conditions	Radioactive concentration (mCi/ml)	Temperature (°C)	Storage time (weeks)	Decomposition (%)[a]
S-Adenosyl-L-methionine-[*methyl*-^{3}H]	1,580	Aqueous sulfuric acid (pH 4)	1	20	1	10
			1	−80	6	8
			1	−196	52	10
Choline-[*methyl*-^{14}C] chloride	1.5	Solid *in vacuo*	—	20	6.6	27
			—	−196	6.6	N.D.
Glucose-[6-^{3}H] 6-phosphate	244	Aqueous solution	1	2	19	25
			1	−196	24	4
L-Leucine-[4,5-^{3}H]	7,100	Aqueous solution	1	2	26	20
			1	−196	26	5
L-methionine-[*methyl*-^{3}H]	8,600	Aqueous solution	5.5	2	6	22
			5.5	−40	6	42[b]
			5.5	−196	7	6

L-Methionine-[^{35}S]	11,500	Aqueous solution + 5 m*M* concentration of mercaptoethanol under nitrogen	4.2	20	20	55
			4.2	−20	20	4
			4.2	−80	20	10
Prostaglandin-[5,6,8,11,12,16,18-^{3}H] E_2	160,000	Ethanol–water (7:3)	0.5	−20	9	12
			0.5	−140	9	9
Thymidine-[6-^{3}H]	6,900	Aqueous solution	3.45	0	25	17
			3.45	−20	26	32^{b}
				−196	25	3
Thymidine-[6-^{3}H]	3,600	Aqueous solution	1	2	64	16
			1	−40	40	20^{b}
Thymidine-[*methyl*-^{3}H]	44,000	Aqueous solution	1	2	5	4
			1	−20	5	17^{b}
			1	−140	5	4
Uridine-[5,6,-^{3}H]	55,000	Aqueous solution	1	2	10	6
			1	−20	10	25^{b}
			1	−140	10	N.D.

[a] N.D., No detectable decomposition.
[b] Note the effect of molecular clustering.

TABLE VII

EFFECT OF RADICAL SCAVENGERS ON THE STORAGE OF RADIOCHEMICALS IN AQUEOUS SOLUTION AT +2°C

Compound	Specific activity (mCi/mmole)	Storage conditions	Radioactive concentration (mCi/ml)	Storage time (months)	Decomposition (%)[a]
DL-Noradrenaline-[7-^{3}H]	3,680	No scavenger	1	3	6
		Plus 1% ascorbic acid	1	3	15
L-Noradrenaline-[7-^{3}H]	2,340	No scavenger	1	3	18
		Plus 1% ascorbic acid in 0.1 *N* acetic acid	1	3	2
L-Cystine-[3,3′-^{3}H]	1,680	0.1 *N* HCl no scavenger	1	20	~18
		Plus 2% ethanol	1	20	~12
L-Dihydroxyphenylalanine-	13,000	No scavenger	1	4.5	50
[2,5,6,-^{3}H]	28,700	Plus 1% ethanol	1	3.5	5
L-Isoleucine-[U-^{14}C]	240	0.1 *N* HCl no scavenger	0.05	6	4
	310	2% aqueous ethanol	0.05	18	2
L-Methionine-[*methyl*-^{3}H]	8,600	No scavenger	5.5	1.5	22
		Plus 1% sodium formate	5.5	1.5	17

L-Phenylalanine-[U-^{14}C]	282	No scavenger	0.18	2	8
		Plus 3% ethanol	0.18	2	<0.5
Thymidine-[6-^{3}H]	3,000	No scavenger	1	6	22
		Plus 1% benzyl alcohol	1	6	13
Thymidine-[6-^{3}H]	20,000	No scavenger	1	2	6
		Plus 10% ethanol	1	3.75	2
Thymidine-[*methyl*-^{3}H]	44,000	No scavenger	1	1.25	4
		Plus 2% ethanol	1	5.75	6
		Plus 10% ethanol	1	5.75	3
L-Tyrosine-[3,5-^{3}H]	48,000	No scavenger	1	3	35
		Plus 1% ethanol	1	3	<1
		Plus 0.1% sodium formate	1	4.5	4[b]
Uridine-[5,6-^{3}H]	55,000	No scavenger	1	4	9
		Plus 2% ethanol	1	4	3
		Plus 10% ethanol	1	5.5	N.D.
Uridine-[G-^{3}H]	2,440	No scavenger	2	13	30
		Plus 1% benzyl alcohol	2	13	8

[a]N.D., no detectable decomposition.
[b]Accompanied by complete racemization.

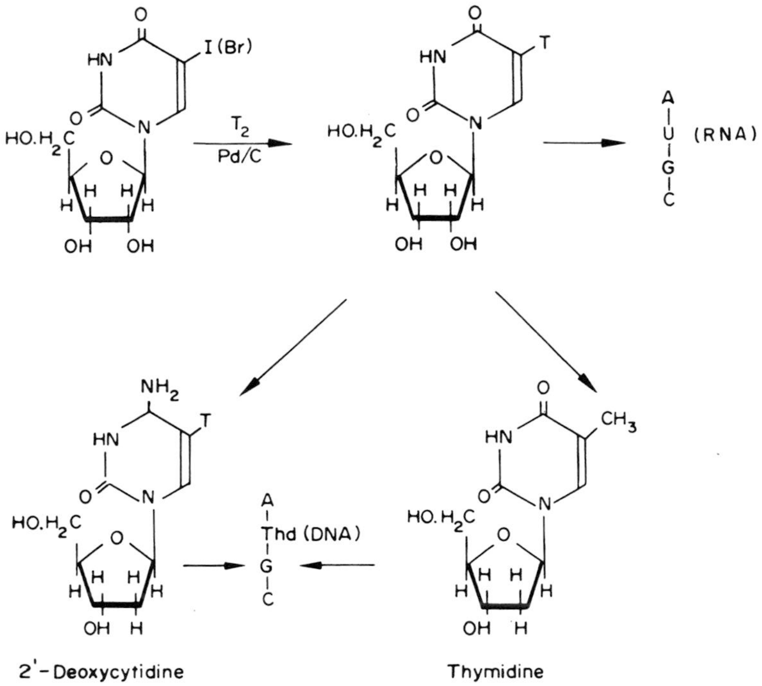

FIG. 4. Biosynthetic pathways of uridine-[5-^{3}H] to RNA and DNA.

xycytidylate (Fig. 4) is blocked. Thus Winter and Yoffey (1966) found DNA labeling in human peripheral mononuclear leukocytes after treatment with RNase, and Comings (1966) found DNA labeling in human (female) fibroblast cells. While no satisfactory explanation was given, it seemed likely from the method used for the preparation of uridine-[5-^{3}H] (Fig. 4), namely, a catalyzed halogen–tritium replacement reaction from 5-bromo- or iodouridine, that some uridine-[6-^{3}H] could be present. This of course as an impurity would lead to DNA labeling. If the reaction conditions are not carefully controlled, up to 20% of the radioactivity can be introduced into the 6-position (Evans, 1972, 1974).

The observed DNA labeling of *Paramecium caudatum* using uridine-[5-^{3}H] (98% tritium at the 5-position) is thought to be due to the deoxycytidylate pathway (Narasimha Rao and Prescott, 1967). Adams (1968), using mouse fibroblast cells, demonstrated that uridine-[5-^{3}H] can label DNA through the deoxycytidylate pathway. Franklin and Granboulan (1965) used uridine-[5-^{3}H] for labeling RNA of *Escherichia coli* and showed that all the tritium label is removed on treatment of the cells with pancreatic RNase. These investigations clearly demonstrate the importance of using RNase for checking that the label is associated with RNA, and the need to use uridine with specificity of the tritium labeling clearly established.

It has been found recently that solutions of tritiated uridine at high specific activity (> 15 Ci/mmole) form a hydrate on storage (Evans *et al.*, 1970)—some of the nonspecific labeling may be due to the incorporation of this hydrate which is not readily detected. It is not yet known whether this hydrate is incorporated into RNA. The hydrate reforms uridine on heating solutions at about 80°C for a few minutes.

B. Biological Hydrogen Transfer Reactions

Stereospecificity of labeling becomes important in biological hydrogen transfer reactions. It has been found, for example, that during the conversion of androstenes to estrogens, the $1\beta,2\beta$-hydrogen atoms are eliminated in the aromatization of the steroid A-ring (Fishman *et al.*, 1969; Brodie *et al.*, 1969). In studies of the localization of these metabolites therefore it is important that the tritium atoms be in the $1\alpha,2\alpha$ configuration, so that they are retained in the metabolites (Osawa and Spaeth, 1971). Similarly, the tritium atom at the 1-position of vitamin D_3 is lost during metabolism of the tritiated vitamin, so in order to pinpoint the site of action of the metabolite (now known to be 1,25-dihydroxy vitamin D_3) (Holick *et al.*, 1971), labeling in positions other than the 1-position is necessary (Lawson *et al.*, 1969; Myrtle *et al.*, 1970). The use of stereospecifically labeled mevalonic acid lactones in the study of steroid and terpene biosynthesis is well known, and there are many examples (Cornforth, 1968).

C. In Metabolism

The incorporation of intact unchanged molecules presents no particular problem, and it does not usually matter in which *stable* position(s) the tritium atom(s) are located. For example, folic-[^{3}H] acid, which was thought to be labeled only in the 3′ and 5′-positions because of the method by which it was prepared, was found later to have up to 60% of the radioactivity in the 9-position. A knowledge of the distribution of the label in this molecule become important for metabolism studies (Zakrzewki *et al.*, 1970).

D. Specific Enzyme Reactions

Autoradiographic techniques are used to locate specific enzymes in tissues. An example is the distribution of folate reductase in tissues by the use of tritiated diisopropyl phosphorofluoridate (DFP) (Darzynkiewicz *et al.*, 1966). While in this case the specificity of the labeling is not important, in the location of other enzymes this may be an important factor. For example, aromatic hydroxylases cause intramolecular hydrogen transfer reactions; the tritium atom of *p*-tritiophenylalanine is retained in the tyrosine formed on reaction with phenylalanine hydroxylase. This is now the well known NIH shift (Guroff *et al*., 1967).

E. Patterns of Labeling

Evidence of patterns of labeling is normally obtained by degradation (Catch, 1971). However, determination of the specificity of labeling is not always easy to perform, as noted by Simon and Floss (1967), and is particularly difficult with tritium compounds in which degradation methods are subject to error because of exchange reactions. Substitution reactions are preferred where applicable, but even then one must be sure that no loss of tritium occurs by exchange reactions with the solvents (Evans, 1974). Physical methods such as infrared and mass spectrometry have many theoretical advantages but unfortunately can rarely be used.

Routine measurements of triton magnetic resonance spectra have been described recently for the first time (Bloxsidge *et al*., 1971), using a microbulb technique to obviate possible radiological hazards. By using a Fourier transform, samples containing as little as 1 mCi of tritium can provide useful scans for examining the specificity of the tritium labeling (J. Bloxsidge, J. A. Elvidge, and J. R. Jones, University of Surrey, Guildford, personal communication, 1973) and is a useful nondestructive technique (Evans, 1974). It provides a method not only for studying the specificity of tritium labeling, but also self-decomposition of tritium compounds and tritium isotope exchange reactions.

VI. Factors Associated with the Autoradiographic Techniques

Although very familiar to those skilled in the art of autoradiographic techniques, there are several factors that can affect the data and the subsequent interpretation of autoradiographs. These factors should not be confused with the problems already discussed, and the more important ones are briefly listed here:

1. The phenomenon of latent image fading (Herz, 1959; Lord, 1963).
2. The correct choice of emulsion and the importance of background measurements, especially in grain counting.
3. The method of sample preparation and its possible influence on the distribution of radioactivity (Levi, 1962).

4. The special techniques required for the autoradiography of diffusible substances (Appleton, 1964; Roth and Stumpf, 1969; Roth, 1971).

5. The reliability of autoradiographic techniques for the study of protain and nucleic acid synthesis depends on differentiating between *incorporated* and *bound* (physical binding) and the correct elimination of precursor pools prior to autoradiography without disturbing the distribution of the incorporated radioactivity (Monneron and Moule, 1969).

6. Reciprocity failure in fluorography (Luthi and Waser, 1965; Randerath, 1970).

7. The problems of *Bremsstrahlung* and light emission associated with the use of tritiated organic polymer reference standards in autoradiography (Evans, 1974; Prydz *et al*., 1970).

8. The possible inhibition of cell growth, and subsequent failure to incorporate radioactive precursors, or cell death due to toxicity of the radioactivity or of the tracer compound (Amici and Paparelli, 1969; Person, 1963, 1968), especially in long-term labeling experiments (Snow, 1973).

VII. Concluding Remarks

It is hoped that by reading this chapter investigators will become more aware of the problems peculiar to radiochemicals and in particular to problems in their use in autoradiography. Two points are clear: First, in order to obtain meaningful results from radiochemical tracer experiments using autoradiography to make the results visible, it is necessary to ensure that the radiochemical is of a *suitable* radiochemical purity and free of impurities likely to give false results. Second, for some studies it may be necessary to establish the pattern of labeling in the tracer compound in order to interpret the autoradiographic data with confidence.

References

Ada, G. L., Humphrey, J. H., Askonas, B. A., McDevitt, H. O., and Nossall, G. J. V. (1966). *Exp. Cell Res.* **41**, 557–572.

Adams, R. L. P. (1968). *FEBS* (*Fed. Eur. Biochem. Soc.*) *Lett.* **2**, 91–92.

Amici, D., and Paparelli, M. (1969). *Ital. J. Biochem.* **18**, 72–77.

Anbar, M., and Neta, P. (1967). *Int. J. Appl. Radiat. Isotop.* **18**, 493–523.

Anet, R. (1965). *Tetrahedron Lett.* pp. 3713–3717.

Apelgot, S., and Ekert, B. (1965). *J. Chim. Phys. Physicochim. Biol.* **62**, 845–852.

Apelgot, S., and Frilley, M. (1965). *J. Chim. Phys. Physicochim. Biol.* **62**, 838–844.

Apelgot, S., and Frilley, M. (1968). *In* "Methods of Preparing and Storing Labelled Compounds," Euratom Rep. EUR 3746 d-f-e, pp. 975–1001. European Atomic Energy Community.

Apelgot, S., Ekert, B., and Tisne, M. R. (1964). *In* "Methods of Preparing and Storing Marked Molecules," Euratom Rep. EUR 1625e, pp. 939–952. European Atomic Energy Community.

Appleton, T. C. (1964). *J. Roy. Microsc. Soc.* [3] **83**, 277–281.

Arrol, W. J., Wilson, E. J., Evans, C., Chadwick, J., and Eakins, J. (1954). *Radioisotope Conf.*, 2, p. 59.

Bartlett, G. M., and Sheppard, G. (1969). *J. Label. Compounds* **5**, 275–280.

Baserga, R., and Kisielski, W. E. (1963). *Lab. Invest.* **12**, 648–656.

Baserga, R., and Malamud, D. (1969). "Autoradiography—Techniques and Application." Harper, New York.

Bayly, R. J. (1962). "Radioisotopes in the Physical Sciences and Industry," Vol II, pp. 305–308. IAEA, Vienna.

Bayly, R. J., and Evans, E. A. (1966). *J. Label. Compounds* **2**, 1–34.

Bayly, R. J., and Evans, E. A. (1967). *J. Label. Compounds* **3**, Suppl. 1, 349–379.

Bayly, R. J., and Evans, E. A. (1968). "Storage and Stability of Compounds Labelled with Radioisotopes," Review Booklet 7. Radiochemical Centre, Amersham, Buckinghamshire, England.

Benes, J. (1967). "Fundamentals of Autoradiography." Iliffe, London.

Blair, J. A., and Dransfield, E. (1969). *J. Chromatogr* **45**, 476–477.

Bloxsidge, J., Elvidge, J. A., Jones, J. R., and Evans, E. A. (1971). *Org. Magn. Resonance* **3**, 127–138.

Brodie, H. J., Kripalani, K. J., and Possanza, G. (1969). *J. Amer. Chem. Soc.* **91**, 1241–1242.

Bryant, B. J. (1966). *J. Cell Biol.* **29**, 29–36.

Castagna, M. (1967). *Biochim. Biophys. Acta* **138**, 598–601.

Catch, J. R. (1961). "Carbon-14 Compounds." Butterworth, London.

Catch, J. R. (1968). "Purity and Analysis of Labelled Compounds," Review Booklet 8. Radiochemical Centre, Amersham, Buckinghamshire, England.

Catch, J. R. (1971). "Patterns of Labelling," Review Booklet 11. Radiochemical Centre, Amersham, Buckinghamshire, England.

Cleaver, J. E. (1967). "Thymidine Metabolism and Cell Kinetics." North–Holland Publ., Amsterdam.

Comings, D. E. (1966). *Exp. Cell Res.* **41**, 677–681.

Cornforth, J. W. (1968). *Chem. Brit.* **4**, 102–106.

Darzykiewicz, Z., Rogers, A. W., Barnard, E. A., Wang, D.-H., and Werkheiser, W. C. (1966). *Science* **151**, 1528–1530.

Diab, I. M., and Roth, L. J. (1970). *Stain Technol.* **45**, 285–291.

Evans, E. A. (1966). *Nature (London)* **209**, 169–171.

Evans, E. A. (1972). *J. Microscopy* **96**, Part 2, 165–180.

Evans, E. A. (1974). "Tritium and its Compounds" 2nd Ed. Butterworth, London.

Evans, E. A., Sheppard, H. C., and Turner, J. C. (1970). *J. Label. Compounds* **6**, 76–87.

Feinendegen, L. E., Bond, V. P., and Huges, W. L. (1966). *Exp. Cell Res.* **43**, 107–119.

Fischer, H. A., and Werner, G. (1971). "Autoradiography." de Gruyter, Berlin.

Fishman, J., Guzik, H., and Dixon, D. (1969). *Biochemistry* **8**, 4304–4309.

Forrest, J. E., Heacock, R. A., Roberts, D. J., and Forrest, T. P. (1970). *J. Chromatogr.* **51**, 525–533.

Franklin, R. M., and Granboulan, N. (1965). *J. Mol. Biol.* **14**, 623–625.

Gahan, P. B., ed. (1972). "Autoradiography for Biologists." Academic Press, New York.

Geller, L. E., and Silberman, N. (1967). *Steroids* **9**, 157–161.

Geller, L. E., and Silberman, N. (1969). *J. Label. Compounds* **5**, 66–71.

Girgis, A. M. (1967a). *C. R. Acad. Sci., Ser. D* **264**, 1800–1802.

Girgis, A. M. (1967b). *C. R. Acad. Sci., Ser. D* **264**, 2401–2404.

Girgis, A. M., and Vieuchange, J. (1966). *Ann. Inst. Pasteur, Paris* **110**, 1–16.

Gude, W. D. (1968). "Autoradiographic Techniques—Localization of Radioisotopes in in Biological Material." Prentice-Hall, Englewood Cliffs, New Jersey.

Guroff, G., Daly, J. W., Jerina, D. M., Renson, J., Witkop, B., and Udenfriend, S. (1967). *Science* **157**, 1524–1530.

Hansen, H. J. M. (1968). *J. Chromatogr.* **35**, 129–133.

Hayhoe, F. G. J., and Quaglino, D. (1965). *Nature (London)* **205**, 151–154.

Herz, R. H. (1959). *Lab. Invest.* **8**, 131.

Herz, R. H. (1969). "The Photographic Action of Ionizing Radiations, in Dosimetry and Medical, Industrial, Neutron, Auto- and Microradiography." Wiley (Interscience), New York.

Holick, M. F., Schnoes, H. K., and Deluca, H. F. (1971). *Proc. Nat. Acad. Sci. U.S.* **68**, 803.

Huggins, A. K., and Moses, V. (1961). *Nature (London)* **191**, 668–670.

Lawson, D. E. M., Wilson, P. W., and Kodicek, E. (1969). *Nature (London)* **222**, 171–172.

Levi, A. (1962). *Science* **137**, 343–344.

Lipton, R., Price, T. D., and Melico, M. M. (1967). *J. Chromatogr.* **26**, 412–423.

Lord, B. I. (1963). *J. Photogr. Sci.* **11**, 342–346.

Luthi, U., and Waser. P. G. (1965). *Nature (London)* **205**, 1190–1191.

Marsh, J. C., and Perry, S. (1964). *J. Clin. Invest.* **43**, 267–278.

Mellish, C. E. (1970). *In* "Analytical Control of Radiopharmaceuticals," pp. 115–125. IAEA, Vienna.

Meyer, R. R. (1966). *Biochem. Biophys. Res. Commun.* **25**, 549–553.

Moffatt, D. J., Youngberg, S. P., and Metcalf, W. K. (1971). *Cell Tissue Kinet.* **4**, 293–295.

Monks, R., Oldham, K. G., and Tovey, K. C. (1971). "Labelled Nucleotides in Biochemistry," Review Booklet 12. Radiochemical Centre, Amersham, Buckinghamshire, England.

Monneron, A., and Moule, Y. (1969). *Exp. Cell Res.* **56**, 179–193.

Morley, C. G. D., and Kingdon, H. S. (1972). *Anal. Biochem.* **45**, 298–305.

Morris, N. R., and Cramer, J. W. (1968). *Exp. Cell Res.* **51**, 555–563.

Myrtle, J. F., Haussler, M. R., and Norman, A. W. (1970). *J. Biol. Chem.* **245**, 1190–1196.

Narasimha Rao, M. V., and Prescott, D. M. (1967). *J. Cell Biol.* **33**, 281–285.

Nauman, L. W. (1968). *J. Chromatogr.* **36**, 398–399.

Nauman, L. W., and Galster, W. (1968). *J. Chromatogr.* **34**, 102–103.

Oldham, K. G. (1971). *Anal. Biochem.* **44**, 145–153.

Ondetti, M. A. (1968). *Sci. Res.* 34.

Opienska-Blauth, J., Charezinski, M., Sanecka, M., and Brzuszkiewicz, H. (1962). *J. Chromatogr.* **7**, 321–328.

Osawa, Y., and Spaeth, D. G. (1971). *Biochemistry* **10**, 66–71.

Person, S. (1963). *Biophys. J.* **3**, 183–187.

Person, S. (1968). *In* "Biological Effects of Transmutation and Decay of Incorporated Radioisotopes," pp. 29–64. IAEA, Vienna.

Phillips. R. F., and Waterfield, W. R. (1969). *J. Chromatogr.* **40**, 309–311.

Pritasil, L., Filip, J., Ekl, J., and Nejedly, Z. (1969). *Radioisotopy* **10**, 525–563.

Prydz, S., Koran, J. F., and Melo, T. B. (1970). *Int. J. Appl. Radiat, Isotop.* **21**, 629–630.

Randerath, K. (1970). *Anal. Biochem.* **34**, 188-205.

Roberts, D. J., and Broadley, K. J. (1967). *J. Chromatogr.* **27**, 407–412.

Rogers, A. W. (1973). "Techniques of Autoradiography," 2nd. Ed. Elsevier, Amsterdam.

Roth, L. J. (1971). *In* "Handbuch der experimentellen Pharmakologie" (O. Eichler *et al.*, eds.), New Series, Vol. 28 Part 1, pp. 286–316. Springer-Verlag, Berlin and New York.

Roth, L. J., and Stumpf, W. E., eds. (1969). "Autoradiography of Diffusible Substances." Academic Press, New York.

Sakai, H., and Kihara, H. K. (1968). *Biochim. Biophys. Acta* **157**, 630–631.

Sala, E., Maier-Hüser, H., and Fromageot, P. (1966). *J. Label. Compounds* **2**, 391–401.
Salpeter, M. M., and Bachmann, L. (1972). *In* "Principles and Techniques of Electron Microscopy. Biological Applications" (M. A. Hayas, ed.), Vol. 2, pp. 221–278. Van Nostrand-Reinhold, Princeton, New Jersey.
Schneider, W. C., and Greco, A. E. (1971). *Biochim. Biophys. Acta* **228**, 610–626.
Sheppard, G. (1969). *J. Chromatogr.* **40**, 312–315.
Sheppard, G. (1972). "The Radiochromatography of Labelled Compounds," Review Booklet 14. Radiochemical Centre, Amersham, Buckinghamshire, England.
Simon, H., and Floss, H. G. (1967). "Bestimmung der Isotopenverteilung in markierter organischer Verbindungen," pp. 247. Springer-Verlag, Berlin and New York.
Snow, M. H. L. (1973). *J. Embryol. Exp. Morphol.* **29**, 601–615.
Stahl, E., ed. (1969). "Thin-layer Chromatography," 2nd ed. Allen & Unwin, London.
Tovey, K. C., and Roberts, R. M. (1970). *J. Chromatogr.* **47**, 287–290.
Waldeck, B. (1973). *J. Pharm. Pharmacol.* **25**, 502–503.
Wand, M., Zeuthen, E., and Evans, E. A. (1967). *Science* **157**, 436–438.
Waterfield, W. R., Spanner, J. A., and Stanford, F. G. (1968). *Nature (London)* **218**, 472–473.
Winter, G. C. B., and Yoffey, J. M. (1966). *Exp. Cell Res.* **43**, 84–94.
Zakrzewski, S. F., Evans, E. A., and Phillips, R. F. (1970). *Anal Biochem.* **36**, 197–206.

Chapter 18

^{125}*I in Molecular Hybridization Experiments*

LEWIS C. ALTENBURG,[1] MICHAEL J. GETZ[2] AND GRADY F. SAUNDERS[3]

I. Introduction . . . 325
II. Iodination Methodology . . . 327
A. Effects of pH . . . 328
B. Ionic Strength . . . 330
C. Nucleic Acid Concentration . . . 330
D. Time and Temperature of the Reaction . . . 330
E. Concentration of Iodide . . . 332
F. Ratio of Cytosine to Iodide . . . 333
G. Reaction Termination . . . 333
III. Applications of ^{125}I-Labeled Nucleic Acids . . . 335
A. Filter Hybridization Experiments . . . 335
B. *In Situ* Hybridization Experiments . . . 337
C. Nucleotide Sequence Analysis . . . 340
D. Secondary-Structure Analysis . . . 341
IV. Discussion . . . 341
References . . . 342

I. Introduction

In the 3-year period following the appearance of Commerford's (1971) procedure for the *in vitro* iodination of nucleic acids, the utility of this label-

[1]Medical Genetics Center, the University of Texas Graduate School of Biomedical Sciences, Health Science Center, Houston, Texas, and the Department of Biology, the University of Texas System Cancer Center, M. D. Anderson Hospital and Tumor Institute, Houston, Texas.

[2]Department of Pathology, Mayo Clinic, Rochester, Minnesota.

[3]Department of Developmental Therapeutics, the University of Texas System Cancer Center, M. D. Anderson Hospital and Tumor Institute, Houston, Texas.

ing method has become apparent by its employment in a wide diversity of applications. These include its use in molecular hybridization experiments at both the molecular and cytological levels, nucleotide sequencing analysis by fingerprinting methods, and nucleic acid secondary structure analysis by iodination of non-base-paired cytosine residues. All these applications required the availability of an *in vitro* labeling procedure that would (1) permit the incorporation of relatively large amounts of a covalently bound, highly radioactive isotope into the molecular structure of nucleic acids, and (2) avoid extensive degradation or other deleterious alteration of the nucleic acid as a consequence of the labeling procedure. When employed under optimal conditions, these are the primary features of the nucleic acid iodination technique.

As described by Commerford (1971), aqueous solutions of denatured nucleic acids can be labeled with iodine by heating in the presence of iodine or thallic trichloride ($TlCl_3$) plus iodide at pH 5. The reaction results in the formation of 5-iodocytosine as the primary product in both DNA and RNA, although trace quantities of iodinated guanine in DNA (Commerford, 1971), iodinated uracil in RNA (Heiniger *et al*., 1973; Robertson *et al*., 1973), and iodinated guanine and adenine in homopolymers (Scherberg and Refetoff, 1974) have been reported. However, the fact that 5-iodocytosine is the predominant, most stable product of the reaction helps to explain the probable mechanism by which iodination of nucleic acids occurs. In the presence of the mild oxidant $TlCl_3$, the iodide ion reacts with the unsaturated bond between C-5 and C-6 of pyrimidines.

As proposed by Yu and Zamecnik (1963) for the bromination of uridylic acid, the introduction of iodine at the C-5 position of pyrimidines probably results in the formation of a transient intermediate with a positive charge carried mainly by the N-1 atom. In aqueous solutions this intermediate is rapidly attacked by a water molecule, resulting in formation of the moderately stable 5-iodo-6-hydroxydihydropyrimidine. As pointed out by Commerford (1971), these compounds should share some of the chemical properties of pyrimidine hydrates which undergo dehydration to reform the 5,6 double bond. However, it is known (Johns *et al*., 1965; Logan and Whitmore, 1966) that dehydration of cytosine hydrate occurs much more rapidly and under milder conditions than dehydration of uracil hydrate. Consequently, dehydration of 5-iodo-6-hydroxydihydrocytosine to 5-iodocytosine occurs rapidly at 60°C and pH 5, whereas complete dehydration of 5-iodo-6-hydroxyuracil requires extensive heating at or above neutral pH (Logan and Whitmore, 1966). Although iodination of RNA yields large quantities of 5-iodo-6-hydroxydihydrouracil (Commerford, 1971; Getz *et al*., 1972), dehydration of the intermediate is apparently accompanied by its deiodination such that only trace quantities of stable 5-iodouracil are formed. The mechanism by which trace amounts of iodoguanine and iodoadenine are formed is not understood.

For several reasons ^{125}I is a highly advantageous isotope for use in molecular hybridization experiments. The decay of ^{125}I to tellurium occurs by electron capture, resulting in the emission of both 27- to 35-KeV γ rays and 2.8- to 23-KeV internal conversion (Auger) electrons (Myers and Vanderleeden, 1960). Thus it is possible to use ^{125}I as a dual label in which the radiation can be detected by both gamma and liquid scintillation counting methods. The soft Auger electrons may also be detected by thin, radionuclide-sensitive emulsions. The principal advantage of ^{125}I, however, is its intrinsically high specific radioactivity due to its relatively short 60-day half-life. Because of this fact it is possible to prepare labeled nucleic acids of high specific radioactivity by the introduction of limited quantities of ^{125}I atoms. This is an important consideration, since high specific radioactivities may be achieved with minimal alterations in the physical properties of the labeled nucleic acid.

II. Iodination Methodology

As described by Commerford (1971), iodination of single-stranded DNA and RNA is a relatively simple, straightforward reaction. In general, a typical iodination reaction mixture consists of (1) the purified nucleic acid, (2) $TlCl_3$, and (3) a mixture of $Na^{125}I$ and carrier KI. These components are assembled at 0°C in acetate buffer adjusted to pH 5.0 and an ionic strength of 0.05–0.10 *M* (Na^+). The reaction mixture is either assembled in or transferred to a tightly sealed container and heated at 60°–80°C for 15–20 minutes, followed by chilling to 0°C. The iodination reaction is terminated by the addition of a reducing agent, raising the pH to 8.5–9.0, and reheating at 60°C for 20 minutes. The iodinated nucleic acid is then recovered from the reaction mixture by gel filtration chromatography, and its specific radioactivity determined by gamma or liquid scintillation counting (Fig. 1).

Following Commerford's (1971) original report on the iodination of nucleic acids, several investigators examined and modified certain parameters of the reaction in order to improve the efficiency of incorporation of iodine and to correct subsequent deiodination of the labeled nucleic acid. These studies have shown that the incorporation and retention of iodine atoms in stable linkage with nucleic acids is strongly affected by (1) the pH of the reaction, (2) the ionic strength, (3) the nucleic acid concentration, (4) the time and temperature of the reaction, (5) the ratio of cytosine to iodide, (6) the concentration of iodide, and (7) the reaction termination step. Because the control of each of these reaction conditions is critically im-

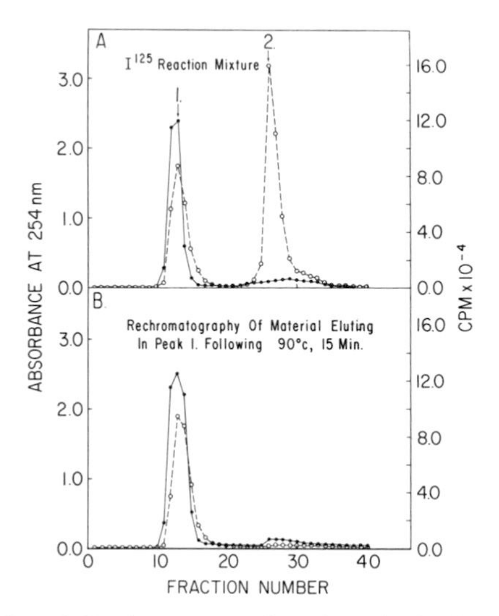

FIG. 1. (A) Sephadex G-25 chromatography of an iodination reaction mixture. Unreacted ^{125}I elutes in fractions 25 to 30. (B) Rechromatography of material eluting in peak I of A, following heating at 90°C for 15 minutes in SSC. Open circles, counts per minute; Solid circles, absorbance at 254 nm. Reprinted with permission from Getz *et al.* (1972).

portant to the successful application of the iodination technique, a review of the most recent information relevant to each is given.

A. Effects of pH

As reported by Commerford (1971), iodination of polycytidylic acid (poly C) proceeds optimally at pH 4 (Table I). However, more recent studies

TABLE I

MAXIMUM AMOUNT OF 5-IODOCYTOSINE FORMED IN POLY C[a,b]

pH	Iodide ($M \times 10^4$)	$TlCl_3$ ($M \times 10^4$)	Cytosine as 5-iodocytosine (%)
5	0.2	6	13
5	0.5	6	27
5	1.0	6	45
5	1.0	0	0
5	1.0	0.7	20
5	1.0	3	34
5	1.0	6	43
5	1.0	15	36
4	1.0	6	96

[a] Reprinted with permission from Commerford (1971).

[b] 10^{-4} *M* poly C was heated with $TlCl_3$ and KI for 1 hour at 80°C.

TABLE II

INFLUENCE OF REACTION CONDITIONS ON LABELING[a,b]

RNA (μg/ml)	Reaction temperature (°C)	pH	Ionic strength (M)	Cytosine/ iodide	Incorporation (%)
Series A					
600	40	5.0	0.05	7.8	3
600	50	5.0	0.05	7.8	6
600	60	5.0	0.05	7.8	12
600	70	5.0	0.05	7.8	20
600	80	5.0	0.05	7.8	23
Series B					
600	70	4.0	0.05	7.8	20
600	70	4.5	0.05	7.8	20
600	70	5.0	0.05	7.8	20
600	70	5.5	0.05	7.8	3
600	70	6.0	0.05	7.8	2
Series C					
600	70	5.0	0.20	7.8	15
600	70	5.0	0.40	7.8	12
Series D					
600	70	5.0	0.05	7.8	32
300	70	5.0	0.05	3.9	15
130	70	5.0	0.05	1.7	12
65	70	5.0	0.05	0.8	14
Series E					
600	70	5.0	0.05	7.8	28
600	70	5.0	0.05	7.8	38
600	70	5.0	0.05	7.8	32

[a]Reprinted with permission from Tereba and McCarthy (1973).

[b]Chicken 28 S rRNA was incubated for 15 minutes at the designated temperature in 50 μl of a solution containing 6 μCi of ^{125}I, 7×10^{-5} M KI, and 8×10^{-4} M $TlCl_3$. Series A illustrates the effect of temperature, B the effect of pH, C the effect of ionic strength varied by addition of NaCl, and D the influence of RNA concentration. In series E the limits of reproducibility are assessed in triplicate reaction mixtures at optimum temperature, pH, and ionic strength. The buffer at pH 4 was sodium formate, at pH 4.5 and 5.0 sodium acetate, and at pH 5.5 and 6.0 sodium maleate. The percentage incorporation of iodide into RNA was assessed by Sephadex gel exclusion chromatography. Series A, B, and C represent experiments carried out in parallel, and the results are directly comparable. The same applies to series D and E. The use of two different radioiodide solutions may account for variation between the two sets of results.

by Tereba and McCarthy (1973) on the iodination of chicken RNA show no appreciable increase in the reaction efficiency below pH 5.0 (Table II). Therefore a reaction pH less than 5.0 appears unwarranted, particularly in view of the increased risk of deamination and depurination of the nucleic acid at lower pH values (Shapiro and Chargaff, 1957). Tereba and McCarthy

(1973) also showed that almost no incorporation of iodine into rRNA occurs at pH 5.5 or higher. Thus the iodination reaction proceeds only under acidic conditions and over an extremely narrow pH range. In this regard it should be noted that the addition of commercial $Na^{125}I$ to the reaction mixture also includes the addition of 0.1 *N* NaOH; therefore the final pH of the reaction mixture must be carefully monitored and corrected if necessary.

B. Ionic Strength

Tereba and McCarthy (1973) examined the effect of ionic strength on the efficiency of iodine incorporation into rRNA (Table II). Their results show that, as the ionic strength of the reaction is increased above 0.05 *M*, the reaction efficiency declines rapidly. As suggested by Tereba and McCarthy (1973), the decline in reaction efficiency is likely due to the induction or maintenance of nucleic acid secondary or tertiary structure. This possibility is consistent with the observation that native DNA is a poor iodination substrate as compared to denatured DNA (Commerford, 1971; Holmes and Bonner, 1974).

C. Nucleic Acid Concentration

Nucleic acid concentration has also been observed (Tereba and McCarthy, (1973) to influence the efficiency of the iodination reaction (Table II). At concentrations below 300 μg/ml, a marked increase in the efficiency of the reaction occurs, which permits the labeling of nucleic acids to very high specific radioactivity (Altenburg and Shaw, 1973). Although this phenomenon has not been adequately explained, Tereba and McCarthy (1973) have proposed that the nucleic acid concentration may affect the extent to which the 5-iodo-6-hydroxydihydropyrimidine intermediate proceeds to 5-iodopyrimidine.

Several investigators (Prensky *et al.*, 1973; Robertson *et al.*, 1973; Tereba and McCarthy, 1973; Dube, 1973; Holmes and Bonner, 1974) have reported the preparation of ^{125}I-labeled nucleic acids of high specific radioactivity when small reaction mixture volumes (50 μl or less) were employed. In constructing these small-volume reaction mixtures, it is advantageous to maintain the nucleic acid stock solutions at relatively high concentrations.

D. Time and Temperature of the Reaction

Commerford (1971) showed that the rate of the iodination reaction is critically influenced by the incubation temperature (Fig. 2). At 70°C incorporation of iodine into poly C is exponential and essentially complete within

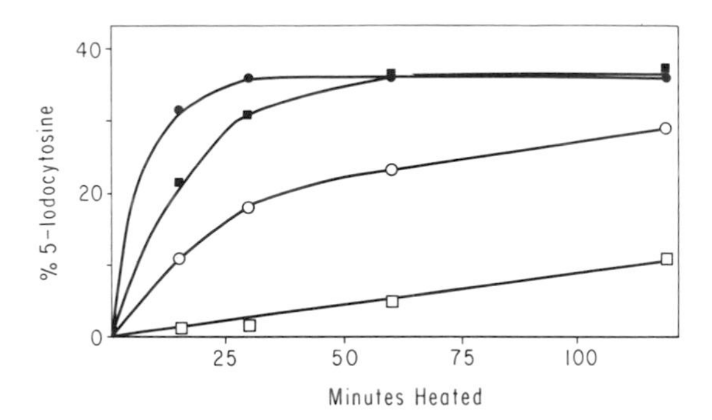

FIG. 2. The effect of temperature on 5-iodocytosine formation. Solutions of poly C were heated at 40°C (open triangles), 60°C (open circles), 70°C (solid triangles), and 80°C (solid circles) at pH 5. The ordinate represents cytosine that reacted to form 5-iodocytosine. Reprinted with permission from Commerford (1971).

20–30 minutes, whereas at 40°C the reaction proceeds slowly and in a linear fashion. Incubation at temperatures greater than 70°C does not result in significant increases in iodine incorporation into either poly C (Commerford, 1971) (Fig. 2) or rRNA (Tereba and McCarthy, 1973) (Table II), although the extent to which degradation of the nucleic acid occurs likely increases. In this respect, Tereba and McCarthy (1973) showed that iodination of 28 S rRNA at 70°C and pH 5.0 constitute sufficiently severe reaction conditions

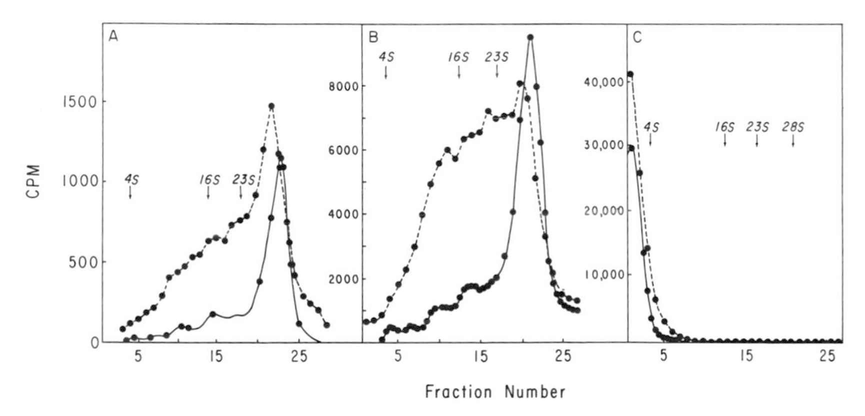

FIG. 3. Sedimentation profiles of labeled and unlabeled 28 S rRNA. 28 S rRNA was sedimented through a 5 × 20% sucrose gradient at 40,000 rpm in a SW 40 Spinco rotor at 20°C for 5.25 hours in 0.1 *M* NaCl–0.001 *M* EDTA–0.01 *M* tris (pH 7.6). *Escherichia coli* RNA was run on an identical gradient for additional standards. Solid line, Optical density of unlabeled RNA; broken line, 100-μl samples of labeled RNA. (A) RNA specific activity 60,000 cpm/μg; (B) RNA specific activity 3 × 10^7 cpm/μg; (C) the same high-specific-activity RNA pretreated with a mixture of T_1 (5 units) and pancreatic RNase (40 μg/ml) at 37°C for 45 minutes. Reprinted with permission from Tereba and McCarthy (1973).

to result in the production of single breaks in approximately half of the RNA molecules (Fig. 3). However, the type of degradation that occurs under the required reaction conditions appears to be limited to phosphodiester bond cleavage, since the extent of mismatching (1–2%) that occurs between iodinated rRNA and DNA in filter hybridization experiments can be attributed entirely to the presence of iodinated bases (Tereba and McCarthy, 1973).

E. Concentration of Iodide

The concentration of iodide has been shown (Commerford, 1971; Getz *et al.*, 1972; Scherberg and Refetoff, 1973) to influence critically the efficiency of the iodination reaction (Tables III and IV). According to Commerford (1971), this phenomenon may be attributed to the relative concentrations of IOH and iodide present in the reaction mixture. However, the reaction itself generates iodide ion which, in concentrations greater than 10^{-4} *M*, strongly inhibits further reaction. Tables III and IV show that the optimum iodide concentration is in the range of 10^{-4}–10^{-5} *M*. When small reaction mixture volumes are employed (50 μl or less), an iodide concentration of 10^{-4}–10^{-5} *M* is readily achieved by the addition of 1–10 mCi of commercially available $Na^{125}I$ (specific activity of approximately 2 Ci/μmole).

TABLE III

EFFECT OF IODIDE CONCENTRATION ON THE FRACTION BOUND TO DNA AS 5-IODOCYTOSINE[a,b]

KI (*M*)	Iodide as 5-iodocytosine	
	Native DNA	Denatured DNA
0	0.9	2.3
1.0×10^{-7}	0.8	6.8
2.5×10^{-7}	0.5	5.8
1.0×10^{-6}	0.1	5.6
2.5×10^{-6}	0.4	5.9
1.0×10^{-5}	1.8	17.1
2.5×10^{-5}	2.8	31.9
1.0×10^{-4}	1.4	21.6
2.5×10^{-4}	0.7	9.0

[a]Reprinted with permission from Commerford (1971).

[b]The DNA concentration of each mixture was 1.44×10^{-4} *M* (as cytosine), and the $TlCl_3$ concentration six times the iodide concentration but not less than 6×10^{-8} *M*. The mixtures were heated 15 minutes at 60°C (pH 5).

TABLE IV
IODIDE DEPENDENCE OF THE CATALYTIC IODINATION OF RNA[a]

Iodide (μmoles × 10^2/reaction)	^{125}I in RNA (cpm)
0	240
0.02	3,310
0.05	7,290
0.10	9,000
0.20	8,920
0.40	7,420
0.80	4,320
1.20	1,715
2.00	570

[a]Reprinted with permission from Scherberg and Refetoff (1973).

F. Ratio of Cytosine to Iodide

Tereba and McCarthy (1973) carefully examined the efficiency of the iodination reaction as a function of the cytosine/iodide ratio (Table II). Their data show that a cytosine/iodide ratio of approximately 8:1 results in maximal efficiency of iodine incorporation into rRNA. At lower cytosine/iodide ratios, the reaction efficiency remains essentially constant; however, as previously noted (see Section II, C), the specific radioactivity of the labeled product is proportionately higher.

G. Reaction Termination

The reaction termination step described by Commerford (1971) employs (1) the addition of Na_2SO_3 to the chilled reaction mixture to reduce $TlCl_3$, and (2) raising the pH of the reaction to 8–9 and reheating at 60°C for 20 minutes to eliminate unstable reaction products. As pointed out by Scherberg and Refetoff (1974), there are several inherent disadvantages in terminating the reaction by these means. First, and perhaps the most critical, is that under mild conditions Na_2SO_3 is known to convert cytosine to uracil by deamination (Hayatsu *et al.*, 1970), thus possibly altering the nucleic acid primary sequence. Second, since Na_2SO_3 treatment likely results in the conversion of substantial amounts of cytosine intermediate to uracil intermediate, deiodination of the nucleic acid is greatly enhanced because of the marked instability of iodine atoms associated with uracil residues. Evidence that simultaneous deamination and deiodination of nucleic acids can occur

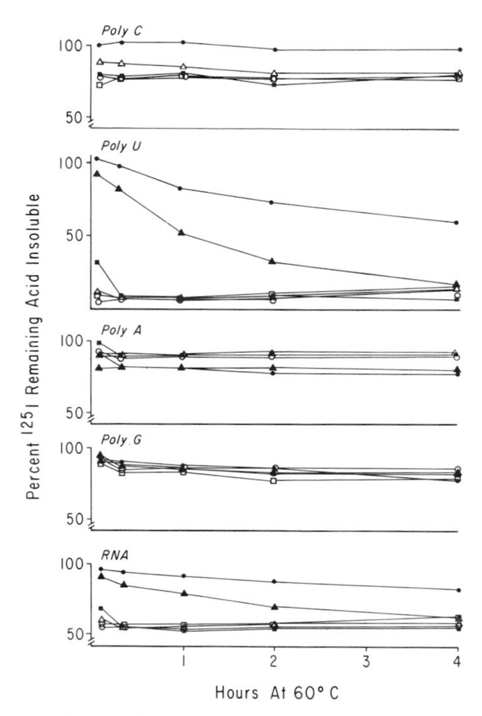

FIG. 4. Stability of iodine adducts in polynucleotides. Isotope was incorporated into polynucleotides in reactions containing 25–50 μg of polymer and 0.1 mCi of ^{125}I (polypyrimidine and RNA reactions) or 0.5 mCi of ^{125}I (polypurine reactions). The final specific activities of polymers used were: poly C, 4.5×10^5 cpm/μg; poly U, 1.7×10^6 cpm/μg; rat polysomal RNA, 0.91×10^6 cpm/μg; poly A, 3.3×10^4 cpm/μg; poly G 4.5×10^4 cpm/μg. At the indicated intervals 0.1-ml aliquots were removed and coprecipitated with trichloroacetic acid in the presence of carrier RNA. The first point was obtained 5 minutes after exposure to buffer. Reprinted with permission from Scherberg and Refetoff (1974).

is demonstrated by an almost quantitative elimination of iodine from iodopoly C treated with Na_2SO_3 at 60°C (Scherberg and Refetoff, 1974).

As Scherberg and Refetoff (1974) also showed (Fig. 4), the stabilities of iodopoly C and iodopoly U at 60°C are quite different over a pH range of 4–9. Over the entire pH range at 60°C, less than 20% of the iodine incorporated into poly C was eliminated after 4 hours, whereas at or above neutral pH over 90% of the iodine incorporated into poly U was eliminated after 5 minutes of incubation. In addition, Scherberg and Refetoff (1974) showed that, immediately following the iodination reaction, the pH of the reaction termination step strongly influences the extent to which deiodination of poly C occurs (Fig. 5). These results support the view that conversion of the iodocytosine intermediate to 5-iodocytosine is not instantaneous, and that terminating the iodination reaction at neutral or higher pH values

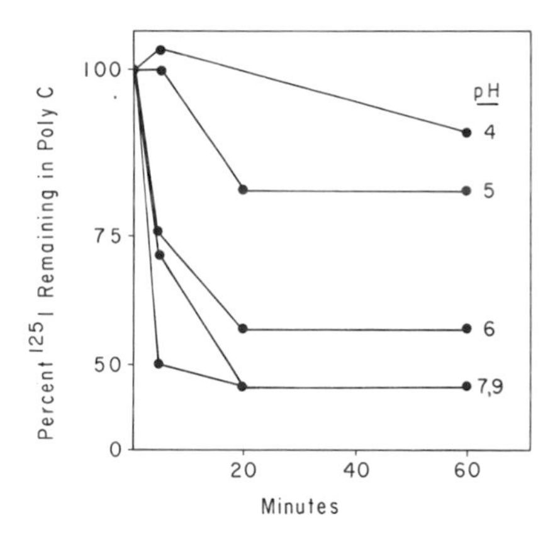

FIG. 5. Early lability of the iodine adduct in polycytidylic acid. ^{125}I was incorporated into polycytidylate in a 0.5-ml reaction containing 2 μmoles of sodium acetate, 2.5 nmoles of KI, 3.1×10^6 cpm of ^{125}I, 13 μg of polycytidylic acid, and 50 nmoles of $TlCl_3$. After 15 minutes at 60°C, 2.5 μmoles of KI were added to quench further incorporation of isotope, and 50-μl aliquots containing 47,000-cpm, acid-precipitable isotope were distributed to buffer pre-warmed to 60°C. Thereafter aliquots were removed at intervals and coprecipitated. Reprinted with permission from Scherberg and Refetoff (1974).

tends to reverse the course of the reaction, i.e., dehydration and deiodination of the pyrimidine.

Termination of the reaction may be very simply accomplished by chilling the reaction mixture to 0°C following the 70°C incubation step. At this temperature conversion of the iodocytosine intermediate to 5-iodocytosine should approach completion in 10–20 minutes. In the case of iodinated RNAs, the pH should then be adjusted to 7.0–7.5, followed by reheating at 60°C for 30 minutes to remove unstable iodine from the uracil residues. The labeled nucleic acids may then be purified by the usual gel filtration procedure.

III. Applications of ^{125}I-Labeled Nucleic Acids

A. Filter Hybridization Experiments

Molecular hybridization experiments are by far the most important current use of ^{125}I-labeled nucleic acids. Their potential usefulness in this application was first demonstrated by Getz *et al.* (1972) and Getz and Saunders (1973) in experiments employing ^{125}I-labeled bulk human RNA hybridized to human placental DNA immobilized on nitrocellulose filters.

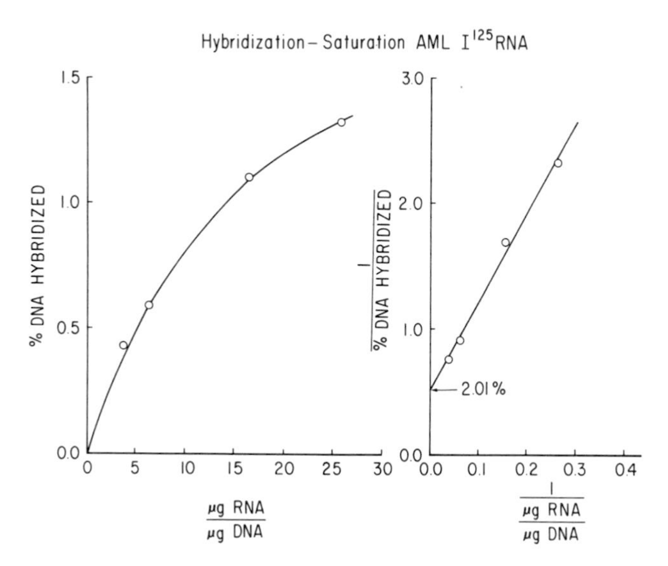

FIG. 6. Hybridization-saturation of ^{125}I-labeled bulk RNA (from leukocytes from patients with acute myeloblastic leukemia) (specific activity 5.2 10^4 cpm/μg). Filters containing 1.5 μg of placental DNA were incubated with increasing amounts of ^{125}I-labeled RNA in 2× SSC and 30% formamide at 45°C for 23 hours. The percentage of DNA that could be hybridized was determined by plotting the data in double-reciprocal form and extrapolating to the point of infinite RNA input by the method of least squares. Reprinted with permission from Getz *et al.* (1972).

The results of these experiments showed that the presence of iodinated bases in the RNA did not interfere with either the annealing kinetics or the saturation values expected for this type of RNA (Fig. 6). Furthermore, thermal elution from filters of the same DNA–RNA hybrids labeled either with ^{3}H or ^{125}I gave identical melting profiles and T_M values (Fig. 7). These experiments therefore demonstrated both the utility and feasibility of employing iodinated nucleic acids in molecular hybridization experiments.

More recent experiments by Scherberg and Refetoff (1973, 1974) have employed ^{125}I-labeled human rRNA to study the reiteration frequency of ribosomal genes in the human genome. The amount of ^{125}I-labeled rRNA observed to anneal to filters containing human DNA in these experiments corresponded to 0.02% of the amount of DNA present. This value is in close agreement with estimates made by other methods (Bross and Krone, 1972). Tereba and McCarthy (1973) employed ^{125}I-labeled rRNA to estimate the reiteration frequency of ribosomal genes in the avian gemone. Their results suggest the presence of 10 to 25 DNA copies per haploid genome, which agrees reasonably well with previous estimates of 15 to 34 copies determined by Merits *et al.* (1966).

The labeling of RNA fractions with ^{125}I for use in molecular hybridization

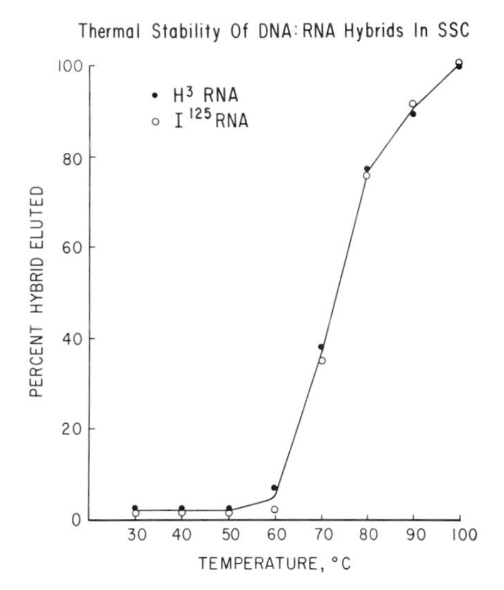

FIG. 7. Thermal stability of DNA–RNA hybrids in SSC. Hybrids were formed by annealing RNA-^{3}H synthesized *in vitro* from human DNA by *E. coli* RNA polymerase or ^{125}I-labeled RNA (from leukocytes from patients with acute myeloblastic leukemia) with 12 μg of placental DNA immobilized on nitrocellulose filters. Hybridization was performed in 2× SSC and 30% formamide at 45°C. Solid circle, RNA-^{3}H; open circles, ^{125}I-labeled RNA. Reprinted with permission from Getz *et al.* (1972).

experiments has been one of the primary goals in the development of the nucleic acid iodination technique. This has now been accomplished by Tonegawa *et al.* (1974), who have reported the labeling of mouse myeloma light-chain mRNA to a specific radioactivity of 10^7 cpm/μg RNA. By using this ^{125}I-labeled mRNA fraction, a reiteration frequency approaching unity for light-chain antibody genes was determined. In a subsequent study, Bernardini and Tonegawa (1974) employed ^{125}I-labeled mouse myeloma heavy-chain mRNA to show that the heavy-chain mRNA molecule may also be coded by a unique or very limited number of DNA sequences.

The iodination technique has also been used to label single-copy and middle repetitive rat ascites tumor cell DNA by Holmes and Bonner (1974). The ^{125}I-labeled DNA fractions were hybridized to giant, heterogeneous nuclear RNA to determine the occurrence of single and repetitive transcripts in this RNA fraction. The results obtained in these experiments suggest that both types of RNA transcripts are present in heterogeneous nuclear RNA.

B. *In Situ* Hybridization Experiments

Several reports (Altenburg *et al.*, 1973; Prensky *et al.*, 1973; Steffensen *et al.*, 1974) have now appeared demonstrating the use of ^{125}I-labeled nucleic

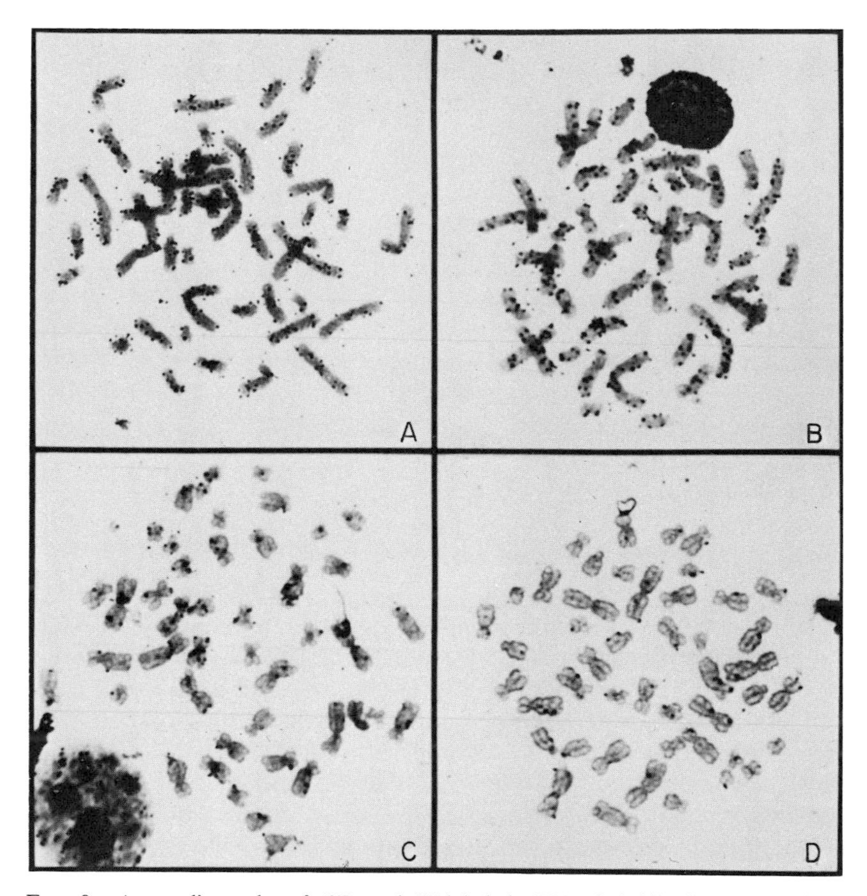

FIG. 8. Autoradiographs of ^{125}I- and ^{3}H-labeled cRNA hybridized *in situ* to human metaphase chromosomes. To air-dried slides prepared for *in situ* hybridization, 34 ng of either (a) ^{125}I-labeled cRNA (1.3×10^6 cpm/μg RNA), or (b) ^{3}H-labeled cRNA (1×10^7 cpm/μg RNA), in 100 μl of 0.30 *M* NaCl–30 m*M* Na sodium citrate (pH 7.0) was added. The slides were incubated at 60°C for 16 hours, treated at 37°C for 30 minutes with 20 μg/ml of RNase in the same buffer, washed with stirring in this buffer at 4°C for 4 hours, and then dehydrated with alcohol before coating with emulsion. Exposure times were (a) 33 days and (b) 21 days. In competition experiments 6.8 μg of either unlabeled bulk human RNA (c) or unlabeled cRNA (d) was mixed with 34 ng of ^{125}I-labeled cRNA (1.3×10^6 cpm/μg RNA) in 100 μl of the above buffer before hybridization. Subsequent steps were the same as (a) and (b). Exposure time for (c) and (d) was 33 days. Reprinted with permission from Altenburg *et al.* (1973).

acids to localize the positions of genetic elements on chromosomes by *in situ* hybridization (Gall and Pardue, 1969). Altenburg *et al.* (1973) demonstrated by competition-hybridization experiments that ^{125}I-labeled nucleic acids anneal to chromosomes in a sequence-specific manner and that the auto-

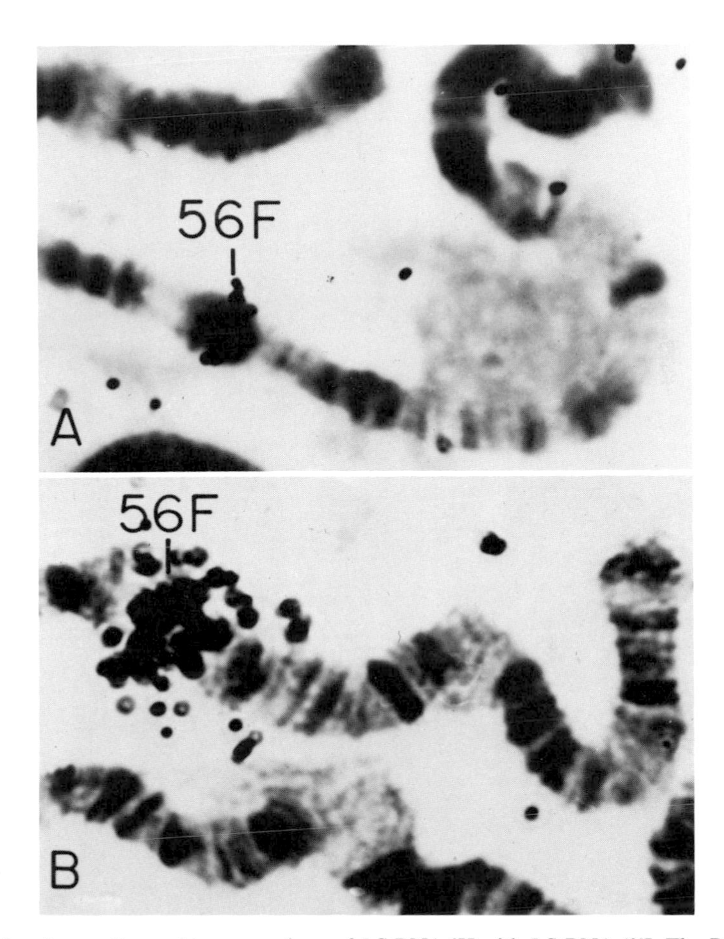

FIG. 9. Autoradiographic comparison of 5 S RNA-^3H with 5 S RNA-^{125}I. The RNA was hybridized to salivary gland chromosomes of *Drosophila melanogaster*. (A) RNA-^3H, about 4×10^6 dpm/μg and a 60-day exposure. The silver grains are located over the bands at 56F, on chromosome 2R. (B) RNA-^{125}I, about 1.16×10^8 cpm/μg, exposure time 4 days. Radioactivity of the RNA probe due to a 3.3% substitution of cytosine with iodocytosine-^{125}I. $\times 3000$. Reprinted with permission from Prensky *et al.* (1973).

radiographic labeling patterns are equivalent to those obtained by the use of ^3H-labeled nucleic acids (Fig. 8). Gene localization was first demonstrated by Prensky *et al.* (1973), who showed that the 5 S RNA cistrons of *Drosophila* are localized at band 56F on chromosome 2R (Fig. 9). By the use of ^{125}I-labeled human 5 S RNA, Steffensen *et al.* (1974) have recently reported the localization of 5 S RNA cistrons on human chromosome 1p (Fig. 10).

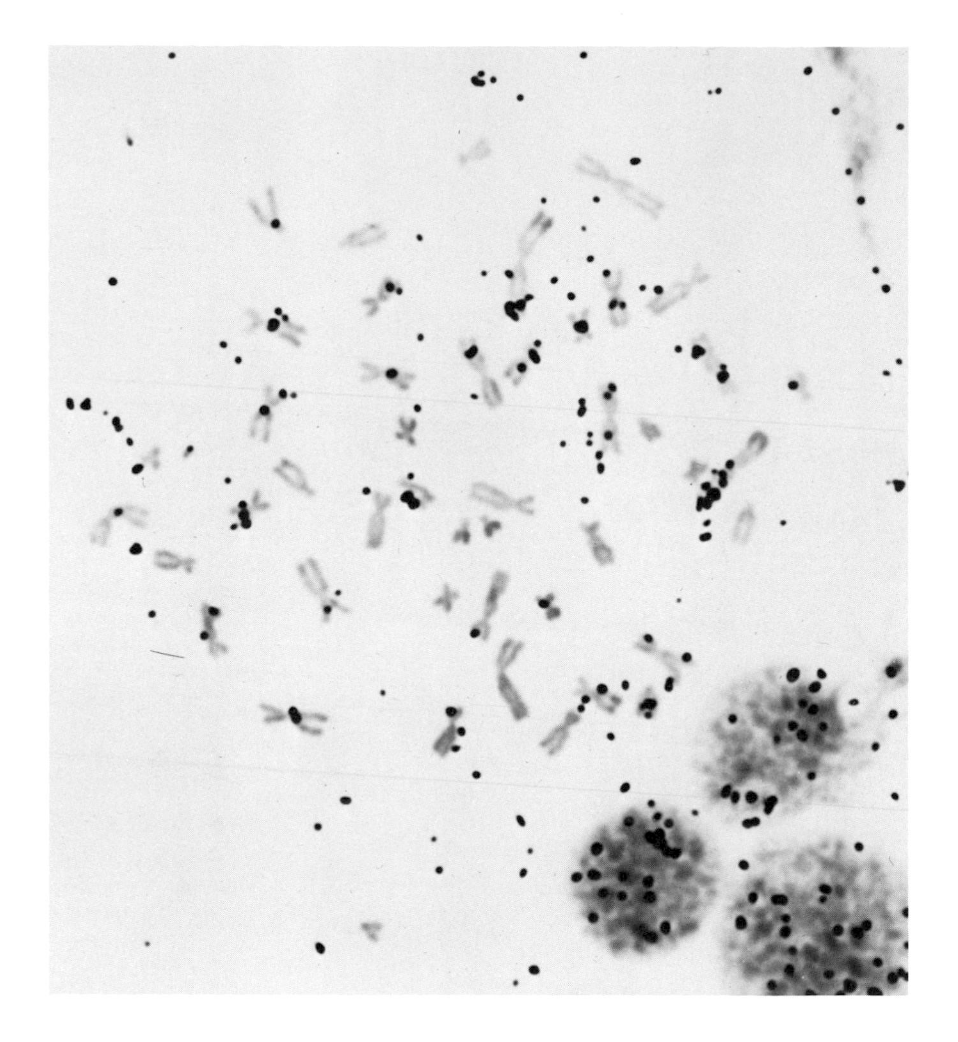

FIG. 10. An autoradiograph using 5 S RNA-^{125}I to hybridize with leukocyte chromosomes of a human male. Reprinted with permission from Steffensen *et al.* (1974).

C. Nucleotide Sequence Analysis

Another important application of ^{125}I-labeled nucleic acids is nucleotide sequencing analysis by fingerprinting methods. Robertson *et al.* (1973) labeled human 5 S RNA and bacteriophage f2 RNA with ^{125}I and, following RNase digestion, showed that the resulting iodinated oligonucleotide fragments are suitable for comparative fingerprinting studies even though the presence of iodinated bases alters their electrophoretic mobility. Similar studies by Dube (1973) show that fingerprints obtained from RNase-digested, ^{125}I-labeled $tRNA_f^{Met}$ are similar to those obtained from *in vivo* ^{32}P-labeled $tRNA_f^{Met}$.

D. Secondary-Structure Analysis

Because the iodination reaction strongly favors denatured substrates, a very useful approach to nucleic acid secondary-structure analysis may be the selective labeling of non-base-paired cytosine residues present in some nucleic acid molecules. Schmidt *et al.* (1973) labeled yeast $tRNA_f^{Met}$ with iodine, under conditions that preserve nucleic acid secondary structure, and then analyzed the ^{125}I-labeled oligonucleotide fragments resulting from RNase digestion. Their results showed that only the non-base-paired regions of the molecule, i.e., the amino acid acceptor end and the anticodon sequence, contained ^{125}I-labeled cytosine residues. Conversely, when the iodination reaction was performed under denaturing conditions, the labeling occurred randomly, as indicated by the presence of iodine in an additional seven oligonucleotide fragments.

IV. Discussion

Although the radioiodination of nucleic acids is a relatively new technique, its potential as a general labeling method in several different areas of cell and molecular biology has already been established. Alternative labeling procedures, such as the *in vivo* incorporation of tritiated nucleotide precursors or $^{32}PO_4$, usually either do not result in sufficiently high specific radioactivities or are limited in their applicability. While the *in vitro* incorporation of tritiated nucleotide precursors by enzymatic transcription of nucleic acid templates does result in high specific radioactivities, the iodination technique permits the direct labeling of the nucleic acid of interest. It should also be mentioned that the cost of labeling nucleic acids with ^{125}I, as compared to labeling with tritiated nucleotide precursors, is trivial.

As with any new technique, *in vitro* labeling of nucleic acids with iodine is not completely straightforward and free of technical difficulties. The two most frequently encountered problems concern either poor incorporation of iodine into nucleic acids during the reaction proper, or deiodination of the nucleic acid following the reaction termination step. As regards the first of these difficulties, i.e., poor incorporation, it has frequently been observed that the pH of the reaction was not properly controlled so that very little, if any, reaction occurred. While this difficulty may be overcome relatively easily, the problem of deiodination may not be. Inherent in the reaction is the conversion of iodinated pyrimidine hydrates to either 5-iodopyrimidine or the pyrimidine itself, disregarding completely the chemistry involving the iodopurines. The problem of deiodination of nucleic acids therefore undoubtedly involves the extent to which the final product of the reaction is formed. Because the iodinated pyrimidine hydrates are relatively unstable compounds, their elimination must precede the use of the labeled nucleic acid in applications requiring low levels of background radiation.

ACKNOWLEDGMENTS

We are indebted to Drs. D. M. Steffensen and W. Prensky for supplying micrographs of their work.

This work was supported in part by grants from the National Institutes of Health (GM-19513), the American Cancer Society (ACS IN-43N), the Robert A. Welch Foundation (G-267), and the Damon Runyon Memorial Fund for Cancer Research (DRG-1061).

REFERENCES

Altenburg, L. C., and Shaw, M. W. (1973). *Abstr. Int. Congr. Genet. 13th,* s5.

Altenburg, L. C., Getz, M. J., Crain, W. R., Saunders, G. F., and Shaw, M. W. (1973). *Proc. Nat. Acad. Sci. U.S.* **70**, 1536.

Bernardini, A., and Tonegawa, S. (1974). *FEBS (Fed. Eur. Biochem. Soc.) Lett.* **41**, 73.

Bross, K., and Krone, W. (1972). *Humangenetik* **14**, 137.

Commerford, S. L. (1971). *Biochemistry* **10**, 1993.

Dube, S. K. (1973). *Nature (London)* **246**, 483.

Gall, J. G., and Pardue, M. L. (1969). *Proc. Nat. Acad. Sci. U.S.* **63**, 378.

Getz, M. J., and Saunders, G. F. (1973). *Biochim Biophys. Acta* **312**, 555.

Getz, M. J., Altenburg, L. C., and Saunders, G. F. (1972). *Biochim. Biophys. Acta* **287**, 485.

Hayatsu, H., Wataga, Y., Kai, K., and Iida, S. (1970). *Biochemistry* **9**, 2858.

Heiniger, H. J., Chen, H. W., Commerford, S. L. (1973). *Int. J. Appl. Radiat. Isotop.* **24**, 425.

Holmes, D. S., and Bonner, J. (1974). *Biochemistry* **13**, 841.

Johns, H. E., LeBlanc, J. C., and Freeman, K. B. (1965). *J. Mol. Biol.* **13**, 849.

Logan, D. M., and Whitmore, G. F. (1966). *Photochem. Photobiol.* **5**, 143.

Merits, I., Schulze, W., and Overby, L. R. (1966). *Arch. Biochem. Biophys.* **115**, 197.

Myers, W. G. and Vanderleeden, J. C. (1960). *J. Nucl. Med.* **1**, 149.

Prensky, W., Steffensen, D. M., and Hughs, W. L. (1973). *Proc. Nat. Acad. Sci. U.S.* **70**, 1860.

Robertson, H. D., Dickson, E., Model, P., and Prensky, W. (1973). *Proc. Nat. Acad. Sci. U.S.* **70**, 3260.

Scherberg, N. H., and Refetoff, S. (1973). *Nature (London), New Biol.* **242**, 142.

Scherberg, N. H., and Refetoff, S. (1974). *J. Biol. Chem.* **249**, 2143.

Schmidt, F. J., Omilianowski, D. R., and Bock, R. M. (1973). *Biochemistry* **12**, 4980.

Shapiro, H. S., and Chargaff, E. (1957). *Biochim. Biophys. Acta* **26**, 596.

Steffensen, D. M., Prensky, W., and Dufy, P. (1974). *Cytogenet. Cell Genet.* **13**, 153.

Tereba, A., and McCarthy, B. J. (1973). *Biochemistry* **12**, 4675.

Tonegawa, S., Bernardini, A., Weimann, B. J., and Steinberg, C..(1974). *FEBS (Fed. Eur. Biochem. Soc.) Lett.* **40**, 92.

Yu, C. T., and Zamecnik, P. C. (1963). *Biochim. Biophys. Acta* **76**, 209.

Chapter 19

Radioiodine Labeling of Ribopolymers for Special Applications in Biology

NEAL H. SCHERBERG AND SAMUEL REFETOFF

Department of Medicine,
University of Chicago,
Chicago, Illinois

I. Introduction 343
II. Special Considerations in Preliminary Preparation of the RNA 344
III. Radioiodination of Polynucleotides 347
A. Substrate and Stabilization 347
B. Parameters in the Forward Reaction 349
C. Standard Procedure for the Radioiodination of RNA 350
IV. Applications of Iodine-Labeled RNA and Nucleotides 351
A. Hybridization 351
B. Molecular Size Distribution 352
C. Structural Analysis 354
D. *In Vitro* Synthesis of RNA 357
References 358

I. Introduction

A complete account of the molecular life cycle of a mRNA will include a description of synthesis, transport, and translation, and the influence of regulatory factors on these processes. To accomplish this objective it is necessary to have quantitative analytical devices capable of measuring the amount of specific nucleic acid in mixtures and purified fractions at a concentration below the level that can be detected by direct measurement. Though the concentration of a specific RNA can in some instances be assayed by indirect means such as determination of the translation product, such analyses may be laborious and are often of low sensitivity. Quantitation of a given messenger RNA through hybridization–competition with a known amount of the same species of RNA is potentially a precise approach. However, quantitative analysis by hybridization has in turn difficult prerequisites; specific mRNA must be available for the preparation of calibration curves relating annealing to input under fixed conditions. While a few messengers have been isolated or highly enriched fractions separated, employment of these in hybridization analysis depends on surmounting a second hurdle. It is necessary to label the ribopolymers to very high specific activity, since the corresponding gene may occupy as little as $10^{-5}\%$ of the whole genome. Polymerase reactions have been used for preparing either highly labeled messenger or, more frequently, a complementary molecule (Pardue and Gall, 1970; Kacian *et al.*, 1972; Melli and Pemberton, 1972; Verma *et al.*, 1972). Radioiodination appears to be a superior alternative, providing simplification as well as the basis for diverse investigations of eukaryote messengers. The information that follows is presented as a guide for applying radioiodination to the objective of marking messengers to high specific activity. Some observations concerning the application of radioiodination of RNAs for other purposes are also discussed.

II. Special Considerations in Preliminary Preparation of the RNA

Many physiological substances can incorporate iodine or inhibit the iodination of nucleic acids. As illustrated in Table I, the addition of iodine to proteins may occur at 40-fold the rate of uptake by an optimal polynucleotide substrate, even though the susceptible aromatic amino acids

TABLE I

COMPARISON OF THE RADIOIODINATION OF HUMAN SERUM ALBUMIN AND POLYCYTIDYLIC ACID AT pH 5[a]

Substrate, 3 μg	^{125}I incorporated (cpm)	Percent
Polycytidylic acid	5,000	0.58
Human serum albumin	196,000	23.05

[a]Radioiodination was carried out in a standard 0.20-ml reaction mixture at 40°C for 5 minutes. The reaction was stopped by rapid chilling at 0°C and addition of trichloracetic acid. The acid-insoluble product was collected by filtration, and the incorporated isotope determined by direct counting of the glass fiber filter in a gamma spectrometer.

constitute only a small fraction of the protein. While it is particularly important to diminish protein contamination, many other cellular substances including DNA, aromatic amino acids, and lipids can incorporate iodine and interfere with subsequent applications of the labeled RNA. Furthermore, since the labeling reaction depends on an oxidation–reduction process, reducing substances such as sugars entrained with the RNA during isolation or added during the preparative steps may inhibit labeling. As a consequence of these considerations, special precautions should be taken in the isolation of RNA when radioiodination is to be used. Although the procedures and problems encountered in the isolation of RNAs have been reviewed in a previous volume of this series (Brawerman, 1973; Kates, 1973; Muramatsu, 1973), preparative methods are reconsidered here in the context of the special problem of radioiodination.

Routine preparation of RNA by extraction with phenol mixtures is a useful first step but does not completely eliminate protein even after multiple extractions. Further purification of the RNA by precipitation at a high salt concentration lessens contamination and may strikingly enhance labeling in a subsequent iodination reaction. The high-molecular-weight RNA obtained in the pellet often contains some protein and rapidly sedimenting polysaccharides such as glycogen. Nevertheless, the amount of contamination generally does not prevent adequate labeling of the nucleic acid for eventual use in sucrose gradient analysis or hybridization. Any iodine-labeled protein in the total reaction product is separated from the RNA during routine processing after the labeling reaction. However, when the highest efficiency of iodine addition to RNA is desired, further treatment of the nucleic acid preparation prior to iodination should be employed.

It is frequently found that increasing the complexity of a procedure for preparing RNA, also increases the tendency to introduce chain breaks. The polymers are sensitive indicators of the presence of endonuclease. We have found that sedimentation of partially purified RNA through cesium sulfate is a simple and effective means of removing contaminants. As illustrated in Fig. 1, high-molecular-weight RNAs are separated from other cellular constituents by the difference in density and molecular weight. RNAs with sedimentation coefficient above 10 are recovered in a narrow band providing centrifugation lasts at least 10 hours in the described protocol. Fractionation of the RNA by molecular weight can be accomplished if centrifugation is interrupted at earlier times. The nucleic acid is subsequently prepared for iodination by exhaustive dialysis, by salt precipitation at 0°C, or by ethanol precipitation after lowering the cesium sulfate concentration.

An alternative preparative route which excludes the use of phenol has also been adopted in our laboratory. A suspension of cells, or subcellular particles, is warmed to 60°C with 3 vol of a buffer composed of 8 *M* guanidinium

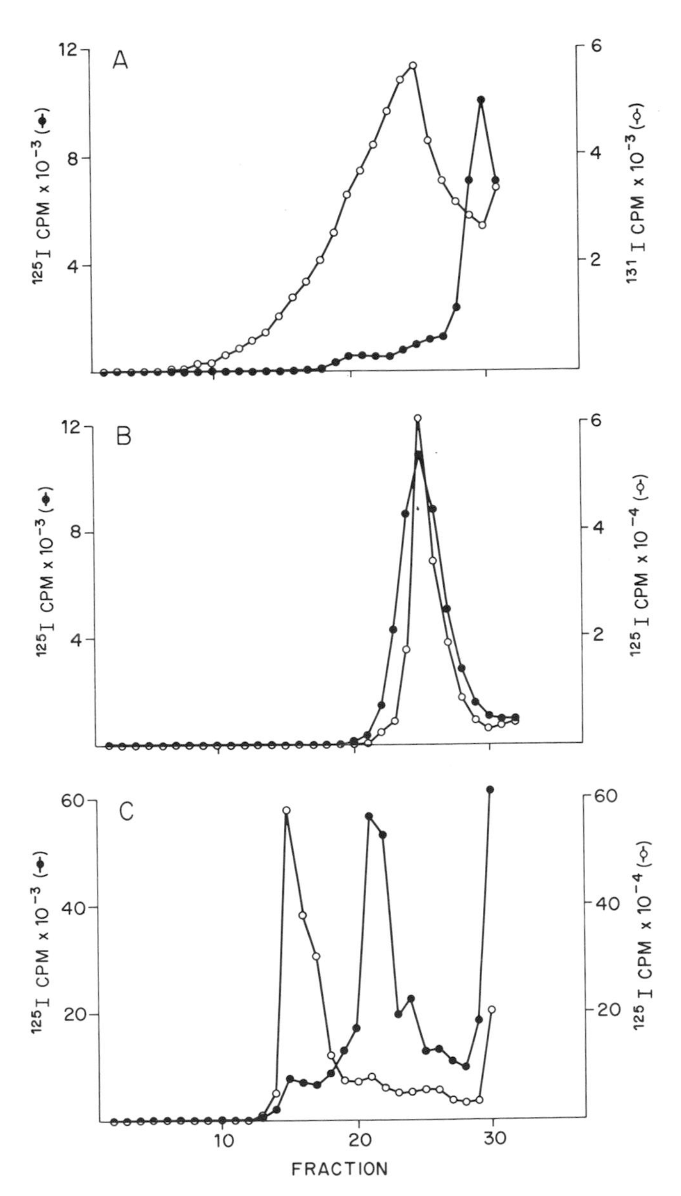

FIG. 1. Separation of iodine-accepting substances commonly associated with partially purified high-molecular-weight RNA. The samples were added at the top of a 4.5-ml cesium sulfate step gradient and spun for 8 hours at 50,000 rpm. Ten-drop fractions were collected from the bottom of the tube. (A) Open circles, iodine; solid circles, protein. (B) open circles, DNA; solid circles, sRNA. (C) open circles, RNA above 8 S; solid circles, unknown by-product of the reaction with RNA purified by phenol alone. From Scherberg and Refetoff (1974a); see for details of preparation.

TABLE II

INHIBITION OF RNASE WITH 6 M GUANIDINIUM CHLORIDE[a]

Condition	Acid-insoluble ^{125}I	Percent of control
Set A		
Complete	5,395	99
6 M urea	980	15
Minus guanidinium	245	4
Set B		
Complete	277,300	96
Minus guanidinium	1,515	0.5

[a]The complete reaction mixture contained in 0.32 ml: 0.02 mmole tris chloride (pH 8) and was 6 M in either guanidinium chloride or urea. Control incubations were without RNase. Set-A reactions contained in addition 6800 cpm of ^{125}I-labeled 28 S RNA (0.7×10^6 cpm/μg), 10 μg of pancreatic and 5 μg of T_1 RNase. Set-B reactions contained 290,000 cpm of poly U-^{125}I (2.4×10^6 cpm/μg) and 60 μg of pancreatic RNAse. Incubation was for 5 minutes at 40°C. The remaining acid-insoluble isotope was determined by filtration.

chloride, 1.3% sodium lauroyl sarcosine, and 1.3 mg/ml heparin. This solution completely disrupts cells and dissociates RNA from organized particles. Unlike urea solutions, guanidinium chloride in concentrated solution is an effective inhibitor of RNase (Sela *et al.*, 1957; Table II). The resulting solution is layered over a cesium sulfate step gradient and centrifuged as already described. RNA is separated as a band and may be collected through a puncture in the bottom of the tube or, when the RNA forms an aggregate in the cesium sulfate, collected from the top after removal of the overlaying solution. Determinations with cells grown in the presence of tritium-labeled nucleoside precursors have shown that high-molecular-weight RNA is separated almost quantitatively (Fig. 2). Since postpreparative labeling of RNA can generate specific activity above 10^7 cpm/μg, fractions with as little as a nanogram of ribopolymer may be characterized in respect to polymer size distribution. In this regard it must be noted that the disrupting buffer does not in all cases eliminate nuclease nicking of the RNA, which is revealed by warming the preparation after recovery from the cesium sulfate.

III. Radioiodination of Polynucleotides

A. Substrate and Stabilization

Iodine is incorporated into both pyrimidine groups of RNA and to a lesser extent into purine residues. The yield in stable reaction product is greater

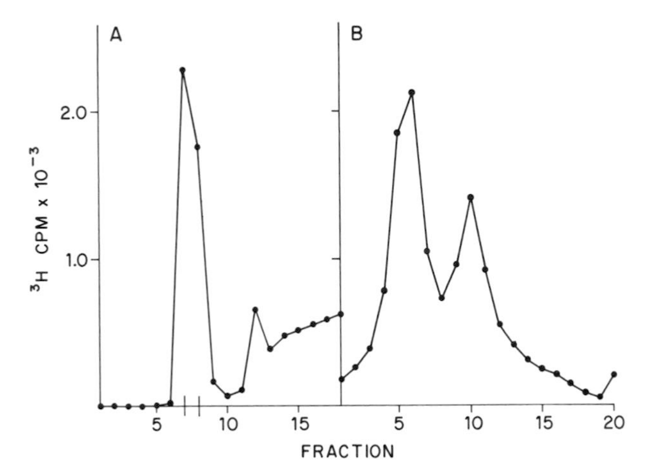

FIG. 2. Effectiveness of a simplified procedure for preparing RNA from the tissue culture cells. A single petri dish of normal human fibroblasts in early log-phase growth was exposed to standard growth media supplemented with 1 mM uridine and 2 μCi/ml carrier-free cytidine-5-^{3}H for 5 days. The cells were washed twice with cold water, suspended in 0.4 ml of disrupting buffer (see text), and warmed 3 minutes at 55°C. The solution was overlaid and centrifuged on a cesium sulfate step gradient (Scherberg and Refetoff, 1974a). Fractions were collected from the bottom of the tube, aliquots removed in capillaries, and the isotope content determined. The distribution of counts is shown in (A). The remainder of the fractions were dialyzed against water and characterized by sedimentation in sucrose gradients. The sedimentation profile of RNA in the combined fraction 7 and 8 is shown in (B). The sedimentation rate and pattern for ^{3}H-labeled RNA was typical of rRNA. The small central peak 12 was composed of 60% DNA and 40% sRNA. Counts in the upper region of the gradient (> 95%) were dialyzable.

with cytidylate than with uridylate, although in both cases the 5′-position of the base is the likely site of iodine substitution and it is probable that the reaction mechanisms are analogous. Commerford has postulated that the reactions proceeds via additions at the 5′- and 6′-positions, producing a saturated product (Commerford, 1971). The resulting compounds undergo further transition, wherein the double bond is reformed through either a reversal, with the release of iodine, or elimination of other substituents at the 5′- and 6′-positions, with retention of the iodine. In a reaction at pH 5, the extent of the first-step addition to polycytidylate may only slightly exceed incorporation into polyuridylate. Quantitatively, the succeeding process is dissimilar. The saturated intermediate in cytidylate is unstable at room temperature and undergoes dismutation, reforming the double bond by one of the mentioned processes. The iodinated intermediate reaction product in the case of polyuridylate has a longer survival and is readily recovered as a component of the total reaction product. The subsequent elimination step occurs predominantly through release of iodine, with the consequence that iodinated RNA is primarily labeled in cytidylate groups. Gradual loss of iodine from unstable intermediates is detrimental to the use of labeled polynucleotides. However, the rate at which release occurs is pH-dependent. As the postiodination pH is adjusted upward, the reversal is accelerated. Exposure of radioiodinated RNA at pH above neutrality at 60°C for 10 minutes is a sufficient treatment to force the reaction to an end point and eliminates slow bleeding of iodine from pyrimidines (see Altenburg, Getz, and Saunders, Fig. 4 this volume).

Addition of iodine to polypurines also occurs. By analogy to other halogenation reactions, the site of substitution in guanylate residues may be the 8-position of the base. While catalytic iodine addition to nucleotides in polyadenylate has been observed, the reaction product has not yet been documented. With the reaction conditions used to date, the contribution of iodinated purine groups to the final level of iodination in RNA is less than a few percent. Iodine present in the final, repurified (see Section III,c) reaction product of polypurines is stable (Altenburg *et al.*, Chapter 18, this volume). Thus, when the thallic trichloride–catalyzed incorporation of iodine into any of the four common nucleotides of RNA is followed by a stabilization treatment at pH above 7, the remaining addition product can be applied in a long-term hybridization reaction at high temperature.

B. Parameters in the Forward Reaction

The composition of the iodination reaction is simple and has few variables aside from the substrate itself. In the reaction condition we adopted, maximal incorporation of isotope occurs in the presence of approximately 10^{-5} *M* iodide (Scherberg and Refetoff, 1973). The fall in incorporation of isotope into polymer as the concentration of carrier iodine is decreased below the optimal level may be partly due to the concentration dependence of the reaction. It may also result from the absorption of iodine to surfaces or removal by combination with small amounts of contaminants present in the solution; we routinely use glass-distilled and Millipore-filtered water in all preparations. The iodine concentration dependence may also be related to the volatility of iodine. The iodination reaction in capillaries described by Prensky *et al.* (1973) effectively eliminates loss of iodine from the reaction solution occurring through volatilization.

The addition of iodine to polynucleotides is pH- and temperature-dependent. As expected from the characteristic stability of the intermediate, incorporation of isotope into polynucleotides decreases rapidly as the pH is raised above 5. Labeling of RNA at pH 4 is more rapid than at pH 5, but carries the hazard of causing depurination or acid hydrolysis of the chains.

The influence of temperature on iodination of polynucleotide was described by Commerford (1971). The relatively poor incorporation at low temperature in comparison to the reaction with protein indicates that the primary barrier to addition is the substrate rather than formation of an active form of iodine. As described in Section IV,C,2 the temperature of the reaction also has an influence through its effect on secondary structure of the nucleic acid.

Thallic trichloride is a convenient catalyst which retains catalytic activity during several months of storage at room temperature in a 0.1 *M* solution. As reported by Tereba and McCarthy (1973), we observed formation of a brown precipitate in dilute solutions of the catalyst. Dilutions made in ethanol have better stability. While the composition of the precipitate is unknown, it may be a factor contributing to background levels of "particulate" isotope (Scherberg and Refetoff, 1974a), since we have found that an insoluble form of iodine arising during incubation in the presence of thallic trichloride is increased by raising the concentration of catalyst. Chloramine T may also be used as catalyst, but has poorer storage qualities and, in our experience, lower catalytic capacity. The effect of thallic trichloride on other functional groups in RNA has not been reported.

C. Standard Procedure for the Radioiodination of RNA

The routine procedure for preparing radioiodinated RNA in our laboratory is as follows. Radioiodine is added to an ice-cold mixture containing 0.1 *M* sodium acetate (pH 5), at least 5×10^{-6} *M* iodide (isotope and stable), and 0.2–20 μg of RNA in a total volume of 0.2 ml. As a last step in composition of the reaction solution, 0.5 ul of 0.1 *M* thallic trichloride is added from a capillary and the tube then sealed with several layers of parafilm. The reaction is carried out at 60°C for the desired period, and then the reaction tube is placed in ice. With protective gloves the Parafilm is removed, discarded as radioactive waste in a sealed tube, and 20 μl *M* tris (pH 8.3) added to raise the pH to 7.5. The tube is resealed and again incubated at 60°C for 10 minutes. At the termination of this incubation period, the vessel is iced and the solution transferred to the top of a Sephadex G-50 column (0.8 × 22 cm) previously heparinized by passage of a solution containing 2 mg of heparin followed by equilibration with 0.1 *M* sodium acetate (pH 5). The polymer is washed through the column with the same buffer and fractions collected until the labeled RNA fraction has been eluted. To this point all steps involving radioiodine are carried out in a well-vented hood. The fractions containing labeled RNA are combined and concentrated to 0.3 ml by flash evaporation. The product is then further purified by sedimentation through cesium sulfate, as described in Fig. 1. After a 40-hour incubation at 70°C, less than 5% of the isotope in a messenger fraction labeled to 5×10^6 cpm/μg and purified by the preceding methods was released from the polymer in an acid-soluble form. DNA and nucleotides may be labeled to high specific activity by similar procedures (Commerford, 1971; Scherberg and Refetoff, 1974b).

IV. Applications of Iodine-Labeled RNA and Nucleotides

A. Hybridization

The preceding discussion has been concerned with practical factors of preparing and iodine labeling of an RNA for application in hybridization. It has been emphasized that it is necessary to eliminate nonnucleic acid substances present in purified RNA in order to label the nucleic acid at optimal efficiency, to establish the correct specific activity, and to eliminate compounds that may contribute to background values in the hybridization reactions. An additional practical concern in annealing experiments is the question whether the presence of iodine in the 5′-position of the base interferes with base pairing.

Several arguments to the effect that iodine is not a severe antagonist of accurate pairing can be presented. First, the position of substitution on the base is a site that is frequently substituted without causing deterioration in pairing. Bulky groups such as methyl or hydroxymethyl are present at the position in DNA. In respect to the chemical as well as the bulk properties of a halogen substituent, Massoulié and co-workers (1966) showed that bromo-substituted cytidylate pairs with inosine in a double-stranded structure of stability exceeding the thermal stability of poly I–poly C. Moreover, in regard to pairing, we have observed that, when iodocytidine nucleotide triphosphate is employed in polymerase reactions, only the corresponding parent nucleotide, as opposed to thymidine or uridine triphosphates, serves as a competitive inhibitor of incorporation of the iodine-labeled nucleotide (Scherberg and Refetoff, 1974b).

In opposition to the foregoing inference of the interchangeability of iodo-cytidylate with cytidylate in base pairing, two observations have indicated that the iodine substituent may interfere with pairing. It has been reported that if more than half of the cytidylate groups of DNA are substituted with iodine, the molecules fail to renature (Commerford, 1971). The blockage of infectivity resulting if phage DNA is extensively substituted with a halogen (bromine) might be due to incorrect base pairing (Goz and Prusoff, 1968). Iodine-containing nucleosides have been effectively applied as antiviral agents (Eidenoff *et al.*, 1959). However, it cannot be concluded that the biological injury derives from inclusions of iodonucleotides into nucleic acids, since inhibition may also arise through interference with a kinase activity (Prusoff *et al.*, 1965). It is of interest in this regard that the possibility that the antiseptic activity of iodine results from its addition to nucleic acids could not be confirmed (Hsu, 1964).

The preceding information showing that the presence of iodine in nucleotides may affect the biological activity is an extreme-case example. The circumstances presented involved conditions in which a large fraction of the nucleotides were substituted. However, specific activities above 10^8 cpm/μg in nucleic acids can be achieved with less than 1% of the bases iodinated (Prensky *et al.*, 1973). Studies depending on the normal *in vivo* incorporation and function of trace amounts of iodine-labeled nucleotide in DNA were carried out, and the parallel catabolism of iodine and tritium-labeled DNA has been documented (Hughes *et al.*, 1964). The strongest evidence on this point is derived from the analysis of hybridization experiments. It has been found that with low amounts of iodine substituent the extent of hybridization of RNA is the same as that occurring with tritium-labeled nucleic acid (Getz *et al.*, 1972). Based on this, and on the data on pairing of oligomers which will be presented, one can conclude that radioiodinated nucleic acids can be a useful tool for hybridization analysis. Careful determination of the effect of iodine on T_m and the upper limit of degree of substitution in respect to annealing of nucleic acid would be of interest.

B. Molecular Size Distribution

Since intact RNA chains are not required in hybridization, cleavage of the molecule during the process of iodination is not of concern. However, if the iodinated product is to be characterized by sedimentation or distribution in electrophoresis, intact polymers are required. This is an application in which the polymerase approach to labeling fails. The effect of iodination is of particular interest for the further reason that very small cell samples may be analyzed for size distribution of RNA. Iodine-labeled RNA has been characterized by sedimentation in sucrose gradients and by electrophoresis (Fig. 3). As illustrated, reaction at pH 5 produces little breakup of the chains. This condition compromises between the increase in iodine incorporation at lower pH and the possibility of acid-catalyzed hydrolysis. When the reaction period is extended to 60 minutes at 60°C there occurs some deterioration in the sedimentation pattern in comparison to the product of the shorter reaction (Scherberg and Refetoff, 1974a).

The distribution of isotope in a particular region of an RNA profile cannot be assumed to reflect directly the relative amount of nucleic acid present.

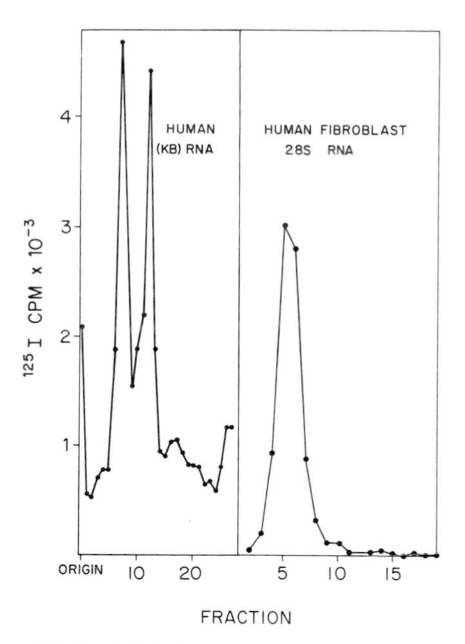

FIG. 3. Distribution of iodine-labeled RNAs by electrophoresis and sedimentation. RNA was prepared from human KB cells by lysis with detergent and phenol extraction, followed by repeated precipitation from 2 *M* lithium chloride. The RNA from human fibroblasts was in addition selected by sucrose gradient sedimentation. Both samples were labeled in the standard reaction, reheated for 10 minutes at 60°C, and dialyzed (KB RNA) or repurified by sedimentation in cesium sulfate (28 S RNA). Acrylamide gel electrophoresis followed the procedure described by Peakcock and Dingman (1967) with 2.6% acrylamide and 0.5% agarose. The sample for electrophoresis was mixed with 30 μg of carrier. The ^{125}I-labeled 28 S RNA recovered from repurification was dialyzed against water, and an aliquot sedimented through a 5–20% sucrose gradient for $2\frac{1}{2}$ hours at 50,000 rpm at 2°C.

The interpretation of the pattern must account for two factors which influence the extent of iodination. As previously discussed, approximately 95% of the label in stabilized iodo-RNA preparations is in the form of iodocytidylate. Consequently, the amount of isotope present in the RNA of a particular zone is expected partly to reflect the proportion of cytidylate in that same region. A second factor which may be influential in regard to the amount of label in a given fraction, is the amount of structure in the RNA. Several types of evidence have indicated that base-paired nucleotides are less susceptible to the addition of iodine. In his original description of thallic thrichloride–catalyzed iodination, Commerford showed that denatured DNA was a better substrate for labeling than was the native polymer. Schmidt and co-workers subsequently found that, in the labeling of yeast methionyl-tRNA, regions arranged in a base-paired relationship in models were poor acceptors for iodination (Schmidt *et al.*, 1973). Finally, in further confirmation of the significance of this factor, we have found that some nucleotides of 28 S RNA iodinated at 80° C apparently are not susceptible to iodination at 40° C. The effect of temperature on iodine addition can be interpreted as resulting from the melting out of double-stranded regions. This finding is further considered in the discussion of radioiodination as a device for the study of structure in RNA.

C. Structural Analysis

1. Feasibility of Radioiodination in the Investigation of the Primary Structure of Nucleic Acid

Radioiodination is a promising tool for the study of nucleic acid structure. The technique is particularly advantageous for the investigation of primary structure because it is possible to work with minute amounts of the original nucleic acid. Iodine-containing RNAs are susceptible to nucleases and can be converted to mononucleotides during treatment with a combined set of RNases. They are therefore suitable for the routines that have been developed for the preparation of fragment fingerprints (Dube, 1973; Robertson *et al.*, 1973). Iodine isotopes provide additional advantages in this application, since the emission is not quenched by paper or other strata used in separating oligomers and thus is comparatively efficiency collected by overlying films. Furthermore, it is possible to make use of two iodine isotopes which may be simultaneously determined by gamma spectrometry.

Important limitations to the application of radioiodinated RNA for the determination of primary structure are also prominent. One drawback follows from the specificity. Detection of oligomers lacking the cytidylate group would be far less sensitive. On the other hand, multiple cytidylate groups in a given fragment might yield several spots during separation of the oligomers. This could occur through the influence of iodine in lowering the pK of the cytosine amine with the consequent anodal shift in electrophoretic mobility and an increase in its binding to anion exchangers. Because of this, different extents of substitution in a given fragment might cause the formation of two separate bands or spots. A further drawback to the use of iodination for sequence determination is the fact that, in the alkaline conditions used to split ribopolymers to single nucleotides, iodine is eliminated from the pyrimidines. This excludes base-catalyzed hydrolysis, as it is ordinarily employed in the sequencing of short oligomers.

2. Assays for Secondary Structure

Recent observations on the chemistry of deiodination in our laboratory have shown that radioiodination can be a valuable tool in assays for secondary structure. As noted previously, several types of evidence have indicated that the pairing of nucleotides hampers the addition of iodine. Thus it seems possible that conditions might be selected through which a direct readout of percent secondary structure might be obtained through quantitation of radioiodination as a fraction of total susceptible bases. While attractive, the feasibility of the proposal has not yet been documented. Indeed, interpretation of net reaction or reaction rate is complicated by the effect of temperature on the reaction rate exclusive of structure and the necessity for coordinate repurification of each labeled preparation. In view of the elaborate preliminary study required to utilize the forward reaction for the objective, dependence of deiodination on structure may be an easier approach. We have found that bisulfite treatment of an iodine-labeled polynucleotide can cause nearly total release of the isotope. The interaction of bisulfite with pyrimidines is well known and probably occurs by reversible addition across the 5,6 double bond (Barrett and West, 1956; Hayatsu *et al*., 1970; Shapiro *et al*., 1970; Kai *et al*., 1971; Sander and Deyrup, 1972). The initial reaction

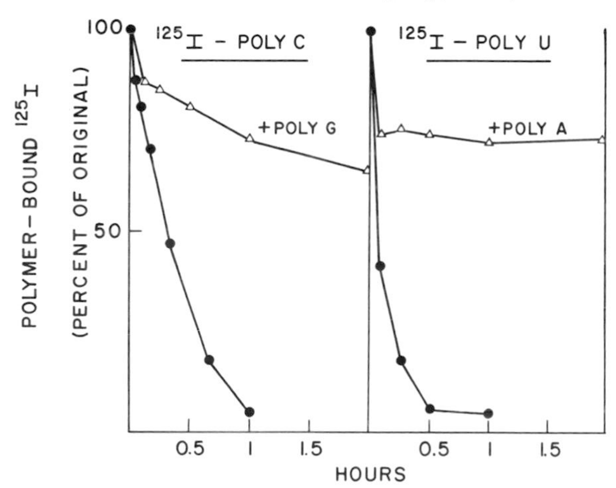

FIG. 4. Pairing protection against bisulfite-induced deiodination. Poly C and poly U were labeled under conditions yielding approximately equal specific activity. The reaction products were heat-stabilized and repurified by sedimentation through cesium sulfate. The final polymer isotope content was 10^6 cpm/μg and 2×10^6 cpm/μg for poly C and poly U, respectively. Reaction mixtures containing 7000 cpm ^{125}I poly C or 9000 cpm ^{125}I poly U were mixed with twice the molar amount of the complementary homopolymer (triangles) or water alone (circles), and then incubated 4 minutes at 60°C. After cooling on ice, sodium bisulfite adjusted to pH 6 was added to a final concentration of 0.5 M; the tubes were closed and incubated at 39°C. At intervals reactions were chilled; carrier RNA was added and precipitated by addition of trichloroacetic acid. The remaining polymer-bound isotope was collected by filtration. The percent of precipitable count represents the fraction of the initial level. Less than 5% of the isotope was converted to an acid-soluble form in the absence of bisulfite.

may be followed by reversal alone, by prior or concomitant deamination. In the case of iodinated polynucleotide, it has been further observed that deiodination, like bisulfite-catalyzed deamination (Goddard and Schulman, 1972), is inhibited by base pairing. Thus if iodinated poly C or poly U is reacted with bisulfite alone rapid deiodination follows (Fig. 4). In contrast, if the complementary homopolymer is added, the reaction is inhibited. A differential in the rate of deiodination of bases in RNA has also been observed. Some of the iodinated nucleotides formed during the iodination of total human fibroblast RNA at 80°C were more resistant to bisulfite than the class of bases labeled at 40°C (Fig. 5). Presumably, in the reaction at elevated temperature, more potential sites for addition are engaged, i.e., dissociated from pairing. Therefore, for deiodination, as for iodination, it should be possible to define conditions allowing differentiation between regions of polynucleotides occurring in base-paired relationships rather than existing as single strands.

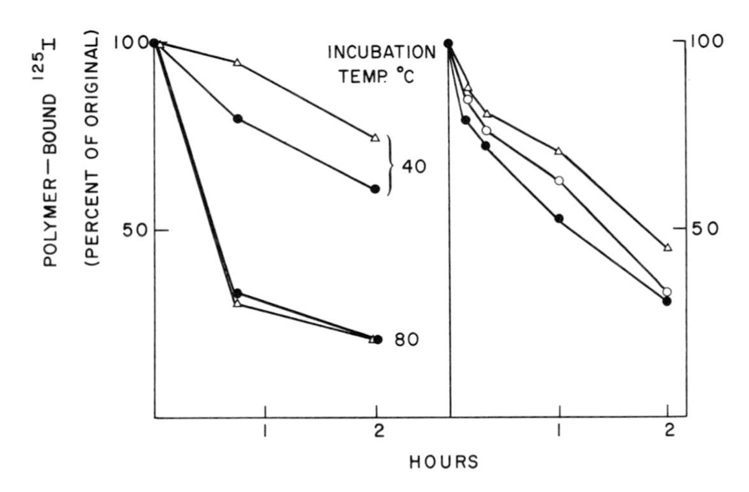

FIG. 5. Differential susceptibility of RNA iodonucleotides to bisulfite-induced deiodination. Human fibroblast RNA was labeled with ^{125}I at 40°, 60°, and 80°C in reactions buffered at pH 6, and also modified in the time and amount of isotope present in order to yield products of similar final specific activity. The products of reaction at 80°C (triangles), 60°C (open circles), or 40°C (solid circles) were stabilized and repurified by sedimentation through cesium sulfate. Aliquots of the high-molecular-weight RNA were incubated with either 0.1 *M* sodium bisulfite at 40°C or 80°C (left), or with 0.5 *M* sodium bisulfite at 40°C (right). Deiodination was interrupted, and the fraction of isotope remaining bound to RNA determined as described in the legend for Fig. 4. Note that incubation with bisulfite under denaturing conditions eliminated the differential susceptibility of the iodine-containing nucleotides. Determination of the labeled iodonucleotide makeup of the RNAs showed identical composition (< 5% deviation in the proportion of isotope present in the several components).

D. *In Vitro* Synthesis of RNA

Observation of the *in vitro* synthesis of high-molecular-weight RNA with nuclei or with purified eukaryote polymerase has been an elusive goal. This is probably attributable both to nucleases and inadequate copying of physiological conditions. In addition, the ability to observe nascent high-molecular-weight chains is limited by the specific activity of the nucleotide precursor. We have reported that either ^{125}I- or ^{131}I-labeled cytidine nucleotide triphosphate can be prepared with specific activity more than 70 times greater than that achieved in tritium-labeled precursors (Scherberg and Refetoff, 1974b). Iodinated nucleotides are suitable for polymerase reactions. Consequently, a product labeled to higher specific activity is obtained, and therefore a smaller amount of the product can be observed. We have found that inclu-

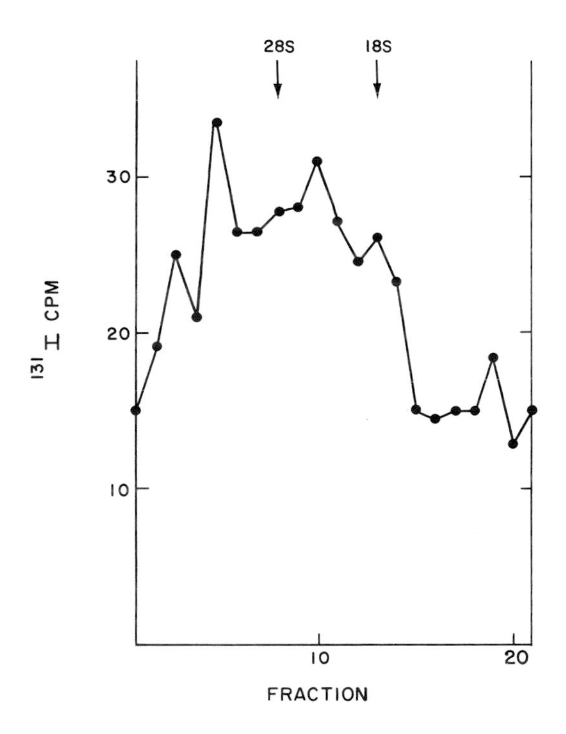

FIG. 6. Sucrose profile of RNA labeled by incorporation of iodocytidylate *in vitro*. Human fibroblasts were lysed with 0.005% NP40. ATP, GTP, and UTP at a final concentration of 0.25 m*M* and carrier-free CTP-^{131}I (5.5×10^7 cpm) were added to the lysate, and the suspension incubated at 37°C for 5 minutes. The reaction was interrupted by the addition of 9 ml of cold 0.1 *M* potassium chloride, and nuclei were collected at low speed. The pellet was resuspended and heated in guanidinium disrupting buffer; high-molecular-weight RNA was separated and analyzed as described in the legend for Fig. 2.

sion of the high-specific-activity precursor in an *in vitro* reaction with nuclei, followed by nuclear disruption and sedimentation through cesium sulfate, results in separation of a small amount of high-molecular-weight product (Fig. 6). It has not been determined whether the high-molecular-weight RNA is representative of all nuclei or selected nuclei, or is derived from a few intact cells with permeable membranes. Whatever the origin of the long-chain products, the combined techniques employing a high-specific-activity precursor and selective isolation result in improved discrimination for the study of physiological processes. Accordingly, these methods may contribute to investigations of the effect of perturbing factors, such as hormones or nuclear proteins, on transcription.

Acknowledgment

The authors are grateful to Mrs. Yolanda Richmond for preparation of this manuscript.

References

Barrett, H. W., and West, R. A. (1956). *J. Amer. Chem. Soc.* **78**, 1612–1615.

Brawerman, G. (1973). *In* "Methods in Cell Biology" (D. M. Prescott, ed.), Vol. 7, pp. 2–22. Academic Press, New York.

Commerford, S. L. (1971). *Biochemistry* **10**, 1993–2000.

Dube, S. K. (1973). *Nature (London)* **246**, 483.

Eidenoff, M. L., Cheong, L., Gambetta-Durpide, E., Benna, R. S., and Ellison, R. (1959). *Nature (London)* **183**, 1686–1687.

Getz, M. J., Altenburg, L. C., and Saunders, G. F. (1972). *Biochim. Biophys. Acta* **287**, 485–494.

Goddard, J. P., and Schulman, L. H. (1972). *J. Biol. Chem.* **247**, 3864–3867.

Goz, B., and Prusoff, W. H. (1968). *J. Biol. Chem.* **243**, 4750–4756.

Hayatsu, H., Wataya, Y., Kai, K., and Iida, S. (1970). *Biochemistry* **9**, 2852–2865.

Hsu, Y.-C. (1964). *Nature (London)* **203**, 152–153.

Hughes, W. L., Commerford, S. L., Gitlin, D., Krueger, R., Schultz, B., Shah, V., and Reilly, P. (1964). *Fed. Proc., Fed. Amer. Soc. Biol.* **23**, 640–648.

Kacian, D. L., Spiegelman, S., Bank, H., Terada, M., Metafore, S., Dow, L., and Marks, P. A. (1972). *Nature (London), New Biol.* **235**, 167–169.

Kai, K., Wataya, Y., and Hayatsu, H. (1971). *J. Amer. Chem. Soc.* **93**, 2089–2090.

Kates, J. (1973). *In* "Methods in Cell Biology" (D. M. Prescott, ed.), Vol. 7, pp. 53–65. Academic Press, New York.

Massoulié, J., Michelson, A. M., and Pochon, F. (1966). *Biochim. Biophys. Acta* **114**, 16–26.

Melli, M., and Pemberton, R. E. (1972). *Nature (London), New Biol.* **236**, 172–174.

Muramatsu, M. (1973). *In* "Methods in Cell Biology" (D. M. Prescott, ed.), Vol. 7, pp. 24–51. Academic Press, New York.

Pardue, M. L., and Gall, J. G. (1970). *Science* **168**, 1356–1358.

Peacock, A. C., and Dingman, C. W. (1967). *Biochemistry* **6**, 1818–1826.

Prensky, W., Stefferson, D. M., and Hughes, W. L. (1973). *Proc. Nat. Acad. Sci. U.S.* **70**, 1860–1864.

Prusoff, W. H., Bakhle, Y. S., and Sekely, L. (1965). *Ann. N.Y. Acad. Sci.* **130**, 135–150.
Robertson, H. D., Dickson, E., Model, P., and Prensky, W. (1973). *Proc. Nat. Acad. Sci. U.S.* **70**, 3260–3264.
Sander, E. G., and Deyrup, C. L. (1972). *Arch. Biochem. Biophys.* **150**, 600–605.
Scherberg, N., and Refetoff, S. (1973). *Nature (London)* **242**, 142–145.
Scherberg, N., and Refetoff, S. (1974a). *J. Biol. Chem.* **249**, 2143–2150.
Scherberg, N., and Refetoff, S. (1974b). *Biochim. Biophys. Acta* **340**, 446–451.
Schmidt, F. J., Omilianowski, D. R., and Bock, R. M. (1973). *Biochemistry* **12**, 4980–4983.
Sela, M., Anfinsen, L. B., Harrington, W. F. (1957), *Biochim. Biophys. Acta* **26**, 506–512.
Shapiro, R., Cohen, B., and Servis, R. (1970). *Nature (London)* **227**, 1047–1048.
Tereba, A., and McCarthy, B. J. (1973). *Biochemistry* **12**, 4675.
Verma, I. M., Temple, G. F., Fan, H., and Baltimore, D. (1972). *Nature (London), New Biol.* **235**, 163–167.

Chapter 20

Autoradiographic Analysis of Tritium on Polyacrylamide Gel

PAOLA PIERANDREI AMALDI

Laboratorio di Biologia Cellulare,
Consiglio Nazionale delle Ricerche,
Roma, Italy

I. Introduction 361
II. Procedure 362
References 364

I. Introduction

After electrophoresis on polyacrylamide gels, the position of compounds labeled with high-energy emitters of radiation is determined very easily by autoradiography. In fact, by properly increasing the length of the exposure time, one can detect minimal amounts of radioactivity which would not be sufficient for standard counting techniques. When working with weak emitters, such as carbon-14, it is necessary to dry the gel before autoradiography. The extreme weakness of emissions of tritium makes almost impossible its detection by autoradiography of most gels (Gordon, 1971). In fact, the thickness of the gel results in self-absorption which does not allow the emission to reach the photographic emulsion (Rogers, 1967).

However, the use of tritium as a labeling isotope is convenient when dealing with small amounts of material and/or with low efficiency of incorporation, because of the high specific activity of the ^{3}H-labeled compounds commercially available.

The technique described here combines electrophoresis on polyacrylamide gel with autoradiography to analyze small amounts of tritium-labeled RNA (Amaldi, 1972). We have increased the sensitivity of the method by working with thin gel slabs dried to a very reduced thickness so as to minimize self-absorption of tritium emissions.

II. Procedure

Electrophoresis of RNA is essentially as described by Loening (1967). Acrylamide and bisacrylamide are used at 2.2 and 0.1% final concentration, respectively. Gel and electrophoresis buffer is 0.02 *M* sodium acetate, 2 m*M* sodium EDTA, and 0.04 *M* tris (pH 7.8); 33 μl of *N*, *N*, *N*′, *N*′-tetramethylethylenediamine and 0.33 ml of 10% ammonium persulfate per gram of acrylamide are added. The solution is rapidly mixed and poured into a polymerization chamber, avoiding air bubbles. Polymerization occurs in about 15–20 minutes.

The polymerization chamber (Fig. 1) consists of two microscope glass slides having very regular surfaces. Slides are previously washed in cleaning acid, rinsed in running water, distilled water, and ethanol, and stored in a dust-free place. One of the two slides is covered with cellophane or Saran wrap, which adheres firmly to the glass surface; the other glass slide has an incomplete frame of 100-μm-thick polyethylene adhesive tape (Fig. 1a). The cellophane-covered slide is placed to face a slide carrying the frame, and by finger pressure the edges of the resulting chamber are made stick to each other. To prevent leakage melted paraffin is brushed on the chamber edges, except the one into which the fluid mixture is to be poured (Fig. 1b). The mixture is gently added through the thin slit between the two slides by a narrow-tipped pipette (Fig. 1c). After polymerization the paraffin is com-

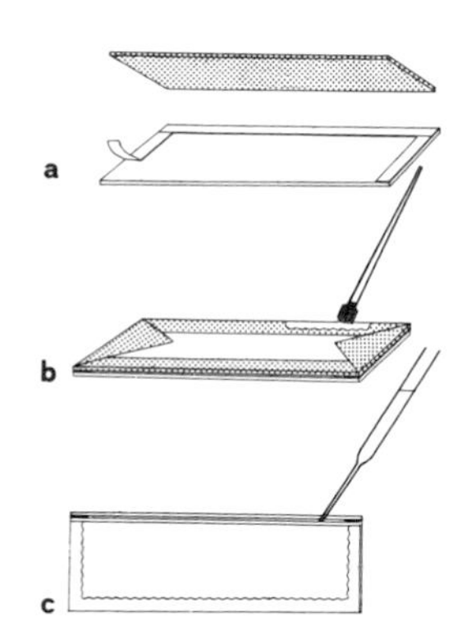

FIG. 1. Polymerization chamber for the preparation of thin slabs of polyacrylamide gel. From Amaldi (1972).

pletely scraped off with the aid of a razor blade; the first slide is removed, leaving the cellophane still adhering to the gel. The cellophane is then carefully peeled off, the frame is removed, and the gel slab lying on the glass is ready for electrophoresis. The surface of the gel is smooth and regular.

The gels are transferred to an apparatus for horizontal electrophoresis kept at 4°C. Two strips of Whatman filter paper are interposed between the gel and the electrophoresis buffer to give electrical contact. The filter paper strips are previously soaked in 5% acetic acid for 1 hour, and then washed in several changes of water and buffer. After a prerun of 20 minutes to remove catalysts and impurities, the sample, containing 5% sucrose, is applied on the surface of the gel in a volume of 0.1 μl by means of a 1-μl Hamilton syringe. A tiny drop of tracking dye in 5% sucrose is applied just above the sample at the starting line. When necessary, a smaller volume can be applied after absorption on a tiny piece of filter paper placed on the gel. Moreover, the radioactivity contained in a larger volume, e.g., 5–10 μl, can be concentrated by evaporation under vacuum on a small piece of filter paper.

Good separation of RNA fractions is obtained in 30 minutes with a current of 10 m*A* and a voltage of 100 V, running 10 gels at a time.

The gels are first dried at room temperature overnight, and then in a warm oven (40°C) for a few hours. The dry gels are so thin that they are visible just as a light opacity on the slides; their thickness is approximately 3–5 μm. Distortion of bands does not occur with drying.

The slides, now ready for autoradiography, are covered with Kodak Ar-10 stripping film, Kodak NTB-2 emulsion, or AR-50 Kodak stripping film, the last giving the most intense blackening. An intensifier spray for autoradiography, spread over the gel before covering with the photographic film, does not improve the blackening. Autoradiographs are stored in dark boxes with a dehydrating agent at −20°C. After an appropriate exposure time, the film-covered slides are processed with Kodak D-19 developer and fixed with Kodak Unifix.

Tiny black spots are visible on the processed film. In order to analyze the pattern of the labeled RNA, the slides can be scanned by means of a microdensitometer such as the Joyce-Loebl. Figure 2 shows an example of electrophoresis obtained with RNA prepared from Chinese hamster cells labeled for 24 hours. A total of about 300 dpm of radioactive RNA was utilized, and the exposure time was 2 months.

It is evident that the method is quite sensitive, allowing the detection of very small amounts of radioactivity.

The same method has been used with satisfactory results to separate proteins on polyacrylamide gels in a sodium dodecyl sulfate–sodium phosphate buffer system, according to Weber and Osborne (1969).

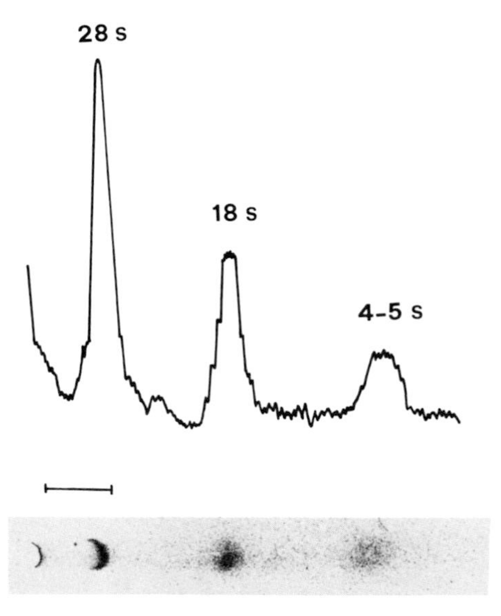

FIG. 2. Densitometric pattern of tritium-labeled RNA extracted by phenol procedure from Chinese hamster cells cultured for 24 hours in the presence of uridine-^{3}H. RNA (0.1 μl) was layered on the gel and run for 30 minutes. The bar represents 2 mm. From Amaldi (1972).

REFERENCES

Amaldi, P. (1972). *Anal. Biochem.* **50**, 439–441.

Gordon, A. M. (1971). "Electrophoresis of proteins in polyacrylamide." North-Holland, Publ. Amsterdam.

Loening, U. E. (1967). *Biochem. J.* **102**, 251–257.

Rogers, A. W. (1967). "Techniques of Autoradiography." Elsevier, Amsterdam.

Weber, K., and Osborne, M. (1969). *J. Biol. Chem.* **244**, 4406–4412.

Chapter 21

A Radioautographic Method for Cell Affinity Labeling with Estrogens and Catecholamines

JOSÉ URIEL

Institut de Recherches Scientifiques sur le Cancer, Villejuif, France

I. Introduction . . . 365
II. Procedure . . . 366
A. Preparation of Specimen . . . 366
B. Cell Radiolabeling . . . 368
C. Versatility of the Method . . . 368
III. Results . . . 369
A. Radioactivity Patterns . . . 369
B. Cell Morphology . . . 370
IV. Concluding Remarks . . . 373
References . . . 374

I. Introduction

The localization of diffusible radiochemicals bound to cell constituents by reversible, noncovalent linkages has been difficult to demonstrate using standard radioautographic procedures. After *in vivo* injection or incubation of tissues with radiochemicals, radioactive material is often lost, or diffusion and translocation occur and reduce the accuracy of the results. Several solutions have been proposed to deal with the technical difficulties of this method (Roth and Stumpf, 1969).

The *in vitro* method described here is based on the high association constants characteristic of some interactions between hormones and their specific cell ligands and receptors. Fixed cell suspensions or histological sections of tissues, mounted on glass slides, are incubated in solutions of tritiated steroids or catecholamines and then washed under experimental conditions which ensure the removal of free or lightly bound hormones, while molecules linked to cell constituents by high-affinity forces are preserved (Uriel *et al.*, 1973a, 1974c)

This radioautographic method was developed primarily for the intracellular localization of rat α-foetoprotein (αFP) on the basis of its estrogen-binding properties demonstrated by biochemical (Nunez *et al.*, 1971; Aussel *et al.*, 1973) and immunochemical (Uriel *et al.*, 1972a) methods. Since the first article reporting the radioautographic localization of αFP in the liver of rat fetuses and neonates (Uriel *et al.*, 1972b), other normal and neoplastic tissues of rat and man have been subjected to cell affinity labeling with tritiated estrogens (Uriel *et al.*, 1973b, 1974a,b) and, more recently, with tritiated catecholamines (Uriel *et al.*, 1974c). We present here a detailed description of the method and some results illustrating the most characteristic patterns of radioactivity obtained with both types of hormones.

II. Procedure

A. Preparation of Specimen

1. Tissue Samples

a. Paraffin Sections. Tissue samples from laboratory animals, killed by decapitation or exsanguination under light ether anesthesia, are quickly removed, cut in small pieces (less than 0.4 cm^3), and fixed overnight at 4°C in absolute ethanol or in either 4–10% formaldehyde or 3% glutaraldehyde in neutral buffers (sodium cacodylate, sodium phosphate, etc.). Specimens fixed in Formol or glutaraldehyde are then washed in several baths of physiological buffers and dehydrated with graded alcohols by hand or automatic processing according to standard techniques. All fixed and dehydrated block tissues are cleared, embedded in paraffin, and stored at room temperature until used. Sections of 4–6 μm are cut, mounted on glass slides (not acid-cleaned) which have been subbed with Mayer's glycerin albumin, deparaffined, and incubated with radiochemicals within 48 hours. Human or animal biopsic specimens are processed in the same manner.

The property of fixation of estrogens and catecholamines seems to be preserved indefinitely when the tissues are embedded in paraffin. However, histological cuts, embedded or not, gradually lose this property in a relatively short time (1 or 2 weeks). (No significant decrease in affinity labeling has been observed in sections cut from block tissues embedded in paraffin for almost 2 years.)

b. Frozen Sections. Alternatively, tissue samples can be frozen by immersion in isopentane cooled with liquid nitrogen, sectioned at 4–6 μm on a cryostat, and mounted on glass slides (not acid-cleaned). After drying in air the sections are immersed one to several hours in absolute ethanol and radiolabeled as soon as possible, or not later than 48 hours after processing. A short fixation (30 minutes) in 4% buffered formaldehyde, followed by rinsing quickly in distilled water can replace the alcohol treatment.

2. Cell Suspensions

Suspensions of cells mechanically or enzymatically dissociated from organs, tissue slices, or cultures obtained by conventional techniques can effectively be used for affinity radiolabeling studies.

Before mounting the cells are centrifuged and the cell pellet resuspended in equal volume of Hanks' medium containing 1% albumin. Samples are withdrawn from the mixture and smeared or, preferably, layered on glass slides with the aid of a Shandon cytocentrifuge. The air-dried slides are immersed for cell fixation in absolute ethanol or in buffered 4% formaldehyde as indicated above (Section II,A, 1b) and radiolabeled within 48 hours.

3. Choice of Fixatives

Fixation of the cells or tissues is a preliminary and critical step in our radiolabeling method. The chosen fixative must not alter the binding properties of the cell constituent(s) involved in the interaction. Experience has shown that the treatment of cells or tissues with acid solutions or strong chelating agents (picrate and chromate salts, heavy metals, etc.) destroys their affinity properties for tritiated estrogens and catecholamines. Thus common histological fixatives such as Carnoy's, Bouin's, or Zenker's fluids must be avoided. Prolonged fixation with formaldehyde or glutaraldehyde also prevents the binding of both types of hormones.

Absolute ethanol, although not a very good tissue preservative, appears to be the best fixative for affinity labeling with tritiated estrogens and catecholamines. The quality and the intensity of radiolabeling seem unaffected by prolonged periods of fixation (several days), which obviously is a great advantage. In addition, alcohol contributes by solvent extraction to the loss of bound endogenous hormones, thus resulting in an increase in

III. Results

A. Radioactivity Patterns

As a rule, the radioactivity appears concentrated in isolated cells or in groups of cells against a moderate or low background of silver grains randomly disseminated over tissue structures (Figs. 1 and 2). With radiolabeled estrogens, the localization is preferentially cytoplasmic, although localizations limited to the cytoplasmic membrane or to the nucleus are also observed. However, intermediate patterns are the most frequent (Fig. 3).

Some extracellular diffusion of radioactivity is apparent when labeling is carried out with catecholamines (Fig. 4b). The artefact seems to appear during and/or after the incubation, but its origin and its exact nature have not as yet been elucidated (Uriel *et al.*, 1974c).

The intensity of the radiolabeling with compounds of similar specific radioactivity varies considerably according to the hormones assayed. The decreasing order of intensity observed for steroids is: estrone > estradiol >

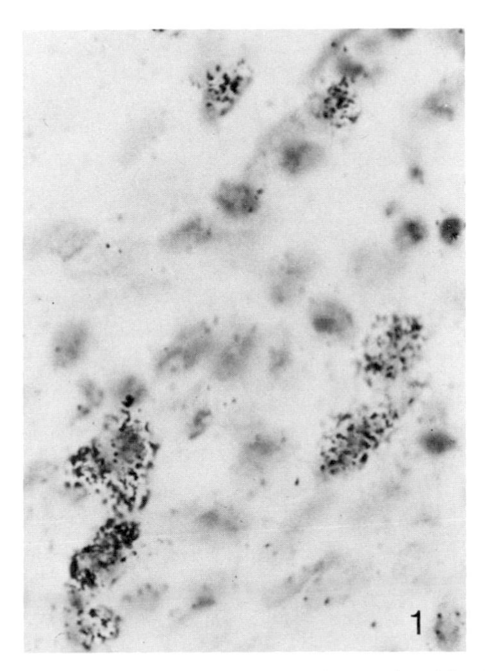

FIG. 1. Radioautograph of fetal liver of baboon (2 months old). Affinity labeling with estradiol-^{3}H (2 μCi/ml). Hematoxylin–eosin. $\times$ 600.

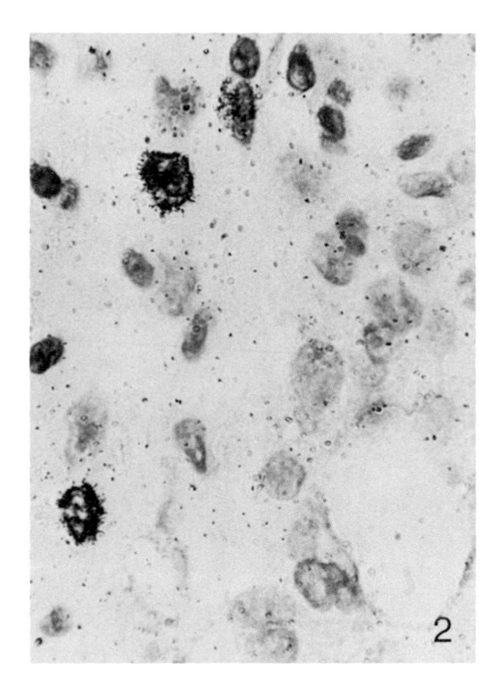

FIG. 2. Affinity labeling with estradiol-^{3}H (2 μCi/ml) of three transitional liver cells proximate to a zone of hyperplastic hepotocytes. Biopsy from a patient with primary hepatoma. ×600. [Reproduced with permission from Uriel *et al.* (1974a).]

estriol; and for catecholamines: noradrenaline > normetanephine > dopamine > dopa. Rat and human tissues labeled by the hormones listed above do not show binding affinity for either adrenaline or the following steroids: testosterone, progesterone, cortisone, and dexamethasone (Uriel *et al.*, 1973b, 1974c).

B. Cell Morphology

The exact morphology of the cells radiolabeled by either estrogens or catecholamines in the different tissues already examined, as well as their identification with known cell phenotypes, has not been clearly established.

The estrogen-binding cells of fetal and neonatal rat liver and of fetal livers of primates seem to represent a transitional phenotype intermediate between small hepatocytes and bile duct cells (Figs. 1 and 4). Transitional liver cells of this type with identical estrophilic properties have also been found in rat and human primary hepatomas (Uriel *et al.*, 1973b, 1974a) and in rat cholangiocarcinomas (Figs. 2 and 5).

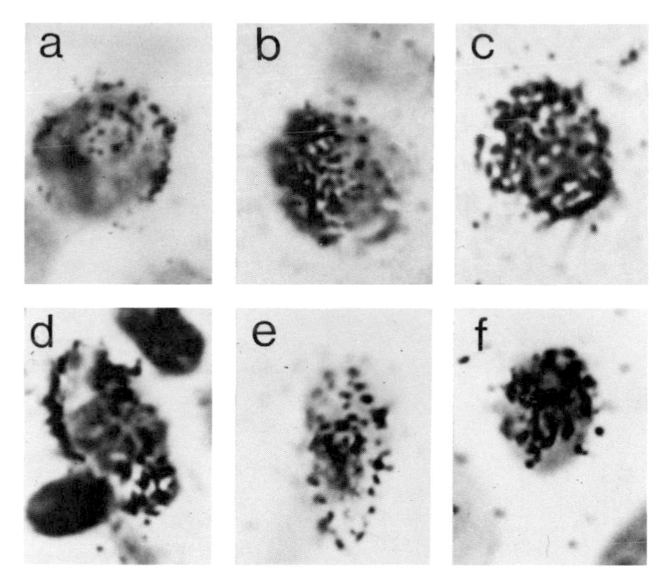

FIG. 3. Intracellular patterns of radioactivity after labeling with tritiated estrogens. (a–c) Rat liver; (d) human liver; (e) and (f) rat uterus. Hematoxylin–eosin.

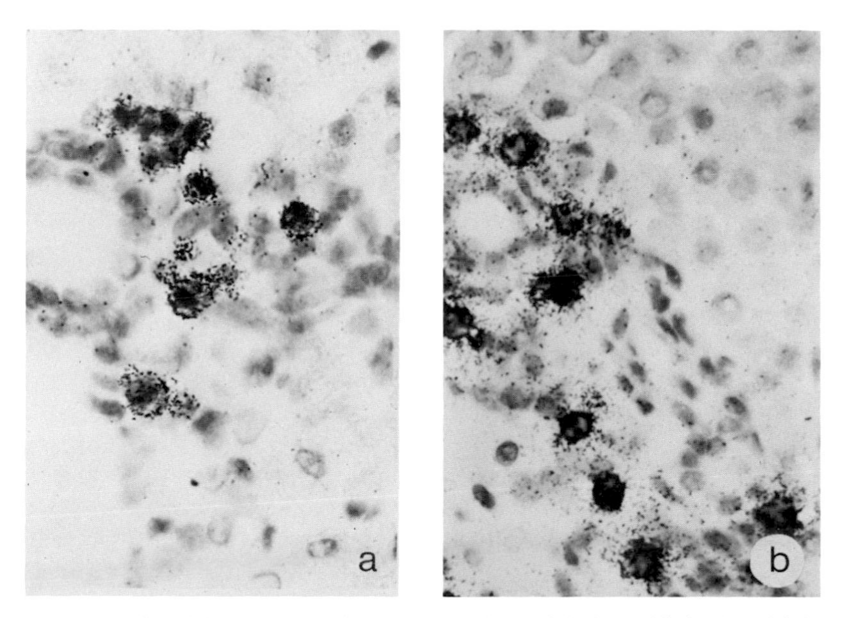

FIG. 4. Radioactivity patterns of neonatal rat livers (10 days old) in the vicinity of portal areas. Labeling after incubation with (a) estrone-^{3}H and (b) dopamine-^{3}H. Hematoxylin–eosin. $\times 600$.

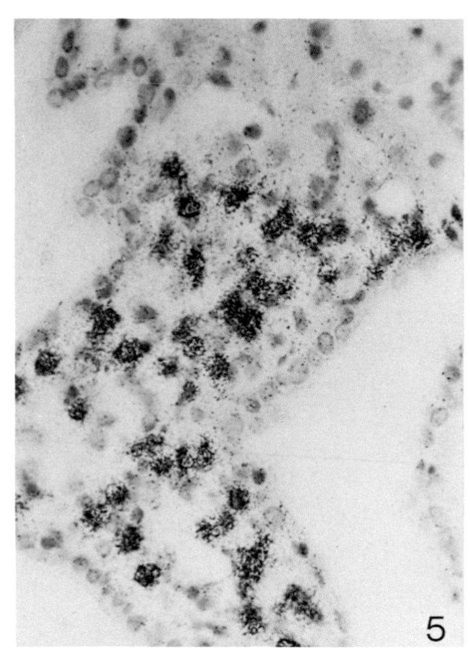

FIG. 5. Radioautograph of a zone of cholangiocarcinoma in a rat liver. Labeling with noradrenaline-^{3}H (1 μCi/ml). Hematoxylin–eosin. ×600.

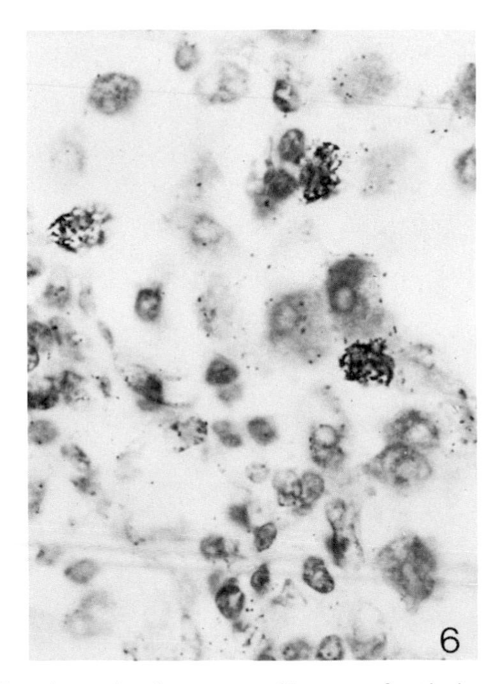

FIG. 6. Area of seminoma in a human teratoblastoma of testicular origin. Three tumoral cells labeled with estrone-^{3}H (2 μCi/ml). Hematoxylin–eosin. ×600.

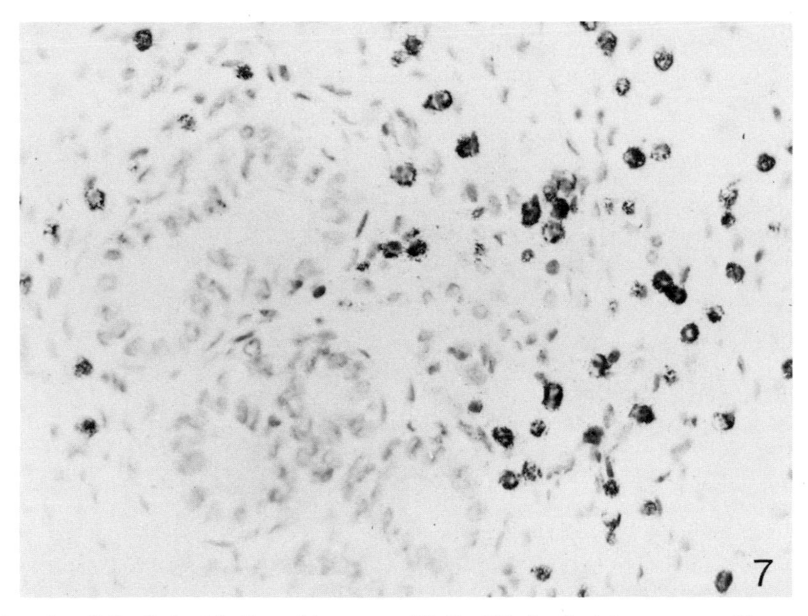

FIG. 7. Cell affinity labeling with estrone-^{3}H (2 μCi/ml) of adult rat uterus. Numerous radiolabeled eosinophylic cells in the lamina propria proximate to epithelial glands. Hematoxylin–eosin. ×163.

In fetal, neonatal (Fig. 4), and neoplastic adult livers, cells with similar morphology bind estrogens and catecholamines. Whether or not the same cell is able to bind both types of hormones remains to be determined (Uriel *et al.*, 1974c). Other estrogen-binding cells of either embryonal or "lymphoid" appearance (Fig. 6) have been observed in human teratoblastomas of testicular or ovarian origin (Uriel *et al.*, 1974b).

Cells with eosinophilic morphology have also been radiolabeled with estrogens (Uriel *et al.*, 1972c) and catecholamines (Uriel *et al.*, 1974c) in the lamina propria around glands and vessels of histological sections of adult rat uterus (Fig. 7). They are probably identical to the "uterine eosinophils" whose estrogen binding properties were demonstrated by Tchernitchin (1967) after *in vitro* incubation of unfixed frozen sections of rat uterus.

IV. Concluding Remarks

The method of cell affinity labeling, relatively simple from a technical point of view and easily modified for diverse applications, provides a wide variety of possibilities and advantages.

The ligands or the cellular receptors for estrogens and catecholamines seem to resist, without damage, the action of certain fixatives or organic solvents utilized during histological processing. Since many of the organic solvents are for the most part denaturing agents of several enzymatic systems, they may help to inactivate the intracellular enzymes implicated in the catabolism of estrogens and catecholamines and thus eliminate possible interference with affinity binding.

This method may also be extended to studies of cell labeling with diffusible radiochemicals other than steroids and catecholamines.

References

Aussel, C., Uriel, J., and Mercier-Bodard, C. (1973). *Biochimie* **55**, 1431–1437.

Bogoroch, R. (1972). In "Autoradiography for Biologists" (P. B. Gahan, ed.), pp. 65–94. Academic Press, New York.

Nunez, E., Engelman, F., Benassayag, C., and Jayle, M. F. (1971). *C. R. Acad. Sci.* **273**, 831–834.

Roth, L. J., and Stumpf, W. E. (1969). "Autoradiography of Diffusible Substances." Academic Press, New York.

Tchernitchin, V. (1967). *Steroids* **10**, 661–666.

Uriel, J., de Néchaud, B., and Dupiers, M. (1972a). *Biochem. Biophys. Res. Commun.* **42**, 1175–1180.

Uriel, J., Rubio, N., and Loisillier, F. (1972b). *C. R. Acad. Sci.* **275**, 1941–1942.

Uriel, J., Rubio, N., Aussel, C., Bellanger, F., and Loisillier, F. (1972c). *C. R. Acad. Sci.* **275**, 2715–2716.

Uriel, J., Rubio, N., Aussel, C., Bellanger, F., and Loisillier, F. (1973a). *Exp. Cell Res.* **80**, 449–453.

Uriel, J., Aussel, C., Bouillon, D., de Néchaud, B., and Loisillier, F. (1973b). *Nature (London), New Biol.* **244**, 190–192.

Uriel, J., Loisillier, F., Szekely, A. M., and Bismuth, M. (1974a). *Biomed. Exp.* **21**, 21–25.

Uriel, J., Loisillier, F., and Aussel, C. (1974b). *Nouv. Presse Med.* **3**, 1213–1215.

Uriel, J., Bouillon, D., Aussel, C., and Loisillier, F. (1974c). *C. R. Acad. Sci.* **279**, 575–577.

Subject Index

A

Acridine orange, 33, 56, 58, 61
Adenovirus type 2, purification of, 101
Aedes albopictus, 185
  cells of, 186
Aedes cells
  fractionation of, 189
  genome size of, 189
  growth rate of, 188
  karyotype of, 188
  medium for, 187
  mitochondrial fraction of, 190
  mRNA from, 190
  polysomes from, 190
  renaturation kinetics of DNA from, 189
  rRNA of, 192
  stocks of, 193
Alizarine navy, 56
Amino acid pool, intracellular, 245
δ-Aminolevulinate dehydratase
  half-life of, 241
  rates of synthesis of, 239
Amniotic fluid cell cultures, 270
Antimycoplasma drugs, 271
Arginine
  content in cell culture media, 268
  determination, 268
Ascites cells, 174
  preparation of, 178
Autoradiographic analysis, of tritium on polyacrylamide gel, 361
Autoradiographic techniques, 320
Autoradiographs
  comparison of 5S RNA-^{125}I, 339
  of ^{125}I- and ^{3}H-labeled cRNA, 338
Autoradiography
  design of experiments using, 293
  high-resolution, 293
  stability of radiochemical tracers in, 291
  staining after, 8
Auxotrophic cells, 214
Auxotrophic mutants, detection of, 216
Auxotrophs, glutamine-dependent, 211
Azure A, 56

B

Balbiani ring, 25, 27, 45
  micromanipulatory isolation of, 26
Baycovin solution, 177
Blood cells, separation of, 98
Blutene chloride, 57
Brilliant chrome, 56
Bromobenzene, 90

C

Carbon-14, physical properties of, 294
Carollia perspicillata, 9
Catacholamine granules, 101
Cell affinity labeling, 365
Cell cycle
  automated analysis of, 157
  calibrating, 168
Cell cycle analyzer, 159–160
Cell dispersion, 95
Cell fusion, 148
Cell generations, formula to determine, 222
Cell kinetic parameters, of ascites cells, 175
Cell separation, 93
Chironomus tentans, salivary glands of, 18
Chlorine-36, physical properties of, 294
Chromomeres, 20
Cilia
  isolation of, 111, 115
  phospholipid composition of, 131
  subfractions of, 115–116
Ciliary membrane, 117

Cleland's reagent, 101
Collagenase, 95
Colloidal silica, 96
  as a gradient medium, 88
  preparation of, 88
  properties of, 87
  self-forming density gradient, 91
  toxic action of, 88
  toxic activity of, 94
Collodial silica density gradient, 101
Colloidal silica density gradient centrifugation, 85
Colloidal silica gradients, 92, 97–98

D

Density gradients
  collection from, 96
  with angle-head rotor, 93
Dextran T40, 86
Dextran T70 gradient, 103
Diethylpyrocarbonate, 66, 177
DNA
  retention on microfilters, 39
  preparation on microfilters, 38
DNA content, distribution of per cell, 170
DNA–DNA hybrids, 6
DNA polymerases, 2
DNA–RNA hybrids, thermal stability of, 336
DNase solution, 39
Drosophila
  5S RNA cistrons of, 339
  rRNA of, 206
Drosophila cell lines, 196
  suspension culture, 196
Drosophila cells
  fractionation of, 203
  karyotype in culture, 200
  macromolecular components of, 201
  RNA extracted from, 204
Dyes, binding of, 54

E

Electrophoresis, 49, 59
Electrophoretic chamber, 32
Enzymes, rate of synthesis of, 243
Eppendorf pipette, 142
Equine abortion virus, 101
Erythrocin B, 94

F

Ficoll, 86
Fluctuation test, 221
Fluorescein, 65
Fluorescence detection, 64

G

Gallocyanine, 56
Genome size, of *Drosophila*, 196
Gentamycin, 271
Giemsa stain, 8, 10
Gradient(s)
  continuous, 92
  discontinuous, 91
  S-shaped, 92–93
  self-forming, 92
Gradient formation, 91
Gradient-forming device, 176
Gradient medium, 93, 97–98
  mixing, 90
  preparation of, 88
Gradient-mixing device, 92, 98
Guldberg-Waage law, 51

H

HAT medium, 215
  definition of, 219
HeLa, growth rate, 94
Hemolymph, in *Chironomas*, 21
Herpes simplex virus, 101
Histone mRNA, 11
Human lung, diploid cells of, 211
Hybrid cells
  analyzing, 152
  cloning, 152
  isolation of, 151
Hybridization, 351
  ^{125}I in molecular, 325
  in situ, 1, 337
  sensitivity of, 41
Hybridization procedure, cytological, 5
Hydrogen transfer reactions, biological, 319
Hypoxanthine-guanine-phosphoribosyl transferase, 215

I

Infectious bovine rhinotracheitis virus, 101
Insect cell lines, 186
Insect cells
  in suspension culture, 185
  methods with, 185, 195

Insect Ringer's, 3, 8, 137
Iodination methodology, 327
Iodination reaction mixture, 328
Iodine-125, physical properties of, 294
Iodine-containing nucleosides, 352
^{125}Iodine-labeled nucleic acids, applications of, 335
Iodine-labeled RNAs, 353
applications of, 351

K

Kinetosomes, isolation of, 120

L

Lactate dehydrogenase, turnover of the isoenzymes, 256
Lactoperoxidose, 255
Linomycin, 286
Lozeman homogenizer, 118
Ludox, 87, 97–98
Ludox TM, 101
Lysosome-gradient medium, 99
Lysosomes, 130
isolation of, 109
separation of, 99–100

M

Macronuclear envelope, 126
Macronuclei, isolation of, 124
Mammalian cells, mutagenesis in, 209
Mast cells, separation of, 98
Membrane, lipid composition of, 130
Membrane components
isolation methods for, 108
isolation of from *Tetrahymena*, 105
Membrane proteins, 82
Mercaptopurine, mutants resistant to, 213
Metabolic cooperation of cells, 212
Methocel, 93
Methyl green, 57
Methylene blue, 56, 58, 61
Methylene blue staining method, 52
Microelectrophoresis, 29–32, 45
Microfilters, 38
DNA-loaded, 42
DNA on, 40
hybridization on, 40
Microhybridization, 34–35, 41, 45
extraction of RNA for, 36
of nucleolar RNA, 42
Micronuclei, isolation of, 125
Microsomes
isolation of, 108, 111
phospholipid composition of, 131
Microsurgical cell fusion, 153
Microsurgical chamber, 149
Microtubules
isolation of, 116
preparation of, 117
Mitochondria
isolation of, 108–109, 111, 121–122
phospholipid composition of, 131
separation of, 100
Mitotic cells
yield and initial synchrony of, 168
yield of, 164–165
yield of WI-38, 166
Mitotic stages, distribution of at selection, 169
Moist chamber, 10
Monocytes, 98
Muntiacus muntjak, 9
Mutagen dose, 227, 230
Mutagenesis
conditions for, 210
induced, 211, 227
proof of spontaneous, 224
selective conditions for, 212
spontaneous, 211, 220
Mutant cell survival, 223
Mutant recovery
effect of cell population density on, 212
optimal conditions for, 213
Mutants
detection of temperature-sensitive (*ts*), 218
drug-resistant, 215
nutritional, 218
Mutation induction, 232
Mutation rate, spontaneous, 221, 226, 228
Mutations
6MP resistance, 231
for resistance to purine and pyrmidine base analogs, 215
forward, 215
kinetics of, 230
reverse, 219
Mycoplasma
and bacterial RNAs, 266

nutrient requirements, 262
RNAs, 265
Mycoplasma arginini, 282, 288
Mycoplasma contamination
biochemical technique for the detection of, 277
detection of, 261
microbiological techniques for the detection of, 268
Mycoplasma detection, sensitivity of techniques, 272
Mycoplasma hyorhinis, 274, 282, 288
Mycoplasma-infected cells, uracil incorporation into, 278
Mycoplasma-infected WI-38 cells, therapy on, 286

N

Nuclear fractions, 77
Nuclear isolation
large-scale, 137
small-scale, 137
Nuclear membranes, 81, 127
densities of, 82
isolation of, 124, 126
phospholipid composition of, 131
purification of, 79
Nuclear purification, in sucrose gradients, 140
Nuclear sap, 25
Nuclear subfractions, 78, 81
Nuclease inhibitors, 74
Nuclei
chemical composition of, 78
cytological analysis of isolated, 141
fractionation of, 69, 77
isolation of, 69, 124, 136
medium for isolation of, 74, 82
sonicated, 75, 82
subfractionated, 73
Nucleic acid hybridization, 1
Nucleoli, isolation of, 136, 140
Nucleoside phosphorylases, 307
Nucleotide sequence analysis, 340
Nyacol, 87
Nylon screen, 137

O

Oil chamber, 24
Oral apparatus, isolation of, 188
Osmometer, 89

P

Paramecium caudatum, DNA labeling of, 318
Paris green, 57
Pellicle ghosts, 112
Pellicles
isolation of, 111, 113, 119
phospholipid composition of, 131
Peroxisomes, isolation of, 109, 130
Phosphorus-32, physical properties of, 294
Phototrophs, glutamine, 214
Plating efficiency, 214
Poliovirus type, purification of, 101
Polyacrylamide-agarose gels, preparation of, 30
Polymorphonuclear cells, 98
Polyribosomes, isolation of, 128
Polytene chromosomes, 14, 20, 45
Polytene nuclei
DNA synthesis in isolated, 143
incubation of, 141
isolation of, 135, 139
manipulation of, 135, 141
mass isolation of, 136
nonaqueous preparation of, 138
of *Chironomus*, 145
RNA synthesis in isolated, 142
Polytron 10-ST, 121
Protease inhibitors, 74–75
Protein(s)
comparison of the half-lives of, 246
half-lives of, 241, 244
rate constants of degradation for, 253, 257
rate of degradation of, 241
rate of synthesis of, 238
Protein degradation
half-life of, 237
in cultured cells, 256
in regenerating liver, 248
measurement of, 249
mechanism of, 255
rate of, 243
relative rates of, 252
Protein turnover, 243
measurement of, 235
Prototrophs, selective death of, 216
Psammechinus milaris, 11
Pyronine G, 57
Pyronine Y, 52, 57–61, 65

R

Radical scavengers, 310
Radioactivity, measurement of in
    chromatogram, 303
Radioautographic method
  for catecholamines, 365
  for estrogens, 365
Radiochemical decomposition, control of,
    309
Radiochemical impurities, 304
  effects of, 306
  uptake of, 307
Radiochemical purity, 296–298
  of fatty (^{14}C) acids, 302
Radiochemicals
  chromatographic analysis of, 297
  effect of radical scavengers on the storage
    of, 316
  effect of temperature on the storage of,
    314
  self-decomposition of, 308
  solvents to minimize self-decomposition
    of, 313
Radioiodination
  of human serum albumin, 344
  of nucleic acids, 341
  of polycytidylic acid, 344
  of polynucleotides, 347
Radioiodination of RNA, procedure for, 350
Radioiodine labeling, of ribopolymers, 343
Radionuclide(s)
  choice of, 293
  physical properties of, 293–294
Radionuclidic (radioisotopic) purity, 296
Radiotracer techniques, 297
Revertants
  detection of drug-sensitive, 219
  detection of prototrophic, 220
  detection of temperature-resistant, 220
Rhodulinorange, 56
Ribosomes, isolation of, 127
Ringer's solution, insect, 3, 8, 137
RNA
  catalytic iodination of, 333
  dyes for staining, 56
  electrophoresis of, 362
  elution of, 29
  extraction of, 28, 36
  in vitro synthesis of, 357
  iodination of, 326
  microanalysis of, 17
  microextraction of, 27
  simplified procedure for preparing, 348
  specific activity determination of, 36–37
RNA detection, 65
  methods of, 63
RNA–DNA hybrids, 6
  microassay for, 34
  thermal stability of, 43
RNA metabolism
  in *Drosophila* cell line, 196
RNA staining, 49
  procedure for, 55
RNase(s), 8
  detection of, 62
  inhibition with 6 M guanidinium chloride,
    347
  inhibitor of, 75
  pancreatic, 4, 8
RNase activity, blockage of, 53
RNase treatment, 40
Roller bottle cultures, 159, 164

S

Saline media, for isolation or fractionation of
    nuclei, 71
Salivary gland chromosomes, 11, 13
Salivary glands, 3, 42
  fixation of, 23
  from *Drosophila hydei*, 136
  labeling RNA in, 20
  microdissection of, 24
Scatchard's equation, 51
Separation of cells, by velocity sedimentation,
    158
Silica, 88
Silica sol, 87
Siliclad, 137
SNB, 8
Soluene, 44
Somatic cell hybrids
  cloning of, 147
  microsurgical production of, 147
Spirogyra grevilleana, incorporation of
    ^{3}H thymidine into, 307
SSC, 8
Stains-all, 57–58, 61
Subbed slides, 8
Subbing solution, 8
Subcellular particles, separation of, 85

Sulfur-35, physical properties of, 294
Suspension growth, adaptation to, 197
Swinnex filter, 151
Swiss blue, 56
Sykes-Moore chambers, 148
Synaptosomes, 101
Synchronized growth, of HeLa cells, 96
Synchronous cell populations, selection of by zonal centrifugation, 173
Synchronous CHO cells, 171
Synchronous culture of HeLa cells, 97
Synchrony
- alignment, 158
- automated, 159
- selection, 158

T

Tetrahymena
- macronuclei from, 124
- membrane studies on, 106

Thionine, 57
Thrombocytes, 98
Thymidine, tritiated impurities uptake by *T. pyriformis*, 306
^{3}H thymidine, self-radiolysis, 305
Tobacco mosaic virus, purification of, 101
Tolazul, 57
Tolonium chloride, 57
Toluidine blue, 54, 57
Toluidine blue 0, 61
Transcription of nucleic acid *in vitro*, 12
Tritium, 293
- physical properties of, 294

Tylosin, 286

U

Ultrabrilliant blue P, 56
Uridine-5-^{3}H
- as a tracer for RNA, 312
- biosynthetic pathways of, 318

V

Viruses, purification of, 101
Vitamin B complex, 21
Vitatron densitometer, 33

W

Whole cells, phospholipid composition of, 131
Wilzbach method, 297

Z

Zonal centrifugation of cells, 180

Cumulative Subject Index Volumes I–X

A

Acanthamoeba, III, 154; IV, 94–95, 389, 417, 433; VI, 190
Acanthamoeba sp., I, 55–83
  aeration of cultures of, I, 73, 79
  cell handling, I, 65–67
  conditions for synchronous encystment in, I, 68–80
  culture media for, I, 60–63, 68, 71–73
  cyst wall components in, I, 56
  ionic requirements of, I, 71–72
  pH effects in, I, 75–81
  pinocytosis in, I, 279
  pure culture of, I, 57–59
  stages in encystment in, I, 57–60
  synchronous encystment of, I, 55–83
  temperature effect on, I, 73–74
Acetabularia
  anatomy of, I, 190
  axenic culture of, IV, 49–69
  bi- and multinucleated cells of, I, 205–206
  biochemical analysis of, I, 207–210
  cell grafting in, I, 110, 189
  cell segments of, I, 202–204
  chloroplastic RNA of, I, 210
  circadian rhythm of, IV, 51
  culture conditions for, IV, 51
  culture techniques for, I, 192–199
  culturing and experimental manipulation of, I, 189–214
  cytoplasmic grafts of, I, 189, 206
  effects of light on, IV, 52
  enucleation of, I, 110, 199, 203–204, 209–210; IV, 49
  gametes of, I, 195–198; IV, 62, 67
  gametogenesis in, I, 195–196
  generation time of, I, 190; IV, 67
  growth medium for, I, 192–194
  growth phase of, IV, 63–64
  growth rate of, IV, 64, 67
  habitat of, I, 190
  injection of materials into, I, 206–207
  internal pressure in, IV, 66
  life cycle of, I, 191; IV, 59
  maturation of, IV, 66
  microsurgery on, I, 199–207
  multinuclear grafting of, I, 199
  nuclear implantation of, I, 206
  nuclear transplantation in, I, 110, 114, 189, 199, 204
  nucleocytoplasmic interactions in, I, 189, 190, 209–211
  protein analysis of, I, 209
  pure culture of, I, 194
  regeneration in, I, 122
  reproduction and early growth of, IV, 62
  RNA synthesis in, I, 210
  spores (cysts) of, I, 192, 195
  sterility in, IV, 54
  stock cultures of, I, 195–198
  studies on, I, 211–212
  synthetic medium for, IV, 57
  taxonomy of, I, 190
  temperature for culture of, I, 195
  working cultures of, I, 198–199
  zygotes of, I, 196
Acetabularia calyculus, I, 195
Acetabularia crenulata, I, 191, 195
Acetabularia mediterranea, I, 194–195, 206; IV, 51
Acetabularia wettsteinii, I, 195
Acetabularium, II, 2

Acetamide, VIII, 345
Acetate flagellates, III, 122
Acetic anhydride reaction, III, 328–329
Acetic-orcein, II, 78, 89–90, 118; IV, 6
Aceto-carmine, II, 270; III, 205; IV, 140, 167
Acetocarmine-fast green, IV, 275
Acetone, as fixative, IX, 201
Acetoorcein, III, 57–58, 91, 205; IV, 167; V, 93; VI, 199; IX, 88
Acetylation procedure, III, 329
Acholeplasma, VI, 149
Acholeplasma laidlawii, VI, 147, 161
Acholeplasma sp., VI, 147
Achromobacter candicans, IV, 253
Acicularia schenckii, I, 194–195, 206; IV, 51
Acid nuclei, composition of, IX, 362
Acid phosphatase, IV, 157, 206
Aconitase, IV, 172, 174
Acrasiales, II, 398; IV, 409, 411
Acridine, IV, 174
Acridine mustard, IV, 304; VIII, 36
Acridine orange, III, 51, 71; V, 337, 340, 343–345, 347, 351–353, 357, 366–368; VI, 85, 366, 368, 371–374; X, 33, 56, 58, 61
  staining of, IX, 208
Acridine yellow, IV, 289; IX, 204
Acriflavin, IV, 187, 276, 289, 399; VI, 84–85, 94; VIII, 194, 202; IX, 203
Acriflavine-Feulgen, VIII, 196–197, 200
Acriflavine-Schiff procedure, VIII, 195
Acrosome, V, 89
Actinomycin D, II, 42, 93, 104, 106, 110; III, 157, 189; IV, 385, 399; V, 270–272; VI, 214; VII, 15, 18, 29, 238–240, 242–243, 372; VIII, 344; IX, 120
  blocking cells in G_2, IX, 86
  chase, VII, 29
  resistance to, VI, 231
  H^3-labeled, III, 52, 54
Adenine, IV, 152
Adenine-C^{14}, III, 189
  incorporation in roots of *Allium cepa*, III, 189
  of *Vicia faba*, III, 189
  of *Zea mays*, III, 189
Adenine nucleotides, VII, 371, 447–448
  pool, VII, 371, 377
Adenine phosphoribosyl transferase, VII, 393, 410
Adenine sulfate, III, 179
Adenocarcinoma in frog renal, IV, 46
Adenosine, III, 30, 33
  prototrophy, VI, 268
Adenosine deaminase, VII, 393, 395
Adenosine diphosphate, IV, 87–88, 93, 106, 109
Adenosine kinase, VII, 393, 396
Adenosine triphosphatase, VI, 241–243
Adenosine triphosphate, II, 13; IV, 85, 93, 98, 100, 106; V, 35; VI, 151
  pool, VII, 378
Adenovirus, VI, 154, 164–167, 169, 172, 176; VIII, 371
  avian, VI, 168
  bovine, VI, 168
  SV40 hybrids, VIII, 415
Adenovirus type 2
  purification of, X, 101
Adenyl cyclase, VII, 393
Adenylate kinase, VI, 298, 303; VII, 400
  assay, VII, 447
Adenylic acid, IV, 14
Adenylosuccinate, VII, 394
Adenylosuccinate lyase, VII, 390, 393–394
Adenylosuccinate synthetase, VII, 393–394
ADP, *see* Adenosine diphosphate
Adrenal cortex, VII, 93, 160
Adrenal glands, VII, 92
Adsorption buffer, VIII, 406
Aedes albopictus, X, 185
  cells of, X, 186
Aedes cells
  cell fractionation of, X, 189
  genome size of, X, 189
  growth rate, X, 188
  karyotype, X, 188
  medium for, X, 187
  mitochondrial fraction of, X, 190
  mRNA from, X, 190
  polysomes, X, 190
  renaturation kinetics of DNA from, X, 189
  rRNA of, X, 192
  stocks of, X, 193
Aerobacter, IV, 254, 366, 373, 397, 415, 418–420, 422–426, 428–431, 435–437, 446, 452–454, 456–457, 467

Aerobacter aerogenes, II, 398, 400; IV, 118–119, 252–253, 289, 350–351, 414–415, 453, 461
Aerobacter cloacae, IV, 253, 266
Aerosolization, VI, 196
Aerosols, VI, 161, 186
African green monkey, VI, 166
  kidney cells, VII, 143, 253
Agar, source of, IX, 161
Agar-Noble, IX, 161–163, 170
Agar plate culture, IX, 158, 171, 174, 177
Agar plates, IX, 164, 172
Agar suspension assay, VIII, 381, 423
Agar suspension method, VIII, 406
Agaropectin, IX, 161
Agarose, V, 42; VIII, 102–104, 394; IX, 162
  cloning method, VIII, 105
Agglutinability, VIII, 79
  of normal and virus transformed cells, VIII, 398
Agglutination, V, 85, 87, 89; VIII, 86
Air filter, particulate, VI, 186
Alanine, V, 14, 63; VI, 6
Albumin, IV, 93
Albumin-kieselguhr, methylated, VII, 117, 124
Albumin powder, bovine, IV, 30
Alcaligens fecalis, VI, 180
Alcian green, V, 51
Alcohol dehydrogenase, IV, 157, 171, 174; VI, 303
  genes for, IV, 171
Alcojet, IV, 23
Alconox, IV, 23; V, 69
Algae, IV, 84; V, 306; *see also* specific organisms
  availability of maintained cultures of, I, 182–183
  blue-green, IV, 85, 111–112, 132
  colorless, IV, 104, 109
  culturing of, V, 374
  mass culture of, I, 181–182
Alginic acid, VI, 2
Alizarine navy, X, 56
Alkaline fast green reactions, III, 312–319
  cytoplasmic staining, III, 204, 316–317
    in amphibian oocytes, III, 317
    in mouse fibroblasts, III, 316
    in onion root cells, III, 317
    in pea seedlings, III, 316
    in *Psammechinus miliaris*, III, 316
    in slug oocytes, III, 316
    in *Stellaria* media, III, 316
    in *Tetrahymena pyriformis*, III, 316
    in unfertilized sea urchin eggs, III, 316
    in nuclei of *Euplotes*, III, 314
    of *Tetrahymena pyriformis*, III, 314
    of *Tradescantia*, III, 315
  procedure after Alfert and Geschwind, III, 203, 317
  in X chromosome of *Rehnia spinosus*, III, 314
Alkaline phosphatase, IV, 157; VII, 31
Alkylating agents, IV, 174
Allantoic fluid, V, 81
Allium cepa, I, *see* onion roots; II, 154, 160–161, 190, 209–212; III, 171, 175, 183, 189, 192, 194, 206; V, 342, 361–367
  Feulgen staining of, III, 199
  turnover of chromosomal proteins, III, 187
Allograft, V, 129
Alpha, IV, 307
  isolation of, IV, 317
Alpha particles, IV, 303
Alumina, IV, 91, 153
Amacronucleate cells, IV, 244, 297–298
α-Amanitin, VI, 214
  cells resistant to, VI, 231
  resistance to, VI, 275
Ambystoma, I, 362; III, 87
Ambystoma mexicanum, II, 19, 44
Amebo-flagellates, IV, 344, 346–347, 349–350, 353, 359–362, 368–370, 393, 403, 406–407, 409–412, 415–416, 421, 425, 440, 456, 463–464, 466
Ameboid flagellate, IV, 349
Ameboid motion, VI, 340
American Type Culture Collection, V, 9, 80; VI, 202; VIII, 25, 65, 368
Amethopterin, V, 256; VI, 44, 214, 233; VII, 368, 386; *also see* Cell synchrony
Amethopterin B, VI, 157
5-Amino-4-imidazole-N-succinocarboxamide ribonucleotide, VII, 390
5-Amino-4-imidazolecarboxamide, VII, 396
5-Amino-4-imidazolecarboxamide ribonucleotide, VII, 390

5-Amino-4-imidazolecarboxamide ribonucleotide transformylase, VII, 390, 392
5-Amino-4-imidazolecarboxylic acid ribonucleotide, VII, 390
5-Aminoimidazole-4-carboxamide, VIII, 36
5-Aminoimidazole-4-carboxylic acid ribonucleotide, VIII, 38
5-Aminoimidazole ribonucleotide, VII, 390
α-Aminoisobutyric acid (AIB)
  kinetics of transport, IX, 274
Amino acid pool, intracellular, X, 245
Amino acids, V, 40
  as inducers of pinocytosis, I, 285–287
  solution of, V, 14
  stock solution, V, 17
Amino acids-H^3, II, 259, 308; III, 171, 187–188
ρ-Aminobenzoate, IV, 58, 171
Aminobenzoic acid, V, 15; VI, 8
α-Aminobutyric acid, VI, 6
δ-Aminolevulinate dehydratase
  half-life of, X, 241
  rates of synthesis of, X, 239
Aminopterin, V, 91; VI, 214, 233–234; VII, 377, 386; VIII, 15, 81
ρ-Aminosalicylate, VII, 112–113, 123
ρ-Aminosalicyclic acid, VI, 309
5-Aminouracil, III, 171
  effects on chromosome structure, III, 194
    on G_2 phase, III, 194
    on S phase, III, 193–194
  inhibition of DNA and protein synthesis, III, 189, 192–193
Amitosis, VIII, 336
Ammoniacal silver reaction, III, 335–338
  procedure after Black and Ansley, III, 338
Amniotic fluid cell cultures, X, 270
Amoeba, II, 1, 65, 138, 400, 404; IV, 95, 182–187, 190–192, 341–344, 346–352, 355–359, 362–367, 369, 371–373, 375, 377–378, 383–384, 386, 388–389, 391–394, 397–409, 412–416, 418, 420–431, 433–434, 436–440, 442, 446, 448, 450–454, 456–457, 459–462, 466–467; VI, 188–193
  binucleate, IV, 191
  enucleated, IV, 185
  marine, IV, 345
  metamorphosis of, IV, 342
  micromanipulation of nuclei in, IV, 179–194
  multinucleate, IV, 387
  nuclei in, IV, 179–194
  transformation of, IV, 342
Amoeba aquatilis, IV, 358
Amoeba diplomitotica, IV, 358
Amoeba discoides, I, 89, 92; IV, 189–190, 192
Amoeba dubia, I, 279, 290
Amoeba gruberi, IV, 352, 356
Amoeba limax, IV, 381
Amoeba lobosa, IV, 342–343, 352
Amoeba proteus, IV, 189–190, 192; V, 282
  amino acid-H^3 labeling of, II, 259–261, 306, 308
  binucleated cells of, I, 93
  cell life cycle of I, 91
  compression of, I, 92–93
  culture of, I, 89–91
  cytoplasmic contaminants in, I, 92
  DNA synthesis in, I, 91, 94–95
  enucleation of, I, 105–106
  experimental uses of, I, 92–95, 97–108
  fixation of, I, 94–95
  inhibition of cytokinesis in, I, 92–93
  isolation of individual nuclei of, II, 132, 134
  isotope labeling of, I, 93–95
  micromanipulation of, I, 98–101
  microtools for, I, 102–105
  micrurgy of, I, 92
  nuclear RNA labeling of, II, 259, 302–305, 307–308
  nuclear transplantation in, I, 92, 97–108
  nuclei of, IV, 232
  nucleus of, II, 133
  operating chambers for, I, 100–101, 103–105
  operating procedure for, I, 103–107
  pinocytosis in, I, 279–281, 285–296, 300
  staining of, II, 283
  thymidine-H^3 labeling of, II, 260, 263–265, 300–301, 307
Amoeba punctata, IV, 352
Amoeba sphaeronucleus, II, 2
Amoeba tachypodia, IV, 412
Amoebida, IV, 411
AMP deaminase, VII, 371, 393–394, 397
AMP (dAMP) kinase, VII, 394
Amphibian cells, I, 98, 110, 114; IV, 20–21, 35

culture conditions for, IV, 21
cultured, IV, 46
Amphibian culture medium, III, 58, 67
Amphibian egg, VII, 48
animal hemisphere of, II, 3–5
animal pole of, II, 6, 15, 22
cortical granules of, II, 6
enucleation of, II, 3, 5–6, 12, 19
jelly coat of, II, 6, 15
nuclear transplantation in, II, 1–34
parthenogenic activation of, II, 3
vitelline membrane of, II, 6, 9, 15, 17, 19–20, 22
Amphibian embryo
blastula stage, II, 6, 9–13, 15, 17–19, 22, 27, 29, 32–33
dissociation of, II, 6–7
gastrula stage, II, 6, 11–13, 15, 17, 25, 33
RNA synthesis in, II, 31–32
Amphibian oocytes, V, 76
nuclei of, V, 168
Amphibian tissues, III, 75–92
cinematography, III, 80
culture technique, III, 76–83
Anuran renal cells, III, 85–87
urodele lung cells, III, 83–85
explantation, III, 77–78
flying coverslip technique, III, 78
hanging drop method, III, 77
incubation temperature, III, 83
monolayer cell culture, III, 78–80
nutrient media for, III, 81–83
Eagle's medium, III, 81
Earle's medium, III, 81
Hank's medium, III, 81
medium 199, III, 81
medium 1066, III, 81
NCTC 109, III, 81–82
sterilization of, III, 77
Amphotericin B, IV, 55–56; VI, 192
Ampicilin, IV, 55
Amylase, VII, 90, 116
α-Amylase, VII, 113, 115, 123
Amytal, IV, 87, 92, 100, 106–107, 110, 112
Anabaena cylindrica, V, 381
Anacystis, V, 373
Anacystis nidulans, V, 374–376, 382
medium for, V, 378
synchrony in, V, 377–380
Anaphase, IV, 3, 73, 76–77, 355, 379–380, 382, 385, 511; V, 112
Anchorage dependence, VII, 393; IX, 168
loss of, VIII, 393–394, 423
Androgenetic haploids, II, 3, 12, 30
Aneuploidy, VI, 218
Angiosperm, III, 189
N^5,N^{10}-Anhydroformyltetrahydrofolate, VII, 392
8-Anilino-1-naphthalene sulfonic acid (ANSA), IX, 212
Animal Cell Culture Collection, VI, 194
Animals
preservation of specimens, V, 6–7
procurement of, V, 2–4
Antechinus flavipes, V, 131
Antechinus swainsonii, V, 130–131, 150, 158–159, 161, 163
Anther, IV, 505–507
corn, IV, 2
of lily, IV, 2
Anti-γ-globulin, VII, 46
Antibiotic treatment, VIII, 244
Antibiotics, IV, 27–29, 31, 41, 54–56, 60–61, 89, 121; V, 16; VI, 162–164, 179, 185–186
Antibodies, fluorescent, II, 88
Antibody, per cell, IX, 230
Antifoam, VIII, 97
Antifoam, B, IV, 90; V, 232
Antifoam SAG, IX, 314
Antifolate drugs, VI, 214
Antifungal medium, VI, 181
Antigen, per cell, IX, 230
Antigenic phenotypes, VIII, 353
Antiglobulin, VI, 175
Antihistamines, II, 217
Antimetabolites, III, 173, 189–195; VII, 401; VIII, 344
Antimycin A, IV, 89, 92, 94, 100, 106–107, 109–110, 112
Antimycoplasma agents, VIII, 239
Antimycoplasma drugs, X, 271
Antineoplastic drugs, VII, 387
Antiviral serum, VI, 175
Anucleate cells
fusion of, VII, 226
properties of, VII, 211
survival times for, VII, 222

Anucleate cytoplasm, IV, 387
Anuran cells, III, 85–87
  culture methods for, IV, 19–47
  embryo, IV, 21
AR-10 stripping film, V, 29–31
1-β-4-Arabinofuranosylcytosine, VI, 214, 246–247, 257
Arboviruses, VI, 174
Arcella, I, 112
Archibold equilibrium method, VIII, 356
Arginine, III, 204; V, 14, 60; VI, 6, 71, 151–152, 154, 156–157
  content in cell culture media, X, 268
  depletion, VIII, 230
  determination, X, 268
Arginine dihydrolase, VI, 151
Argon ion laser beam, VI, 341
Argyrol, IV, 55
Asarum europaeum, II, 178
Asbestos filter pads, V, 9
Ascaris embryos, VII, 110
Ascites carcinoma, VII, 160
Ascites cells, VII, 80, 87, 96, 350, 387, 389, 394, 397, 416; X, 174
  preparation of, X, 178
Ascites hepatoma, IV, 214; V, 181
Ascites tumor cells, I, 282–284, 289–290; IV, 212; V, 79, 92; VII, 97, 110, 224
Ascomycetes, IV, 133
Ascorbate-TMPD, IV, 88–89
Ascorbic acid, V, 15; VI, 7
Asparagine, II, 223–224; V, 14, 62; VI, 6
Asparate transcarbamylase, IV, 157; VII, 378–379, 400
Aspartic acid, II, 222–224; V, 14, 63; VI, 6
Aspergillin, V, 43
Astasia, IV, 103; *see also* specific organisms
Astasia klebsii, IV, 104
Astasia longa, II, 218; IV, 101–104
  continuous culture techniques for, I, 148–155
  culture apparatus for, I, 148, 151
  dry weight content of, I, 152–154
  growth medium of, I, 144
  protein content of, I, 152–153
  pure culture of, I, 86
  synchrony stability and repetitiveness of, I, 151–155
  temperature-induced synchrony of, I, 147–155
Asterias rubens, oocyte RNA in, I, 433
Astrantia major, II, 178
Asynchronous cultures, IV, 161
ATP, *see* Adenosine triphosphate
ATPase, V, 192, 194–195; VIII, 226; IX, 261, 272, 275–276
Auramine, O, IX, 203
Aureomycin, IV, 55, 314, 324; VIII, 238–239
Autogamy, IV, 243–245, 273–279, 281–283, 290, 298–299, 301–304; VIII, 347
  induction of, IV, 276, 331
Autolytic enzymes, VIII, 154
Autoradiographic analysis, of tritium on polyacrylamide gel, X, 361
Autoradiographic emulsions, gelatins in, V, 296
Autoradiographic techniques, X, 320
Autoradiographs, IV, 151; V, 36
  comparison of 5S RNA–^{125}I, X, 339
  degraining of, V, 33
  of ^{125}I- and ^{3}H-labeled cRNA, X, 338
  of nuclei, IV, 10, 12
  resolution of, VIII, 268
Autoradiography, III, 50, 52–56; V, 29, 31, 35, 278; VII, 340, 350, 352, 442; *see also* Autoradiography of water-soluble materials, Electron microscope autoradiography, High-resolution autoradiography, Quantitative autoradiography
  application of, II, 255–257
  background determinations, II, 336–338
  crossfire in, I, 307–309
  design of experiments using, X, 293
  film stripping for, I, 309, 324, 365
  grain density in, I, 315–317, 353–355
  high resolution, X, 293
  history of, I, 4–5
  image spread, *see* Quantitative autoradiography
  in electron microscopy, V, 289
  Kodak AR-10 stripping film, use of in, III, 52
  latent image in, I, 306, 309
  light microscope techniques, II, 257–268, 326–328, 362–363
  liquid emulsions in, I, 309, 324, 330, 334–335, 365–370, 388; III, 52
  of *Amoeba proteus*, II, 260, 263–264

of isolated lens epithelium, III, 54
of lampbrush chromosomes, II, 52–53
of macronuclear anlagen, II, 136–137
of mammalian chromosomes, I, 381–384
for measuring interphase periods, *see* Mitotic cycle measurements
of mouse epidermis, II, 359–384
of *Stentor* macronucleus, II, 262–263
preparing filters for, IX, 89
stability of radiochemical tracers in, X, 291
staining cells after, IX, 90; X, 8
staining cells before, IX, 88
wire loop technique in, I, 332–333, 336, 373
Autoradiography of water-soluble materials, I, 371–380
detection of insoluble radioactivity in, I, 374–375
detection of soluble radioactivity in, I, 374–375
experimental examples of, I, 375–378
procedures for, I, 372–375
Auxins, III, 171, 178, 192; IV, 57
as growth factors, III, 195–196
effects on cell division, III, 196
3-indoleacetic acid, III, 178, 196–197
induction of lateral roots by, III, 197
α-naphthaleneacetic acid, III, 196
Auxotrophic analysis, VIII, 27
Auxotrophic cells, X, 214
Auxotrophic lines, VI, 211, 214, 225
cross feeding in, VI, 254
Auxotrophic markers, VI, 212, 217, 227, 250
drug resistance, VI, 263
Auxotrophic mutants, V, 91, 113; VI, 220, 224–225, 247, 250, 276; IX, 177
detection of, X, 216
isolation of, VI, 246; IX, 174
Auxotrophic mutations, VI, 216, 249
Auxotrophic phenotypes, VI, 250
Auxotrophs, VI, 214, 252; VIII, 31; IX, 172
glutamine-dependent, X, 211
induction of, VI, 268
isolation of, VI, 250
nutritional, VI, 214, 218–219, 249–250
Auxotrophy, VI, 251, 268, 270, 273
glycine, VI, 252
nutritional, VI, 267
Avian erythroblastosis, VIII, 370
Avian leukemia virus, assays with, VIII, 422
Avian leukosis-sarcoma complex, VIII, 370
Avian leukosis viruses, VI, 174, 176; VII, 320
Avian lymphomatosis virus, VIII, 370
Avian myeloblastosis virus, VII, 314–315, 327–328; VIII, 370
assay, VII, 319–320
transformation of cells by, VII, 319
Avian myelocytomatosis, VIII, 370
Avian sarcoma, VI, 168
Avian viruses, assays for, VIII, 381, 421
Axenic culture techniques, VIII, 342
Axenic cultures, VIII, 342, 349
Axolotl, II, 19, 23–25, 27, 43–44
Azaguanine, IX, 172
resistance to, IX, 174
8-Azaguanine, V, 91; VI, 213, 228, 231–232, 235–236, 238, 244; VII, 396
cells resistant to, VI, 231–232, 235, 237; VII, 256
cells sensitive to, VI, 237
resistance to, VI, 214, 216–217, 226, 230–231, 234, 270, 273, 275; VIII, 42
resistant clonal lines, VI, 233–234, 237
resistant mutants, VI, 231, 235, 240, 271
resistant phenotypes, VI, 271–272
8-Azaguanosine, VI, 231–233
8-Azahypoxanthine, VI, 231–232, 235
Azahypoxanthine resistance, VI, 267
Azaorotic acid, VII, 380
Azaserine, VII, 380, 382, 392
Azauridine, VII, 380
Azide, IV, 87, 92, 106, 112
Azure A, X, 56
Azure A staining, IV, 306
Azure B, II, 51, 270, 274–275, 279, 282, 284–285; III, 200
Azure B bromide, VIII, 336
Azure B- eosin Y
procedure, III, 342
reaction, III, 341–342
Azure C, IV, 205, 216–217, 226–227
Azurophilic granules, VII, 322

B

Babesia rodhaini, I, 282
Bacillaris, II, 218
Bacillus, VI, 180
Bacillus cereus, IV, 433

Bacillus fluorescens, I, 11
Bacillus megaterium, III, 232
Bacillus stearothermophilus, VII, 89
Bacillus subtilis, I, 339, 347–350, 356, 358; II, 257
  chromosome replication in, VII, 148
Bacitracin, V, 81
Bacterial L forms, VI, 161
Bacteriocidal agents, IV, 54, 56
Bacteriocides, IV, 55
Bacteriophage, I, 351, 354, 356, 377
Bacto-agar, IV, 124; IX, 162
Bactogen, I, 142
Bacto-peptone, IV, 420
Bacto-tryptone, V, 224, 226
Bacto tryptose phosphate broth, VII, 316
Badhamia foliicola, I, 13
Badhamia utricularis, I, 10, 13, 15, 35
Baked lettuce, IV, 289
Balamuth's medium, IV, 431
Balanced growth, I, 142–144
Balanced salt solution, IV, 27, 40, 43, 77
Balbiani rings, II, 81; X, 25, 27, 45
  micromanipulatory isolation of, X, 26
Bandicoots, V, 130–131
Banding techniques, VIII, 183, 188
Barley root tips, fixation of, I, 219–220
Barley seeds, germination of, I, 217
Basal bodies, IV, 375–376, 379, 395–396, 405, 448, 464
Basal medium, IV, 73
Basic fuchsin, V, 23; VI, 83
Basic proteins, IV, 141, 211
Basophil granulocytes, VII, 323
Basophils, VI, 335
Batch culture, I, 142
Bats, V, 24
Baycovin solution, X, 177
BBL Micro test II, V, 20
Bean seeds, germination of, I, 217
Beer-Lambert law, V, 314
Behavioral mutants, VIII, 345
Bellevalia, root tips, I, 225
Bentonite, VII, 9, 33
Benzoflavine, IX, 204
Benzopyrene, VIII, 373
Berberin, V, 340
Berkefeld filter, IV, 331
Beta emitters, IV, 303
Betaprone, VII, 229
Bettongia cuniculus, V, 135
Bettongia lesueur, V, 135
Bettongia penicillata, V, 135
BHK cell line, VI, 78, 261
BHK cells, VII, 375, 411, 453; VIII, 77
  polyoma-transformed, VIII, 77
  transformants of, VIII, 77
BHK-21 cells, VII, 215, 234, 266; VIII, 181
BHK 21/13 cells, VII, 406
Biebrich scarlet
  procedure after Spicer and Lillie, III, 335
  reaction, III, 334–335
Bile canaliculi, IV, 206
Binucleate cells, IV, 135, 190; V, 87, 100–101, 120; *see also* Caffeine
  technique, III, 102
Biohazard and Environmental Control, VIII, 369
Biohazards in Cancer Research, VIII, 369
Biological clocks, IV, 245
Biotin, IV, 122–123, 143, 170–171; V, 15, 60, 226; VI, 8
1,3-Bis(2-chloroethyl)-1-nitrosourea, VII, 392
Bizarre virus, VI, 189
Blastomeres, V, 88
Blastula stage, IV, 29
Belpharisma japonicum, II, 138–139
Blood, V, 4, 149; VI, 182
Blood agar plates, VI, 179–180, 184
Blood cells, IV, 478
  separation of, X, 98
Blood cultures, IV, 477–495
  human peripheral, VI, 218
Blue-green algae, V, 373–374, 380, 382
  synchrony in, V, 373
Blutene chloride, X, 57
Body cavity fluid, frog, II, 14
Bomford's solution, IV, 327
  for culture for *Chlorella*, IV, 331
Bone marrow, V, 12–13, 23–24, 59, 133; VII, 26, 323, 392
  cells, VIII, 101
  cultures, VII, 320
  preparations of, V, 23, 27
Borate methods, VII, 440
Bouin's fluid, III, 266; VI, 191
  Hollande's modification, III, 265–266
Bovine
  diarrhea virus, VI, 167, 170–171
  embryonic kidney, VI, 178

n of nuclei, V, 173

, 60, 91, 115, 199,

9
IV, 64–65
9
50

, X, 101
93

8, 352, 409, 464

or small numbers
oscopy; *see also*
e also specific types
nique in, II, 312–317

ding, II, 319
2–353

ique in, II, 317–319

00–101, 103–106,

00, 104, 106

ities in, I, 320

d disrupting, VII,

nizing, IV, 72
of, II, 319–320

mitotic, V, 252
monkey kidney, VI, 12, 29
mouse liver, VI, 15
mouse lung, VI, 15
mouse lymphoma, VI, 45
multinucleate, V, 95, 104–105
murine leukemia, VI, 3
rat ascites hepatoma, VI, 12
rat liver parenchymal, VI, 12–13, 19, 24, 26, 34
rat spleen plasma, VI, 12
rat thymus reticulum, VI, 12
reticuloendothelial, VI, 3
single, V, 56
squash preparation of, II, 89–90
storage of, IX, 129
synchronized populations of, V, 96–98, 107, 113, 116
synchrony in, VI, 44
3T6, VI, 106–107

Cell affinity labeling, X, 365
Cell clones, V, 45
Cell collection, IV, 144
Cell counting, V, 234
methods for, VII, 334
Cell counts, VIII, 69–70
Cell Culture Collection Committee, VI, 186, 202
Cell cultures, synchronous, VI, 43
Cell cycle, III, *see also* Mitotic cycles; IV, 133, 149, 158, 161, 263–264; V, 75, 87, 97, 99, 104, 109, 144–146, 148, 152, 378; VI, 46–47, 82–97, 100–101, 115, 259; VII, 271, 333, 346, 376
age gradients, III, 221–225
analysis, III, 28–29, 150–157, 213–257; V, 280; IX, 76, 182, 218
application of computer results to, III, 246–251
automated analysis, X, 157
calibrating, X, 168
cell division, III, 152
computer program characteristics in, III, 244–246
continuous labeling, III, 246–250
definition of G_1, S, G_2, M phases, III, 26, 96, 98
determination of G_1, S, G_2 phases, III, 152–153
effect of G_2 duration on mitotic cycle, III, 241

embryonic lung, VI, 178
enteroviruses, VI, 171
herpes virus, VI, 171, 178
parainfluenza type 3 virus,
rhinotracheitis virus, VI, 1
serum albumin, II, 14
trachea, VI, 178
tracheal cells, VI, 170–171
viruses, VI, 167
Boyden chamber technique,
Boyle-Mariotte's law, II, 157,
Bradysia mycorum, II, 75–76
Brain, VII, 374, 387
Brain-heart infusion, VI, 157
Brain-heart infusion agar, VI,
Brain-heart infusion broth, VI
Bräun MSK homogenizer, VII
Breaking pipettes
description of, I, 128–129
for amoeba, I, 91
for *Euplotes*, I, 89
Brilliant chrome, X, 56
Bromobenzene, X, 90
5-Bromodeoxyuridine, V, 45, 9
213, 228, 244, 246; IX, 1
cells resistant to, VI, 231, 23
resistance to, VI, 216, 226, 2
substitution with, IX, 121
5-bromo-2-deoxyuridine, VIII,
24–25, 29, 21–35, 44–45,
88–90
cells containing, VIII, 28
incorporation, VIII, 26
5-Bromouracil, VII, 388; VIII, 8
Bromthymol blue, II, 9
Bronchitis virus, avian, VI, 167
BSC-1 cells, VIII, 140
BT-88 basal medium, VII, 315
Buccal apparatus, VIII, 332
Buccal parts, isolation of, VIII,
Buccal structures, VIII, 332
isolated, VIII, 333
Budding yeast, IV, 132, 135, 141
BUdR, VIII, 411, 414, 416, 420
Buffer solutions, *see also* specific
preparation of, V, 370
Bufo, II, 25, 27
Bufo americanus, IV, 42
Bufo marinus, III, 52, 69
Bull sperm, V, 86

Carcinoma
cultures of, VI, 165
human, VI, 170, 173
human uterus, VI, 12
Carcinoma cells, isolati
Carnoy, V, 24
Carnoy's fluid, III, 48,
309
Carollia perspicillata, X
Cartesian diver balanc
Cartilage, V, 5, 50, 53,
Cartilagenous tissue,
Caryopsis, III, 175
Casein, VII, 92–93
Catacholamine granul
Catalase, IV, 206; VI,
Cathepsins, IX, 261
Cats, VI, 177
Cavostelium, IV, 347–
Celanol 251, VI, 319
Celite, VII, 422
Cell(s), II, preparatio
for electron m
Giant cells; V
agar embedding te
AH-130, VI, 2
blastula, IV, 30
capillary tube em
cultivation of, III
dog kidney, VI,
embryonic, IV, 3
epithelial, VI, 3
extracted, I, 372-
feeder layer, IV,
flat embedding t
frog embryo, IV,
fused, V, 92, 119
gastrula, IV, 30
heterophasic, V,
108–109
homophasic, V,
human, VI, 4
L-929, VI, 2
mastocytoma,
measurement o
methods for, II
methods for ly
110
methods for sy
microcentrifug

experimental methods for obtaining data on, III, 217–220
G_0 phase, III, 221
G_1, S, G_2, M phases, III, 26, 28, 97, 105, 110, 120, 148, 153, 155, 178, 181, 184, 214, 220, 221, 223
generation time, III, 150–152
distribution and determination of, III, 233–241
Mak technique, III, 223
morphological and biochemical events, III, 27
of paramecium, VIII, 340
of *Tetrahymena*, V, 276
pulse labeling, III, 250–251
reproductive cycle, III, 26–27
synchrony of, V, 380
volume through, VII, 270
Cell cycle analyzer, X, 159–160
Cell degeneration, VI, 39
Cell density, IV, 157
turbidimetric estimation of, V, 237
Cell dispersal, IX, 241, 249; X, 95
enzymes for, IX, 198
Cell disruption, VII, 2
methods for, IX, 286
Cell division(s), IV, 33, 134, 149, 155, 188–189, 191, 263
synchronization of, IX, 51
synchronous, V, 376–377
Cell DNA content, VII, 453
Cell fixation, IX, 199
Cell fractionation, IV, 197
Cell freezing, VIII, 377
Cell fusion, IV, 347; V, 75–79, 83, 85, 87, 90, 95, 97; VII, 228–229, 252, 259, 308; VIII, 37; X, 148
Cell generations, formula to determine, X, 222
Cell grafting, I, 110–112
Cell growth, IV, 157
Cell hybridization, VI, 216, 219, 226, 239, 263, 376; IX, 171
Cell hybridization technique, VI, 227
Cell hybrids, VI, 216
Cell isolation, VII, 309
Cell kinetic parameters, of ascites cells, X, 175
Cell life cycle, I, *see* specific organisms; Mitotic cycle measurements, III, IV, *see* cell cycle
measurements, III, IV, *see* cell cycle
Cell lines, IV, 35, 42; V, 19, 21, 42, 45–48, 53, 91–92, 148, 150; VI, 217–218
amphibian embryo, IV, 46
anuran embryo, IV, 33
BHK-21/C13, IX, 159
BHK-21/py6, IX, 159
BHK-21/SR, IX, 159
BHK-21/SR reverted, IX, 159
characteristics of, IV, 33
Chinese hamster ovary, V, 47
CHO-K1, IX, 159
drug-resistant, VI, 228
embryo, IV, 32, 42
frog, IV, 38, 45
FM3A, IX, 159
HeLa, V, 47
HeLa S3, IX, 159
JTC-15, IX, 159
L5178Y, IX, 159
L929, IX, 159
mouse L, V, 47
mutant, VI, 5
procurement of, V, 8–9
propagation of, IV, 32
temperature-sensitive, VI, 214
V79, IX, 159
Yoshida sarcoma, IX, 159
YSC, IX, 159
Cell markers, III, 108–110, 112–116
colchicine, III, 108–110, 112–116
thymidine, tritiated, III, 108–110, 112–116
Cell membrane(s), I, 277; IV, 185, 212, 370–371, 373, 375, 395, 404; V, 13, 43
Cell migration, IV, 30
Cell monolayers, adult kidney, IV, 21
Cell movement, role of nucleus in, IV, 187
Cell plate(s), IV, 134, 136, 149–150, 155–156, 159, 161
index, IV, 160
Cell populations, II, *see also* G_1 populations and G_2 population
analysis of, II, 323–357
cell types in mouse epidermis, II, 360–361
determination of progenitor cell cycle, II, 331–335
double labeling of, II, 332–335
G_1 and G_2 cell populations in, II, 361–379
handling of, for autoradiography, II, 326–328

history and terminology of, II, 324–326
measurement of tissue turnover and transit times, II, 346–348
over-all concept of, II, 390–394
patterns of cell divisions in, II, 359–395
physiological subpopulations, II, 361, 380–394
procedure for detection of G_2 in, II, 379–380
proliferation and specialization, II, 348–355
synchronous, VI, 48
thymidine-H^3 labeling of, II, 335–346
Cell preservation, V, 21
freezing, V, 21
thawing, V, 21–22
Cell proliferation and specialization, relationship between, II, 351–355
Cell protein content, VII, 453
Cell protein distributions, IX, 214
Cell separation, X, 93
Cell size, IX, 230
Cell size distribution, VI, 52, 61
Cell spectrometer, III, 353–354
Cell surface area, IX, 232
Cell surface area distribution, IX, 233
Cell suspensions
fixation of, IX, 199
methods for preparing, IX, 196
Cell synchronization, VI, 117
Cell synchrony
aminopterin, III, 30
amethopterin block
of DNA synthesis, III, 30–33
thymidine release and, III, 33–35
cold shock, III, 42
deoxyadenosine, III, 30
deoxyguanosine, III, 30
5-fluorodeoxyuridine (FUDR), III, 30, 33, 35
inhibition and release of DNA synthesis, III, 29–35
loss of, III, 39
multiple synchronization, III, 41
selection of mitotic cells, III, 35–36
Cell volume, VII, 373, 453; IX, 192, 223, 226, 230, 232
analysis of, IX, 221, 228, 230
analysis of DNA, IX, 229
determination of, VII, 453
plotter, VII, 274
Cell volume distributions, IX, 211, 214
of HeLa cells, IX, 209
Cell wall, IV, 157, 163
vital staining of, V, 342, 366–367
Celloidin, use in slide preparation, III, 205
Cellophane, IX, 21
Cellophane sheets, V, 18
Celloscope, III, 2, 6–7, 11–13, 17–19; IV, 101, 269, 418; V, 237
Cellular dry mass, IV, 136
Cellulose, IV, 389
Cellulose powder, VII, 62
Cemulsol NPT 10, VI, 319
Centrifugation, sucrose gradient, VI, 43
Centrifuge
Foerst plankton, V, 239
Sharples continuous-flow, V, 238
Centriole(s), IV, 141, 380–381, 383–384, 386, 388, 396; VII, 243–244; VIII, 134, 140
Centriole replication, III, 121
Centromeres, IV, 141; VIII, 183
Centromeric areas, V, 34
Cephaloridine, IV, 55
Cephalothin, IV, 55
Cercaertus, V, 135
Cercaertus concimnus, V, 134
Cercaertus lepidus, V, 134
Cercopithecus aethiops, VI, 166; VIII, 44
Cerebellum, VII, 375
Cerebral cortex, VII, 90, 160
Cerophyl, IV, 255, 332
infusion, IV, 119, 121, 250
Cervix, VI, 165, 170
Cesium chloride, VII, 123
Cesium sulfate, VII, 123
Cetyl-trimethyl ammonium bromide, VII, 123
Chaikoff's homogenizer, IV, 208, 213
Champy's solution, V, 283
Chang's liver cells, VI, 118, 125, 152
Chaos chaos, I, 86, 89, 279, 285, 289–291, 294, 296, 300–301; IV, 95
Characean cell, II, 168, 170
Charcoal
activated, IX, 115
as adsorbent for thymidine, IX, 116
Charcoal adsorption, VII, 441
of ^{32}P, VII, 452
Chauveau centrifugation, IV, 221

Chauveau method, IV, 197–198, 202–206, 208–209, 211–212, 215, 220; V, 173, 188; VII, 7; VIII, 155–156
Chelating agent
  EDTA, II, 219
  HEDTA, II, 219
  histidine, II, 219
  sodium citrate, III, 123
  Versene, II, 6–7
Chemical mutagens, VI, 214
Chemoautotrophs, II, 227
Chemostat, I, 142
Chemotactic agents, VI, 326–328, 334–335, 341
Chemotaxis, VI, 325, 337–339, 341
Chemotherapeutic agents, VI, 101
  effects of, IX, 227
Chemotherapy, VI, 100
Cherenkov counting, VII, 443
Chi-square test, III, 8–12
Chick cells, transformation of, VIII, 421
Chicken, VI, 177
  embryo extract, V, 40, 50
  embryo fibroblasts, VI, 174
  embryonic cells, V, 38; VI, 169
  embryos, V, 54; VI, 167
  fibroblasts, VI, 92, 107, 110
  serum, VII, 315
Chicken viruses, VI, 167
Chinese hamster(s), V, 7–8; VI, 218, 346
Chinese hamster cell(s), V, 91, 106–107, 113; VI, 70, 72–73, 94, 114, 132, 135–137, 140–141, 151, 211, 214, 231–232, 235, 275–276, 287, 292, 300, 361, 374; VII, 148
  chromosome extraction from, II, 115–118, 120, 124–129
  Don-C, III, 26–27, 37–38, 40–41
  Don lines, VI, 132
  fibroblasts, III, 39–40
  from lung, III, 350
  from ovary, III, 352
  generation time of, II, 114
  isolation and fraction of chromosomes, III, 285
  lines, III, 152, 278, *see also* Colcemid, VI, 218
  metaphase arrest in, II, 114–115
  metaphase chromosomes of, III, 279
  properties of isolated chromosomes, II, 118
  RNA, DNA, and protein content of chromosomes, II, 119
Chinese hamster cell cycle, VI, 102
Chinese hamster chromosomes, VI, 311
Chinese hamster fibroblasts, VI, 286, 299, 308
Chinese hamster ovary cells, VI, 71–90, 94, 97, 100–102, 104, 107–108, 214, 219, 222–223, 227, 235–238, 240, 242, 246, 248–250, 252–254, 257–260, 265, 267–269, 271, 274, 276; VII, 332, 340, 344, 346, 375, 411, 419–420, 425, 429, 438, 453
  analysis of genome, VIII, 197, 200
  auxotrophic mutants of, VI, 250
  cell cycle, VI, 101
  colchicine-resistant lines of, VI, 244
  Feulgen-DNA distributions of, IX, 213
  growth, VII, 333
  karyotypes, VIII, 201
  line, V, 47, 90, 110; VII, 330
  mutant frequencies in, VI, 268
  mutant lines of, VI, 256, 262, 275
  mutations in, VI, 263
  subclone of, VI, 253
Chinese hamster V79 cells, VII, 374
Chironomids, II, 73, 79
Chironomus, II, 85, 87; IX, 377
  nuclei and chromosomes of, IX, 378
Chironomus tentans, II, 67
  salivary gland RNA in, I, 433
  salivary glands of, X, 18
Chironomus thummi, II, 67, 72, 74–75, 78, 81–83
Chitinase, IX, 378
Chlamydomonas, IV, 375
  culture conditions for
    aeration, III, 125
    agitation, III, 126
    lighting, III, 125
    media, III, 123–125
    temperature, III, 125
  induction of synchrony, 128–145
  light and dark cycles, 141–143
Chlamydomonas reinhardi, II, 227; III, 122
Chloealtis conspersa, I, 232
Chloramphenicol, IV, 56; VI, 163; VIII, 345; IX, 336–337
Chloramphenicol resistance, VIII, 335
Chlorella, III, 120; IV, 245, 255, 326; V, 305, 315

as symbiotic organism, I, 112
autospores of, I, 161, 175–177
autotropic growth of, I, 161–162, 165
bursting division of, I, 174, 178, 180
carbon sources of, I, 164–165
continuous culture of, I, 172–173
culture apparatus for, I, 169–172
culture of, I, 163–182
growth of, I, 160–162
handling and culture of, I, 159–188
light for culture of, I, 168–169
media for, I, 161–162, 174–175
medium for axenic cultures of, IV, 332
nutrient media for, I, 165–167
photosynthesis in, I, 159
stock cultures of, I, 163–164
synchronous culture of, I, 173–181
taxonomy of, I, 162–163
temperature for culture of, I, 169
Chlorella ellipsoidea, I, 162, 174
Chlorella pyrenoidosa, I, 162–163, 175–176; V, 308, 316, 320, 322
cell sizes of, V, 317–318
photosynthetic reaction in, V, 319
size of daughter cells, V, 321
synchronous cultures of, V, 317
Chlorella vulgaris, I, 162–163
Chlorine-36, physical properties of, X, 294
Chlorobenzene, IV, 142; V, 226
1-Chlorobutane, IV, 142; V, 226
Chlorocide, V, 9
Chloroform, VII, 11, 14–17
Chlorogloea fritschii, synchrony in, V, 382
Chlorogonium elongatum, IV, 253–254
p-Chloromercuribenzoate, VII, 410, 418
Chlorophen, VI, 187
m-Chlorophenylhydrazone, IV, 94
Chlorophyceae, III, 121
Chlorophyll, VI, 183
Chloroplast(s), II, 218; IV, 50
Chloroplast fragments, IV, 96
Chlorpromazine, IV, 87
Chlortetracycline, VI, 163, 192
Cholesterol, VI, 9
Choline, IV, 171; V, 15, 60; VI, 8
Choline kinase, VII, 399
Cholinesterase, VII, 357
Chondrocytes, V, 42, 49–50
Chorioallantoic membrane, VII, 228
Choroid, V, 52
Choroid cells, V, 52
Choroid coat, V, 52
Chorthippus curtipennis, I, 232
Chortophaga viridifasciata, I, 231–233, 238
see also Grasshopper, Neuroblast, 1
Chortophaga viridifasciata australior, I, 231
Chrom-alum gel, VIII, 336
Chromatids, IV, 492
Chromatin(s), IV, 139–141, 198, 205, 213, 379, 381, 386, 462; VII, 30, 129–130, 140–141, 143, 145
contamination by, IV, 223
disruption of, VIII, 161
DNA of fractions, VIII, 169
fractionation of, VIII, 163
integrity of, VIII, 159
isolation of fractions, VIII, 160, 172
nucleolus-associated, IV, 219
reconstitution of, IX, 351, 372
Chromatin fibers, VII, 140, 143, 145
Chromatin fraction II, IX, 372
Chromatin fraction II proteins, IX, 368–369
Chromatography solvent, VIII, 234
Chromatography standard solution, VIII, 234
Chromium, radioactive, VIII, 404
Chromocenters, VIII, 166
Chromogene, V, 327
Chromomeres, X, 20
Chromosomal analysis, VI, 199
Chromosomal fibers, IV, 79; VII, 141, 143
Chromosomal mapping, in *S. pombe*, IV, 172
Chromosomal nondisjunction, IX, 213
Chromosomal proteins, VII, 143; IX, 389
Chromosomal segments, removal of, II, 80–82
Chromosomes, IV, 1–3, 9–10, 15, 20, 33, 42, 45, 77, 79–80, 139, 141, 167, 172, 292, 354, 356, 379, 381, 383–384, 388, 447, 459, 479, 488, 511; V, 1–4, 6, 12, 20, 22–23, 26–27, 29, 33–34, 45, 76, 87, 94–95, 104, 106–113, 116–122, 128, 130, 133, 135–140, 142–143, 152, 155, 157–158; VII, 120
aberrations in arrested blastulae, II, 31, 33
acrocentrics, V, 6, 34
autoradiographs of, VII, 147
autoradiography of, II, 52–53
banding, VI, 218
Chinese hamster, VI, 291–292

complement of, VI, 227
composition of, VI, 293
contents of, VI, 294
diploid number of, VI, 218
DNA content of, I, 2, 6; II, 117, 119
effect of deoxyribonuclease and trypsin on, II, 119
effect of pronase on, II, 119, 128–129
effect of sodium lauryl sulfate on, II, 119, 128–129
electron microscopy of, II, 54–56
expulsion of, VII, 306
extraction of, VII, 305
fractionation of, VI, 292
giant, II, 62, 66
HeLa, VI, 284, 286
human, VI, 213, 216, 219
in meiotic metaphase, II, 40
isolation of, II, 74, 113–129; VI, 283–315
lampbrush, I, 3, 419, 427; II, 37–58
manipulation of isolated, IX, 386
markers on, VI, 213
metaphase, IV, 1, 14, 478, 486
number of, IV, 16
of *Drosophila* spermatocytes, II, 38
of mammals, I, *see* Mammalian cells
permanent preparations of, II, 117–118
polytene, II, 61–91
premature condensation of, V, 77, 87, 105, 107–108, 110, 113, 116–121
preparations of, V, 1–36
properties of, II, 118–119, 127
purification of, II, 116
removal of, II, 80–82
RNA and protein content of, II, 119, 127
RNA synthesis in, II, 42, 52
semiconservative replication in, I, 5
sizes of, II, 67
structure of, VII, 147
submetacentrics, V, 6
suspensions of, II, 116
transplantation of, II, 84–85
Chromosome aberrations, VI, 152–154, 156, 239
Chromosome analysis, IV, 43
Chromosome fragments, IV, 79
Chromosome loss, VI, 241; VIII, 13, 15, 38
Chromosome mapping, V, 99
Chromosome morphology, IV, 509
Chromosome number analysis, IX, 216
Chromosome number histograms, IX, 216–217
Chromosome numbers, V, 131, 133–136, 141; VI, 31–34, 218, 250, 253
variability of, VI, 212
Chromosome replication, III, 181–184; IV, 489
Chromosome segregation, IV, 511
Chromosome spreads
from amphibian leukocytes, III, 89–91
of *Rana catesbeiana* lens, III, 56–58
Srinivasan and Harding technique, III, 57
Chromosome structure, IX, 387
Chrysoidine, V, 343, 353, 365–366, 368
Chrysomonads, IV, 411
α-Chymotrypsin, VI, 361
Cilia, V, 222; VIII, 328; IX, 282
isolation of, V, 262–265; X, 111, 115
microtubules of, V, 263, 272
phospholipid composition of, X, 131
subfractions of, X, 115–116
Ciliary antigens, VIII, 355
Ciliary beating, VIII, 345
Ciliary membrane, X, 117
Ciliates, I, see also specific organisms; IV, 310, 313
experimental techniques with, I, 109–125
pinocytosis in, I, 281, 290
Cinemicrography, I, 388–389, 397–400
Cinephotomicrography, III, 147–148, 152, 157
Circadian rhythm, IV, 283, 286
Circaea lutetiana, II, 178
Circotettix verruculatus, I, 232
Circular dichroic spectra, VII, 82
Citrate solution, V, 13
Citric acid, V, 22
Citric acid procedure, IV, 196, 209
Citric acid-sucrose procedure, IV, 211
Cladosporium, II, 227
Cleavage histones, III, 330
Cleland's reagent, X, 101
Clonal colonies, IV, 41–42
Clonal cultures, V, 39, 43, 49
Clonal derivatives, V, 20
Clonal growth, V, 38–40, 42–43, 46–47, 49, 52–54, 57–59
medium for, V, 40
Clonal inoculum, V, 45
Clonal line, V, 45
isolation of, IV, 40

Clonal morphology, V, 92
Clonal populations, V, 44–46, 55–56
  selection of, V, 44
Clonal strains, V, 20–21
Clone(s), IV, 43; V, 20, 38
  drug-resistant, IX, 172
  growth of, VI, 138–139
  hybrid, V, *see* Hybrid clones
  isolation of, IX, 158, 164
  isolation of UV-sensitive, IX, 174
  manual imprinting of, IX, 170
Clone isolation, VI, 223; VIII, 27, 29, 43
Clone selection, V, 45
Cloned cells, V, 55, 58
Cloning, IV, 40; V, 20, 38, 46; VI, 223
  by agar suspension, IX, 165, 167
  dishes, IV, 40
  in liquid medium, IX, 165, 167
  on agar plates, IX, 167
  procedure, IV, 42
  techniques for, IX, 165
Cloning cylinder(s), V, 54, 56
Cloning efficiency, VIII, 105, 377
Cloning medium, VI, 50
Clostripain, IX, 197
Cobalt-60, II, 257
Cocarboxylase, VI, 8
Coccus cacti, V, 327
Cockerel plasma clot, III, 78, 81
Coconut milk, II, 14
Coelomic fluid, II, 44
Coenzyme A, VI, 8
Coenzyme Q, VIII, 213, 216, 224
Colcemid, II, 363, 381; III, 36, 38–39, 157, 220; IV, 74, 481–482, 492; V, 4, 11–12, 23–24, 30, 94, 96, 103, 105, 108–109, 113–117, 150, 252; VI, 93–94, 100–102, 114, 116, 199, 219, 287–289, 355; IX, 51, 55, 72
  arrest of mitosis, III, 36, 351
  Chinese hamster cells Don-C, III, 37
  concentrations of, IX, 73, 79, 81–82
  effect reversal of, IV, 74
  effects of, IX, 56
  HeLa lines, III, 39
  injections of, IX, 61
  reversibility of, IX, 73
  synchronization, III, 36–39; VIII, 114
  time of incubation in, IX, 82
Colchicine, II, 115, 347, 363, 365, 368, 372, 374–375; III, 57–58, 98–99, 105, 109, 171, 190, 220; IV, 43, 191, 481, 488, 491–493; V, 252, 272–273; VI, 114, 199, 228, 231; VII, 410, 418; VIII, 5; IX, 55
  as cell marker in mitosis, III, 108–110
  induction of octoploids and polyploids, III, 186, 190–191
  induction of tetraploids, III, 98–102, 108–109, 112, 190
  metaphase accumulation, III, 190
  mitotic inhibitor, III, 189
  preparation of solutions, III, 190
  resistance to, VI, 244–245, 268
Colchicine-^{3}H, V, 270; VI, 245
Colistimethate, IV, 55
Collagen, V, 42
Collagenase, V, 43, 50, 53, 67–68, 152; VI, 319; VIII, 53, 67; X, 95
  in tissue dispersal techniques, IX, 197
s-Collidine, VI, 300
Colloidal silica, X, 96
  as a gradient medium, X, 88
  preparation of, X, 88
  properties of, X, 87
  self-forming density gradient, X, 91
  toxic action of, X, 88
  toxic activity, X, 94
Collodial silica density gradient, X, 101
Colloidal silica density gradient centrifugation, X, 85
Colloidal silica gradients, X, 92, 97–98
Colony counting, VIII, 13
Colony formation, V, 47, 55
Colony growth, V, 58
Colony morphology assay, VIII, 412
Colony morphology method, VIII, 406, 409
Colorimeter, photoelectric, V, 238
Colpidium colpidium, IV, 308
Colpidium colpoda, IV 308
Complement, V, 149
Complement fixation, VIII, 403–404
Complementation, VIII, 34
  analysis, V, 99
  groups, VIII, 16
  interallelic, IV, 173
  tests, VIII, 12, 16
Concanavalin A, VI, 214; VIII, 77, 85–86, 90, 397–398, 424, 428; IX, 6, 198

binding of, IX, 231
monovalent, VIII, 397
resistance to, VIII, 79
selection of resistant cells, VIII, 86
Conditional lethal mutants, VI, 246, 250, 260
isolation of, VIII, 425
temperature-sensitive, VI, 246–247, 255–256
Conditional-lethality, VI, 219
markers, VI, 217, 268
Conessine, IV, 450
Conference on Cell Cultures for Virus Vaccine Production, VI, 164
Congo red, V, 344
Conjugation
chemical induction of, IV, 289
incomplete, IV, 277
Contact inhibition, V, 18
of growth, VII, 355
Contaminants, V, 18, 20
Contamination
bacterial, V, 18
fungal, V, 12, 18
microbial, IV, 35
yeast, V, 18
Contractile vacuole, IV, 343–345, 355, 357, 359, 365, 371, 373, 378, 395, 404; VI, 189; VIII, 339
Convallaria majalis, II, 178
Convolvulus arvensis, III, 177–179, 185
Coomassie brilliant blue, VII, 79, 94
Coors filter, IV, 331
Copper sulfate, IV, 55
Copromastix prowazeki, IV, 359
Coriphosphine, IX, 204
Cornea, V, 52
Corneal epithelium, V, 133
Cortical granules, IV, 25
Corynebacterium sp., VI, 151, 180
Cot values, VIII, 172–173
Cotyledons, VIII, 157
Coulter counter, III, 2, 5–9, 13, 16, 33–34, 163; IV, 150, 154, 269, 416–418; V, 237; VII, 274, 317, 333–335; VIII, 70
standardization of, VIII, 71
Countercurrent distribution, IX, 38
of epithelial cells, IX, 46
Counting chamber
Levy, V, 235–236
plankton, V, 234
Coupled transport, VII, 402
Coverbond, VIII, 263
Coxsackie virus, VI, 165
Craigia hominis, IV, 349
Cream separator, IV, 125, 129, 267; V, 238; VIII, 348
Creatine kinase, VI, 303
Cresyl blue, V, 342, 345, 347, 351, 357, 368
Cricetids, V, 7
Cricetulus griseus, IV, 16; VI, 346
Crithidia fasciculata, IV, 89–90, 92–93
Crossing over, somatic, VIII, 12
Cryoform, VIII, 252
Cryostat, VIII, 255
Cryptic transformants, VIII, 390, 412
Crystal violet, IV, 150; V, 49, 53, 95; VI, 50
α-Crystalline lens proteins, VII, 97
CsCl gradients, IX, 104, 106, 109, 112
CTP, V, 35
deamination of, VII, 382
pool, VII, 414
synthetase, VII, 381–382
Culture growth, IV, 154
Culture of roots and root cells
adenine sulfate, III, 179
media, III, 176–178
myoinositol, III, 179
nicotinic acid, III, 177, 179
root calluses, III, 178–179
thiamine-HCl, III, 177
Torrey's basal medium for pea roots, III, 177
Culture tube borosilicates, II, 219
Cultured cells
monolayer of, IV, 28
Cultures
axenic, V, 221
clonal, V, see Clonal cultures
homocontinuous, V, 303
long-term, V, 19
monoxenic, V, 221
primary, V, 55
synchronized, V, 319, 321
synchronous, V, 302–304
Cuscuses, V, 135
CV-1 cells, VII, 143
Cyanide, IV, 99, 106, 109–112
Cyanogen bromide peptide patterns, VII, 82
5-Cyanouracil, VII, 387

Cyclic AMP, V, 120; VI, 326, 328, 330, 336–342; VII, 377, 399, 449–450; VIII, 77–78, 90
radioimmunoassay for, VII, 449
Cyclic AMP phosphodiesterase, VII, 393
Cyclic GMP, VII, 449
Cyclic nucleotides, assay of, VII, 449
Cyclodeaminase, VII, 385
Cycloheximide, IV, 459; V, 273–274; VI, 89, 106; VIII, 132–134, 345; IX, 65, 336–337
cell cycle block, IX, 86
transition point for, IX, 100
Cyclohexylamino propane sulfonic acid, VI, 310
Cyclohydrolase, VII, 385
Cycloserine, IV, 55
CYDAC, IX, 181
Cynomolgus monkey, VI, 166
Cysteine, V, 62; VI, 6
Cystine, VI, 6
Cysts, IV, 342, 344
Cytidine (CdR)-H^3, II, 303, 305, 307–308; III, 193, 201
Cytidylic acid, IV, 14
Cytoanalyzer, IX, 181
Cytocenters, VII, 304–305
Cytocentrifuge, V, 93–94
Cytochalasin A, VII, 213
Cytochalasin B, VII, 189, 191–192, 203–207, 212–215, 217–221, 223–226, 233, 242, 245, 410; VIII, 78, 124, 146–147
cytotoxic effects of, VII, 222
enucleation with, VIII, 142, 145
Cytochalasin C, VII, 213
Cytochalasin D, VII, 213
Cytochalasin E, VII, 214
Cytochalasin F, VII, 214
Cytochemistry, historical survey, I, 1–8
Cytochrome(s), IV, 100, 102, 107, 111–113; V, 192; IX, 273, 275–276
Cytochrome *a*, IV, 104
and a_3, IV, 89, 109–110
Cytochrome *b*, IV, 100, 109–110
Cytochrome b_5, VIII, 226; IX, 262, 272–273; 275, 277
reductase flavoprotein, VIII, 226
Cytochrome *c*, IV, 89, 104, 109–110
Cytochrome c_1, IV, 109
Cytochrome *c* oxidase, IV, 110; VIII, 227; IX, 261, 272, 275
Cytochrome *o*, IV, 112
Cytochrome oxidase, IV, 89, 102, 105, 110, 112, 205; V, 192, 195; VIII, 226
Cytochrome P-450, VIII, 224, 226; IX, 262, 273, 275
Cytofluorometry, IX, 221
Cytogenetic analysis, II, 43
Cytokinesis, III, 120, 138, 214; IV, 380, 386–387
inhibition of, VII, 225
lesion in, VI, 215
Cytokinins, III, 171, as growth factors, III, 195–196
Cytology, automated, IX, 180
Cytolyze, II, 12
Cytomegalic virus, VI, 164–166
Cytomegalo-like virus, VI, 168
Cytopathogenicity, VI, 170
Cytophotometry, III, 73; IX, 180
Cytoplasm
microinjection into, II, 86–88
transplantation of in *Chironomus*, II, 82–84
Cytoplasmic debris, II, 114, 117, 127, 140
Cytoplasmic exchange
during conjugation, IV, 294
Cytoplasmic organelles
autonomy of, VII, 243
Cytoplasmic symbionts, IV, 294
Cytosine arabinoside, VI, 99; VII, 377, 384, 396, 412, 417; VIII, 5, 44, 81; IX, 51, 54, 60
effects of on mitotic activity, IX, 56
injection of, IX, 61
Cytosine arabinoside-Colcemid
effects of, IX, 59
Cytostatic drugs, IX, 120
Cytostome, IV, 344–345, 408–409
Cytotoxic antibody test, VI, 197
Cytotoxic tests, VIII, 402–403

D

Dalton's law, II, 157, 181
Dansyl chloride, IX, 213
Darco, VII, 442
Dasyuridae, V, 128, 130–131
Dasyurids, V, 135
Dasyurus hallucatus, V, 131

Dasyurus maculatus, V, 131
Dasyurus quoll, V, 131
Datura stramonium, II, 178
Deamination procedure after Van Slyke, III, 329
Deciliation of *Tetrahymena*, V, 282
Deer fibroma virus, VIII, 371
Defano fixative, V, 283–284
Dehydrogenases
  lactate, II, 71–72
  malate, II, 72
Delta, IV, 307
Deltotrichonympha, IV, 349
6-Demethyl-7-chlortetracycline, VI, 163
Densitometry, use in quantitative autoradiography, I, 321
Density-dependent inhibition of growth, VIII, 384
  mechanism of, VIII, 386
Density gradient(s), III, 73
  collection from, X, 96
  in purification of chromosomes, III, 286–291
  resolution of with angle-head rotor, X, 93
Density gradient centrifugation, II, 100, 119; III, 291
Density inhibition of DNA synthesis, VIII, 390
Deoxyadenosine kinase, VII, 393, 396
Deoxycholate, IV, 212, 224; VII, 5–6, 8, 16
Deoxycytidine kinase, VII, 381, 388–389
Deoxycytidylate deaminase, VII, 365–366, 381, 383–384, 399
Deoxyribonuclease, II, 201, 310, 312; III, 152; IV, 13, 214, 224–225, 513; V, 11; VI, 153, 299, 303, 317; VII, 15–16
  digestion with, VIII, 215
  electrophoretically purified, VIII, 215
  kinetics of, II, 42
  pancreatic, II, 119
  solution, I, 2, 435; VII, 342; X, 39
Deoxyribonuclease, I, electrophoretically purified, VII, 33
Deoxyribonucleic acid, *see also* specific organisms, Autoradiography, IV, 10, 13–14, 31, 143, 151–154, 157, 161, 163, 203, 211, 213, 379, 385, 394, 411, 413, 460, 462, 478; V, 34–35, 76–77, 107, 120–121, 130, 149, 152–155, 157–159, 163; VI, 51, 70, 86, 94, 98, 149, 151–153, 156, 171, 183
  analysis of, VII, 434; IX, 221, 234
  cell content of, II, 119
  chloroplast, VII, 107
  chromosome content of, II, 119
  circular, VII, 107, 118
  colorimetric determination of, VII, 433
  constancy, VIII, 182, 203
  content of individual cells, VIII, 190
  cytochemical detection of, I, 1–2
  cytoplasmic, III, 316
  demonstration of, II, 255, 307
  depurinated, VII, 107
  determination of, I, 29, 418–419; VII, 336
  distribution of values, VIII, 190
  extraction, basic methods for, VII, 111
  fractionation techniques, VII, 119
  from macronuclei, IV, 226
  from micronuclei, IV, 226
  high molecular weight, VII, 120
  in individual cells, VIII, 195
  in protein synthesis, I, 4, 6
  in *Schizosaccharomyces pombe*, IV, 132
  in spermatids, I, 431–432
  in thymocytes, I, 431–432
  incorporation of BUdR into, IX, 121
  incorporation of ^{32}P, IX, 106
  inhibition of, VII, 377
  inhibition of synthesis of, I, 53
  isolation and characterization of, VIII, 347
  kinetics of labeling, mammalian, IX, 116
  labeling of *in vivo*, IX, 115
  labeling with ^{32}P, IX, 106
  length of, VI, 308
  mass increase of, IX, 110
  melting temperature, VII, 108
  microextractions of, VII, 122
  microphotometric determination of, III, 199
  microspectrophotometry of, I, 2
  mitochondrial, VII, 107
  molecular weights of, VII, 107
  mycoplasma, VI, 159
  of chromosomes and nuclei, III, 171, 198–200
  of *Naegleria gruberi*, IV, 461
  physical-chemical characteristics of, VII, 106
  preparation on microfilters, X, 38
  procaryotic, VI, 151, 153
  purification and fractionation of, VII, 116

quantitative determination of, III, 199–200
quantitative staining procedures for, IX, 202
rates of, VII, 330
repetitive, VIII, 172–173, 176
replication in phytohemagglutinin stimulated cells, III, 316
replication of, IV, 195
retention on microfilters, X, 39
semiconservative replication of, I, 5
semiconservative segregation in, II, 256
sequence repetition in, VIII, 174
sequences in, II, 113
shear degradation, VII, 107
single-stranded, VII, 107
spreading of, VII, 149
staining reagents for, III, 200
azure B, III, 200
methyl green, III, 200
synthesis of, I, 2, 5–6, 87–88, 387–401; II, 68, 131, 256, 260, 263, 329, 331, 340, 369–370, 373; V, 29, 77, 99–102, 104–105, 121, 149–150; VII, 278–279, 337–338, 340, 344, 349–352, 355, 358–359, 362–363, 367–370, 372, 378
in *C. moewussi*, III, 144
in cell cycle, III, 155
synthetic index of, II, 326, 338–339, 347, 378–379; III, 262, 273
diurnal rhythm of, III, 273
thymidine H^3 labeling of, I, 88
X-ray effects on, VII, 378
Deoxyribonucleoside kinase, VII, 396
Deoxyribonucleoside triphosphates
enzymatic determination of, VII, 445
levels, VII, 376
pools, VII, 376
Deoxyribonucleosides, arresting properties of, VII, 383
Deoxyribonucleotides
complex mixtures of, VII, 439
pools, VII, 376, 444
Deoxyuridine triphosphatase, VII, 381
DePex, VIII, 190
Deplasmolysis, II, 146, 151, 153, 187
Deproteinizations, VII, 55, 114
of DNA, VII, 112
with phenol, VII, 30
Desferal, VII, 383
Desiccator, V, 20
Desicote, II, 273; VII, 291
Detergent-disinfectant solutions, VI, 187
Dexamethasone, VI, 15, 215
Dextran, VI, 2
Dextran gradients, IV, 160
Dextran sulfate, VII, 9, 33
Dextran T40, X, 86
Dextran T70 gradient, X, 103
Dextran T500, IX, 29, 35
Dextrose, VI, 157
Diakinesis, IV, 3
Dialysis membrane, III, 83
Dialysis tubing, III, 80
Diaminopurine, IV, 313; VI, 214
Diamond knife, II, 271, 273
Diamond object marker, Leitz, IV, 40
Diamond pencil, II, 241, 270
6-Diazo-5-oxo-L-norleucine, VII, 380, 391
5-Dizaouracil, VII, 387
Dibutyryl cyclic AMP, VII, 344, 389, 401, 410, 425
2,4-Dichloro-α-naphthol, in Sakaguchi reaction, III, 204, 207, 321
Dichlorodifluoromethane, VIII, 252
Dichloroethane, IV, 142; V, 226
2,4-Dichloraphenoxyacetic acid, III, 178
see also Auxins, III
Dichroism, V, 344
Dicoumarol, IV, 87
Dictyostelium, I, 10; II, 401; IV, 459–460
Dictyostelium discoideum, II, 398–399, 403–407; IV, 433; VIII, 313
Dictyostelium mucoroides, II, 398
Dictyostelium purpureum, II, 398
Dicyclohexylcarbodiimide, VII, 61–62
Didascalus, IV, 408
Didascalus thorntoni, IV, 358, 408
Didelphidae, V, 128, 141–142
Didelphis, V, 129, 146, 148–150, 152–153, 155, 158
Didelphis azarae, V, 141
Didelphis marsupialis, V, 6
Didelphis marsupialis aurita, V, 141
Didelphis marsupialis virginianus, V, 141–142, 144–145, 152
Didelphis virginiana, V, 6
Didinium, IV, 308, 322
Didinium nasutum, IV, 310
Didymium, IV, 406

Didymium difforme, I, 13
Didymium nigripes, I, 15; IV, 347, 379, 392, 405–406
Didymium squamulosum, I, 13
Dienes stain, VI, 160
Diepoxybutane, IV, 174
Diethylpyrocarbonate, VII, 34; X, 66, 177
Diethyl sulfate, IV, 174
Differential centrifugation, II, 79, 115
Differentiation, of spermatids, IV, 3
Diffraction effects, IX, 188
Diffraction intensity, IX, 189
Diffusion, II, 151
  constants, VII, 416
  facilitated, VII, 402–403, 406–408, 416
  simple, VII, 402–404, 407–408, 411, 413, 416
Diffusion chambers, intraperitoneal, VIII, 312
Diffusion coefficient, VI, 330, 332; VIII, 272
Diffusion constant, VI, 331
Diffusional uptake, VII, 406
Digitonin, V, 260, 263
Dihydrofolate, VII, 384
Dihydrofolate reductase, VII, 385
  inhibition of, VII, 384
Dihydroortase, VII, 379
Dihydroortic acid, VII, 380
Dihydroorotic acid dehydrogenase, VII, 379
Dihydrostreptomycin sulfate hydrochloride, IV, 27
Dihydrouracil, VII, 387
Dihydrouracil dehydrogenase, VII, 381, 386–387
Diisopropylfluorophosphate (DFP), IX, 352
Dileptus sp., IV, 310, 322
Dimastigamoeba, IV, 353, 358, 412
Dimastigamoeba gruberi, IV, 353, 355–356
Dimastigamoeba tachypodia, IV, 352
N^6-Dimethyladenosine, VII, 405
4-Dimethylaminoazobenzene, VI, 13
Dimethylbenzanthracene, VIII, 78
Dimethylnitrosamine, VIII, 373
Dimethylsulfoxide, IV, 38–39; VI, 36–38, 183, 222; VII, 42
Dimorpha radiata, IV, 353
Dinitrofluorobenzene procedure
  after Bloch, III, 325
  after Danielli, III, 325
Dinitrofluorobenzene reaction, III, 323–325
Dinitrofluorobenzene-Sakaguchi procedure after Bloch and Brack, III, 327
Dinitrofluorobenzene-Sakaguchi reaction, III, 325–327
Dinitrophenol, III, 191
Dinoflagellates, staining of, II, 283
Diphenylamine, IV, 110, 112
Diphenylamine technique, IX, 218
Diphtheria toxin bottles, IX, 314
Diphtheroids, VI, 179
Diploid cell lines, VI, 169, 194; VIII, 3
Diploid human fibroblasts, VI, 214, 231–232
Diploid karyotypes, VI, 217–218
Diploid lung culture, VI, 286
Diplotene, IV, 3, 502
Dipteran cells, II, 62, 66
Disc electrophoresis, III, 71
Disinfectants, V, 4, 10
Disintegrator
  Mickle, IV, 135
  ultrasonic, IV, 150
Dissosteira caoline, III, 152
Dissosteira carolina, I, 232
Division, IV, 149, 191
  nuclear, IV, 344
  rate, in ameba, IV, 188
"Division" triggers, III, 121
DNA, *see* Deoxyribonucleic acid
DNA-containing tumor viruses, VI, 176
DNA contents, VI, 83–85, 94, 96, 98, 101, 115–116; IX, 218, 227, 230
  abnormal, IX, 236
  analysis, IX, 216, 228
  distribution of per cell, X, 170
  measurements of diploid and heteroploid cells, IX, 213
  of mammalian cells, VIII, 203
  per cell, VIII, 202
  variations in, IX, 213
DNA content distribution, IX, 233–234
  for methylcholanthrene-induced mouse tumor, IX, 236
DNA distribution patterns
  of CHO cells, IX, 228
DNA distributions, IX, 208, 216–217, 220
  of CHO cells, IX, 208, 215
  of L5178Y, IX, 222
DNA-DNA hybrids, X, 6
DNA-histone complexes, IV, 224
DNA mass, VI, 82, 84, 87

DNA-nonhistone protein complex, VII, 141
DNA polymerase(s), IV, 196; V, 257–259; VI, 106, 298, 303; VII, 366, 381, 393, 445–446; X, 2
assay, VII, 424
DNA polymerase I, VII, 134, 136–137
DNA-protein complexes, VII, 130–131
isolation of, VII, 129
DNA-protein content distributions, IX, 226
DNA provirus, VIII, 426
DNA repair, IV, 172
DNA replication, synchronization of, IX, 111
DNA–RNA hybridization, VII, 27
DNA–RNA hybrids, thermal stability of, X, 336
DNase, *see* Deoxyribonuclease
DNA-specific staining
quantitative, IX, 210
DNA synthesis, IV, 158, 188, 304, 492–493; VI, 44, 46–47, 58, 62–63, 69, 71, 75–76, 78, 85, 89, 92–94, 98, 100–104, 106, 117, 122–124, 153
determined by hypoxanthine-^{3}H, VIII, 112
inhibitors of, IX, 71, 120
initiation of, VI, 69, 92, 105–106, 108
measurement of, IX, 108
mitochrondrial, VI, 125
rate of, VI, 108
synchrony of, VI, 57
DNA-to-cell-protein ratio, IX, 225–226
DNA-to-cell volume, IX, 225
DNA tumor viruses, VIII, 369
DNA values, IX, 221
DNA viruses, V, 78; VIII, 371
DNP, IV, 87, 99
Dog vaccine, VI, 170
Dominant gene D, II, 23
Domiphen bromide, IV, 55
Don, V, 19–20
Don C, V, 19
Don cell clones, VII, 268
Don Chinese hamster cells, VII, 267, 367, 391–392
Double diffusion technique, VIII, 356
Double labeling
application and limitations of, II, 334–335
principles of, II, 332–334
with colchicine and thymidine-H^3, III, 112–116
with thymidine-H^3 and -C^{14}, III, 184
Double-staining of cells, IX, 221
Doubling time, III, 26
Dounce homogenizer, II, 115; VII, 3, 5, 165, 167
Down's syndrome, VI, 350–351
Drinking habits of mice, III, 266–267
Drosophila, II, 72–74; VI, 216, 276; VII, 114, 149, 236
embryos, VII, 115
rRNA of, X, 206
5S RNA cistrons of, X, 339
Drosophila busckii, II, 74–75
Drosophila cell lines, X, 196
suspension culture, X, 196
Drosophila cells
fractionation of, X, 203
karyotype in culture, X, 200
macromolecular components of, X, 201
RNA extracted from, X, 204
Drosophila hydei, II, 67
Drosophila melanogaster, II, 67, 69, 73, 75; III, 329–330
cleavage histones, III, 330
Drosophila sp., II, 38
Drug resistance, VI, 219, 227, 243; VII, 400
markers, VI, 216–217
phenotypes, VI, 213–214
Drug-resistant cells, VI, 135, 234; VII, 364
Drug-resistant lines, VI, 212
Drug-resistant phenotype, IX, 172
Drug-sensitive cells, VI, 230
Dry mass, IV, 157, 163
of cells, IV, 152
Dryl's salt solution, IV, 275, 286, 303, 329, 331, 333–334; VIII, 353
Duck viruses, VI, 167
Dulbecco's medium, VII, 252, 254
Dunaliella marina, IV, 253–254
Duodenal progenitor cell population, II, 349–351
Duodenal villi, VII, 185
Duponol, VII, 147
Dyes
as inducers of pinocytosis, I, 295–296, 299–300
binding of, X, 54

E

Eagle's minimal essential medium, III, 57, 278, 350; IV, 73, 144, 150, 154–156; V, 40, 150–152; VII, 162, 252, 288, 315, 350
for amphibian tissues, III, 81–82

Eagle's spinner medium, IV, 74
Earle's balanced salts, VII, 163–164
Earle's medium, III, 58
  amphibian tissues, III, 81–82
Earth medium, IV, 252, 254
Eaton press, IV, 153
Echinostelium minutum, IV, 379, 393, 403, 405–406
ECHO virus, VI, 165, 168, 174
Edinburgh minimal medium, IV, 142–143
Eggs, VII, 110
Ehrlich ascites cell, V, 175, 180, 182
Ehrlich ascites tumor, IV, 208, 224; VIII, 138
  cells, VII, 374
    RNA from, VII, 57
Elastase, V, 43; VIII, 53, 67; IX, 198
Electric sensing zone instruments, III, 1–2, 9–10, 19, 23, *see also* Celloscope, Coulter Counter
  causes and remedies of distortion, III, 13–23
  comparison of, with electron microscope measurements, III, 7–13
  for counting and measuring of bacteria, III, 1–2
    of protozoa, III, 2
    of red and white blood cells, III, 1–2
    of tissue culture cells, III, 1–2
Electric sensing zone transducer, III, 3–5
Electrode
  Clark, V, 200–205
  Joliot, V, 212
    micro, V, 215
    ultramicro, V, 205
Electromigration
  apparatus, IV, 247
  of *Paramecium aurelia*, IV, 118, 126
Electron microscope autoradiography, III, 293–305
  background testing for, II, 295–296
  comments on, II, 296–299
  development for, II, 249; III, 297–299
    chemical, III, 299
    fine grain, III, 299
    physical, III, 299
  electron microscope techniques in, II, 285–308
  emulsion application for, II, 287–293
  emulsion coating for, II, 245–248
  examples of, II, 307–308
  high-contrast thin sections for, II, 229
  intermediate layer in, II, 245
  light microscopic techniques in, II, 257–285
  obtaining thick sections for, II, 268–274
  optimum emulsion layer for, II, 321
  photographic processing for, II, 293–295
  resolution, III, 295–296
  section mounting of, II, 241
  sensitivity of emulsions, III, 299–304
  staining in, II, 274–285
  staining procedure for, II, 243–245
  stains for, II, 276–284
  storage in, II, 248–249
  stripping and mounting for, II, 250–251
  usefulness of thick sections, II, 257–268
Electron transfer, IV, 87, 89
Electronic cell counter, I, 391; VIII, 55, 70
Electronic particle counting, VII, 334
Electrophoresis, X, 49, 59
  acrylamide gel, II, 42
  analysis of, II, 134
Electrophoretic chamber, X, 32
Electrophoretic mobility, of cells, IX, 33, 40
Elodea sp., II, 155
Emacronucleation, V, 281
Embedding medium
  Araldite, II, 268, 275, 281–282
  Epon, I, 268, 275, 282–284, 303, 316
  methacrylate, II, 247, 268, 275–280, 317–319
  polyester, II, 268
  Vestopal W, II, 268, 280–281, 312, 316
Embedding technique, II, 312–320
Embryo(s), IV, 24, 27; VII, 110
  androgenetic, IV, 25
  blastula, IV, 29
  diploid, IV, 25, 29
  gastrula, IV, 29
  haploid, IV, 25–26, 29
    frog, IV, 20
Embryo line(s), IV, 42, 44
Embryonic induction, III, 45
Embryonic lung, VII, 288
Emetine, IV, 450
Emulsion, *see* Photographic emulsion
Encephalitis, VI, 165, 167
Encephalomyelitis virus, VI, 167
Encephalomyocarditis virus, VII, 96–97; IX, 162
  polypeptides, VII, 97

Endocytosis, VII, 225, 234
Endomitosis, IV, 170
Endomycetaceae, IV, 133
Endomycetoideae, IV, 133
Endonucleases, VII, 108
Endosperm, V, 76
Endosymbiotes, IV, 121
Endoxan, IX, 120
Energy-regenerating system, II, 79
Entamoeba invadens, I, 281
Enteroviruses, II, 100; VI, 166–167
Enucleate(s), IV, 65
Enucleated cells, IV, 65
  BHK-21, VIII, 134–136, 140
  cell division in, VIII, 140
  CV–1, VIII, 150
  fusion between, VIII, 148
  growth of, IV, 66
  longevity of, VIII, 137
  monkey, VIII, 138, 146
  possible uses of, VIII, 138
  protein synthesis in, VIII, 131
  responses of, VIII, 139
  ultrastructure of, VIII, 134
  uses and properties of, VIII, 127
  viability of, VII, 197
  virus activities in, VII, 197
Enucleation, IV, 66, 187, 244; V, 282
  effects of colchicine on, VII, 226
  of ameba, IV, 184
  of amphibian eggs, II, 3, 5–6, 19, 22, 24
  mechanism of, VII, 224
  percentage of, VIII, 139
  spontaneous, VIII, 124
  technology of, VIII, 125
Enzymatic digestion, I, 2–3
Enzyme activities, subcellular distribution of, IX, 272
Enzymes, rate of synthesis of, X, 243
Eosin, II, 328, 362, 383; V, 333; VI, 191
Eosin, Y, V, 44
Eosin Y-fast green procedure after Bloch, III, 340
Eosin Y-fast green reaction, III, 338–340
Eosinophil granulocytes, VII, 323
Eosonophils, VI, 335
Epichlorohydrin, IV, 197
Epidermal cells, IV, 27
  of haploid embryos, IV, 26
Epidermis, III, 45; IV, 27
  frog, staining of, II, 281
  G_1 and G_2 cell populations in, II, 361–372
Epidermoid carcinoma, VII, 161, 184
Epiphyses, V, 24
Epithelium, *see also* epidermis
  cell division in, II, 378
  functional compartment of, II, 325, 339
Epon, III, 62
Eppendorf pipette, X, 142
Epstein Barr virus, VIII, 108, 116–117, 371
Equine abortion virus, X, 101
Equine encephalomyelitis, VII, 12
Equine herpes, VIII, 371
Equine kidney cell cultures, VI, 168
Erd-Schreiber medium, IV, 50, 65
Erythrobacillus prodigiosus, IV, 253
Erythroblasts, VII, 225–226, 233, 323
Erythrocebus patos, VI, 166
Erythrocin B, X, 94
Erythrocytes, III, 45; V, 79, 90, 92, 117–118; VI, 154, 198, 341; VII, 97, 141, 143, 233, 238, 253, 318, 371, 373–374, 395, 402, 404, 428
  aldehyde-fixed, IX, 48
  electrophoretic mobilities of, IX, 33
  ghosts, VII, 230, 233, 237
  laser-damaged, VI, 326, 342
  partition of, IX, 41
Erythromycin resistance, VIII, 335
Erythiopoietic stem cells, VIII, 101
Erythrosin, V, 353; VI, 38
Erythrosin B, V, 44
Escherichia coli, I, 334, 340, 343; II, 257, 398, 400, 407; III, 6–7, 232; IV, 242, 253, 364, 414–415, 420, 433, 463; V, 35; VI, 44, 180; VII, 12, 136, 141
  chromosome of, VII, 147–148, 150–151, 153
  dATP pools in, VII, 370
  dTTP pools in, VII, 373
  electron microscopic measurement, III, 10–11
  nucleotides from, VII, 431
Eserine, VII, 357
Esterase activity, in viable cells, IX, 238
Estradiol-17B, V, 120
Ethanol, as fixative, IX, 201
Ethidium bromide (EB), VII, 107, 118, 124; VIII, 400; IX, 200, 204
Ethidium bromide protocol, IX, 242

Ethidium bromide stained cells, IX, 206
Ethoxyethanol, VII, 124
Ethyl methane sulfonate, IV, 174, 303; VI, 243, 245, 247–248, 251, 253, 258, 267–270; VIII, 4, 25–26, 31–32, 36
Ethylenediaminetetraacetic acid (EDTA), V, 19; VII, 124
  in culturing roots and root cells, III, 176
L-*O*-Ethylthreonine, VI, 105
Euchromatin, isolation of, VIII, 160
Euglena, II, 218–221, 223–224, 226–227; IV, 86, 97–101, 103; VII, 111
  chloroplast-free, IV, 96
Euglena gracilis, III, 239; IV, 96, 101–103, 113
  autotrophic media for, II, 219–222
  bleached strains of, II, 217–218
  continuous culture techniques, I, 146–147
  culture media for, II, 217–227
  general comments on, II, 226–227
  growth medium of, I, 144–145
  heterotrophic acid medium, II, 222–224
  light-dark cycle of, I, 144
  light-induced synchrony of, I, 144–147
  maintenance of, II, 218–219
  neutral and alkaline media for, II, 225–226
Euglenoid cells, IV, 101
Eukaryotes, III, 307
Eukaryotic cells, IV, 140
Eukaryotic unicells, IV, 242
Euparol, VII, 280
Euplotes, III, 314; IV, 286
  emicronucleation of, I, 120
  viable fragments of, I, 120
Euplotes eurystomus, I, 85–89
  cell life cycle of, I, 87
  culture of, I, 86–87
  DNA synthesis in, I, 87–90
  experimental uses of, I, 88–89
  generation time of, I, 87
  isolation of macronuclei from, II, 132–134
  isotope labeling of, I, 88–90
  macronucleus of, I, 87–89
  micronucleus of, I, 87
  micrurgy of, I, 88
  protein synthesis in, I, 88–89
  replication bands of, II, 133, 265–268
  RNA synthesis in, I, 88
  staining of, II, 283
  thymidine-H^3 labeling of, II, 265–268
Euplotes minuta
  isolating macronuclei, IX, 297
  procedure for isolating macronuclei, IX, 284
Euplotes patella, I, 86
Euplotes woodruffi
  isolation of macronuclear anlagen from, II, 135–137
  isolation of micronucleus from, II, 138
  thymidine-H^3 labeling of, II, 136
Eutherian mammals, V, 145, 148
Eutherians, V, 128
Exconjugants, II, 135, 137
Exocytosis, VII, 225
Exponential culture, IV, 162
Exponential growth, IV, 139, 154
Exponential phase, IV, 150, 158

F

F-12 salts, VII, 331
Fat cells, VII, 160
Feedback inhibition, VII, 366
Feeder cells, VIII, 101–102, 105
Feeder layer technique, IV, 42
Feeder layers, IV, 40
  cloning with, IV, 42
Feline kidney cells, VI, 168
Feline rhinotracheitis virus, VI, 168
Feline syncytia-forming virus, VI, 168
Fernbach flasks, V, 231–232
Ferritin, VII, 90
Feulgen-eosin-fast green procedure, III, 341
Feulgen-eosin Y-fast green reaction, III 340–341
Feulgen-fast green procedure after Bloch and Goodman, III, 318
Feulgen procedure(s), III, 52, 57–58, 113, 115, 198–200; IV, 447; VI, 83–85, 94, 199; VIII, 195; IX, 202, 241
  hydrolysis and staining of root cells, III, 199
  microspectrophotometry, II, 371, 373; III, 52
  preparation of Schiff's reagent, III, 115
Feulgen reaction, II, 51, 270, 328; IV, 380, 383
Feulgen reagent, IV, 387
Feulgen stain, IV, 151, 305, 509

Feulgen staining, V, 241
Fibrin, V, 42
Fibroblasts
  contact-inhibited, VI, 114
  cultures of, VI, 217
  diploid, VII, 184, 264
  establishing human cultures, IX, 14
  growth properties of diploid, IX, 14
  human diploid, IX, 13
  mass-scale culturing of, IX, 19
  mouse, VI, 12, 15, 35
  transformed, VII, 161
Ficin, IV, 276
Ficoll, VI, 322, 324; VII, 159, 220, 284; X, 86
Ficoll gradient, IV, 98, 197
Ficoll-isopaque solution, IX, 8
Filter(s)
  Millipore, IV, 22, 24, 52, 64, 124, 145
  oxoid, IV, 145
Filter holder, Millipore PVC, IV, 147
Filtration, Millipore, IV, 26–27, 30–31
Filtration systems, VI, 183
Fingerprinting, RNA, VII, 25
Fission yeast, IV, 131
Fixation, formaldehyde, IX, 200
Fixation and staining of root cells, III, 198–208
  for DNA of chromosomes and nuclei, III, 198–200
  for proteins, III, 202–208
  for RNA, III, 200–202
Fixation of cells, in formalin, IX, 200
Fixative, for cells, IX, 201
Flagella, IV, 342, 344–349, 353–354, 356–358, 365, 369, 375–379, 387, 395–396, 398–400, 404–406, 409–412, 444–446, 450–452, 459–460, 464, 468
Flagellar apparatus, IV, 375–377, 379, 405, 448, 464
Flagellar protein, IV, 452
Flagellate(s), IV, 96, 101, 326, 342–346, 348–349, 351, 353–354, 356–359, 361–362, 364–365, 367, 369–370, 374–378, 384, 387, 392, 395, 398–416, 424–425, 428–440, 442, 445–446, 448, 450–451, 454–457, 459–460, 464, 466–467
Flagellum formation, IV, 467
Flat revertants, VIII, 390–391
Flavobacterium brunneum, IV, 253
Flavophosphine N, VIII, 194–195; IX, 204
Flavoprotein, IV, 109
Fleischmann's dry yeast, VI, 157
Flow meters, VIII, 96
Flow microfluorometric analysis, VIII, 190, 192
  spectra, VIII, 196
Flow microfluorometry, VIII, 180, 192, 195, 197, 200
Fluctuation test, X, 221
Fluorescamine, IX, 212–213
Fluorescein, V, 354; VI, 196; X, 65
Fluorescein-conjugated immunoglobulins, VI, 175
Fluorescein isothiocyanate (FITC), IX, 207, 212
Fluorescein-labeled antiglobulin, VI, 175
Fluorescein mercuric acetate, VII, 158–159
  method, VII, 184
Fluorescein-tagged antiviral serum, VI, 175
Fluorescence, V. 339–340
  Feulgen, VIII, 197
  of HeLa, VIII, 196
Fluorescence detection, X, 64
Fluorescence emission spectra, IX, 212
Fluorescence microscopy, III, 51
Fluorescent antibody procedure, VI, 174
Fluorescent dyes, VI, 83–85; IX, 200, 202–203
Fluorescent enzyme assay, IX, 238
Fluorescent stains, IX, 185
2-Fluoroadenine, VI, 214
Fluorochrome(s), III, 71; V, 339–340, 355, 366–367; VI, 358
Fluorodeoxyuridine, IV, 173; VII, 151, 387
5-Fluorodeoxyuridine, III, 131, 171, 195; IX, 105
  radiomimetic action, III, 195
5-Fluoro-2-deoxyuridine, VIII, 5–7, 9, 16, 20, 32, 79, 81, 88–90, 108
  selection with, VIII, 8
  survivors, VIII, 11
  treatment, VIII, 10
5-Fluoro-2′-deoxyuridine, VI, 44, 246
Fluorodinitrobenzene, VII, 418
Fluorophenylalanine, III, 157
Fluorouracil, IV, 173; VII, 387–388; VIII, 82
5-Fluorouracil, VII, 388; IX, 62, 120
Foamy agent virus, VI, 164, 166

Focus assays, VIII, 381, 413, 421–422
Fogging machine, V, 9
Folate antagonists, VII, 386
Folate metabolism, VII, 391
Folic acid, IV, 122–123; V, 15, 61, 226, 230; VI, 8, 233
  analogs, VII, 384
  antagonist, VI, 233
  metabolism, VII, 385, 389
Folic acid reductase, VIII, 15
Folin reagent, VIII, 352
Folin-Ciocalteau reagent, VII, 336
Folinic acid, VI, 8
Foot and mouth disease virus, VI, 171
Formaldehyde
  buffered, II, 48
  commercial, II, 51
  in fixing proteins, III, 202, 204, 309
  neutral, II, 48
  root cell fixation, III, 199
Formalin, VI, 191
  definition of, IX, 200
5-Formamido-4-imidazolecarboxamide ribonucleotide, VII, 390
Formycin A, VII, 396
Formycin B, VII, 416
Formyl tetrahydropteroylglutamic acid, VI, 258
Formylglycinamide ribonucleotide, VII, 390–391
Formylglycinamide ribonucleotide amidotransferase, VII, 390, 392
1-Formylisoquinoline thiosemicarbazone, VII, 383
Freeze-substitution, III, 310
Freezing ampoule, V, 21
French press, IV, 320
French pressure cell, IV, 91, 153, 214, 221, 223; VIII, 154, 171
Freund's complete adjuvant, VIII, 354
Frog cell(s), IV, 22, 27, 39
  cultured, IV, 31, 43
  freezing of, IV, 38
Frog embryos, VI, 217
Frog saline, IV, 39
Frog virus, IV, 35
Frogs, IV, 35
Frontonia, I, 119, 122
Fucose, plasma membranes and, VII, 184
Fucose-^{3}H, incorporation of, IX, 255
Fulgio septica, I, 13
Funaria, V, 343
Fungal contamination, VI, 160, 179
Fungi, IV, 346; VI, 179, 181–185
Fungizone, V, 231; VI, 181, 185, 188
Furadantin, IV, 55–56
Fusidic acid, VI, 163
Fusion(s)
  heterophasic, V, 113, 116
  homophasic, V, 113, 116
Fusion factor(s), V, 84; VII, 251
Fusion index, V, 95

G

G bands, VI, 347, 351–352, 354–356, 361–362, 364, 366–376; VIII, 185–187, 189–191, 201
G_0, VI, 68, 100, state, VI, 107
G_1, *see* Mitotic cycle measurements
G_1 arrest, VI, 68–69, 71–72, 76–78, 80, 82, 86, 89–91, 93–95, 97–101, 105, 107, 109; IX, 228, 230
G_1 cells, VI, 95
G_1 period, IV, 158, 488–491, 494; VI, 46–47, 58, 63–64, 70–73, 75–76, 79, 82–85, 89–91, 93–95, 97–103, 105–107, 114–115, 121–124
  length of, VI, 114
G_1 population, II, 324, 334, 340–341, 353, 355, 360–361, 365, 368–375, 378–380, 389–394
G_2, see Mitotic cycle measurements
G_2 period, IV, 158, 488–489; VI, 46–47 58, 72–73, 75, 83–84, 97, 100, 102–103, 105, 114, 116, 122–124
G_2 population, II, 324, 330, 332, 340–341, 360–361, 366–369, 371–375, 378–381, 390–394, 398
  history of, II, 255–257
Galacteric acid, VI, 7
Galactose repressor, VII, 136
Gallocyanine, X, 56
Gametes, phototactics of, IV, 53
Gamma, IV, 307
Gamma globulins, VI, 196
Ganglion cells, I, 419, 431
Gas-flow planchet counter, VII, 345
  windowless, VII, 338
Gastric enzymes, dessicated mixture of, VIII, 67

Gastric mucin, VI, 191
Gastrula stage, IV, 29
GDP-sugar pyrophosphorylase, VII, 393
Gel electrophoresis, VII, 59
  one-dimensional, VII, 75, 87
  preparative, VII, 83
  two-dimensional, VII, 76–77, 79–81, 86–87, 93
  two-dimensional polyacrylamide, IX, 349
Gelatin, as inducer of pinocytosis, I, 301
Gelatine capsule, II, 54
Gene activity, II, 42, 62, 73
Gene-centromere distances, IV, 172
Gene conversion, IV, 173
Gene dosage, VI, 217
Generation time, III, 26; IV, 38, 155, 190; V, 19
  determination of, VIII, 117
  methods of estimating, III, 216–217
Genes, mapping in human cells, VIII, 38
Genetic diseases, VI, 215
Genetic linkage groups, II, 113
Genetic markers, VI, 216
Genital secretions, VI, 182
Genome size, of *Drosophila*, X, 196
Gentamicin, VI, 163
Gentamicin sulfate, IV, 55, 68
Gentamycin, X, 271
Geotropism, IV, 268
Germicidal lamp, VIII, 83
Germicidal tube, V, 82
Germinative zone, III, 47
Gevaert NUC, III, 294, 301, 304–305, 307; V, 290, 307
  sensitivity of, III, 299–305
Gey's balanced salt solution, IV, 72, 74
Ghosts, isolation of, VII, 171
Giant cells
  culture from *Diptera*, II, 72–73
  induction of in HeLa, II, 94–96, 110–111
  induction of in monkey kidney cells, II, 96
  method of anucleation, II, 96–99
  viral synthesis in anucleates of, II, 100–111
Giemsa, IV, 151; V, 22, 25, 32, 34, 90; VII, 205, 256
Giemsa banding, VI, 359, 362–363, 371
  analysis, VI, 274
Giemsa powder, V, 23
Giemsa solution, IV, 10, 140, 167, 168
Giemsa stain, II, 118; III, 278–279, 281; IV, 305, 448, 482; VI, 152, 156, 199, 346–347, 358; 360–361, 364–366, 368, 374, 377–378; VIII, 183, 185–187; X, 8, 10
  alkaline-saline method, VIII, 191
  basic, IV, 485
  blood, IV, 45, 481
  buffer, IV, 482, 485
Giemsa stock solution, V, 22
Gigantomonas, IV, 349
Glass beads, growing cells on, IX, 16
Glass-cutting wheel, micrurgy with, II, 96–101
Glaucoma pyriformis, V, 279
Glial cells, IX, 123
Globin, VII, 97
  β-chains, VII, 97
  mRNA, VII, 20, 85, 96
Glucosamine, VI, 7; VII, 378
Glucose, V, 15, 62, 64, 66–67
Glucose 1–6 diphosphatase, VIII, 226
Glucose-6-phosphate, IV, 206
  of nuclear membrane, V, 195
Glucose-6-phosphate dehydrogenase, VI, 194, 200, 303
α-Glucosidase, IV, 163
β-Glucuronidase, IV, 206
Glucuronolactone, VI, 7
Glutamate dehydrogenase, IV, 206
Glutamic acid, II, 222–223; III, 162; V, 14, 63; VI, 6
Glutamine, V, 14, 17, 62, 150–152; VI, 6, 70–72
Glutamine-requiring lines, VI, 214, 250
Glutaraldehyde-fixed monolayers, preparation of, VIII, 87
Glutathione, II, 13; V, 15; VI, 9
Glycerol phosphate dehydrogenase, VI, 303
Glycerophosphate, II, 226
Glycinamide ribonucleotide, VII, 390–391
Glycinamide ribonucleotide transformylase, VII, 390–391
Glycine, V, 14, 63; VI, 6
  mutant, VI, 252
  synthesis, VI, 233
  prototrophy, VI, 268
Glycine-requiring mutant, VIII, 27–28, 32–33
  growth responses of, VIII, 33

Glycolic acid, II, 223
N-glycolylneuraminic acid, IX, 48
Glycolytic enzymes, IV, 206
Glycolytic pathway, IV, 89
Glycoproteins, VII, 159–160, 175, 179–181
  fucose-containing, VII, 161
  precursors, VII, 180
GMP (dGMP) kinase, VII, 394
GMP reductase, VII, 371, 393, 395, 397
GMP synthetase, VII, 393–394
Gold colloid, as inducer of pinocytosis, I, 289, 296–298, 301
Golgi apparatus, IV, 141, 163
Golgi bodies, II, 69
Gonadotrophic hormone, II, 48
Gonads, V, 5
Gonyaulax, III, 122
Gradients
  continuous, X, 92
  discontinuous, X, 91
  discontinuous density of, IV, 4
  discontinuous sucrose density of, IV, 5–6, 16
  Ficoll, VI, 45, 63
  linear sucrose density of, IV, 13
  S-shaped, X, 92–93
  self-forming, X, 92
  sucrose, VI, 44–45, 49, 51–52
Gradient apparatus
  for generation of, VII, 272
  to fractionate, VII, 273
Gradient formation, X, 91
Gradient-forming device, X, 176
Gradient medium, X, 93, 97–98
  mixing, X, 90
  preparation of, X, 88
Gradient-mixing device, X, 92, 98
Gramicidin, IV, 92
Granulocytes, V, 42; VI, 334, 336–337, 342; VII, 324; VIII, 152
  velocities of, VI, 340
Granulocytic leukocytes, VII, 384
Grasshopper, *see also* Neuroblast
  collecting of, I, 233–234
  coloring of, I, 232
  development of, I, 238–240
  diapause in, I, 231–232
  eggs of, I, 236–238
  food for, I, 234–236
  fungal infection of, I, 236
  life cycle of, I, 231
  nymph stages of, I, 239
  rearing of, I, 234–236
  temperature for eggs of, I, 236
  testes of, IV, 2
Grid preparation, VII, 133
Grids, platinum, II, 54
Group translocation, VII, 133
Growth, *see* specific kinds
  effect of serum on, VIII, 389
  of transformed cells, VIII, 389
Growth curve(s), IV, 37
  single-cell, IV, 65
Growth factors, III, 195–197; VI, 220
  see also Auxins, Cytokinins
Growth medium, V, 13–14, 18–22, 25, 38, 50, 52–53, 58
Growth-promoting substances, VIII, 388
Growth rate, IV, 156
Growth rate measurements, I, 390–392
  by cell number, I, 390–391
  by chemical constitution, I, 391–392
  by physical dimensions, I, 391
Guanazole, VII, 377, 383
Guanidinium chloride, VIII, 351
Guanidium chloride, VII, 17
Guanine analogs, VI, 234
Guanine deaminase, VII, 393, 395, 397
Guanine nucleotides, VII, 371
Guanosine kinase, VII, 396
Guanyl cyclase, VII, 393
Guldberg-Waage law, X, 51
Gum arabic, V, 175; IX, 319
Gurr's buffer, VI, 360
Gyrotory shaker, IV, 90

H

Habrobracon juglandis, II, 74
Haemanthus, III, 152
Haematoxylon campechianum, V, 327
Halistuara cellularia, III, 152
Haliteria, I, 117
Ham's F-10 medium, III, 280, 352; V, 13, 39–40
Ham's F-12 medium, VII, 331–332
Hamster(s), IV, 2; V, 116; VI, 166
  Chinese, IV, 10, 15–16
  golden, VI, 34
  isolated chromosomes of, I, 367, 382–384

kidney cells, V, 80
Syrian, IV, 3, 10, 13–14, 16
testis
Chinese, IV, 14, 16
Syrian, IV, 14
Hamster cells, VI, 234, 237–238, 241, 250, 261
clones, VI, 270
cultures, VI, 267
mutants, VI, 233
Hamster embryo, VI, 220
Hamster sarcoma virus, VII, 215
Hank's basal salt solution (BSS), IV, 27; V, 10, 81–82; VII, 252
Haploid idiogram, IV, 45
Haploid line(s), IV, 35, 45; VI, 217; VIII, 3
Haploid syndrome, II, 30
Haplopappus gracilis, III, 186
Hares, VI, 167
Harris' hematoxylin, III, 51, 266
root cell staining, III, 199
Hartmanella, VI, 188–190, 192–193
Hartmanella agricola, VI, 192
Hartmanella astronyxis, IV, 390
Hartmanella castellanii, VI, 190
Hartmanella glebae, VI, 192
Hartmanella hyalina, IV, 369
Hartmanella rhysodes, I, 72, 82; VI, 192
Hartmannellidae, VI, 190
Harvard press, IV, 199, 211, 215
Harvesting of cells, VIII, 234
HAT medium, VIII, 15; IX, 173; X, 215
definition of, X, 219
Hedera helix, II, 177–178
HeLa cells, I, 281, 301, 314, 412; IV, 72–73, 214, 224; V, 47, 86, 89, 96–98, 105, 107–112, 117–118, 121; VI, 4, 12, 15, 18, 28, 39, 106, 114, 116, 120, 124, 128, 132–133, 152–153, 164, 170, 173, 181–182, 194, 196, 214, 218, 220, 223, 239, 242, 250, 323; VII, 4–5, 35, 48, 59, 63, 91, 121, 148, 158, 215, 234, 238, 242, 288, 350, 353, 355–359, 375, 383, 404, 414, 453; VIII, 36, 181, 202, 233, 237, 239, 241–242, 244; IX, 216
cell ghosts, VII, 169
cell growth cycle of, VI, 118
cell life cycle in, VI, 122, 125
cell volume, IX, 209
chromosome extraction from, II, 115–119, 121
clones, VI, 133–134
cutting of, II, 96–104, 108–109
DNA content of, IX, 218
DNA distributions of, IX, 206, 208
DNA and protein content of, IX, 225
effects of exposure to thymidine-H^3, II, 330
enucleated, VII, 199
growth of, VI, 119
growth rate of, X, 94
induction of giants, II, 94–96, 110–111
isolation of nuclei from, II, 132, 140–142
metaphase arrest in, II, 114–115
metaphase chromosomes, III, 278
poliovirus RNA synthesis in, II, 104, 107–110
properties of purified chromosomes, III, 283–285
reticulocytes, VII, 2
ribosomes of, VII, 62
RNA synthesis in, II, 104–106
RNA, DNA, and protein content of chromosomes, II, 119
separation of mitotic cells, III, 352
survival after micrurgy, II, 97
synchronization by amethopterin, III, 30–33
synchronization of, VI, 121
synchronized, VI, 118–119
virus infection of anucleates, II, 101–104
HeLa sublines, IX, 217
Helianthus annuus, II, 178
Helix aspersa, III, 330
Helminthosporium dematiodium, VII, 203, 212–214; VIII, 124, 145
Helper viruses, VI, 172–173, 176
Hemadsorption, II, 102–103; VI, 39, 174; VII, 232; VIII, 403–404
Hemagglutinating activity, V, 85
Hemagglutinating factors, VI, 198
Hemagglutinating units, VII, 228
Hemagglutination, VI, 198
assay, V, 81; VII, 228
titer, V, 81–82; VII, 228
Hemagglutination test, VI, 197
Hemagglutinin, viral, V, 84
Hematopoietic tissues, VII, 314
Hematoxylin, II, 270, 276–280, 284, 328, 362, 383; V, 327; VI, 191

Hemin, IV, 90
Hemocytometer, I, 390–391; III, 33; IV, 132, 149, 161; V, 49, 53, 89; VI, 117, 197; VII, 317, 335
standardization of, VIII, 71
Hemoglobin, VII, 96
biosynthesis, VII, 45
mRNA, VII, 19, 64
Hemolymph, II, 68, 73, 82, 86
in *Chironomus*, X, 21
Hemolytic activity, V, 84
Hemorrhagic disease agent, VI, 166
Henrica sanguinolenta, V, 170
HEp-2 cells, VI, 173; VII, 215, 220, 230, 232–234
HEPA filters, VI, 186
Heparin, III, 88; IV, 289, 480; V, 120; VII, 9
Hepatectomy, IX, 53
Hepatitis virus, VI, 165, 168
Hepatocytes, VII, 344
Hepatoma(s), V, 183; VII, 26, 43, 80, 87, 99, 160, 367, 387, 400, 403, 404–406, 408, 411–412, 414, 419, 422, 429, 453
ascites, VII, 29
cells, VI, 215, 318, 323
isolation of nuclei, V, 173
minimal deviation, IV, 209; VIII, 157
rat ascites, 2, 21, 32
HEPES (*N*-2-hydroxyethylpiperazine-*N′*-2-ethane sulfonic acid), V, 41; VII, 180, 337
Herpes ateles, VIII, 371
Herpes B virus, VI, 166
Herpes saimiri, VIII, 371
Herpes simplex type 2, VIII, 371
Herpes simplex virus, VI, 164–165; X, 101
Herpes sylvilagus, VIII, 371
Herpes-type virus, VI, 174
Herpes virus(es), V, 78; VI, 165, 167–168, 170; VII, 184; VIII, 371
canine, VI, 168
Herpes zoster virus, VI, 165
Heteramoeba, IV, 349, 352, 369, 405, 409, 456, 464, 467, 468
Heteroamoeba clara, IV, 345
Heterochromatin, III, 154, 184; V, 34
centromeric, VIII, 175, 183
constitutive, V, 33–36; VIII, 152–153, 161
distribution of, VIII, 170
facultative, VIII, 152
isolation of, VIII, 160
separation of, VIII, 171
staining procedure, V, 25
Heterokaryotypes, VII, 252, 255–258
Heterokaryon(s), V, 79, 86, 90, 96, 110, 116, 118–119; VIII, 226–227, 230–233, 243, 245
Heteromita, IV, 349
Heteroploid cell lines, VI, 212, 220, 232
Heteroploid human D98 cells, VI, 231
Heteroploid mouse cells, VI, 231
Heterosynkaryon, V, 96
Heterothallic cells, III, 121
Heterozygosis, VIII, 12
Hexokinase, IV, 93, 98, 106, 171, 174; VI, 303
genes for, IV, 171
Hexylene glycol, VI, 284–285, 287, 289, 299, 310, 313
Hexylresoccinol, IV, 55
High-resolution autoradiography, I, 327–363
background eradication for, I, 330–331
chemical developers in, I, 344
electron microscope techniques in, I, 332–334, 360–361
emulsion preparation for, I, 330, 332–333, 336–339
experimental uses of, I, 361–362
exposure for, I, 331, 333, 341–342
light microscope techniques in, I, 329–332, 360
photographic processing for, I, 331, 333, 342–346
physical developers in, I, 344–346, 349–350
resolution in, I, 348–360
slide preparation for, I, 329
specimen preparation for, I, 330, 332, 335–336
staining in, I, 333
Histomonas, IV, 349
Histomonas gigantomonas, IV, 409
Histone(s), II, 113, 118, 127, 406; IV, 141, 204, 211; VI, 90–91, 103, 106, 109; VII, 93, 131, 141, 143, 145, 449
cytological and cytochemical methodology, III, 203–207, 307–343
cytoplasmic staining, *see* Cytoplasmic staining

definition, III, 307
eosinophilia, III, 309
fixation, III, 205–207, 309–310
isolation of, IV, 196
message, VII, 54
mRNA for, VII, 19, 53
recovery of, V, 257
staining procedures
acetylation, after Monne and Slaughterback, III, 329
alkaline fast green, after Alfert and Geschwind, III, 203, 317
ammoniacal silver, after Black and Ansley, III, 338
azure B-eosin Y, III, 342
Bebrich scarlet, after Spicer and Lillie, III, 335
deamination, after Van Slyke, III, 329
dinitrofluorobenzene, after Bloch, III, 325
after Danielli, III, 325
dinitrofluorobenzene-Sakaguchi, after Bloch and Brack, III, 327
eosin Y-fast green, after Bloch, III, 340
Feulgen-fast green, after Bloch and Godman, III, 318
Millon, after Baker, III, 311–313
Naphthol-yellow S, after Deitch, III, 320–321
picric acid-bromophenol blue, after Bloch and Hew, III, 331–334
picric acid-eosin Y, after Bloch and Hew, III, 331–334
picric acid-Schiff, after Dwivedi and Bloch, III, 318–319
Sakaguchi, after Deitch, III, 322–323
staining reactions
alkaline fast green III, 312–319
ammoniacal silver, III, 335–338
acetic anhydride, III, 328–329
azure B-eosin Y, III, 341–342
Biebrich scarlet, III, 334–335
dinitrofluorobenzene, III, 323–325
dinitrofluorobenzene-Sakaguchi, III, 325–327
eosin Y-fast green, III, 338–340
Feulgen-eosin Y-fast green, III, 340–341
Millon, III, 310–312
Naphthol-yellows, III, 319–321
picric acid-bromophenol blue, III, 329–334
picric acid-eosin, III, 329–334
Sakaguchi, III, 203, 205, 321–323
Sakaguchi-Feulgen, III, 327–328
Sakaguchi reagent, III, 204–205
Van Slyke, III, 328–329
Histone mRNA, X, 11
Histone phosphorylation, VI, 101
Hoagland's nutrient medium, III, 112, 114, 190
Hog cholera antibodies, fluorescein-labeled, IX, 230
Hog cholera virus (HCV), IX, 230–231
Homofolate, VII, 386, 392
Homokaryocytes, VII, 256, 258
Homokaryon, V, 96
Homoserine dehydrogenase, IV, 157
Homosynkaryon, V, 96
Hordeum vulgare, III, 176
Hormone-dependent cells, VIII, 36
Hughes press, IV, 105
Human adenoid tissue, VI, 164
Human adenovirus, VIII, 12, 414
Human amnion cells, I, 284, 392, 404, 412; III, 232; IV, 72–73; VI, 156, 173
simian virus 40 transformation of, IX, 72
Human amnion cultures, III, 152
Human bone marrow, VI, 220
Human cells, VI, 219
cultures, VI, 168
Human diploid cells, V, 40; IX, 80
Human diploid fibroblast, VI, 152, 154, 233–234, 238, 267; IX, 81
Human embryonic fibroblasts, VI, 169
Human embryonic liver, VI, 152
Human embryonic lung, VI, 152
Human heteroploid cell lines, IX, 218
Human lung, diploid cells of, X, 211
Human lymphoblasts, VI, 231, 238
Human tumor cells, III, 39; VI, 177
Human tumors, IV, 214
Human viruses, VI, 165, 178
Humidity, control of, V, 71
Humidity chamber, I, 439, 441
Hyacinthus orientalis, III, 175
Feulgen staining of root cells, III, 199
Hyalodiscus simplex, I, 279
Hyaluronic acid, VI, 80
Hyaluronidase, VI, 80–81; IX, 198

Hybrid(s), V, 96, 106, 113, 116–117
Hybrid cell(s), V, 45, 85, 87–88, 90–92, 104, 109, 116, 120; VI, 217, 219, 227, 233, 241, 250; VII, 226, 231–233, 243, 245, 252–253, 256, 308; VIII, 12–15, 37–38
  analyzing, X, 152
  cloning, X, 152
  colony, VII, 256, 258
  isolation of, X, 151
  production of, VII, 227
  regulation in, VII, 227
Hybrid clones, V, 79, 87; VI, 226; VIII, 15
Hybrid colonies, V, 92
Hybridization, V, 34, 36, 88; X, 351
  among mutants, VIII, 14–16, 38
  applications of, VIII, 400
  ^{125}I in molecular, X, 325
  in situ, VI, 314; X, 1, 337
  in vitro, VI, 346
  of cells, V, 78; VI, 216, 219, 226, 239, 263, 376
  of somatic cells, VIII, 12, 78
  sensitivity of, X, 41
Hybridization experiments, with viral nucleic acids, VIII, 401
Hybridization procedure, cytological, X, 5
Hydra, IV, 415, 442
Hydra littoralis, IV, 361
Hydrocortisone, VII, 92
Hydrocortisone hemisuccinate, VI, 15
Hydrogen peroxide, IV, 55
Hydrogen transfer reactions, biological, X, 319
Hydrogenomonas, II, 221
Hydroxyapatite chromatography, VIII, 400
Hydroxylamine, IV, 174, 303; VI, 268
Hydroxylapatite, VII, 117–119, 124
8-Hydroxyquinoline
  in Sakaguchi reaction, III, 321
5-Hydroxytetracycline, VI, 163
Hydroxyurea, VI, 44, 98–99; VII, 368, 373, 377, 383; IX, 120
  dose, IX, 63
  response of intestinal epithelial cells to, IX, 64
Hygromycin B, VI, 163
Hypertonic solution, definition of, II, 151
Hypodermic adaptor, Swinney, IV, 23
Hypotonic buffers, VII, 169
Hypotonic disruption, VII, 171
Hypotonic shock, IV, 223
Hypotonic solution(s)
  definition of, II, 151
  effect of on mitotic recovery, IX, 76
  sodium citrate, III, 89
Hypotonic swelling of cells, III, 280–281; VII, 165
Hypotonic treatment and cell rupture, III, 285–286
Hypoxanthine, V, 63; VI, 8, 232–234; VII, 371
  analogs, VI, 234
Hypoxanthine-guanine phosphoribosyltransferase (HGPRTase), V, 91; VI, 213, 215, 232, 234–235; VII, 393, 396; IX, 173; X, 215
  cells lacking, VI, 233
  deficiency, VI, 270
Hypsiprymnodon moschatus, V, 135

I

Ilford G5, IV, 233, 235–236, 239
  liquid emulsion, IV, 237
Ilford K2, IV, 235, 239
  liquid emulsion, IV, 232–234
Ilford K5, IV, 233, 236, 239
  liquid emulsion, IV, 238
Ilford L4, III, 294, 301, 304–305; IV, 235–236, 239
  data for monolayers, V, 295
  liquid emulsion, IV, 233, 238; V, 290, 293, 296–298
Ilford nuclear research emulsion K5, V, 33
Immobilization antigens, VIII, 353
Immunodiffusion test, VI, 197
Immunofluorescence, II, 93; VI, 174–175, 196; VIII, 403–404
Immunofluorescence technique(s), VIII, 116–117; IX, 230
Immunoglobulin light chain, VII, 97
Immunoglobulins, VI, 215
  production of, IX, 7
IMP dehydrogenase, VII, 393–394
Impatiens nolitangere, II, 178
Impatiens parviflora, II, 178
Impermeability, of cell wall, II, 155
Indoleacetic acid (IAA), III, 178, 192
  effect on cell division, III, 196
Infectious bovine rhinotracheitis virus, X, 101
Influenza virus, VI, 165; VII, 242

Infusion(s)
cerophyl, IV, 250
grass, IV, 251
hay, IV, 250
lettuce powder, IV, 252
baked, IV, 251
Inosine kinase, VII, 396
Inosinicase, VII, 390
Inositol, IV, 143, 170–171; V, 15; VI, 5, 7, 17, 39
Insect cell lines, VI, 197; X, 186
Insect cells, V, 87
in suspension culture, X, 185
methods with, X, 185, 195
Insect larvae, VII, 111
Insect Ringer's, X, 137
Insulin, VI, 9
Insulin-^{14}C, VII, 454
Interference contrast microscopy, III, 68–70
Interferon, VI, 174; IX, 3
Intermitotic time measurements, I, 388–390
see also Mitotic cycle measurements
Intestinal crypts
mitotic activity in, IX, 60–61
Intestinal epithelium, V, 133
Intestine epithelium cells
lifetime of, IX, 117
Intestinal microvilli, VII, 160
Intestinal mucosa, VII, 160, 387
Intracameral injection, III, 52, 55
Inverted microscope, I, 404–406
Iodination methodology, X, 327
Iodination reaction mixture, X, 328
Iodine-containing nucleosides, X, 352
Iodine-125, physical properties of, X, 294
^{125}Iodine-labeled nucleic acids, applications of, X, 335
Iodine-labeled RNAs, X, 353
applications of, X, 351
Iodoacetate, III, 72; VII, 164
Iododeoxyuridine-H^3, III, 52
5-Iododeoxyuridine, IV, 115
Ionic coupling, II, 73
Ionic detergents, IV, 212
Ionizing radiation, VI, 178
Iridectomy scissors, II, 64
Irradiation chamber, II, 22
Isocitric dehydrogenase, VI, 303
Isogametic haploid cells, III, 121
Isogenic stocks, IV, 245
Isoleucine, V, 14, 60; VI, 6, 68, 70–72, 75, 85–87, 89, 100–105
Isoleucine deficient medium, VI, 72–74, 87–88, 107
Isoodon, V, 133
Isoodon macrourus, V, 133
Isoodon obesulus, V, 133
Isopentane, VIII, 252
Isoproterenol, IX, 53
Isopycnic banding, VIII, 402
Isothiocyanate, VI, 196
Isoton, VIII, 70
Isozymes, III, 71
IUdR, VIII, 411, 414, 420

J

Janus green, V, 329
Jensen sarcoma, IV, 211
Jeweler's tweezers, II, 45–46, 48, 56, 292–293
Joyce Loebel gel slicer, VII, 59

K

Kanamycin, IV, 27, 55–56; VI, 152, 162–163, 192; VIII, 238–239
Kangaroo(s), V, 135, 150
great grey, V, 137, 149
rat, V, 130, 135, 138–139
red, V, 137
tree, V, 135
Kappa, IV, 244, 294, 307, 310, 313–319, 321, 325
aureomycin-resistant, IV, 324
killing activity of, IV, 320
methods for purifying, IV, 319
mutant, IV, 324
mutations of, IV, 323
special methods for, IV, 318
transplantation of, IV, 326
Karakashian's medium, IV, 327
Karyokinesis, IV, 344, 355, 359, 365, 380–381, 383, 385, 388, 398, 406, 410, 462, 466
synchronous, IV, 387
Karyotype(s), V, 3, 6, 128, 130, 132, 135–144, 161–162; VI, 252–253
diploid *R. pipiens*, IV, 44
evolution of, V, 158–159
haploid, IV, 33

Karyotyping, V, 28
KCl, hypotonic, IV, 43
KCN, IV, 87, 92, 107
Kel-F halofluorocarbon oils, II, 65
Kidney tissue
giant cell induction *in vitro*, II, 96
staining of, II, 276–280, 282–284
Kinetin, III, 171, 179, 197
effects on mitosis, II, 197
Kinetodesmal fibers, IV, 315–316
Kinetoplasts, IV, 91
Kinetosome(s), V, 222, 250–260, 267; VIII, 328
DNA in, VIII, 329
isolation of, V, 261–262, 269; X, 120
microtubules of, V, 272
Kinety patterns, VIII, 331
Knop solution, IV, 332
Koala, V, 135
Kodak AR10, IV, 239
stripping film, IV, 10; VII, 256
Kodak NTB, IV, 233, 235–236, 239
liquid emulsion, IV, 235
Kodak NTB2, IV, 233, 235, 239; VII, 340
liquid emulsion, IV, 235
Kodak NTB3, IV, 232–233, 235, 237, 239
liquid emulsion, IV, 236
Kodak NTE, III, 294, 301, 304–305
sensitivity of, III, 299–305
Krebs-2 tumor cells, IV, 213
Krebs-cycle substrate, II, 223–224

L

L cells, I, 284, 381; II, 114, 122–123; IV, 223; VI, 71, 78, 152, 164, 193, 212–214, 218–220, 232, 239–240, 242–243, 247, 257, 261, 274, 276; VII, 3–4, 35, 159, 161, 183, 190, 196–197, 212, 220, 232, 253, 270, 288, 330, 377, 391–392, 414, 419
anucleate, VII, 232–233
enucleated, VII, 195; VIII, 138
mutant lines of, VI, 256, 259
mutants, VI, 262
mutations in, VI, 263
L-4 emulsion, VII, 150, 154
L929 cells, VII, 204, 214–215, 217, 222–224, 230, 234, 237–238, 242, 372, 375–377, 453
L929 line, VII, 373
L1210 cells, VII, 392, 411
LS cells, VII, 272, 274–275, 278–279, 281, 284
synchronization of, VII, 277
volume distribution in, VII, 276–277
Labeled compounds, degradation of, VII, 451
Labeling, continuous, 122
Labyrinthula, I, 10, 12
Labyrinthules, IV, 411
Lac repressor, VII, 137
Lactalbumin hydrolyzate, V, 16–17, 150; VI, 2, 4–5, 39
Lactate dehydrogenase, VI, 201
turnover of the isoenzymes, X, 256
Lactic dehydrogenase, IV, 206; VI, 303
Lactic-orceine, II, 89–90
Lactoorcein, IV, 306
Lactoperoxidose, X, 255
Lactuca sativa, III, 197
effects of kinetin on mitosis, III, 197
Lagorchestes hirsutus, V, 134
Lagostrophus fasciatus, V, 134
Lambda, IV, 307, 311, 315–316, 322; VI, 216
Lambda DNA, VII, 136–137
Lambda repressor, VII, 136
Laminar flow hood, VI, 186
Lamium purpureum, II, 194
Lampbrush chromosomes, I, see chromosomes
as experimental material, I, 40–42
autoradiography of, I, 52–54
electron microscopic techniques for, I, 54–56
history of, I, 37–38
occurrence of, I, 38
preparation for microscopy, I, 45–49
techniques for, I, 37–60
Lasiorhinus latifrons, V, 131, 135
Latent image fading, VIII, 284
Latex bead method, IX, 276
Latex beads, IX, 262, 276–277
ingestion of, IX, 274
kinetics of uptake, IX, 266
Latex spheres
in electron microscopy, III, 2, 6–8, 12–13
in evaluation of particle counters, III, 6–7
in X-ray diffraction, III, 3
polystyrene, III, 6
polyvinyltoluene, III, 6

n-Lauroyl sarcosine, VII, 149, 151
Laurylsulfate, IV, 212
Lectins, VIII, 77–78, 85–86, 397, 399; IX, 1–2
  susceptibility to, VIII, 90
Leibovitz medium, IV, 22, 27–33, 38–43
Leishman's stain, VI, 360
Leitz marker objective, VIII, 103
Lemna minor, II, 200–201
Lens capsule, III, 47
Lens cells, VII, 45, 97
Lens epithelium, III, 45
  autoradiography, III, 52–56
  whole mount, III, 48–52
Leptomonas karyophilus, IV, 326
Leptotene, IV, 3, 502, 510–511
Lesch-Nyhan cells, VII, 288
Lesch-Nyhan syndrome, VI, 215, 231, 233
Lettuce juice medium, IV, 252, 254, 289, 313
Lettuce powder, IV, 251, 254
Leucovorin, VI, 258
Leukemia, VI, 156
Leukemia cells, VII, 398, 402, 412, 447
Leukemia virus, VI, 168
Leukemic bone marrow, VII, 324
Leukemic leukocytes, VII, 368–370, 412, 415
Leukemic tissue, VII, 389
Leukocyte counting, differential, IX, 236
Leukocyte nuclear staining, IX, 236
Leukocytes, V, 87, 92, 133; VI, 114, 328, 333–335, 338; VII, 160, 223, 368, 392 404, 416, 427, 437
  cultures of amphibians, III, 87–91
    Ambystoma, III, 87
    Necturus, III, 87, 89
  differential analysis, IX, 236
  G_1 and G_2 *in vitro*, II, 374–375, 377
  migration of, VI, 327
  partition of, IX, 41
  polymorphonuclear, VI, 325–326, 336, 347
  sorting of human, IX, 236
Leukosis virus, VI, 167; VII, 314
Licea flexuosa, I, 13, 15
Life cycle analysis, VIII, 190
Light-scattering, IX, 188
  of Chinese hamster cells, IX, 195
Light-scattering properties of cells, IX, 187
Liliaceous plants, IV, 497–498
Lilium hollandicum, IV, 499
Lilium longiflorum, III, 152; IV, 499, 502, 510
Lilium maximowiczii, IV, 499
Lilium speciosum, IV, 500, 502–503, 511
Lilium tigrinum, IV, 499–503, 511
Lily (lilies), IV, 498–504, *see also* specific species
  culture medium for meiotic cells of, IV, 507
Linbro trays, VI, 218, 251, 258; VII, 267; VIII, 104
Lincomycin, IV, 55; X, 286
Linoleic acid, V, 64; VI, 9
Linum usitatissimum, III, 180
Lipase, VIII, 67
Lipoic acid, V, 63; VI, 8
Lipomyces, IV, 140
Lipopolysaccharide, V, 84
Lipovirus, VI, 190
Liquid emulsion, IV, 151; V, 33
Lithium chloride precipitation, VII, 56
Lithium dodecyl sulfate, VII, 11, 124
Liver, VII, 26, 36, 39, 70, 74, 79–80, 83, 85–87, 89–90, 92–93, 99, 110, 141, 143, 160–161, 372–375, 389, 400, 453
  homogenization of, VIII, 208
Liver cells
  countercurrent distribution of, IX, 43
  electrophoretic mobility of, IX, 42–43
Liver concentrate, IV, 431–432
Liver extract, IV, 433; V, 224
Liver fraction, IV, 90
Liver fraction L, V, 225
Liver nuclei, IV, 200–201; V, 173
Loefer's salt solution(s), IV, 327
  for *P. busaria*, IV, 333
Logeman homogenizer, X, 118
Logeman press, V, 257
Lowry method, VII, 335
Luciferase, VII, 448–449
Luciferase-luciferin reaction, VII, 448
Lucke virus, VIII, 371
Ludox, X, 87, 91, 97–98
Ludox TM, X, 101
Luer-Lok adaptor, VII, 300
Lugol's iodine, IV, 354, 356, 376, 391, 444, 446, 455
Lung(s), V, 4, 23
  cells, VII, 215, 375
  cultures, V, 18

Lupinus albus, III, 183
Lutreolina crassicandata, V, 141
Lymph node, V, 129
cells, VII, 45
Lymphatic system, V, 129
Lymphatic tissues, V, 128
Lymphoblast cultures, synchronization of, VIII, 107
Lymphoblastic cell, VII, 429
Lymphoblastic leukemia, VIII, 119
Lymphoblasts, VI, 232, 267
Lymphocyte(s), IV, 478–480, 485, 487–488, 490–494; V, 87, 117–118, 128–129, 144–150, 152–153, 155–156; VI, 68, 154, 354, 356, 366, 373; VII, 160–161, 215, 398, 405; VIII, 152
bovine, IX, 1
contact cooperation, IX, 9–10
cultures, VII, 350
glass fiber purification of, IX, 4
primary culture of, IX, 2
stimulation by mercaptoethanol, IX, 6
thymus, IX, 2
Lymphocyte cultivation
serum-free, IX, 8
Lymphocyte cultures, V, 4
Lymphocyte stimulation, IX, 7, 10
Lymphocytic choriomeningitis, VI, 165
Lymphocytic leukocytes, VII, 384
Lymphocytoid cells, VI, 218
lines of, VI, 215
Lymphoid cells, VII, 160, 254, 257–258
Lymphoid leukosis viruses, VI, 167
Lymphokines, IX, 3
Lymphoma, VI, 14
cells, VI, 241
Lymphosarcoma, IV, 168
cells, VI, 188
Lyon hypothesis, V, 155
Lysolecithin, V, 79–80, 84, 90; VII, 251
cytotoxicity of, VII, 253
preparation of, VII, 252
Lysolecithin-induced fusion, VII, 254
Lysolecithinase, VII, 252
Lysosomal enzymes, IX, 274
Lysosome-gradient medium, X, 99
Lysosomes, VI, 172, X, 130
isolation of, X, 109
separation of, X, 99–100
Lysozyme, IV, 112; VII, 147, 149
Lyzosomal RNase, VII, 4
Lyzosomes, VII, 3–4, 7–8

M

Macaca irus, VI, 166
Macaca mulatta, V, 166
Macaloid, VII, 33
MacIlvaine's buffer, VI, 166
Macronuclear anlagen, IV, 275
of *Euplotes woodruffi*, II, 135–137
Macronuclear breakdown, IV, 276
Macronuclear envelope, X, 126
Macronuclear fragmentation, IV, 274
Macronuclear fragments, IV, 275, 297–298
Macronuclear regeneration, IV, 292, 297–298
Macronuclear reorganization, IV, 298
Macronuclei (macronucleus), IV, 226–227, 313, 326; VIII, 335; IX, 281–282
composition of, IX, 298
DNA in, IX, 297
dry weight of, VIII, 337
isolation of, VIII, 337; IX, 282, 311, 316; X, 124
of *Blepharisma japonicum*, II, 138–139
of *Blepharisma undulans*, II, 138–139
of *Euplotes eurystomus*, II, 132–134, 265–268
of *Euplotes woodruffi*, II, 138
of *Paramecium caudatum*, II, 137–138
of *Spirostomum ambiguum*, II, 138–139
of *Stentor*, II, 2, 134–135, 262
of *Tetrahymena*, II, 138–140, 314–315
polytene giant chromosomes in, II, 66
protein content of, VIII, 337
scheme for isolating, IX, 292
volume of, VIII, 336
Macrophages, VII, 214–215, 223, 230, 233, 235, 237, 324; VIII, 124
anucleate, VII, 232, 237
virus transformed, VIII, 237
Macropodidae, V, 128, 134–136
Macropods, V, 134, 137
Macropus, V, 136, 150
Macropus major fulignosus, V, 134
Macropus major major, V, 134, 136–138
Macropus major malanops, V, 134
Macropus major ocydromus, V, 134
Macropus major tasmaniensis, V, 134
Macropus robustus, V, 134, 137–138

Macropus rufus, V, 134, 136–138
Magnetic stirrer, Teflon, IV, 90
Maize, VII, 111
Malpighian tubules
  culture of, II, 72
  isolation of nuclei and chromosomes from, II, 74
  polytene chromosomes in, II, 66–67
  structure of, II, 69
  transplantation of cytoplasm from, II, 82–84
Malt extract
  agar, IV, 144
  broth, IV, 141, 155
Mammalian cell genetics, IX, 171
Mammalian cells, IV, 71–72; VI, 103, 105, 109
  chromosomes of, I, 381–385
  cloning of, V, 37–74
  cycle, VI, 107, 109
  in culture, VI, 97
  mutagenesis in, X, 209
  pinocytosis in, I, 282–284, 289–290, 297–299, 300
  synchronization of, VI, 44
Mammary carcinoma, DNA synthesis and mitotic activity in, IX, 63
Mammary explants, VII, 91
Mammary glands, VII, 160
Marek's disease, VI, 167; VIII, 371
Marmosa, V, 129, 141, 143, 146–147, 150
Marmosa elegans, V, 141, 144
Marmosa giganteus, V, 149
Marmosa mexicana, V, 141
Marmosa mitis (robinsoni), V, 141, 144, 146, 148–149, 151–152, 163
Marmoset, VI, 166
Marsupiala, V, 128
  cell types, V, 146
Marsupials, V, 128–130, 133, 139, 141, 147, 149
  cells, V, 127, 144
Mass culture of *Paramecium aurelia*, IV, 125
Mast cell leukemia, VI, 168
Mast cells, V, 42
  separation of, X, 98
Mastigophora, IV, 343, 345, 406
May Grünewald giemsa stain, V, 90; VI, 156, 191
MBET medium, IV, 86–87
McCoy's 5a medium, V, 13, 21
Measles, VI, 166
Measles virus, V, 78; VI, 165
Medium, VI, 4, 8–12; VII, 135, 199
  acetate-*n*-butanol, II, 225
  autotrophic, II, 219–222
  Barth's, II, 14–15
  basal, VI, 3
  chemically defined, VI, 2, 5
  conditioned, V, 20
  Cramer-Myers, II, 219
  dissociating, IV, 26; V, 50
  "Drosophila-Ringer", II, 73
  Eagle's minimum essential, IV, 21; VI, 6, 8, 10, 38–39, 136
  for chromosome isolation, II, 49–50
  for culture of nuclei, II, 74–78
  for nuclear transplantation, II, 13
  HeLa maintenance, VI, 4
  heterotrophic acidic, II, 222–224
  Holtfreter's solution, II, 13
  list of synthetic, VI, 3
  "Medium-199", II, 73
  minimum essential, VI, 3
  modified Eagle's, VI, 3
  neutral and alkaline, II, 225–226
  Niu-Twitty solution, II, 6–7, 14
  preparation, V, 17
  protein-free, VI, 1, 5, 12–15, 17, 32, 34, 38
  Ringer's solution, II, 2–3, 6
  Stearn's, II, 14
  Steinberg's solution, II, 7, 14, 22
  synthetic, VI, 5, 17–18, 20, 22–23, 25, 28, 32–39
  wash, IV, 30, 32–33, 38, 40
Medium F10, V, 39–40
Medium F12, V, 39–40, 64, 66, 73, 89–90
Medium F12M, V, 40, 47, 50, 52, 59, 64–66, 72
Megaloblasts, V, 129
Meiosis, IV, 3, 12, 16, 347, 499, 508; V, 28
  in *S. pombe*, IV, 171
  prophase of, IV, 1–3
Meiotic cells, culture of, IV, 497–513
Meiotic cycles, IV, 498
Meiotic division, IV, 141
Melanocytes, IV, 28
Melanomas, IX, 62, 216
Melanoplus differentialis, I, 232, 235
Melanoplus femur-rubrum, I, 232

Melinex, VIII, 93
Membrane(s)
adhesive factors on, IX, 196
lipid composition of, X, 130
Millipore filter of, IV, 23
salic acid, IX, 34
surface charge, IX, 32
Membrane charge
inner vs outer surface, IX, 47
nature of, IX, 33
Membrane components
isolation of from *Tetrahymena*, X, 105
isolation methods for, X, 108
Membrane filters, IV, 52, 146–147; VI, 161
nucleopore, VI, 288
Sartorius, IX, 9
Membrane filtration, VI, 178, 184, 187
Membrane potential, VIII, 309
Membrane proteins, X, 82
Membrane resistance, VIII, 307
determination of, VIII, 300
Membrane resistance capacitance, VIII, 306
Menadione, VI, 8
Mengovirus RNA synthesis, VII, 372
Meningoencephalitis, VI, 192
2-Mercapto-1-(β-4-pyridethyl)-benzimidazole, VII, 344, 410, 421
Mercaptoethanol, V, 275
β-Mercaptoethanol, VII, 73, 89
Mercaptopurine
mutants resistant to, X, 213
6-Mercaptopurine, VI, 231; VII, 377; IX, 120
Merthiolate, IV, 26, 29
Mesocricetus auratus, IV, 2
Messenger RNA
extraction of, VII, 13, 15
isolation of, VII, 18
lifetimes of, VII, 240
purification of, VII, 60
purified, VII, 96–97
species of, VII, 95
Metabolic cooperation of cells, X, 212
Metachromasy, V, 337–339, 345, 348, 352, 355
Metamyelocytes, V, 129
Metanil yellow, V, 51
Metaphase(s), IV, 1, 3, 33, 35, 72, 77, 79, 139, 380, 382–383, 385, 387, 479, 487, 491–493; V, 12, 26–27, 35, 104–105, 109–111, 118, 149
isolated, IV, 73
Metaphase arrest, II, 115; III, 280
reversibility of, III, 74, 81
Metaphase chromosomes, isolation of, II, 113–129
Metaplasia, IV, 394
Metarrhizium anisopliae, VII, 213
Metatherians, V, 128
Methacrylate, tritiated, VIII, 278
Methanol, as fixative, IX, 201
Methanol-acetic acid, II, 117, 119
Methionine, V, 14, 60; VI, 6, 71
Methocel, VIII, 69, 90, 393, 408; X, 93
medium, VIII, 84–85
Methotrexate, V, 256; VIII, 108
Methyl green, IX, 316; X, 57
Methyl green-pyronin, II, 132–134; V, 241
Methyl methanesulfonate, IV, 174; VI, 267
N-methyl-*N*′-nitro-*N*-nitrosoguanidine, VI, 247–248, 251, 253, 258, 261, 267–269; VIII, 4, 16, 31–32, 36, 44; IX, 174
N-methyl-*N*-nitroso-*N*′-nitroguanidine, IV, 303; VIII, 347
Methyl oleate, VI, 9
Methyl red, V, 335, 355
Methyl violet, V, 353
N^6-Methyladenosine, VII, 395, 405
Methylazoxymethanol acetate, VII, 357–359
α-Methylbutyryl-CoA, VI, 104
Methylcellulose, II, 82; V, 42; VIII, 393
Methylcholanthrene, VIII, 79, 373
3-Methylcholanthrene, VI, 35
Methylcytosine, VI, 8
Methyldeoxycytidine, VI, 8
Methylene blue, II, 276; IV, 38, 151; V, 342; VI, 236; 327; X, 56, 58, 61
Methylene blue staining method, X, 52
N^5, N^{10}-Methylenetetrahydrofolate, VII, 384
1-Methylguanine, VII, 396
6-Methylmercaptopurine ribonucleoside, VII, 377
1-Methyl-6-thioguanine, VII, 396
6-Methylthiopurine ribonucleotide, VII, 391
Metofane, V, 3
Metrizamide, IX, 325
Metronidazole, VI, 192–193
Michaelis-Menten equation, VII, 404
Mickle tissue disintegrator, VIII, 333

mitotic phases, III, 251–253
use of caffeine, III, 102–104
use of colchicine, III, 98–102, 104
use of colchicine and tritiated thymidine, III, 108–110
use of tirtiated thymidine, III, 105–107
effects of 5-aminouracil, III, 192–195
mitotic time, III, 224
anaphase, III, 224
metaphase, III, 224
prophase, III, 224
telophase, III, 224
response to TdR-H^3, III, 185–187
Mitotic cycle measurements, I, 387–401
cultures for, I, 388–389
interphase times in, I, 5, 392–400
photography and instruments for, I, 389
Mitotic delay, radiation-induced, IX, 85
Mitotic detachment, VI, 260
Mitotic extract, V, 117, 119
Mitotic incidence in *Necturus* leukocyte cultures, III, 89
Mitotic index(es), II, 346; III, 57, 222, 262, 273; IV, 74, 487, 489, 492–494
definition, III, 222
diurnal rhythm, III, 273
Mitotic inducer(s), V, 107, 121–122
Mitotic inhibitors, III, 189
Mitotic movement, IV, 195
Mitotic scoring, III, 353
Mitotic shake-off, VII, 333
Mitotic spindle, V, 107
Mitotic stages, distribution of at selection, X, 169
Mitotic stimulant, II, 384
Mitotic synchrony, V, 104
see Physarum polycephalum, I
Mixed agglutination test, VI, 198
Miyake solution, IV, 287–289
Miyake's balanced physiological salt solutions, IV, 33
Moist chamber, X, 10
Mold growth, V, 49
Molds, VI, 181, 185
broth, VI, 184
contaminations, VI, 179
infections, VI, 188
Molluscum contagiosum, VIII, 371
Monas, IV, 411
Monkey kidney cells, VI, 164, 169, 173–174, 188, 193, 196, 214
Monoamine oxidase, V, 192, 195; VIII, 216
Monocytes, VI, 3; X, 98
Monodus, III, 122
Monolayers, IV, 32
confluent, IV, 31
dispersing cells from, IX, 197
Mononucleosis, infectious, VIII, 109, 119
Monophosphokinase, VII, 394
Monosomy, VI, 217
Monoxenic cultures, VIII, 342
Morphalloxis, I, 122
Morphological assay, VIII, 424
Morris hepatoma, IV, 209–210, 214, 218
isolated nucleoli of, IV, 219
nucleolar preparation from, IV, 218
Mosquito, V, 86, 107, 110, 121
cells, V, 92
Mougeotia, II, 197
Mouse
cell lines, V, 78
cell harvest, V, 35
marsupial, V, 130–131, 158, 161
Mouse bone marrow, VI, 220
Mouse fibroblasts, III, 157
cytoplasmic staining of, III, 316
Mouse lines, VI, 222
Mouse mammary tumor, VIII, 370
Mouse myeloma cells, VII, 8
Mouse osteosarcoma, VIII, 370
Mouse sarcoma 180 ascites cells, VII, 3–6, 15
Mouse sarcoma virus, VIII, 77–79, 89
Movement of protozoa, IV, 328
MS-222 (tricaine methane sulfonate), III, 52, 77
MSV-NRK, VIII, 79
revertants of, VIII, 87
MSV revertants, VIII, 89
MSV-3T3, VIII, 79, 87
Mu, IV, 308, 316
Mucopolysaccharides, IX, 211
Mucoproteins, IX, 211
Multinucleate cells, IV, 191; VII, 257
formation, VII, 255
Multiple synchronization, III *see also* Cell synchrony prerequisites, III, 42
Multiplication rates of individual cultures, IV, 36
Multipolar mitosis, III, 36
Mumps, V, 78; VI, 168
Mumps virus, VI, 165

Muntiacus muntjak, X, 9
Murids, V, 7
Murine leukemia virus, VIII, 89, 370
  assays with, VIII, 424
Murine leukosis, VIII, 370
Murine lymphoma, VIII, 181
Murine mastocytoma cells, VI, 45
Murine sarcoma, VI, 168
Murine sarcoma viruses, VI, 176; VIII, 370
  assays for, VIII, 381
Murine tumor cells, V, 79
Murine viruses, assays for, VIII, 422
Mus musculus, V, 7
Muscle, V, 5
Muscle cells, VII, 73, 90, 92–93, 97, 160, 236, 374, 387
  mRNAs, VII, 96
Mushroom, VI, 183
Mustard ICR-170, IV, 174
Mutagen does, X, 227, 230
Mutagenesis, VI, 128, 217, 226, 250, 259, 270, 272; VIII, 4
  conditions for, X, 210
  induced, X, 211, 227
  proof of spontaneous, X, 224
  selective conditions for, X, 212
  spontaneous, X, 211, 220
Mutagenic agents, VI, 211; VIII, 34, 44
  detecting, VI, 269
  effects of, VI, 263, 267
Mutagenic chemicals, VIII, 373
Mutagenic treatment, VIII, 25, 32
Mutagenicity, VI, 269
Mutagens, V, 130; VI, 212, 226, 235, 243; IX, 174
Mutant cell survival, X, 223
Mutant cells, VI, 230; IX, 172
  isolation of, VI, 238
Mutant clones
  radiation-induced, VI, 267
  spontaneous, VI, 213
Mutant frequencies, VI, 263–265, 267–268, 273
  in CHO cells, VI, 268
Mutant isolates, VI, 225
Mutant lines, VI, 220, 227, 243
  cold storage of, VI, 221
Mutant markers, VI, 213, 215, 217
Mutant phenotypes, VI, 215, 258, 261, 264–265, 276
Mutant recovery
  effect of cell population density on, X, 212
  optimal conditions for, X, 213
Mutants, V, 113–114, 116, see also specific types; VI, 137–139, 214, 216, 224, 241, 243
  auxotrophic, VIII, 23–24, 26, 32, 34–37, 43
  cell division, VI, 260
  colchicine resistant, VI, 269
  conditional lethal, VIII, 41
  detection of temperature-sensitive (*ts*), X, 218
  drug-resistant, VI, 227–228; X, 215
  frequency of, VI, 258
  frequency of spontaneous, VI, 264
  glutamine-requiring, VI, 141
  hypoxanthine-auxotorphic, VI, 140
  in BHK cells, VI, 261
  in *E. coli*, VIII, 36
  isolation of, IX, 171
  non-leaky, VIII, 36
  nutritional, X, 218
  nutritionally deficient, IX, 174
  of Chinese hamster cells, VI, 135
  ouabain-resistant, VI, 269, 271
  radiosensitive, IX, 222
  selection for *ts*, VIII, 5
  selection methods for, VIII, 24
  somatic cell, VI, 274
  spontaneous, VI, 234
  temperature-sensitive, VI, 215, 221, 224–225, 227, 261
  temperature-sensitive conditional lethal, VI, 214
  thymidine-auxotrophic, VI, 140
  viability of, VI, 253
  X-ray induced, VI, 234
Mutation induction, X, 232
Mutation rates, VI, 211, 213, 264, 275
  measurement of, VI, 266
  spontaneous, X, 221, 226, 228
Mutations, VI, 215, 217, 226, 239
  conditional lethal, VI, 216
  for resistance to purine and pyrimidine base analogs, X, 215
  forward, X, 215
  kinetics of, X, 230
  leaky, VIII, 11
  missense, VIII, 4

rates of, VIII, 42–43
reverse, X, 219
6MP resistance, X, 231
suppressor, VI 270
temperature-sensitive, VI, 255
Mycetozoa, IV, 346
Mycetozoan amebae, IV, 407
Mycifradin, V, 16
Mycological media, VI, 183
Mycoplasma, VIII, 394
and bacterial RNAs, X, 266
cultivation of, VIII, 230
nutrient requirements, X, 262
RNAs, X, 265
Mycoplasma arginini, VI, 147–148, 161; VIII, 239–240, 243; X, 282, 288
Mycoplasma arthritidis, VI, 147, 154–155; VIII, 240
Mycoplasma broth, VI, 179–180
Mycoplasma canis, VI, 147
Mycoplasma contamination
biochemical technique for the detection of, X, 277
detection of, VIII, 229; X, 261
microbiological techniques for the detection of, X, 268
Mycoplasma cultures, axenic, VIII, 239
Mycoplasma detection, sensitivity of techniques, X, 272
Mycoplasma DNA, VIII, 231
Mycoplasma fermentans, VI, 147, 152; VIII, 240
Mycoplasma gallinarum, VI, 147
Mycoplasma gallisepticum, VI, 147, 154–155; VIII, 240
Mycoplasma hominis, VI, 147, 151–153, 155–156; VIII, 240, 243
Mycoplasma hyorhinis, VI, 82, 147–148, 153, 155, 157, 161; VIII, 231, 238–240, 243; X, 274, 282, 288
Mycoplasma-infected cells, uracil incorporation into, X, 278
Mycoplasma-infected WI-38 cells, therapy on, X, 286
Mycoplasma laidlawii, VI, 147, 149, 152–153, 155; VIII, 239–240, 243
Mycoplasma neurolyticium, VI, 154; VIII, 240
Mycoplasma orale, VI, 147, 152, 156; VIII, 240, 243
Mycoplasma pneumoniae, VI, 151, 154, 157
Mycoplasma pulmonis, VI, 147, 154–155
Mycoplasma RNA, VIII, 231–232, 243
Mycoplasma salivarium, VI, 155; VIII, 240
Mycoplasma sp., VI, 147–148, 152
Mycoplasmacidal factors, VI, 160, 164
Mycoplasmalike organisms, VIII, 244
Mycoplasmas, VI, 144–149, 151, 153–154, 156–157, 159–164, 176, 178, 185, 188, 194
broth, VI, 184
classification of, VI, 149
contamination, VI, 82
detection of, VI, 156
Dienes method of staining, VI, 159
elimination of, VI, 162
eradication of, VI, 161
genome size, VI, 149
infection, VI, 107
media, VI, 185
resistance to, VI, 164
resistant to antibiotics, VI, 149
Mycoplasmataceae, VI, 149
Mycoplasmatales, VI, 149
Mycostatin, III, 64; IV, 55; V, 12, 18; VI, 181, 188
Myelin, VII, 160
Myeloblasts, VII, 316, 324, 327–328
transformed, VII, 316
Myelocytes, V, 129
Myeloma cells, VI, 215; VII, 45
Myoblasts, IV, 28; V, 87; VII, 45
Myoinositol, III, 179; V, 63
Myosin, VII, 97
Myotubes, V, 87
Myromecobius fasciatus, V, 131
Myxamoeba, IV, 346–347, 369, 378–379, 388, 405–406
Myxoma virus, VI, 167
Myxomycete(s), IV, 346–350, 352, 368–369, 378, 388, 392–393, 403, 405–411, 434, 456, 461–462, 464–468
Myxomycophyta, II, 398
Myxovirus(es), II, 96, 100–101, 104; V, 78; VI, 174
Myxovirus-like virus, VI, 168

N

NAD, IV, 92–93
NAD glucosidase, IV, 206
NAD pyrophosphorylase, IV, 206; VII, 234
NADH, IV, 88, 92, 100, 106, 109, 113
NADH-cytochrome b_5 reductase, VIII, 227

NADH-cytochrome *c* oxidoreductase, IV, 100, 104
NADH-cytochrome *c* reductase, IX, 261, 275
NADH oxidase, IV, 105, 107, 109–112; VIII, 226–227
NADPH, IV, 100
NADPH-cytochrome *c* reductase, V, 195; VIII, 216, 224, 226; IX, 261, 272–273, 275–276
Naegleria, IV, 341, 344, 347–348, 350, 353–354, 360–363, 366–368, 370–371, 373–375, 377–381, 383, 385–386, 388–392, 398–399, 401–402, 404–409, 411–412, 414–416, 419–421, 423–431, 433–439, 441–442, 446–450, 452–455, 457, 459, 461–468; VI, 190, 192
  encystment of, IV, 388
  excystment of, IV, 388
  multinucleate, IV, 387
  taxonomy of, IV, 352
  transformation of, IV, 350
Naegleria bistadialis, IV, 358, 361, 409–410
Naegleria gruberi, IV, 343, 345, 357, 349–354, 357, 359, 361, 364–368, 370, 372, 376, 379–382, 384, 393, 407–410, 412–414, 423, 425, 432, 434, 436, 454, 456, 459, 462, 464–466
  transformation of, IV, 391
Naegleria limax, IV, 353
Naegleria mutabilis, IV, 358–359
Naegleria punctata, IV, 353, 355
Naegleria thorntoni, IV, 358, 408
Nagisa culture, VI, 5, 12–13, 19, 24, 26, 34
α-Naphthol, VII, 204
Naphthol yellow S, VIII, 337
  procedure after Deitch, III, 320–321
  reaction, III, 319–321
Nasopharyngeal carcinoma, VIII, 371
Necturus, II, 43, 48; III, 87, 89
Necturus maculosus, III, 90
Needle puller, II, 15
Neff's medium, composition of, V, 227
Neomycin, IV, 55–56; V, 11 16–17; VI, 163, 185, 192
Neomycin A, IV, 121
Neomycin sulfate, IV, 124; V, 81, IX, 87
Neoplastic transformation, VIII, 372
  spontaneous, VIII, 383
NETS buffer, composition of, V, 267
Neuramicin acid, VI, 154
Neuraminidase, IX, 33–34, 47, 198
Neuroblast(s), I 229–276
  autoradiography of, I, 263–265
  carbon dioxide tension for, I, 256–257
  chemical effects on, I, 271
  culture media for, I, 240–242
  culture techniques for, I, 240–257
  effect of osmotic pressure on, I, 254–257
  electron microscope techniques for, I, 265
  enzymatic digestion of, I, 262
  fixation of, I, 261
  fixed cell techniques for, I, 259–261
  gas effects on, I, 271
  hanging-drop cultures of, I, 248, 254
  microdissection on, I, 272–274
  microinjection of, I, 274
  mitosis of, I, 230
  multiplication of, IX, 248
  oxygen tension for, I, 256–257
  pH effect on, I, 257
  radiation effect on mitosis of, I, 270–271
  radiation techniques for, I, 265–271
  sectioning of, I, 261
  separated cell techniques for, I, 257–259
  staining of, I, 262
Neuroblastoma, IX, 248
Neurons, V, 87, 92; VII, 160
Neurospora, IV, 84
Neurospora crassa, II, 2
Neutral red, IV, 285, 373; V, 333, 335–336, 342, 345, 347–352, 355, 357, 360–364, 368; VI, 182
Neutrophils, VI, 335, 339
Newcastle disease myxoviruses, V, 80
Newcastle disease virus, II, 101–104; V, 78; VI, 167; VII, 242
Newts, II, 45, see also specific species
Niacinamide, V, 60
Nictoinamide (niacinamide), V, 15; VI, 8
Nicotinic acid (niacin), III, 177; IV, 143, 170–171; V, 15, 226; VI, 8
Nigrosin, V, 44, 241
NIL2E hamster cells, VII, 223
Nile blue sulfate, IV, 285; V, 342
Nitella flexilis, II, 197
Nitex monofilament screen, V, 44
Nitroblue tetrazolium, III, 72
Nitrogen mustard, IV, 187, 303–304, 313; IX, 66
Nitroquinoline oxide, VIII, 415

4-Nitroquinoline 1-oxide (4NQO), VI, 13, 35–36
Nitrosodimethylamine, VIII, 36
N-Nitroso-*N*-ethylurethane, IV, 174
Nitrosoguanidine, IV, 303–304, 457; VIII, 347
N-Nitroso-*N*-methyl-*N'*-nitroguanidine, IV, 174
Nitrosomethylurethane, VIII, 36
N-Nitroso-*N*-methylurethane, IV, 174
Nitrous oxide, V, 96–97, 109, 112
Niu-Twitty solution, IV, 26
NKM solution, IX, 352
Nondisjunction
  colcemid, VIII, 202
  colcemid induced, VIII, 181, 196; IX, 213, 215
  defect, VIII, 197
Nondisjunctive mechanism, VIII, 203
Nonidet P40, IV, 121, 225–226; V, 256, 258; VII, 3, 8, 40, 55
Nonionic detergents, IV, 212
Norit, VII, 442
Norit A, VII, 452
Nostoc, V, 374
  synchrony of heterocyst production in, V, 380–381
Notoryctidae, V, 128
Novikoff hepatoma, IV, 208, 214, 216, 223; VII, 368, 372, 374–375
Novobiocin, IV, 55; V, 163
NP-40, VIII, 158
NTB3 emulsion, VII, 149–150, 153
NTE emulsion, V, 290
Nu, IV, 307
Nuclear-cytoplasmic compatibility, IV, 189, 192
Nuclear-cytoplasmic hybridization, IV, 190
Nuclear-cytoplasmic hybrids, IV, 189
Nuclear-cytoplasmic incompatibility, IV, 190
Nuclear-cytoplasmic interactions, IV, 189; VIII, 123, 125, 142
Nuclear-cytoplasmic relations, IV, 187, 191
Nuclear cytoplasmic size, IX, 226
Nuclear diameters, IX, 226–227
Nuclear division, IV, 134, 139, 149, 157, 159, 386, 388, 398
  in *Acetabularia*, IV, 67
Nuclear electron transport, site of, VIII, 227
Nuclear envelope(s), IV, 213, 370, 380, 383, 386; V, 168, 190
  enzymatic activities of, V, 193
  from ascites tumors, V, 176
  from rat liver, V, 176
  isolation of, V, 167, 169
  of phospholipid, V, 191
Nuclear fractions, X, 77
Nuclear isolation, VIII, 348
  large-scale, X, 137
  small-scale, X, 137
Nuclear manipulations, in ameba, IV, 193
Nuclear membrane(s), IV, 3, 141, 195, 198, 209, 211, 344, 355, 382, 384, 388, 410; V, 107, 121, 168, 171, 173, 176; IX, 281; X, 81, 127
  chemical composition of, VIII, 225–226
  comparison with endoplasmic reticulum, VIII, 225–226
  composition of, V, 189–190
  contamination in isolated, VIII, 225
  densities of, X, 82
  electron transport system in, VIII, 227
  enzymatic equipment of, V, 192
  fractionation of, VIII, 227
  isolation of, VIII, 213; X, 124, 126
  of *A. proteus*, IV, 418
  outer, IV, 225
  phospholipid content of, VIII, 225; X, 131
  purification of, X, 79
Nuclear pore(s), V, 168
  complexes, VIII, 206, 220
Nuclear preparations, IV, 197
Nuclear protein(s), IV, 204; IX, 389
  of ameba, IV, 204
  recovery of, VIII, 225
  sulfuric acid-extractable, IX, 365
  sulfuric acid-soluble, IX, 358
Nuclear purification, in sucrose gradients, X, 140
Nuclear RNA, IV, 201
  extraction of, VII, 15
Nuclear sap, X, 25
Nuclear staining, IV, 151
Nuclear subfractions, X, 78, 81
Nuclear symbionts, IV, 294
Nuclear-to-cytoplasmic ratio, IX, 226–227
Nuclear transfer, IV, 20, 46
Nuclear transplantation, I, *see* Micrurgy, specific organisms; IV, 180, 183, 187, 191–192; V, 76; VII, 240
  abnormal embryos from, II, 29

apparatus for, II, 8
applications of the methods of, II, 25–33
chromosomal analysis after, II, 29
differences in methods in, II, 23–25
effects of liver fractions on, II, 31
effects of nitrogen mustard on, II, 30
in *Amoeba*, II, 1–2; IV, 179, 186
in amphibia, II, 1–33
in anurans, II, 2–18
in urodeles, II, 18–23
interspecific, II, 28
media for, II, 12–15
nuclear marker in, II, 18, 21, 23
procedure for, II, 2–23
tumor nuclei, II, 32–33
Nuclease inhibitors, X, 74
Nucleases, VII, 3–4, 10–11
Nuclei (nucleus), IV, 9, 25, 39, 37, 66, 79, 87, 91, 125, 132, 134–136, 139–141, 151–152, 154, 158, 163, 172, 181, 184, 186–192, 202, 235–238, 286–287, 305–306, 342–345, 348, 355, 370–373, 378–380, 386–388, 392, 404, 410, 447–448, 510; VII, 7
ameba, IV, 179–180
ascites tumor, V, 172
bulk isolation from oocytes, II, 43
chemical composition of, VIII, 225–226; X, 78
composition of, IX, 356
contents of, VI, 294
cultivation of from *Drosophila*, II, 75–78, 90–91
cytological analysis of isolated, X, 141
disruption of, VIII, 161, 171–172
enzymatic activities of, V, 193
fractionation of, V, 170; X, 69, 77
in discontinuous gradient, IV, 6
in testicular homogenate, IV, 7, 10
insertion of, IV, 185
isolated, IV, 6
from Morris hepatoma, IV, 217
isolation from *Drosophila*, II, 74–82
isolation of, II, 131–142; IV, 3, 195–230, 232; V, 172; VI, 283–306; VIII, 153–154, 157, 207; IX, 315, 349; X, 69, 124, 136
from rat liver, IV, 196
isolation of salivary gland, IX, 380–381
labeled with tritiated uridine, IV, 14
liver, V, 172
manipulation of, IV, 186
manipulation of isolated, IX, 384
mass isolation of pachytene, IV, 14
mass preparation of, IV, 2–3
maternal, IV, 25
medium for isolation of, X, 74, 82
meiotic prophase, IV, 2
methods for isolation of, IX, 352
microinjection of, II, 131–142
phospholipid of, V, 191
pachytene, IV, 2
preparation of, VIII, 160
prophase, IV, 2
protein content of pachytene, IV, 14
protein exchange between, IV, 191
purification of, VII, 5, 40
recovery of, VIII, 159
sizes of, II, 67
sonicated, X, 75, 82
sonication of, VIII, 163
sperm, IV, 6
spermatid, IV, 6, 9
subfractionated, X, 73
testicular, IV, 14–16
thymocyte, IV, 205
thymus, V, 172
transplantation in amphibia, II, 1–34
transplantation in honey bee, II, 2
tritium-labeled, IV, 233
viability of, VII, 200
Nucleic acid hybridization, X, 1
Nucleic acids, I, *see also* DNA, Microelectrophoresis of nucleotides, Microextraction of nucleotides, RNA
analysis of, I, 417–447; IV, 14
base composition in, I, 417–418
calculations of base ratios in, I, 444
calculations of total amounts of, I, 444–445
determination of, I, 29, 417–447
electrophoresis of, I, 424–427
hydrolysis of, I, 423–424
Nucleocapsid protein, VII, 181
Nucleochromatin material, II, 113
Nucleocytoplasmic feedback, II, 62
Nucleocytoplasmic interactions, V, 76, 96
Nucleohistone fibers, VII, 143
Nucleolar apparatus, IV, 2
Nucleolar division, IV, 386–387

Nucleolar protein, IV, 213
Nucleolar vacuole, IV, 373
Nucleoli, II, 38, 81, 133; IV, 2, 16, 29, 135–136, 139–140, 202, 342, 344–345, 352, 355, 357, 363, 365, 371, 379, 381, 383–386, 388, 390, 396, 399, 410–411, 454, 466
  division of, IV, 409
  enlargement of, VIII, 16
  isolation of, IV, 195–230; VIII, 163; IX, 353, 373; X, 136, 140
  Penman-type, IX, 374
Nucleolo-chromosomal apparatus, IV, 224
Nucleolonema, IV, 219
Nucleolus, II, 18, 32; III, 154–155, 171, 201–202
  staining, Rothwell's modification of Rattenbery-Serra procedure, III, 202
Nucleolus-associated chromatin, IX, 358
Nucleolus organizer, IV, 383
Nucleolytic activities, IV, 211
Nucleolytic enzymes, IV, 209, 221
Nucleoplasmic RNA, extraction of, VII, 16
Nucleopore filtration, IX, 318
Nucleoproteins, IV, 192
Nucleoside deaminase, VII, 381
Nucleoside diphosphokinase, VII, 381–382, 394
Nucleoside kinases, VII, 376, 388, 398, 407, 409–410, 415–417
Nucleoside monophosphokinase, VII, 382, 386
Nucleoside phosphorylase(s), VII, 387–388; VIII, 231–232; X, 307
Nucleosides
  inhibition of, VII, 416
  transport, VII, 416, 421
  uptake, VII, 415
5′-Nucleotidase, IV, 206; VII, 158–160, 184, 366, 381, 386, 393, 395, 409; IX, 261, 272, 275–276
Nucleotide kinases, VII, 394
Nucleotide sequence analysis, X, 340
Nucleotides
  metabolism, VII, 378
  microelectrophoresis of, I, *see* Microelectrophoresis of nucleotides
  microextraction of, I, *see* Microextraction of nucleotides
  pools, VII, 378, 414, 420, 423, 429
    changes in, VII, 378
    four-factor model for, VII, 366
    sizes of, VII, 373, 376
  techniques for separation of, VII, 444
Nucleus, *see also* Nuclei
  dry mass of, II, 43
  isolation from oocytes, II, 46–48
  role in differentiation, II, 25–28
Nucleus isolation media, IX, 315, 318
Nyacol, X, 87
Nylon knives, V, 174
Nylon screen, X, 137

O

Obligate photoautotroph, III, 120–121
O.C.T. compound, VIII, 252
Ocular adnexa, III, 48
Oil chamber, X, 24
Oil injections, VII, 293
Oleandomycin, VI, 192
Oligomycin, IV, 87; VII, 418
Omadine, IV, 54–56, 65
Omnifluor, VII, 338, 345
Omnimixer, IV, 201, 209, 214
  grass bead-Sorvall, IV, 102, 105, 110
OMP decarboxylase, VII, 379–381
OMP pyrophosphorylase, VII, 379–381
Oncogenic agents, VI, 173
Oncogenic viruses, VI, 166, 176–177; VIII, 77
Onion roots
  fixation of, I, 219–220
  germination of, I, 216–217
Onychogalea unguifer, V, 134
Oocytes, VII, 110, 212, 236
Oogenesis, IV, 2
Operating chambers, I, 100–101, 103–105, 407–409
Ophthaine, III, 53
Opossum(s), V, 6, 141, 148–149, 155
  four-eyed, V, 142
  Philander, V, 141–142
  pouchless, V, 142, 144
  rat, V, 141
  Virginia, V, 128–129, 141–142, 147, 154–155
  Woolly, V, 129, 142–143
Oral apparatus, isolation of, V, 267–269; X, 118
Oral papilloma virus, VI, 167

Oral structures, VIII, 332
Orange G, V, 334
Orcein, IV, 510; V, 22, 25–26, 93
Orcein stain, VI, 156, 358, 373
Orcinol reaction, IV, 10
Orcinol technique, VIII, 351
Ornithine, VI, 6
Ornithine transcarbamylase, IV, 157
Ornithosis virus, infection by, I, 415
Oronite polybutene-128, II, 65
Orotic acid
  permeability of cells to, VII, 380
Oryotolagus cuniculus, III, 69
Oscillating mechanical shaker, III, 335
Osmic acid fixation, III, 198
Osmometer, X, 89
Osmoregulation, II, 155
Osmosis, II, 145
Osmotic ground value, II, 149, 152–154, 156–163, 168–169
Osmotic pressure, II, 147–148, 163–166, 168, 220
Osmotic properties, II, 144
Osmotic quantities, II, 144–145, 147, 149, 167
Osmotic value, II, 147, 186
Osterhaut's solution, IV, 333
Ouabain, VI, 214, 228, 242
  resistance phenotype, VI, 244
  resistance to, VI, 241–242, 265, 268, 274
  resistant cell lines, VI, 243
  resistant mutants, VI, 269
Ouchterlony double-diffusion, IV, 455
Ova, V, 88, 129
Ovalbumin, VII, 97
Oviduct, VII, 97
Ovulation, IV, 25
  in frogs by FSH, II, 22
  induction of in frogs by frog pituitaries, II, 2
Oxalate, III, 72
Oxamate, III, 72
Oxidative phosphorylation, IV, 86–87, 89, 92–93, 95, 97, 99–100, 102, 107, 112, 204
  measurement of, IV, 85
Oxygen consumption by reticulocytes, V, 209
Oxygen electrodes, IV, 87, 91, 106; V, 199
  apparatus, IV, 88
  flush type, V, 200
  needle type, V, 200
Oxygen sensor, IV, 106
Oxytetracycline, VI, 192

P

Pachytene, IV, 3, 502, 511
Pachytene nuclei, mass preparation of, IV, 16
Pachytene stage, isolation of nuclei of, IV, 1–17
Pancreatin, V, 43; VI, 361
Panheprin, IV, 481
Panthothenate, IV, 143
Pantothenate, IV, 122, 170–171; V, 15, 60, 226
Pantothenic acid, VI, 8
Papain, V, 43
Papilloma viruses, VIII, 371
Papio sp., VI, 166
Papova viruses, VIII, 371
  transformation by, VIII, 405
Papovirus, VI, 176
Parabiosis, II, 26
Paraformaldehyde, IX, 200
Parainfluenza, VI, 167–168, 171
Parainfluenza I, V, 79–80
Parainfluenza virus type, VI, 3, 172
Parainfluenza viruses, V, 78; VII, 260
Paramecia, swimming behavior of, VIII, 346
Paramecium, II, 132, 138; IV, 242, 244, 250–251, 253–254, 257, 262, 264, 266, 268–270, 276, 278–279, 286, 291, 295–296, 304–307, 309–315, 324, 326–327, 329–331
  amicronucleate clones of, I, 119–120
  available stocks of, IV, 248
  binucleated cells of, I, 118–119
  Chlorella-bearing, IV, 327
  clonal life cycle of, IV, 272
  culture of, I, 122–124
  endoplasm of, I, 120, 122
  experimental uses of, I, 117–124
  fission rate of, IV, 261, 274
  fixation of, VIII, 325
  generation time of, I, 132
  geographical distribution of, IV, 246, 280
  growth cycle of culture of, IV, 252
  isogenic stocks of, IV, 245
  kappa-bearing, IV, 318
  killer of, IV, 322
  killer trait in, I, 117
  lambda bearers of, IV, 322
  macronucleus of, I, 118–119, 122

mailing cultures of, IV, 249
mass cultures of, IV, 265
mating types of, I, 117
methods in research in, IV, 241–339
micronuclei of, I, 118–119
nucleocytoplasmic ratio of, I, 119
nutritional abnormalities in, I, 122–123
predators of, IV, 322
regeneration of, I, 118, 122
sectioning of, I, 135
symbiont-bearing, IV, 322
clones, IV, 323
symbiont-free, IV, 294
symbionts of, IV, 321
synchronization of, IV, 263
viable fragments of, I, 120, 122
Paramecium aurelia, I, 119, 122; IV, 118–121, 244–248, 254, 262, 265, 269, 272–277, 279–284, 286, 288–290, 299, 301, 303–304, 307–308, 311, 315, 321, 331, 334; VIII, 324
axenic cultivation of, IV, 117–130
genetic analysis in, IV, 291
genetics of, IV, 245
growth curve for, IX, 296
isolation of nuclei from, IX, 281
lethal genes in, IV, 245
membranes of, VIII, 326
senescence, IV, 273–274
silver-impregnated specimens of, VIII, 321
symbionts of, IV, 294, 304
Paramecium bursaria, I, 117, 119; IV, 244–246, 248–249, 254, 259, 263, 267, 272, 280, 284–285, 287, 291, 326–327, 331–332
balanced salt solution for, IV, 332
Loefer's salt solutions for, IV, 333
Pringsheim's solutions for, IV, 334
symbionts of, IV, 326
Paramecium calkinsi, IV, 245, 249, 254–255, 277, 280, 287
Paramecium caudatum, I, 118–120; IV, 244–246, 248, 254, 276–278, 280, 282, 287–291, 296, 308; VI, 35; VIII, 324
DNA labeling of, X, 318
induction of conjugation in, IV, 291
isolating macronuclei from, IX, 297
isolation of micro- and macronuclei from, II, 137–138
Paramecium jenningsi, IV, 246, 249, 254, 274, 281, 289
Paramecium multimicronucleatum, I, 122; IV, 244–250, 254–255, 262, 276, 281–284, 286–287, 289–291, 298, 308; VIII, 324, 350
Paramecium polycaryum, IV, 246, 254, 274–275, 280, 326
Paramecium putrinum, IV, 249, 278, 281–282, 287
Paramecium traunsteineri, IV, 289–290
Paramecium trichium, IV, 245–246, 249, 254–255, 267, 281, 291, 326; VIII, 323
Paramecium woodruffi, IV, 245–246, 249, 254, 291
Paramoeba, IV, 349
Paramoeba eilhardi, IV, 348
Paramyxoviruses, V, 79–80, 84–85, 90; VI, 166, 174
Pararosaniline-Schiff, IX, 221
Parasite contamination, VI, 188
Paris green, X, 57
Parlodion, II, 287, 295
Paromomycin, VI, 163
Parr Bomb, IX, 286–289, 304–305
Particle counter, electronic, IV, 150; V, 237
PAS, *see also*, Periodic acid-Schiff, II, 270, 276, 278–281
Patapar paper, VIII, 51
Patas monkey, VI, 166
Patulin, VIII, 345
Pea roots, culture medium for, I, 217–218
Pea seedlings, nuclei of, IV, 214
Pectinase, III, 207
Pellicle(s)
isolation of, V, 260; X, 111, 113, 119
phospholipid composition of, X, 131
Pellicle ghosts, X, 112
Penicillin, II, 140; III, 64; IV, 27, 121, 480; VI, 9, 148, 185, 192
Penicillin G, IV, 55, 314, 481
Penicillin method, VIII, 24
Pentachlorophenol, IV, 87
Pentaerythritol, II, 219–221
Peptidyl-tRNA, VII, 74
Perameles, V, 133
Perameles bourganvillei, V, 131
Perameles gunni, V, 133
Perameles nasuta, V, 132–133
Peramelid(s), V, 132–133, 155
Peranema, II, 217

Perchloric acid solutions, preparation of, VII, 433
Perinuclear cytoplasm, IV, 212
Periodic acid-Schiff, *see also* PAS, II, 270, 328
Periodic acid-Schiff (PAS) reaction, IX, 211
Perlacell, IX, 16
Perlacell apparatus, IX, 18, 21
Perlacell culture vessel, IX, 17
Permeability
  absolute constant for, II, 151, 180, 188, 192–195
  determination of, II, 179–195
  differential, II, 144–145
Permeability series, II, 151
Permease, VII, 402
Permelidae, V, 128, 130–131
Permissive hosts, transformation in, VIII, 413
Permount, V, 95; VIII, 190, 280
Perognathus, V, 8
Perognathus baileyi, V, 8
Peromyscus, V, 8
Peroryctes longicauda, V, 133
Peroxisomes, isolation of, X, 109, 130
Persantin, VII, 410, 416
Persian gazelle, VIII, 185
Perspex hemagglutination trays, VII, 228
Petaurus breviceps, V, 134, 136
Peter's solution, IV, 333
Petrogale penicillata pearsoni, V, 134
Petrogale rothschildi, V, 134
PHA, IV, 485, 490, 492
Phage M13, VII, 130, 138
  structure, VII, 139
Phage-membrane association, VII, 138
Phage T7, VII, 136
Phagomyxa, IV, 411
Phalangeridae, V, 128, 134–135
Phalangerids, V, 134
Phalangers, V, 135
Phascolarctos cinereus, V, 134
Phascolomidae, V, 128, 130–131, 135–136
Phascolomis ursinus, V, 131, 135
Phaseolus vulgaris, II, 177, 374
Phenol, IV, 55
  extraction, VII, 14, 16
  procedure, VII, 12, 13
Phenol nuclei, IV, 224
Phenol red, IV, 27, 30; V, 11, 16–17, 48, 61, 63, 67, 83; VI, 9; VIII, 194
Phenol red pH indicator set, VIII, 56
Phenol solution, V, 23
Phenoloxidase dehydrogenase, II, 72
Phenosafranine, IX, 204
Phenotype, drug-resistant, VI, 227
Phenylethyl alcohol, VII, 418
Phenylhydrazine-treated rabbits, VII, 71, 94
Philander, V, 141
Phisohex, V, 5, 10
Phleum pratense, IV, 250
Phlorizin, VII, 418
Phoma, VII, 213
Phomin, VII, 213
Phosphatase(s), VII, 365, 446; VIII, 67; IX, 261, 272, 274, 276
  acid, VI, 303
  alkaline, VI, 303
  fluorescent assay of, IX, 240
Phosphodiesterase, Vi, 339; VII, 447
Phosphoglucomutase, VI, 194, 303
6-Phosphogluconate dehydrogenase, VI, 303
Phosphohydrolases, VI, 234
Phospholipase B, VII, 252
Phosphoprotein phosphatase, IX, 261
Phosphoribosylamine, VII, 391
5-Phosphoribosyl-1-amine, VII, 390
5-Phosphoribosyl-1-pyrophosphate, VII, 387, 390
Phosphorus-32, physical properties of, X, 294
Phosphorylase, VII, 397–398, 395–396, 398, 401
  activity, VIII, 242–243
Phosphorylation, IV, 87, 91, 94, 97–100, 102–103, 105–106
  role of in translational control, VII, 95
Phosphorylcholine, VII, 421
Phosphoserine, VII, 92
Phosphothreonine, VII, 92
Phosvitin, VII, 93
Photographic developer
  Dektol, II, 236–239, 249
  Elon-ascorbic acid, II, 236, 239, 250
  gold Elon-ascorbic acid, II, 238, 247, 249
  Kodak D-19, II, 362
  Microdol-X, II, 237, 239, 249, 252, 288
  p-phenylenediamine, II, 239, 249
Photographic emulsion
  application of, II, 230–236, 245–248, 287–293, 296–299

background in, II, 295–296
criteria for thickness of layers of, II, 233–236
gelatin removal, II, 299, 301, 303, 305
Gevaert, II, 207, 231, 234–236
Ilford L4, II, 230–237, 239, 248, 251–252, 286, 288, 292, 295–298, 301, 307
Kodak NTB, II, 362
Kodak NTB-2, II, 327
Kodak NTB-3, II, 327
Kodak, NTE, II, 230–237, 240, 246–248, 251–252, 286
processing of, II, 293–295
resolution of, II, 240
sensitivity of, II, 237–240
Photosynthesis, II, 217–218; IV, 50
Phototrophs, glutamine, X, 214
Physarum cinereum, I, 11; IV, 406
Physarum didermodies, I, 35
Physarum flavicomum, IV, 406
Physarum polycephalum, I, 10–54; IV, 392
aeration of cultures of, I, 24
carbon sources for, I, 17–18
coalescence of plasmodia, I, 43–53
culture techniques for, I, 22–28
DNA synthesis in, I, 52–53
glucose utilization in, I, 17, 21
growth measurements of, I, 17, 21, 24, 38–40, 44
isolation of, I, 11–13
life history of, I, 10, 46–47
light effects on growth of, I, 26, 34–35
macroplasmodia of, I, 27–28
microplasmodia of, I, 14, 23–24, 27–28, 43–53
mitotic cycle in, I, 50–52
mitotic synchrony in, I, 27, 43–54
motility of, I, 10
nucleic acids of, I, 29, 50, 52–53
organic requirements of, I, 20–22
pH effect on growth of, I, 25, 35
pigment in, I, 29, 34
pure culture of, I, 10, 13–40, 47–49
sclerotia formation in, I, 36–37, 47
semidefined media for, I, 15–18
spherule formation in, I, 36–37
sporulation medium for, I, 27
sporulation of, I, 29–36, 38–40
submersed culture of, I, 23–27, 44
surface culture of, I, 22–23
synchronous growth in, I, 27
synthetic media for, I, 18–20
temperature effect on growth of, I, 25–26, 35
Physostigmine, VII, 357–359
Phytohemagglutinin, IV, 478, 480–481, 488; V, 149–150; VI, 68, 154; VII, 388, 401, 409; VIII, 109; IX, 1, 6–7, 10
Phytohemagglutinin M (PHA-M), III, 88–89
Phytomastigina, III, 121
Phytomonad, IV, 108
Picofarads, III, 4
Picornaviruses, VI, 179
Picric acid-bromophenol blue procedure after Bloch and Hew, III, 331
Picric acid-bromophenol blue reaction, III, 329–334
Picric acid-eosin Y procedure after Bloch and Hew, III, 331–334
Picric acid-eosin reaction, III, 329–334
Picric acid-Schiff procedure after Dwivedi and Bloch, III, 318–319
Piericidin A, IV, 109
Pinocytosis, I, 277–304, III, 85
definition of, I, 278
demonstration of, I, 279–283
in enucleated fragments, I, 415
in mammalian cells, I, 282–283, 289–290
in plant cells, I, 283, 290
in protozoa, I, 279–282, 285–290
inducers of, I, 283–290
pH effect on, I, 292–294
quantitative measurements of, I, 290–301
temperature effect on, I, 292–294
Pinocytotic vesicles, VIII, 131
Piperazine-*N*,*N'*-bis(2-ethane sulfonic acid) monosodium monhydrate ("PIPES"), VI, 287, 295–296, 300, 304, 313–314
Pipette
braking, II, 132, 134, 319
holder, II, 15
Pasteur, II, 139–140, 314–315
Pipette puller, VIII, 295
automatic, IV, 329
Pi-Pump, VII, 288
Pisum, I, see pea roots
Pisum sativum, III, 171, 174, 182–183, 185, 187, 190–191
PK-15, VIII, 181
Plagiothecium, II, 207

Planimeter, I, 139
  polar, V, 210
Plankton counting chamber, V, 234
Plant cells, V, 90
  nucleoli from, IV, 214
  pinocytosis in, I, 282–284, 289–290
Plant hormones, IV, 57
Plant protoplasts, V, 90
Plasma cell tumor, VII, 97
Plasma extenders, II, 13
Plasma membrane(s)
  composition of, IX, 253
  composition of brain, IX, 252
  damage to, IX, 277
  ghosts, VII, 171, 176–177, 179, 183
  isolation by latex bead method, IX, 262
  isolation by $ZnCl_2$ method, IX, 263
  isolation methods, IX, 249, 259
  neuron of, IX, 248
  of KB cells, IX, 259
  permeability, VII, 407
  precursors, VII, 179, 181
  properties of, IX, 259
  purification of, IX, 247
  purity of, IX, 274
  synthesis of proteins and glycoproteins, IX, 254
Plasma membrane markers, IX, 248
Plasma membrane proteins, IX, 256
Plasmacytoma, VII, 79–80
Plasmodia, IV, 347, 388, 407, 464
Plasmodiophorales, IV, 349, 410–411
Plasmodium gallinacium, III, 152
Plasmolysis, II, 144–149, 151–158, 160–161, 163, 168–169, 184–188, 204–212
Plasmometric method, II, 146, 152, 159–161, 179, 189
Plastic coverslips, VIII, 147
Plastic films, IX, 21
Platelets, VII, 392
Plating efficiency, IV, 40–42; IX, 167, X, 214
  effect of conditioned medium on, IX, 130
  of FM3A cells, IX, 162
  on agar plates, IX, 165
  single cell, VIII, 68
Platinum-iridium wire, IV, 416
Pleomorphism, III, 154
Plethodontids, II, 48
Pleurodeles, II, 23, 25, 27, 48
Pleurodeles waltlii, II, 23
Plus X reversal, III, 150
Podophrya paramecium, IV, 326
Poikilothermic animals, III, 82
Poikilothermic tissues, III, 76, 81
Pokeweed mitogen, VIII, 309; IX, 6
Polarizing microscope, III, 148; IV, 81
Poliomyelitis, VI, 169
Poliovirus, II, 104–110; VI, 164–165; VII, 199, 242; VIII, 137, 146
  cell cycle dependency, VI, 125
  RNA replication, VIII, 137
Poliovirus type, purification of, X, 101
Polonium-210, II, 257
Poly(A)-binding, VII, 46
Poly-ADP-ribose-synthesizing enzyme, IV, 206
Poly(dT), VII, 20
Poly(dT) cellulose, VII, 60
  columns, running of, VII, 62
Poly(L-lysine), VII, 120
Poly-L-lysine-kieselguhr, VII, 124
Poly(U), VII, 20
Poly(U) fiber glass filters, VII, 57, 62
  for RNA binding, VII, 58
  preparation of, VII, 58
Polyacrylamide-agarose gels, preparation of, X, 30
Polyacrylamide gel electrophoresis, VII, 19
Polyadenylic acid [poly(A)], VII, 21, 46, 57
  ribonuclease resistance of, VII, 56
Polyamines, V, 120, 122; VII, 435
Polyarginine, VII, 435
Polycarbonate, IX, 21
Polychromatophilic erythroblasts, III, 339
Polyethylene glycol, IX, 29–30, 36
Polyethylene polyester, IX, 21
Polyethylene tubes, II, 319
Polykaryocytes, VII, 229
Polylysine, VII, 435
Polymixin, IV, 55
Polymorphonuclear cells, X, 98
Polymorphonuclear leukocytes, IX, 40
Polynucleotide kinase, VII, 31
Polynucleotide phosphorylase, VII, 64
Polyoma transformed cells, VIII, 89
  enucleated, VIII, 134–136
Polyoma virus, VII, 215; VIII, 77–78, 86, 371
  transformed BHK, VIII, 78–79
  transformed Syrian hamster cells, VIII, 87
  tumor antigen, VIII, 78

Polypeptone, IV, 86
Polyphenylalanine, VII, 70
  synthesis, VII, 72–74, 89
Polyploidy, induction of, III, 190–191
Polyporus frondosus, I, 13
Polypropylene, IX, 21
Polyribosomes, II, 406
  isolation of, X, 128
Polysaccharide staining, procedure for, IX, 211
Polysomal RNA, VII, 12, 14
Polysomes, VI, 86–89, 108–109; VII, 2–4, 6–7, 15, 19–20, 33–34
  isolation of, VII, 5, 8
  preparation of, VII, 7, 9
  preparation of membrane-bound, VII, 8
Polysphondylium, II, 401
Polysphondylium pallidum, I, 10; II, 378, 400
Polystyrene "tissue culture" petri dishes, V, 42
Polytene chromosomes, II, 66, 68–69, 79–80, 89
  IX, 378, 387; X, 14, 20, 45
  isolation of, IX, 377, 384
  isolated, IX, 385
Polytene nuclei
  DNA synthesis in isolated, X, 143
  incubation of, X, 141
  isolation of, X, 135, 139
  manipulation of, X, 135, 141
  mass isolation of, X, 136
  nonaqueous preparation of, X, 138
  of *Chironomus*, X, 145
  RNA synthesis in isolated, X, 142
Polytoma uvella, IV, 102, 105, 109–112
Polytomella caeca, IV, 108–109, 113
Polytron 10-ST, X, 121
Polyvic, IV, 145
Polyvinylpyrrolidone (PVP), VI, 2, 17–18; VIII, 62
Polyvinylsulfate, IV, 224; VII, 9, 33
Pools
  analysis, chemical methods for, VII, 444
  definitions of, VII, 364
  extraction, VII, 430
    kinetics of, VII, 437
  general properties of, VII, 364
  growth-rate dependence of, VII, 376
  intracellular concentration of, VII, 453
  intracellular location of, VII, 372
  sizes, VII, 373
    in animal cells, VII, 374
POPOP, VII, 352
Pores, IV, 141
Porter-Blum MT-2 ultramicrotome, III, 62
Possum(s), V, 135
  bushtail, V, 136
  dormouse V, 135
  gliding, V, 136
Postneurula, IV, 26
Potorous tridactylus, V, 130, 135, 137–140, 145–146, 148, 150, 152, 155, 157–158, 163
Potter-Elvehjem homogenizer, VII, 32, 39–40, 71; VIII, 156, 158
Potter homogenizer, VII, 3, 5
Pox viruses, V, 78; VIII, 371
PPLO, VI, 146, 162; VII, 162, 387
PPLO contamination, VI, 163
Precursors
  incorporation, measurement of, VII, 336
  purity of, VII, 342
Primary cultures, IV, 26, 32, 40
Primary root meristem, III, 96
  model, III, 97–98
Pringsheim's solutions, for *P. bursaria*, IV, 334
Procaryotic contaminants, VI, 159
Proflavine, IV, 303
Progenitor cell cycle
  analysis, of, II, 342–344
  determination of, II, 331–335
  division of, II, 326
  duration of G_1 in, II, 334
  estimates of, II, 345–346
  measurement of, II, 339–342
Progenitor cells, II, 325–354
Progenitor compartment, II, 325–355
Progesterone, IV, 25
Prokaryotes, IV, 132
Proline
  auxotrophy, VI, 269
  prototrophy, VI, 265, 268
Proline-requiring mutants, VIII, 33
Pronase, II, 79, 119; V, 10–11, 24, 43, 52–53, 152; VI, 361; VII, 92, 115–116, 124
  solution, V, 13, 19, 24, 67
  treatment, V, 24
Pronucleus, inactivation by radiation, II, 18
Prophase, IV, 2, 379–380, 385
  meiotic, IV, 2, 502

Prophase nuclei, meiotic, IV, 10
Propidium-diiodide, VII, 121, 124
Propidium iodide (PI), IX, 200, 207
Propidium iodide protocol, IX, 242
β-Propiolactone, V, 82–83, 87–88; VI, 178–179; VII, 228
Propionyl-CoA, VI, 104
Prostaglandins, production of, IX, 131–132
Protamines, III, 309; VII, 93
Protease inhibitors, X, 74–75
Protease treatment, VIII, 397
Proteases, IV, 86; VII, 160
  liberation of, VIII, 237
Proteins, II, 79, 85; V, 152
  active material for, II, 31
  analysis of, IX, 221
  as inducers of pinocytosis, I, 285, 288–291, 293–294, 298
  autoradiography of, III, 207
  comparison of the half-lives of, X, 246
  cytochemistry of, I, 3
  cytoplasmic staining of, III, *see* Cytoplasmic staining
  determination of, I, 21, 26, 28, 34; VII, 335
  fixation of, III, 205–207
  half-lives of, X, 241, 244
  histones, III, *see* Histones
  ^{131}I-labeled, VI, 2
  in meristematic cells, III, 202–207
  location of, I, 3
  Millon reaction of, III, 202
  phosphorylation, VII, 94
  rate constants of degradation for, X, 253, 257
  rate of degradation of, X, 241
  rate of synthesis of, X, 238
  staining in root cells, III, 202–207
  staining with fluorescent dyes, IX, 212
  two-dimensional electrophoretic analysis of, IX, 362
Protein content of cells, IX, 226
Protein degradation
  half-life of, X, 237
  in cultured cells, X, 256
  in regenerating liver, X, 248
  measurement of, X, 249
  mechanism of, X, 255
  rate of, X, 243
  relative rates of, X, 252
Protein-free growth medium, VIII, 54
Protein kinases, VII, 92–94, 449
  cyclic AMP-dependent, VII, 93, 95
Protein staining techniques, IX, 212
Protein synthesis, I, 4, 6, 88–89; II, 31, 101, 131, 405; VII, 339, 351, 355
  in cell cycle, III, 157
  in *C. moewusii*, III, 144
  inhibition of, VI, 152
  mutations affecting, VI, 259
Protein-to-cell volume, IX, 225
Protein turnover, X, 243
  measurement of, X, 235
Protemnodon, V, 136
Protemnodon agilis, V, 134
Protemnodon bicolor, V, 130, 134, 136, 139–140, 150, 155, 157–159, 162–163
Protemnodon dorsalis, V, 134
Protemnodon eugenii, V, 134
Protemnodon irma, V, 134
Protemnodon parryi, V, 134
Protemnodon rufogrisea, V, 134
Proteomyxids, IV, 349, 410–411
Proteose peptone, IV, 123, 331, 431–432; V, 223–224, 227–228, 233
  medium, V, 225
Proteus vulgaris, IV, 461
Protista, IV, 381, 384
Protoplasts, IV, 141, 152–153, 163
Protostelida, IV, 347, 409, 411
Prototheca, IV, 106–107
Prototheca zopfi, IV, 84, 104
Prototrophs, selective death of, X, 216
Protozoa, I, *see also* specific organisms; IV, 84, 346, 348; VI, 144
  availability of maintained cultures of, I, 182–183
  continuous synchronous culture of, I, 141–158
  culture of, I, 85–86
  electrophoretic analysis of, I, 209
  flagellated, IV, 264
  light-induced synchrony in, I, 144–147
  pinocytosis in, I, 279–282
  selection for synchronized cells, I, 142–143
Provirus, VIII, 416
  C-type, VIII, 420
  formation and integration of, VIII, 417
PRPP amidotransferase, VII, 390–391

PRPP synthetase, VII, 390–391
Psammechinus miliaris, III, 316; X, 11
Pseudocherius peregrinus, V, 134
Pseudomonas aeruginosa, VI, 180–182
Pseudomonas fluorescens, VI, 180
Pseudopodia, IV, 342
Pseudorabies, VI, 168
Pseudospora, IV, 411
Pteris, II, 197
Puck and Steffen equation, III, 32
Puck's saline, VIII, 194
Puffballs, VI, 183
Puffing phenomenon, II, 62, 69, 74
Pullularia, II, 227
Pulse amplification measurement, III, 5
Pulse labeling, III, 250
Purine analogs, VI, 213
Purine mutants, VIII, 36
Purine nucleoside kinases, VII, 396
Purine nucleoside phosphorylase, VII, 396–398
Purine nucleotide
  pools, VII, 371
  synthesis, VII, 390
Purine nucleotide phosphorylase, VII, 393–394
Purine phosphoribosyl transferases, VII, 388, 396–398
Purine ribonucleoside kinase, VII, 396
Purine synthesis, VI, 233–234; VII, 377, 380
Purkinje nerve cell bodies
  RNA extracts of, I, 429
Puromycin, II, 93; III, 157; V, 275; VI, 213; VII, 72, 74, 80, 85, 90–91, 410, 416; VIII, 345
Puromycin-H^3, III, 52
Puromycin-resistance phenotype, VI, 214
Puromycin-resistant frog cells, VI, 212
Putrescine, V, 62, 120
Putrescine dihydrochloride, VI, 9
PyBHK, VIII, 85
Pyridine carboxaldehyde thiosemicarbazones, VII, 383
Pyridine nucleotides, IV, 100
Pyridoxamine, V, 226
Pyridoxamine HC1, IV, 122–123
Pyridoxine, IV, 143, 171; V, 15, 60; VI, 9
Pyrimidine(s)
  fluorinated, VII, 386
  synthesis, VII, 379
Pyrimidine analog, VI, 213
Pyrimidine deoxyribonuclease, VI, 153
Pyrimidine nucleoside deaminase, VII, 386
Pyrimidine nucleoside phosphorylase, VII, 381, 386, 388
Pyrimidine phosphoribosyl transferase, VII, 381, 387
Pyronine G, V, 241; X, 57
Pyronine Y, X, 52, 57–61, 65
Pyrophosphorylase, VII, 387–388, 396, 401
Pyruvate kinase, IV, 206

Q

Q bands, VI, 347, 351, 353–356, 358–359, 365, 367, 371, 373, 375–376; VIII, 186
Quadrol, II, 226
Quantitative autoradiography, I, 305–326
  absolute measurements in, I, 306, 322–324
  adjustment of grain count for, I, 320–321
  arrays of grains in, I, 322–323
  assay methods in, I, 321–322
  calculation for tritium in, I, 310–315
  coincidence in, I, 306, 315–316
  factors influencing measurements in, I, 306–308
  grain count distribution in, I, 317–318
  image spread in, I, 306–310
  matrix theory in, I, 309–310
  number of grains for, I, 315–317, 320–321
  relative measurements in, I, 306
  self-absorption corrections in, I, 311–315, 318–320
  theoretical considerations for, I, 308–318
  track counting in, I, 321–322
  visual grain counting in, I, 321
Quinacrine dihydrochloride, VI, 346–347, 351, 358–359, 373, 376
Quinacrine fluorescence, VI, 365, 373, 375; VIII, 182, 185
Quinacrine mustard, VI, 346–347, 358; VIII, 182

R

R bands, VI, 347, 353–354, 356, 365–368, 371–372, 375
Rabbit cell cultures, VI, 168
Rabbit heart line, VIII, 58, 60–63
Rabbit lens, III, 48
Rabbit lymph node, VI, 152
Rabbit myxoma-fibroma, VIII, 371

Rabbit reticulocytes, VII, 2, 4
Rabbit viruses, VI, 167
Rabbitpox virus, VI, 167
Rabies, VI, 165, 168
Radiation damage, III, 219; VI, 249
  to chromosomes, II, 330
Radical scavengers, X, 310
Radioactive compounds, purity of, VII, 450
Radioactive phosphate, problems with, VII, 451
Radioactivity
  measurement of, VII, 338, 352
  measurement of in chromatogram, X, 303
Radioautographic efficiency, VIII, 286
Radioautographic method
  for catecholamines, X, 365
  for estrogens, X, 365
Radiochemical decomposition, control of, X, 309
Radiochemical impurities, X, 304
  effects of, X, 306
  uptake of, X, 307
Radiochemical purity, X, 296–298
  of fatty [^{14}C] acids, X, 302
Radiochemicals
  chromatographic analysis of, X, 297
  effect of radical scavengers on the storage of, X, 316
  effect of temperature on the storage of, X, 314
  self-decomposition of, X, 308
  solvents to minimize self-decomposition of, X, 313
Radioimmune assay, VIII, 403
  of surface antigens, VIII, 355
Radioiodination
  of human serum albumin, X, 344
  of nucleic acids, X, 341
  of polycytidylic acid, X, 344
  of polynucleotides, X, 347
Radioiodination of RNA
  procedure for, X, 350
Radioiodine labeling, of ribopolymers, X, 343
Radiolysis, VII, 26, 450
Radiomimetic chemicals, III, 189
Radiomimetic compounds, I, 225
Radionuclide(s)
  choice of, X, 293
  physical properties, X, 293–294
Radionuclidic (radioisotopic) purity, X, 296
Radiotracer techniques, X, 297
Rag doll germination, I, 217
Rainbow trout, V, 92; VII, 155
Rana, II, 3, 12, 18, 22–27, 29
Rana arvalis, II, 12
Rana catesbeiana, III, 52, 56, 61–64, 69, 71
  chromosome spreads, III, 56–58
  isozymes, III, 71
  lens epithelium electron microscopy, III, 61–63
  malate dehydrogenase, III, 72
Rana clamitans, III, 69
Rana nigromaculata, II, 12; III, 87, 90
Rana pipiens, II, 2–3, 5–6, 10–12, 17, 23–25, 28–31; III, 47, 52, 69; IV, 24–25, 31 35, 38, 42, 45; IV, 170
  eye vertical section, III, 47
  haploid embryos of, IV, 33
  isozyme activity, III, 71
  lactate dehydrogenase, III, 71
  nictitating membrane, III, 47, 53
  optic nerve, III, 47
  "skinning procedure" in mounting lens, III, 50
Rana sylvatica, II, 28–29
Rana temporaria, II, 12, 28
Randomization table, VIII, 58
Rat liver
  isolated nuclei of, IV, 215–216, 220
  isolated nucleoli of, IV, 220
  nuclei from, IV, 198, 203
Rauscher leukemia virus, VII, 183
Razor-blade cutter, rotary, I, 120–122
Reaction chamber
  plexiglass, IV, 92
Receptor-destroying enzyme, VIII, 407, 414
Recombination, IV, 173
Red blood cell(s), V, 53
  different ages of, IX, 45
  enzyme treatment of, IX, 47
  partition coefficient of, IX, 45
  stored, IX, 47
Refraction, index of for biological samples, IX, 190
Refractive index, IX, 188
Refractive indices, of liver cells, IX, 192
Rehnia spinosus, III, 314
Reithrodontomys fulvescens, V, 34–35
Renaturation kinetics, VIII, 400

Renucleation, IV, 187
Reovirus I, VI, 168
Reoviruses, VI, 166, 168
Repipet, VII, 332
Replica plating, VI, 223–224; VII, 261; IX, 338
efficiencies of, IX, 170–172, 345
fidelity of, V, 267–268; IX, 169
of mammalian cells, IX, 157, 168
techniques, V, 58
Replicate coverslip cultures, VII, 333
preparation of, VII, 331
Replicate culture planting device, VIII, 49–50, 55, 67
Replicate culture unit, VIII, 57
Replicating ring, IX, 338
Replicating units, VII, 148
Replication forks, VII, 148, 150
Resolution-limiting errors, II, 240
Respiration, IV, 88, 97, 100, 104
Respiratory rate, IV, 88
Respiratory syncytial virus, V, 78, 84
Respirometer, differential, IV, 93
Reticularia, IV, 406
Reticulocyte standard buffer, VIII, 158
Reticulocytes, VII, 9, 19–20, 45, 71–73, 76, 79–80, 83–86, 92–94, 96–97, 371, 395; IX, 45
lysis of, VII, 4
standard buffer, VII, 3, 164, 234
Reticuloendothelial system, IV, 478
Retina, V, 52
Retinal epithelium, V, 52
Reverse transcriptase, VI, 176
virion-associated, VIII, 416
Reversion
gratuitous, VIII, 89
in saturation density, VIII, 90
rate, VIII, 11
selective assays for, VIII, 78
spontaneous, VIII, 11, 34
Reversion frequencies, VI, 267; VIII, 10, 43, 45
Revertant cell lines, VIII, 78–81
cAMP in, VIII, 90
preexistent, VIII, 88
properties of, VIII, 88
Revertant phenotypes, VIII, 87–88
Revertants, VIII, 11, 34, 80
anchorage, VIII, 87, 90
BUdR-induced, VIII, 98
con A, VIII, 89
definition of, VIII, 78
density, VIII, 89
isolation of, VIII, 78, 86–87
detection of drug-sensitive, X, 219
detection of prototrophic, X, 220
detection of temperature-resistant, X, 220
negative selection of, VIII, 80
phenotypic, VIII, 202, 386
selection of, VIII, 85
selection of density, VIII, 82, 90
selection of serum, VIII, 82
serum, VIII, 90
spontaneous morphological, VIII, 87
Revertants in agar, selection of, VIII, 84
Reynolds lead citrate, VII, 171
Reynolds number, III, 17–18
Rhesus monkey, VI, 166
Rhinoviruses, VI, 169, 174
Rhizomastigada, IV, 411
Rhizomes, IV, 498
Rhizoplast, IV, 375–376, 379, 395, 448
Rhodamine B, V, 334, 343, 346, 348, 352–353, 355, 365, 368
Rhodamine B isothiocyanate (RITC), IX, 212–213
Rhodamine-tagged antiviral serum, VI, 175
Rhodulinorange, X, 56
Rhoeo discolor, II, 186–187
Rhynchosciara angelae, II, 67, 75–76
Riboflavin, IV, 122–123; V, 15, 62, 226; VI, 9; VIII, 25
Ribonuclease, IV, 151, 388
activity, IV, 17
Ribonuclease inhibitor, VII, 7, 9
Ribonuclease solution (RNase), I, 2, 435
Ribonucleic acid (RNA(s)), I, *see also* Ribonucleic acid analysis, specific organisms, Nucleic acids, Nucleotides; II, 85, 113, 119; IV, 10, 13–14, 17, 43, 132, 151–152, 157, 163, 195, 198, 205, 207, 211, 213, 388, 399; V, 34–36, 152, 155
accumulation of, IV, 2
acrylamide gel fractionation, III, 291
base compostion of, IV, 14
of pachytene nuclei of, IV, 14
base ratio of, II, 42
catalytic iodination of, X, 333
chromatin-associated, IX, 373
chromosomal, II, 119, 127

cytoplasmic, II, 18; VII, 28
degradation of messenger, VII, 46
deproteinization of, VII, 26
determination of, I, 29
dyes for staining, X, 56
electrophoresis of, X, 362
elution of, X, 29
extraction of, I, 421–423; V, 265–267; X, 28, 36
extraction of nuclear, VII, 31
extranucleolar nuclear, IV, 216
fixation and staining of, III, 200–202
fractionation of, VII, 63
half-life of ribosomal, VII, 29
hemoglobin messenger, VII, 45
heterogeneous, IV, 208
heterogeneous nuclear, VII, 45–46
high molecular weight, VII, 32–35, 40, 42
high molecular weight nuclear, IV, 206
in glia, I, 432
in neurons, I, 432
in protein synthesis, I, 4
in vitro synthesis of, X, 357
intramitochondrial, VII, 48
iodination of, X, 326
isolation of, II, 406; IV, 196
isolation of messenger, VII, 45–46
of pure messenger, VII, 45
of ribosomal, VII, 39, 41
labeling of, VII, 25
metabolism of, II, 32, 42, 407
microanalysis of, X, 17
microextraction of, X, 27
nuclear, IV, 204–205, 211, 224; VII, 24, 26–30, 32–36, 40, 56
nuclear origin of, I, 432
nuclear synthesis of, I, 5
nucleic acid content in, I, 428–430
nucleolar, IV, 204, 206, 209, 214, 216, 221, 224; VII, 26–29, 35–36, 38
nucleolar 45 S, VII, 35–37
phenol extraction, VII, 10
polymerase, II, 79, 407
precursors of, II, 42, 307
precursors to ribosomal, VII, 30, 34
preparation of messenger, VII, 44
of mitochondrial, VII, 47
of ribosomal, VII, 38
of transfer, VII, 47
presence of, II, 255
profiles of, IV, 13
purification
of 5S ribosomal, VII, 43–44
procedures, VII, 17, 54
separation of ribosomal and, VII, 42
ribosomal, II, 18, 32
ribosomal precursor, VII, 35–36
ribosomal synthesis of, IV, 2; VII, 30
SDS-hot phenol-ethanol precipitation procedure of, IV, 13
sedimentation of, IV, 14
characteristics of, IV, 13
simplified procedure for preparing, X, 348
specific activity determination of, X, 36–37
specific activity of, IV, 13
staining with azure B, III, 200–201
with methyl green-pyronin (Unnappenheim-Brachet), III, 200
with methylene blue, III, 200–201
synthesis, V, 149–150
in root meristems, III, 171, 200–202
synthesis in *C. moewusii*, III, 144
in cell cycle of, III, 155
synthesis of, II, 31, 79, 101–102, 110, 131, 255–256, 307; IV, 1, 13, 16
in nucleoli of, IV, 2
movement and, IV, 188
Ribonucleic acid analysis
in cytoplasm, I, 432
in Dipteran giant chromosomes, I, 419, 432
in lampbrush chromosomes, I, 419, 427
in nucleus, I, 432
in spinal ganglion cells, I, 419
Ribonucleoprotein matrix, II, 48
Ribonucleoprotein particles, IV, 221, 224
Ribonucleoside diphosphate reductase, VI, 234
Ribonucleoside triphosphates, VII, 421
Ribonucleotide reductase, VII, 365–366, 381–383, 393–394, 399–400
Ribonucleotides
complex mixtures of, VII, 439
enzymatic analysis of, VII, 447
periodate oxidation of, VII, 440
pools, VII, 372–373, 376, 444–445
Ribosomal proteins
analysis of, VII, 75
extraction of, VIII, 352
heterogeneity of, VII, 78, 85
molecular weight distribution and number of, VII, 78
molecular weights for, VII, 79–80

number of, VII, 80
phosphorylation of, VII, 92, 95
split proteins, VII, 89
two-dimensional separation of, VII, 83–84
Ribosomal RNA, extraction of, VII, 13
Ribosomal subunits, preparation of, VII, 72
Ribosomes, IV, 2, 141, 152–153, 163, 205, 207, 370–371
extraction of RNA from, VII, 40, 42
heterogeneity, VII, 98
isolation of, IV, 204; V, 265–266; VII, 6, 39; X, 127
lyophilized, VII, 72
membrane-bound, VII, 89–91, 99
membrane-bound vs free, VII, 87
normal vs tumor, VII, 87
nuclear, IV, 224
preparation of, VII, 70
preparation of total protein from, VII, 74
preservation, VII, 72
protein-deficient, VII, 88
Ringer's solution, III, 79; IV, 25, 334, 419
insect, X, 3, 8
Rinsing solution, V, 19, 24
rRNA
precursors, VIII, 19
processing, VIII, 19
RNA, *see* Ribonucleic acid
RNA-containing tumor viruses, VI, 176
RNA detection, X, 65
methods of, X, 63
RNA-directed DNA polymerase, VI, 177
RNA-DNA hybrids, X, 6
microassay for, X, 34
thermal stability of, X, 43
RNA labeling kinetics, VII, 372–373
RNA metabolism, in *Drosophila* cell line, X, 196
RNA polymerase, IV, 206; V, 35, 257; VI, 298, 303, 319; VII, 136, 141, 381, 393, 447
RNA polymerase II, VI, 275
RNA staining, X, 49
procedure for, X, 55
RNA synthesis, VII, 29, 35, 339, 343–344, 349–351, 355, 362, 372
mutations affecting, VI, 259
nucleolar, IV, 225
rates of, VII, 330
RNA tumor viruses, VII, 21; VIII, 369
cells transformed by, VIII, 87
RNA virus(es), II, 101; V, 87
transformation by, VIII, 416
RNase, I, *see* Ribonuclease solution; II, 308; IV, 13, 204, 212, 214; V, 11, 34, 36; VI, 363–364; X, 8
detection of, X, 62
inhibition with 6 *M* guanidinium chloride, X, 347
inhibitors of, VII, 2, 4, 10, 33–34, X, 75
pancreatic, X, 4, 8
solution, VII, 342
RNase activity
blockage of, X, 53
RNase treatment, X, 40
Rodents, breeding of, V, 7–8
Roller bottle cultures, X, 159, 164
Roller bottles, IX, 19–21
plastic, IX, 21
Root calluses, III, 178–179
cultures from growing roots, III, 179
of carrot, III, 178–179
Root tips
autoradiographic methods for, I, 222–223
cell cycle in, I, 224–225, 375–377
cell growth and differentiation in, I, 225–226
DNA synthesis in, I, 225
effects of ionizing radiation on, I, 225
of metabolic inhibitors, I, 226
of ultraviolet radiation on, I, 226
fixation of, I, 218–220
germination of seeds, I, 216–218
growing of roots, I, 216–218
isolation of nuclei of, I, 223–224
meristematic region in, I, 215–216
middle lamella of, I, 219
mitotic process in, I, 225
nuclear proteins of, I, 223
permanent slides for, I, 221–222
pinocytosis in, I, 283
pre-DNA synthetic period in, I, 216
protein synthesis in, I, 225
RNA synthesis in, I, 225
semiconservative replication in, I, 225
somatic chromosome studies on, I, 216
staining of cells of, I, 220–221
studies on, I, 225–226
Roots
techniques for growing
Allium cepa, III, 175
grasses, III, 175–176

available free ligands or receptors. Chemographic artefacts were never observed after alcohol fixation of cells and tissues.

B. Cell Radiolabeling

1. Incubation

Just before use, tritiated steroids and catecholamines of high specific radioactivity (5–50 Ci/mmole, according to commercial preparations) are brought to a concentration of 1–2 μCi/ml in buffered isotonic saline (pH 7.4) containing 5×10^{-3} M β-mercaptoethanol.

The slides, mounted with cell monolayers or tissue sections and treated as indicated above, are placed cell side up on a horizontal stay. The radiochemical solution is carefully pipetted over the biological material, and the preparation left at room temperature in a humidified box to avoid water loss by evaporation. After incubation the slides are washed by immersion in two successive baths (20–30 minutes each) of buffered isotonic saline and air-dried.

2. Liquid Emulsion Radioautography

After incubation, the slides are transferred to a darkroom, briefly rehydrated in distilled water, and dipped in a liquid photographic emulsion (Kodak NTB or Ilford K series). After storage in appropriate boxes at 4°C for 21–23 days, the emulsion is developed and fixed, and the cell monolayers or tissue sections stained with hematoxylin-eosin, toluidine blue, Giemsa or any other adequate stain. Detailed descriptions of liquid emulsion radioautography have already been published (Roth and Stumpf, 1969; Bogoroch, 1972).

C. Versatility of the Method

This method can be readily manipulated for different purposes. Before or after the radiolabeling steps, the slides can be subjected to different treatments with chemicals and natural products capable of modifying cell affinity properties for the steroids and catecholamines of given biological material. Also, by using serial sections from the same block tissue, competition assays with unlabeled compounds or comparative tests with different radiolabeled steroids, catecholamines, and their derivatives and analogs can be carried out simultaneously. In all cases valuable information on the nature and/or the specificity of the interaction studied can be obtained.

Hordeum vulgare, III, 176
Hyacinthus orientalis, III, 175
Pisum sativum, III, 174
Vicia faba, III, 173
Zea mays, III, 175
Rosa sp., II, 175
Rose multipurpose chamber, III, 65, 68, 80
Rosellina necatrix, VII, 214
Rotenone, IV, 92, 100, 106–107, 109–110, 112
Rotocompressor, IV, 330; VIII, 323
Rous-associated virus, VI, 156, 172
Rous-sarcoma-helper virus, VI, 172
Rous sarcoma virus-transformed cells
revertant from, IX, 167
Rous sarcoma viruses, VI, 156, 167, 174; VII 215, 314–315; VIII, 2, 158, 370
assay, VII, 320
mutants of, VIII, 426
Roux flasks, IV, 52–53, 63–64
RSV-BHK, VIII, 79
RSV-transformed NRK cells, VIII, 88
Rubber policeman, VIII, 235
Rubella, VI, 165, 169
Rubella virus, VI, 174
Rubeola virus, VI, 165
Rupture of hypotonically swollen cells, III, 281
Ruthenium red, VIII, 395
Ryan virus, VI, 189

S

S-100 protein
production of in glial cells, IX, 131
S phase, II, *see also* Progenitor cell cycle, Thymidine-H^3, Thymidine-C^{14}, Double labeling, Deoxyribonucleic acid, DNA; IV, 158, 489–491, 494; VI, 73, 75, 84, 88–90, 93–95, 98–103, 105–106, 108, 114–115, 119, 121–124, 151, 249
biochemical and genetical methods in, II, 397–410
cell constituent analysis of, II, 407–408
maintenance and preservation of, II, 398–404
nucleic acid isolations in, II, 407
preparation of cells for, II, 404–407
SA 7, VI, 166
Sabourand dextrose broth, VI, 181, 184
agar, VI, 184
Saccharomyces, IV, 140–141
Saccharomyces cerevisiae, I, 12; II, 374; IV, 132, 141, 153
Saccharomyces ellipsoideus, I, 12
Saccharose
and osmotic pressure, II, 163–166
table of concentration, II, 163–165
Sakaguchi-Feulgen procedure after Bloch and Brack, III, 327–328
Sakaguchi-Feulgen reaction, III, 327–328
Sakaguchi procedure after Deitch, III, 322–323
Sakaguchi reaction, III, 203, 205, 309, 321–323
improved by Messineo, III, 322
Salamander, II, 43, *see also* under species names
Saline D_2, II, 142
Saline G, VIII, 26, 30
composition, VIII, 31
Saline media, for isolation or fractionation of nuclei, X, 71
Saline solutions, VII, 316
Saliva, VI, 182
Salivary gland, isolation of nuclei and chromosomes from, IX, 377
Salivary gland chromosomes, X, 11, 13
Salivary glands, II, 66; X, 3, 42
fixation of, X, 23
from *Drosophila hydei*, X, 136
labeling RNA in, X, 20
microdissection of, X, 24
Salmine sulfate, VI, 9
Salmonella typhi, V, 129
Salt precipitation, VII, 18
Sambucus ebulus, II, 173
Sanicula europaea, II, 178
Sanitizer, V, 9
Saponin, II, 115–117
Sapphire mortar, IV, 153
Saprospira, IV, 85, 112
Saquinus sp., VI, 166
Sarcinia lutea, IV, 253
Sarcodina, IV, 343, 406, 411
Sarcolysin, IX, 120
Sarcoma cells, VII, 92
Sarcoma 180 cells, VII, 374, 377
Sarcomas, VI, 168
Sarcophaga bullata, II, 67, 75
Sarkomycin, IV, 450
Sarkosyl, IV, 389, 391; VI, 125; IX, 106

Satellite DNA(s); VIII, 153, 166–168, 170, 172–175, 183
detection of, VIII, 176
mouse, VI, 346
Satellite viruses, VI, 172
Satellites, V, 133
Saturation densities
of normal and virus transformed cells, VIII, 385
of normal and virus transformed 3T3 cells, VIII, 385
Scanning interference microphotometry, VIII, 337
Scatchard's equation, X, 51
Scenedesmus, V, 305, 313–314, 321
Scenedesmus obliquus, V, 306, 315–316, 320, 322
cell sizes of, V, 317–318
photosynthetic reactions in, V, 319
synchronous cultures of, V, 317
Schiff's reagent, III, *see* Feulgen procedure; IV, 510; IX, 211
Schiff's solution, IV, 509
Schizomycete(s), IV, 326; VI, 149
Schizosaccharomyces octosporus, IV, 133
Schizosaccharomyces pombe, IV, 132–133, 136, 141, 144, 149, 152–154, 162–163
genetical methods for, IV, 169–177
physiological and cytological methods for, IV, 131–165
staining nucleus of, IV, 167–168
Schizosaccharomyces versatilis, IV, 133, 140
Schmidt-Thannhauser technique, VI, 82
Schoinobates volans, V, 134
Schuster's medium, IV, 430
Sciarids, II, 73–74
Scilla campanulata, III, 182–183, 206
Scintillation counting procedures, III, 58–61
dimethyl popop, III, 60
NCS solubilizer, III, 60
omnifluor, III, 60
Scotch grass infusion, IV, 266
Scotch grass medium, IV, 251, 266
Sedimentation coefficient, IV, 14
Seeds, germination of corn, lettuce, pea, and wheat, II, 372–374
Seitz filter, V, 10
Selection of mitotic populations, III, 355–370
cell collection protocol, III, 367–370
conditions affecting yield and quality, III, 362, 366
of suitable subclones, III, 366–367
process of, III, 355, 362
Semi-solid suspension culture, VIII, 392
Sendai virus, V, 77–80, 82–83, 86–89; VI, 226; VII, 227–231, 233, 235, 251–252, 256, 258–260; VIII, 13, 138, 140, 148, 150, 418
treatment with, VIII, 14
Separation of cells, by velocity sedimentation, X, 158
Sephadex, VI, 220
Sepharose, VII, 20
Serine hydroxymethyl transferase, VII, 385
Serine hydroxymethylase, VIII, 37
Serotype antigens, VIII, 356
Serotypes, VIII, 353
Serratia marcescens, IV, 253, 373
Serum
fractionation of, VIII, 29, 392
requirements for growth, VIII, 388
reductions in, VIII, 389
Serumless medium, IX, 123
adaptation of cells to, IX, 124, 127
care of cells in, IX, 124
growth of cells in, IX, 136
Setonix brachyurus, V, 134
7X (detergent), V, 68–69
Sex vesicle, IV, 16
Sigma, IV, 307
Silica, X, 88
Silica sol, X, 87
Siliclad, VI, 292; VII, 332; IX, 379; X, 137
Silicone oil, II, 85, 87
Silicone RD emulsion, VIII, 97
Siliconized slides, preparation of, II, 89
Silk fibroin, VII, 45
Silk gland cells, VII, 45
Silkworm, VII, 45
Silver impregnation, VIII, 320
Silver impregnation technique, VIII, 331
Silver staining, V, 283–284
Simian adenovirus, VIII, 7, 414
Simian viruses, VI, 166
Sindbis virus, VII, 198, 201
Single cell growth, VIII, 25, 30
Single cells, preparation of, IX, 163
Single clones, isolation of, VII, 263
Slime mold(s), IV, 346–347, 350, 410, 420, 435; VII, 110–111, 114

Sminthopsis crassicaudata, V, 131
Sminthopsis macrura, V, 131
Smog, extracts of, VIII, 373
Snail enzyme, IV, 172
Snail gut enzyme, IV, 152
SNB, X, 8
Sodium chloride-sodium citrate, VII, 108, 125
Sodium deoxycholate, VII, 39–40, 71–72
Sodium dodecyl sarcosinate, VII, 30–33, 40–42, 125
Sodium dodecyl sulfate, II, 406; VII, 11–12, 14–17, 20, 24, 30, 32–35, 38–41, 43, 46, 48, 54–55, 76, 125
Sodium lauroyl sarcosine, IV, 389
Sodium lauryl sulfate, II, 119; IV, 54–55, 61; VII, 125
Sodium tetraphenylboron, V, 43
Sodium triisopropylnaphthalene sulfanate, VII, 33
Soja hispida, II, 197
Soluene, X, 44
Solution, antibiotic, IV, 27
Somatic cell hybrids, VI, 213
  cloning of, X, 147
  microsurgical production of, X, 147
Somatic cell mutants, VI, 213, 215, 227
Somatic cell phenotypes, VI, 213
Somatic cell variants, VI, 211
Somatic nuclei transplant
  ectodermal, II, 25
  endodermal, II, 25–26
  mesodermal, II, 27
Sonchus laciniatus, II, 190, 192
Sonic oscillation, IV, 213, 221
  methods with, IV, 214
Sonic oscillator, IV, 91; VI, 322
  Raytheon, IV, 215
Sonication, IV, 202; VIII, 163
  conditions of, VIII, 170
  procedure, IV, 220
Sonicator, IV, 105; VIII, 332
  Branson, V, 257
Sonifer cell disrupter, V, 185
Sonifer type L667, V, 180
Sonifier, Branson, VIII, 163
Sorenson buffer, VI, 360
Soybean, III, 180
  suspension medium for cells, III, 180
Spectrophotofluorometer, IX, 202, 212
Spermatids, IV, 3
Spermatocytes, IV, 2–3
Spermatogenesis, IV, 2, 16; V, 28
  meiotic prophase of, IV, 1
Spermatogonia, IV, 3
Spermatozoa, III, 45; IV, 2–4; V, 87, 89–90, 117, 119, 129; VII, 108
Spermatozoan nuclei, VII, 258
Spermidine, II, 131–132, 134, 137–138; IV, 212, 232; V, 117, 120, 258, 265
Spermine, II, 131, 140, 142; IV, 212, 221; V, 120–122
Spermiogenesis, V, 28
Spheroplasts, VII, 120, 130, 149, 151
  yeast, VII, 111
Spiramycin, VI, 163
Spirogyra, II, 199
Spirogyra grevilleana
  incorporation of ^{3}H thymidine into, X, 307
Spirostomum ambiguum, II, 138–139
Spleen, V, 5, 59, 129, 133, 152; VII, 26, 110, 322
Spongioblasts, multiplication of, IX, 248
Spores, IV, 347; VI, 188
Sporulation agar, IV, 170
Sporulation medium, IV, 170
Spring water, artificial, IV, 24
Squirrel fibroma virus, VIII, 371
SSC (standard saline citrate), V, 12; X, 8
STAFLO system, VII, 284
Staining, methods for, IX, 202
Staining artifacts, IX, 210
Stainless steel mesh, VIII, 103
Stains, V, 22–23, *see also* specific stains
  acidic, V, 334
  amphoteric, V, 334
  basic, V, 341
  disturbant, V, 330
  inturbant, V, 330
  metachromasy of, V, 337
  orthochromatic, V, 338
  perturbant, V, 330
  supravital, V, 329
  turbant, V, 330
  vital, V, 325, 331–332
Stains-all, X, 57–58, 61
Staphylococcal filtrate, IX, 6
Staphylococcus aureus, VI, 180
STAPUT apparatus, VII, 283
Steinberg's solution(s), IV, 29–30
Stellaria media, III, 316

Stemonitis, I, 11
Stemonitis axifera, I, 13
Stentor, II, 2, 134–135, 138
alteration of nuclearcytoplasmic ratio in, I, 114–116
cell grafting of, I, 110–112
chimeras of, I, 112
culture of, I, 117
endoplasm of, I, 111, 113–114
enucleation in, I, 112–114
experimental uses of, 86, 110–117
interracial grafts of, I, 112
interspecific grafts of, I, 112
intraclonal grafts of, I, 111–112
macronucleus of, I, 112–114
micronuclei of, I, 112
micrurgy of, I, 110–116
nuclear beads in, I, 113–114
nuclear transfer in, I, 112–114
regeneration of, I, 116–117, 122
synchronous mass regeneration in, I, 116–117
Stentor coeruleus, I, 110, 112
macronucleus isolation of, II, 134–135
thymidine-H^3 labeling of, II, 260, 263
Stentor polymorphus, I, 112
Sterility testing, IV, 56
Stigmasterol, IV, 122–123
Stools, VI, 182
Streptomycin, II, 140, 217, 225, 404–405; III, 64; IV, 55–56, 190, 480–481; VI, 9, 148, 185
sensitivity, IV, 190
Stripping film, IV, 151; V, 33; VII, 147–148, 150
Strophanthin G, VI, 241
Student *t* test, VII, 359
Stylonychia, VII, 136
Stylonychia mytilus, IX, 303
Styrofoam rest, II, 63
Subbed slides, I, 329, 368; II, 132, 134; V, 33; X, 8
Subbing solution, I, 134; VII, 149; X, 8
Subcellular components, isolation of, VII, 2
Subcellular particles, separation of, X, 85
Subcloning, VI, 225
Subculturing, V, 18–20
Succinate-cytochrome *c* oxidoreductase, IV, 100, 104
Succinate-cytochrome *c* reductase, VIII, 216, 224; IX, 261, 272, 275–276
Succinate dehydrogenase, IV, 100, 104, 206; V, 195
Succinate oxidase, IV, 104
Succinate-PMS reductase, VIII, 216, 224
Succinoxidase, IV, 106, 109; VIII, 216, 224
Sucrase, IV, 157
Suction potential, II, 151, 163, 167–179, 183–184
Suctorian, IV, 326
Sulfadiazine, VI, 192
Sulfaflavine, IX, 212–213
Sulfated polysaccharides, VIII, 394
Sulfonamides, VI, 192
Sulfur-35, physical properties of, X, 294
Superinfection experiments, VIII, 418
Suppressor genes, IV, 172
Suppressor sensitive mutants, IV, 174
Surgical suture, II, 45
Suspension growth, adaptation to, X, 197
SV3, VI, 166
SV3T3, VIII, 79, 82–83, 86, 88–89, 180, 244
revertants of, VIII, 89–90
temperature-sensitive, VIII, 89
SV3T3 cells, VII, 215
SV5, VI, 166
SV20, VI, 166
SV33, VI, 166
SV34, VI, 166
SV37, VI, 166
SV38, VI, 166
SV40, V, 78; VI, 166, 172, 194; VIII, 77–78, 89, 138, 150, 158, 371
rescue of, VIII, 146
SV-40 virus, II, 114; VII, 215, 243, 253
Swelling of the cell wall, II, 155
Swinnex filter, X, 151
Swiss blue, X, 56
Sykes-Moore chamber, VII, 309; X, 148
Synaptosomes, X, 101
Synchronization, V, 96, 375; VI, 125, 287
by cold, V, 250
by hypoxia, V, 253
characterization of, VIII, 115
in melanoma, IX, 63
in the intact animal, IX, 62
in vivo, IX, 53
maintenance of, IX, 58
methods of, VI, 219
of crypt cells, IX, 65
of DNA synthesis, V, 255
repetitive, VIII, 118

S phase, IX, 103
thymidine, VI, 119, 121
thymidine method of, IV, 74
Synchronized cell populations, analysis of, IX, 219
Synchronized cells, VII, 332
DNA replication kinetics of, IX, 107
Synchronized growth, of HeLa cells, X, 96
Synchronous cell populations, selection of by zonal centrifugation, X, 173
Synchronous CHO cells, X, 171
Synchronous cultures, IV, 157–162; VI, 50–51
of HeLa cells, X, 97
of *Schizosaccharomyces pombe*, IV, 132
Synchronous division, IV, 154; VII, 282
Synchronous growth, VII, 275, 279–280
Synchronous populations
establishment of, IX, 80
Synchrony, IV, 264; V, 99, 105
alignment, X, 158
automated, X, 159
degree of, VI, 45, 61–62; VII, 279
factors affecting, VI, 60
in *Anacystis*, V, 374
index, VII, 283
loss of, VI, 97
maintenance of, VI, 62
methods, IV, 163
of cell division, V, 375
selection, X, 158
Syncytical plasmodium, IV, 347
Syncytium, IV, 347
Synthetic medium
composition of, IX, 124–125
for *Acetabularia*, IV, 58
preparation of, IX, 124
storage of, IX, 124
Syrian hamster cells, II, 114, 123; VI, 71, 78
Syrian hamster fibroblasts, VIII, 2
Syrian hamsters, VI, 156
Syringe replicators, VII, 262–264
Systrophe, II, 206–208

T

T test, III, 7
T2-bacteriophage, II, 257
T2 DNA, molecular weight, VII, 121
T4, VI, 216
Taricha torosa, III, 84
Tasmanian devil, V, 130
Tasmanian wolf, V, 130
Taurine, VI, 7
TCA solutions
preparation of, VII, 435
Technicon mounting medium, II, 362
Technicon slide holder, VIII, 279
Teflon-coated slides, VIII, 255, 259
Teflon tubing, V, 13
Tegenaria domestica, oocyte RNA in, I, 433
TEM-4T (diacetyl tartaric acid esters), IV, 122
TEM-4T/stigmasterol, IV, 123, 125
TEMED (N,N,N^1,N^1-tetramethylethylenediamine), VII, 59
Temperature-sensitive phenotype, VIII, 89
Temperature sensitivity, VI, 227
Terramycin, IV, 313
Testis (testes), IV, 2–3, 10, 12–13, 25; V, 133; VII, 108, 110
Syrian hamster, IV, 1–2
Tetracycline, VI, 162–163, 192
Tetrahydrofolate, VIII, 35
Tetrahydrohomofolate, VII, 386, 392
Tetrahymena, III, 145, 161–162, 164–166, 169, 235, 239, 248, 250; IV, 87–89, 232, 250, 263
amicronucleate strains of, IX, 323
as food organism, I, 86–88, 90–91, 94, 291
drug resistant mutants of, IX, 336
enzymatic digestion of, I, 134–135
experimental manipulations of, I, 133–134
fixation of, I, 134
generation time of, I, 132
growth on agar medium, IX, 330
handling of cells of, I, 130–139
isolation of macro- and micronuclei from, IX, 282
labeling of, I, 88, 94, 133, 135–138
lysis of, I, 133
macronuclear measurements in, I, 138
macronuclear synthetic period in, I, 132, 375–378
macronuclei from, X, 124
membrane studies on, X, 106
methods for culturing, IX, 313
micronuclear synthetic period in, I, 132, 375–376
normal cell cycle of, I, 127–140
pinocytosis in, I, 281, 290, 299–300
pure stock cultures of, I, 85–86, 130–131
replica plating, IX, 343

spectrophotometric measurements on, I, 134, 138
staining of, I, 137–138
synchronously dividing cells of, I, 85, 117, 127, 131–133
tryptone glucose medium for, III, 162–163
uptake of thymidine-H^3 by, I, 375–378
viable fragments of, I, 120, 122
Tetrahymena pyriformis, I, 85, 128, 132, 375–378; II, 314; III, 161–162, 219, 232, 235, 248, 314, 316; IV, 84, 86, 226, 359; V, 220; VII, 141, 149, 330, 344–345, 389
amicronucleate, V, 222, 254
cell cycle of, V, 279
cloning of, IX, 330
composition of synthetic medium for, V, 228
conditions for heat synchronization of, V, 249
cytidine-H^3 labeling, II, 308
efficiency of plating, IX, 334, 342
electron microscopy of, II, 312–317
generation times of, V, 245
effect of pH on, V, 246
growth rate of, V, 245
inorganic medium for, V, 229
isolated macronuclei of, IV, 227
isolated micronuclei of, IV, 227
isolation of micro- and macronuclei from, IV, 225; IX, 311
isolation of organelles from, V, 256–269
macronucleus of, V, 222
manipulations on solid medium, IX, 329
mass isolation of nuclei from, II, 138–140
media for, V, 223
micronucleate strains, V, 223, 254
micronucleus of, V, 222
mutant strain CHXF-3, IX, 340
optimal pH for growth, V, 245
optimal temperature for growth, V, 245
replica plating, IX, 340
scanning electron micrograph of, V, 221
starvation of, V, 253
synchronization of, V, 246–256
by colchicine or colcemid, V, 252
by vinblastine, V, 254
thymidine-H^3 labeling, II, 263
thymidine-H^3 pools, III, 219
ultraviolet irradiation of, V, 276
uridine-H^3 labeling, II, 259, 308
Tetramitus, IV, 341, 347, 350, 360, 362, 368–370, 377–378, 387, 392–393, 401–402, 404–405, 407, 409, 411–412, 415, 420, 423–424, 429, 433, 435, 437, 440–442, 448, 455–456, 459, 465–468
taxonomy of, IV, 352
Tetramitus rostratus, IV, 344–345, 347, 349, 359, 361, 392, 405, 407–410, 413–414, 429, 464, 466
Tetramitus spiralis, IV, 359
Tetraphenylboron, use of, IX, 198
Thallium acetate, VI, 157, 160
Thermal shock, III, 351
Thermosensitivity of cell mutants, VIII, 3
Thiglycollate broth, VI, 179, 184
Thiobacillus thioöxidans, II, 227
Thioctic acid, IV, 122–123; V, 226
Thioglycolic acid, IV, 101
Thioguanine, VI, 235
6-Thioguanine, VI, 231; VII, 396; VIII, 81
Thioguanine-resistant cells, VI, 234, 267
Thioguanine-sensitive cells, VI, 234
6-Thioinosinic acid, VII, 394
Thionine, II, 270, 281; V, 342, 352; X, 57
Thionine-azure-fuchsin technique, II, 274, 280
Thiouracil, IX, 54
Thorotrast, VIII, 326
3T3 cells, VI, 220, 222–223, 242; VII, 215, 219, 223, 230, 374–376, 397; VIII, 3, 76–78, 81, 83, 86, 89, 180, 233, 241–242, 244
ATP pool in, VII, 370
contact inhibition of enucleated, VIII, 135
enucleated, VIII, 129, 131, 134–136
nuclei from, VIII, 158
origin of, VIII, 377
polyoma transformed, VII, 422
SV-40 transformed, VIII, 77, 81, 89
virus transformed, VII, 161
3T6 cells, VI, 128, 130–133; VIII, 180
Thrombocytes, X, 98
Thylogale billardierii, V, 134
Thylogale stigmatica, V, 134
Thylogale thetis, V, 134
Thymectomy, V, 129
Thymidine, III, 33; IV, 151; V, 63
biological effects of, II, 329–330

effect on cells, III, 186
pool, VII, 365
prototrophy, VI, 268
tritiated impurities uptake by *T. pyriformis*, X, 306
Thymidine block, III, 351; V, 96, 98; VI, 119; VII, 383; VIII, 108
cell viability during, VIII, 115–116
Thymidine blockade, IX, 220
Thymidine-C^{14}, III, 52, 289
continuous labeling with, IX, 117
double labeling with thymidine-H^3, II, 332–335
radiation damage by, II, 330
Thymidine cellulose, synthesis of, VII, 60
Thymidine-H^3, III, 28, 30, 49, 52–55, 57–58, 61, 97, 101, 104–106, 108–110, 112, 152, 171, 177, 181, 183–187, 189, 193–195, 216, 218–220, 222, 242–244, 250, 262–264, 268–269, 272–274, 290, *see also* Cell cycle, Mitotic cycle; IV, 31, 332; V, 30
advantages of, II, 330–331
analysis of cell cycle with, II, 360, 363–384, 389–394
chromosome replication, III, 181–184
clearance from blood, II, 329
double labeling with thymidine-C^{14}, II, 332–335
effects on cells, III, 186
epithelial labeling with, II, 335–355
in determining mitotic cycle and periods, III, 105–107
intracameral injection, III, 55
labeling of *Amoeba proteus* with, II, 261, 263–265, 307
macronuclear labeling, II, 136–137, 265–268
mitotic cycle, III, 184–185
precautions and criticisms, III, 218–220
radiation damage by, II, 330
responses of primordia, III, 185–186
self-radiolysis, X, 305
specific activity of, II, 258
specificity of, II, 329
uptake by root cells, III, 186
use of in cell renewal studies, III, 263
in mice, III, 263–266
Thymidine kinase(s), III, 185–186; V, 91; VI, 213, 239–241; VII, 366, 381, 388–389, 397, 399, 413–414
cells lacking, VI, 240
deficiency in BHK cells, VIII, 14–15
deficient mutants, VIII, 32
induction of, II, 329
mitochondrial, VII, 372
viral gene for, VIII, 416
Thymidine 5′-monophosphoric acid, anhydrous pyridine solution of, VII, 61
Thymidine phosphorylase, VI, 153; VIII, 230
Thymidine-5′-phosphotransferase, VI, 239
Thymidine shock, VI, 117, 122
Thymidine synchronization, VIII, 114
Thymidine synthesis, VI, 233; VIII, 32
Thymidine triophosphate, III, 181
Thymidylate kinase, VII, 381, 386, 399
Thymidylate synthetase, VII, 365, 367–368, 381, 384–385, 392, 399; VIII, 9, 81
Thymidylic kinase, stimulation of, II, 329–330
Thymine nucleotides, VII, 366–368
pool, VII, 367–368, 377, 412, 419, 437
Thymine starvation, VIII, 20
Thymineless death, III, 35
Thymocytes, VII, 160; VIII, 152
DNA in, I, 431–432
Thymus nuclei, V, 185
TICAS, IX, 181
Tight junctions, V, 77
Time-lapse cinematography, III, 33, 220; VI, 115
Time-lapse photomicrography, III, 147
Timothy hay, IV, 250
Timothy roots, cell growth and differentiation in, I, 225–226
Tissue(s)
culture(s), V, 8–9, 23
primary culture, V, 4, 17–19, 24
Tissue grinder, Tenbroeck, IV, 60
Tissue turnover
estimates of, II, 347–348
Tissumizer, IX, 357
TKM buffer, VII, 40
composition of, V, 265
TKM solution, VIII, 208
Tobacco cells, III, 149
Tobacco mosaic virus, VII, 10
purification of, X, 101
Tobacco pith tissue, G_1 and G_2 cell populations in, II, 371–373

Tocopherol phosphate, VI, 9
Tolazul, X, 57
Tolmach method, for collecting mitotic cells, III, 350
Tolonium chloride, X, 57
Toluidine blue, II, 140, 270, 275, 285, 328, 381–383; IV, 151, 205; V, 342, 345, 347, 352, 355, 357, 363, 368; VII, 341; VIII, 279; X, 54, 57
Toluidine blue 0, IV, 25; X, 61
Topinhibition, VIII, 391
Torula sp., VI, 180
Toxoplasma gondii, III, 152
Trace metal toxicity, II, 220
Tradescantia, II, 330; III, 315
Tradescantia guianesis, II, 162
Tradescantia ohiensis, III, 183
Tradescantia paludosa, III, 182
Transamination, VI, 104
Transcription of nucleic acid, *in vitro*, X, 12
Transfer RNA, alterations of, VIII, 2
Transformants
  spontaneous, VIII, 408
  temperature-sensitive, VIII, 89
Transformation, IV, 344, 351, 370, 387, 396–404, 409, 412–414, 420, 429, 433–434, 436–443, 445, 450–452, 457–459, 464–465, 467–468, 487; V, 76; VI, 154, 164, 173–174
  abortive, VIII, 390, 408, 411, 426
  blastoid, IV, 478
  by tumor viruses, VIII, 375
  cryptic, VIII, 390
  delayed, VIII, 408
  detection of, VIII, 413
  efficiency of, VIII, 393, 407
  efficiency of virus in, VIII, 409
  factors affecting the efficiency of, VIII, 409
  kinetics for, VIII, 422
  of ameba-to-flagellates, IV, 350, 394
  of *Naegleria*, IV, 350
  of *Naegleria gruberi*, IV, 391
  quantitative assays of, VIII, 414
  quantitative studies of, VIII, 406
  selective assay for, VIII, 76
  stable, VIII, 390, 408
  synchronous, VIII, 421
  to serum independence, VIII, 411
  viral functions required for, VIII, 425
Transformation assays, VIII, 381
Transformation frequency, VIII, 411, 414–415
Transformed cells
  doubly, VIII, 411
  general properties of, VIII, 378
  selective growth of, IX, 167
  specific properties of, VIII, 399
Transformylase, VII, 392
Transit time, II, 325
Transition points, in the cell cycle, IX, 92
Transplantation of chromosomes, II, 84–85
Transplantation of cytoplasm, II, 82–84
Transplantation of nuclei, II, see Nuclear transplantation in amphibia
Transport
  facilitated, VII, 411
  inhibitors of, VII, 416
  kinetics, VII, 418
  systems, VII, 417
Traps, V, 3
  Havahart for larger animals, V, 3
  Sherman type for rodents, V, 3
Tributylin chloride, IV, 87
Tricarboxylic acid-cycle, IV, 89, 109
Trichloroacetic acid, III, 61, 203, 206, 312
Trichoblast, III, 176
Trichocysts, IV, 125, 267, 305, 315, 329; VIII, 328–330; IX, 282
Trichomoniasis, VI, 192
Trichomycin, IV, 450
Trichosurus caninus, V, 134
Trichosurus vulpecula, V, 134, 136
Triethanolamine, IV, 90
Triethylene melamine, IV, 303–304
Tri-iso-propylnapthalene, VI, 309
Trillium, IV, 498–499, 508
Trillium erectum, IV, 511; VI, 346
Trimastigamoeba, IV, 353, 409, 467
Trimastigamoeba philippinensis, IV, 347–348, 352, 357
Trimerotropis maritima, I, 232
Tri-*n*-butylamine, III, 205
Trinucleate, V, 94, 120
Tris-glycine buffer, III, 72
Tris-HCl, II, 14, 224, 226
Tris-HCl buffer, III, 203
Tritium, X, 293
  interaction with matter, I, 310–311
  physical properties of, X, 294

quantitative calculations, I, *see* Quantitative autoradiography
Tritium efficiency
for detection of, IV, 233
of counting with radioautographic emulsions, IV, 231–240
Triton, II, 19, 24
Triton alpestris, II, 19
Triton N-101, VII, 5
Triton palmatus, II, 19
Triton-X, III, 19
Triton X-100, II, 131–132, 134, 138, 140; IV, 212, 232; V, 256, 263; VII, 3–6, 8 130, 163, 352; VIII, 158, 233
Triturus, II, 40, 43–44
Triturus cristatus, II, 43–44, 46, 52; V, 169
back-crosses in, II, 43
F_1 interracial hybrids in, II, 43
Triturus cristatus carnifex, II, 46, 56
Triturus viridescens, II, 38, 44, 46, 52–53, 56; V, 170
Trizma base, VII, 164
Trophozoite, VI, 189, 191
Truxalis brevicornis, I, 232
Trypan blue, IV, 39; V, 44; VI, 197; VII, 255
exclusion test, VII, 254
Trypanocidal drugs, IV, 94
Trypanosoma lewisi, IV, 112
Trypanosoma mega, I, 282
Trypanosome, IV, 89
Trypsin, II, 6–7, 17, 94, 140; III, 37; IV, 38, 40–43, 513; V, 10–11, 13, 18–19, 21, 43, 48, 53, 55, 67, 89, 150, 152; VI, 161
inhibitor of, IV, 32
monolayer cultures of, IV, 30
Trypsin-EDTA-saline preparation, VIII, 66
Trypsin solution, IV, 31–33
Trypsinization of monolayers, III, 33, 36
cold method, III, 79
Trypticase, IV, 90, 122–123
Trypticase soy broth, VI, 184
Tryptone, IV, 112
Tryptone medium, V, 230
composition of, V, 226
Tryptophan, staining, III, 202
Tryptophan synthetase, IV, 157
Tryptose phosphate broth, VI, 184
Tuberculosis bacilli, V, 84
Tubericidin, VII, 416
Tubifex, II, 44
Tulips, IV, 498–499
Tumor(s), V, 20; VI, 33–35, 156, 166, 168, 176; VII, 26
malignant, IV, 216
murine mast cell, VI, 46
Tumor cells, IV, 212, 214; V, 20; VII, 39
isolation of nuclei from, IV, 196, 208
Tumor formation, VI, 173
Tumor lines, genetic homogeneity of, VIII, 180
Tumor nodules, VI, 35
Tumor tissues, IV, 216; V, 20
Tumor viruses
and their hosts, VIII, 370
brief survey of, VIII, 369
changes induced by, VIII, 374
Tumorigenicity, VI, 32; VIII, 77–79
Tungsten needle, II, 64, 76
Turbidostat, I, 142
Turgidity, II, 145, 150–151, 209–212; II *see also* Plasmolysis and deplasmolysis of plant cells
Turgor pressure, II, 147–148, 150, 159
Turnover time, II, 325
Tween-40, IV, 212, 224; VII, 5, 16; VIII, 158
Tween-80, III, 281, 283, 286; IV, 212; V, 173–174; VI, 9
Tylocine, VIII, 238–239
Tylosin, X, 286
Tylosine, VI, 163
Tyrosine, staining, III, 202
Tyrosine aminotransferease, VI, 215

U

Ubiqumone, VIII, 213
UDP-glucose dehydrogenase, VII, 447
UDP-N-acetylhexosamine, VII, 377
UDP-sugars, VII, 373
synthesis of, VII, 378
Ulcers, V, 148
Ultrabrilliant blue P, X, 56
Ultrasonic disintegrator, IV, 105
Ultrasonic generator, Kubota, IV, 215
Ultrasonic probe, IV, 320
Ultra-Turrax, VI, 319
Ultraviolet irradiation, II, 18, 22, 24; VI, 178; VIII, 6
Ultraviolet microbeam irradiation, III, 148; IV, 330; VIII, 331

Ultraviolet microscopy, I, 440, 442
Ultraviolet radiation, IV, 304
Ultraviolet-sensitive cells, VIII, 36
Ultraviolet-sensitive clones, IX, 175
Ultraviolet sensitivity, IV, 163
Unbalanced growth, III, 35; VIII, 9
Unna methylene blue, II, 118
Uptake, facilitated, VII, 404, 407
Uracil arabinoside, VII, 384
Uracil nucleotides, VII, 372–373, 447–448
Uranin, V, 341, 353, 355, 367
β-Ureidopropionic acid, VII, 387
Uridine-5-^{3}H
  as a tracer for RNA, X, 312
  biosynthetic pathways of, X, 318
Uridine kinase, VII, 381, 388, 399
Uridine nucleotides, VII, 378
Uridine phosphorylase, VIII, 237–241, 244–246
Uridine phosphorylase assay, VIII, 233–234
Uridine triphosphate, VI, 8
Uridylate kinase, VII, 381–382
Uridylic acid, IV, 14
Urine, VI, 182
Urodele lung cells, III, 83–85
Urostyla, VIII, 342
UTP pool, VII, 449

V

V79 cells, VII, 343
Vaccines, VI, 169
Vaccinia RNA, VII, 62
Vaccinia virus, VI, 156, 164–165; VII, 64, 151, 154, 190, 197, 200, 242; VIII, 137
  messenger RNA, VII, 63
Vacuolar contraction, II, 155
Vacuolating virus, VI, 164
Vacuum sample-handling device, VIII, 253
Vahlkampfia, IV, 353, 356, 360, 407–409, 411
Vahlkampfia fragilis, IV, 358
Vahlkampfia gruberi, IV, 353
Vahlkampfia salamandrae, IV, 358
Vahlkampfia soli, IV, 352, 358
Vahlkampfia tachypodia, IV, 355
Vancomycin, IV, 55
Van Slyke reaction, III, 328–329
Van't Hoff's Law, II, 157, 181
Varicella virus, VI, 165
Variola virus, VI, 165
Vegenite, VIII, 349
Vegemite medium, IV, 252, 254
Velban, V, 4, 12, 23–24
  solution of, V, 24
Vermiculite, III, 173
Veronal acetate, VI, 300
Vervet monkeys, VI, 166
Vesicular stomatitis virus, VII, 181–183
Vicia faba, I, see Bean seeds; II, 162, 168; III, 171, 173–174, 182–185, 187, 189, 192–193, 205; VI, 346
  effects of 5-aminouracil on, III, 192–193
  Feulgen staining, III, 199
  induction of polyploidy, III, 191
  radiation effects, III, 197–198
  turnover of chromosomal proteins, III, 187
  uptake H^3-TdR, III, 186–187
  var. *major*, III, 173
  var. *minor*, III, 173
Vicia sativa, III, 183–184
Vinblastine (VLB), III, 162, 164–165, 167; V, 254, 276; VIII, 5
  accumulation of cells in metaphase arrest, III, 280
  in blocking cell division, III, 163–165
  recovery from inhibition, III, 167–169
Vinblastine sulfate (VELBAN), II, 115; III, 163, 280, 351; V, 12
Viokase, V, 43
Viola silvatica, II, 178
Viral antigens, VI, 175
Viral diseases, V, 177
Viral DNA
  transformation by, VIII, 412
  uptake of, VIII, 412
Viral genetic material, detection of, VIII, 399
Viral RNA, VII, 21
Viral synthesis
  detection of by hemadsorption, II, 101–104
  detection of poliovirus RNA, II, 104–110
Virus III, VI, 168
Virus adsorption, VIII, 406, 410, 421
  and penetration of, VIII, 416
Virus cell interactions, VIII, 416
Virus contamination, VI, 165
  of cultured cells, VI, 164

Virus-specific antigens
detection of, VIII, 402
in tumor or transformed cells, VIII, 403
Virus susceptibility, VIII, 376
Virus titrations, IX, 151
Virus vaccines, VI, 166
Virus-virus interactions, VIII, 418
Viruses, III, 26; VI, 144, 153, 157, 169–170, 175–176, 189, 224; see also specific types
detection of, VI, 173
hemagglutinating, V, 79–80
inactivation, V, 82
oncogenic, V, 130
particles, V, 88
purification of, X, 101
replication, VII, 241
rescue of, VII, 252
Viscosity, II, 144, 146, 155, 195–204
Visna virus, V, 78
Vital stain, IX, 209
Vitamin A, VI, 9
Vitamin B_{12}, II, 218, 221, 224; IV, 57; V, 15, 17, 62; VI, 9
Vitamin B complex, X, 21
Vitamin requirements, VI, 4
Vitamins, V, 15; VI, 5
stock solution, V, 17
Vitatron densitometer, X, 33
Vitreoscilla, IV, 85
Volume-concentration relationship, II, 182
Vortex-type mixer, IV, 79

W

Walker 256 carcinosarcoma, IV, 208, 214, 216
Walker tumor, IV, 209, 211, 218
tissue of, IV, 223
Wall attachment, II, 144, 146, 155, 195–204
Wall pressure, II, 147–148, 150, 183
Wallabia agilis, V, 138
Wallaby (wallabies), V, 129–130, 135, 159
black-tailed, V, 139–140
Wallaroos, V, 135
Washing solutions for cells, VII, 452
Wasielewskia, IV, 353
Wasielewski limax, IV, 352
Wasserman tubes, VI, 197
Water saturation, II, 148–149
Wescodyne, VI, 187; VII, 289
Wheat germ agglutinin, VIII, 77, 85, 397–398, 428
Whole cells, phospholipid composition of, X, 131
WI38 cells, VII, 204–205, 221, 223, 238, 253; VIII, 181; IX, 216
anucleate, VII, 222
DNA distributions of, IX, 206, 208
Feulgen-DNA distribution of, IX, 204
fluorescence distributions for, IX, 213
Wickerham medium, IV, 171
Wilzbach method, X, 297
Wombats, V, 130, 135
Wright's giemsa, VI, 361
Wright's stain, VI, 199, 361; VII, 322; VIII, 165; IX, 89

X

Xanthine oxidase, VII, 393, 395–397
Xenopus, II, 17–19, 23–27, 29, 32
egg ribosomes, VII, 99
eggs and ovaries, VII, 110
nucleus, V, 110
ribosomes from VII, 75
Xenopus laevis, II, 28–29, 32; V, 86, 107, 169
kidney cells from, IV, 31
Xenopus tropicalis, II, 29
Xeroderma pigmentosum cells, IX, 175
X-irradiated cells, VI, 116
X-irradiation, III, 220; VI, 115
effects of, IX, 221
synchronization action of, IX, 54
X-rays, II, 19, 94, 96, 104, 111; IV, 303
transition point for, IX, 100

Y

Yaba monkey virus, VIII, 371
Yeast, IV, 84, 140–141, 144; VI, 180–181, 183, 185
autolysate, IV, 90
broth, VI, 184
budding, IV, 132, 135, 141; VI, 44
contamination, VI, 179
extract, IV, 86, 90, 253, 318, 332, 420, 430–432; V, 226–227, 230; VI, 2
fission, VI, 44
infection, VI, 188
juice, IV, 255
media, IV, 141, 318

Yeast cells, VII, 120
  chromosome from, VII, 107
Yeast-peptone-liver medium, IV, 431, 433
Yeast-peptone medium, IV, 402
Yolk platelets, IV, 27, 32, 35
Yolk sac
  cells, VII, 318
    transformation of, VII, 327
  cultures, VII, 315–319, 325–326
  membrane, VII, 321

Z

Zea mays, III, 171, 183, 189
Zebrina pendula, II, 161, 190; V, 363
Zephrin chloride, IV, 55
Zonal centrifugation, of cells, X, 180
Zygonema, IV, 507
Zygote nucleus, II, 32
Zygotene, IV, 3, 502, 511
Zygotene stage, IV, 510

A 5
B 6
C 7
D 8
E 9
F 0
G 1
H 2
I 3
J 4